Computation as Logic

C. A. R. Hoare, Series Editor

BACKHOUSE, R. C., *Program Construction and Verification*
DEBAKKER, J. W., *Mathematical Theory of Program Correctness*
BARR, M. and WELLS, C., *Category Theory for Computing Science*
BEN-ARI, M., *Principles of Concurrent and Distributed Programming*
BIRD, R, and WADLER, P., *Introduction to Functional Programming*
BORNAT, R., *Programming from First Principles*
BUSTARD, D., ELDER, J. and WELSH, J., *Concurrent Program Structures*
CLARK, K, and McCABE F. G., *Micro-Prolog: Programming in logic*
CROOKES, D., *Introduction to Programming in Prolog*
DAHL, O-J., *Verifiable Programming*
DROMEY, R. G., *How to Solve it by Computer*
DUNCAN, E., *Microprocessor Programming and Software Development*
ELDER, J., *Construction of Data Processing Software*
ELLIOTT, R. J. and HOARE, C. A. R. (eds), *Scientific Applications of Multiprocessors*
FREEMAN, T. L. and PHILLIPS, R. C., *Parallel Numerical Algorithms*
GOLDSCHLAGER, L. and LISTER, A., *Computer Science: A modern introduction (2nd edn)*
GORDON, M. J. C., *Programming Language Theory and its Implementation*
GRAY, P. M. D., KULKARNI, K. G. and PATON, N. W., *Object-oriented Databases*
HAYES, I. (ed.), *Specification Case Studies*
HEHNER, E. C. R., *The Logic of Programming*
HENDERSON, P., *Functional Programming: Application and implementation*
HOARE, C. A. R., *Communicating Sequential Processes*
HOARE, C. A. R. and GORDON, M. J. C. (eds), *Mechanized Reasoning and Hardware Design*
HOARE, C. A. R., and JONES, C. B. (eds), *Essays in Computing Science*
HOARE, C. A. R., and SHEPHERDSON, J. C. (eds), *Mathematical Logic and Programming Languages*
HUGHES, J. G., *Database Technology: A software engineering approach*
HUGHES, J. G., *Object-oriented Databases*
INMOS LTD, *Occam 2 Reference Manual*
JACKSON, M. A., *System Development*
JONES, C. B., *Systematic Software Development using VDM (2nd edn)*
JONES, C. B. and SHAW, R. C. F. (eds), *Case Studies in Systematic Software Development*
JONES, G., *Programming in occam*
JONES, G. and GOLDSMITH, M., *Programming in occam 2*
JOSEPH, M., PRASAD, V. R. and NATARAJAN, N., *A Multiprocessor Operating System*
KALDEWAIJ, A., *Programming: The derivation of algorithms*
KING, P. J. B., *Computer and Communications Systems Performance Modelling*
LALEMENT, R., *Computation as Logic*
LEW, A., *Computer Science: A mathematical introduction*
MARTIN, J. J., *Data Types and Data Structures*
McCABE, F. G., *High-Level Programmer's Guide to the 68000*
MEYER, B., *Introduction to the Theory of Programming Languages*
MEYER, B., *Object-oriented Software Construction*
MILNER, R., *Communication and Concurrency*
MITCHELL, R., *Abstract Data Types and Modula-2*
MORGAN, C., *Programming from Specifications*
PEYTON JONES, S. L., *The Implementation of Functional Programming Languages*
PEYTON JONES, S., and LESTER D., *Implementing Function Languages*
POMBERGER, G., *Software Engineering and Modula-2*
POTTER, B., SINCLAIR, J. and TILL, D., *An Introduction to Formal Specification and Z*
REYNOLDS, J. C., *The Craft of Programming*
RYDEHEARD, D. E. and BURSTALL, R. M., *Computational Category Theory*
SLOMAN, M. and KRAMER, J., *Distributed Systems and Computer Networks*
SPIVEY, J. M., *The Z Notation: A reference manual (second edition)*
TENNENT, R. D., *Principles of Programming Languages*
TENNENT, R. D., *Semantics of Programming Languages*
WATT, D. A., *Programming Language Concepts and Paradigms*
WATT, D. A., *Programming Language Processors*
WATT, D. A., WICHMANN, B. A. and FINDLAY, W., *ADA: Language and methodology*
WELSH, J. and ELDER, J., *Introduction to Modula 2*
WELSH, J. and ELDER, J., *Introduction to Pascal (3rd edn)*
WELSH, J., ELDER, J. and BUSTARD, D., *Sequential Program Structures*
WELSH, J. and HAY, A., *A Model Implementation of Standard Pascal*
WELSH, J. and McKEAG, M., *Structured System Programming*
WIKSTRÖM, Å., *Functional Programming using Standard ML*

COMPUTATION AS LOGIC

René Lalement
Ecole Nationale des Ponts et Chaussées, Paris

Translated by John Plaice
Université Laval, Québec

Masson

Prentice Hall

First published in French by Masson, Paris
under the title *Logique, réduction, résolution*,

First published in English as a co-publication between
Masson, Paris and Prentice Hall, Hemel Hempstead.

Printed and bound in Great Britain by
the University Press, Cambridge

Library of Congress Cataloging-in-Publication Data

Lalement, René.
[Logic, réduction, résolution. English]
Computation as logic / René Lalement; translated by John Plaice.
p. cm. – (Prentice Hall international series in computer science)
Includes bibliographical references and index.
ISBN 0-13-770009-1
1. Logic programming. 2. Functional programming (Computer science). 3. Logic, symbolic and mathematical. I. Title.
II. Series.
QA76.63.L3515 1993
005.1–dc20 92-45893
CIP

British Library Cataloguing in Publication Data

A catalogue record for this book is available from
the British Library

ISBN 0-13-770009-1 (hbk)

1 2 3 4 5 97 96 95 94 93

In memory of
Jean-Bernard Saint

Contents

Foreword

René Lalement was kind enough to ask me for a preface. I told him that I was incompetent, but he replied that his taste for logic began with a computer algebra seminar on the Knuth–Bendix algorithm that I gave here at the Ecole Polytechnique. Since I required no competence to set off the chain of events which resulted in this book, I was in no position to require any for the preface!

The few lines that follow are those of a non-specialist, a mathematician who has always[1] been fascinated by computer science, but, in keeping with French tradition, very cautious with respect to logic. It is by spending time with the mathematical and logical objects that are computers and programs, and also with computer scientists and logicians, that I have become more reasonable, although I am still irritated by the form of monographs in logic and theoretical computer science.

Mathematicians, particularly French ones, are known for writing important and erudite books which are unreadable by the non-specialist. Each of us has attempted to do otherwise, but very few of us have succeeded in preventing formal rigour from meaning a 'glaciation' of form. It is when science is at its most formal and the technical syntax the most strict that the syntax should obscure the explanations the least. What is true for mathematics is even more true for logic, be it considered a part thereof or a younger sibling, where the need for formal rigour is greater. And here begins the irritation I was referring to: we sit in front of books which we feel must be talking about important things, but we cannot grasp the ideas, as they are hidden in perfectly adequate, hence completely unreadable, networks of definitions and systems of notations. One feels that one is missing the 'folklore' and that certain relations, probably obvious to those familiar with the field, are unattainable. But, as everyone knows, it is precisely this 'folklore' that is the most difficult to pass on.

It should be clear that, in my mind, the basic strength of René Lalement's work is that it avoids these common problems. It is not the work of someone who 'knows' and who has come to tell us what is the best or unique way of writing and

[1]Since the prehistoric age of the CAB 500 and the Gamma 3 (early French machines).

understanding, why we must forget everything that we have already learned and that we must stop asking the questions that we have been asking because they are the wrong questions. Although well structured, it will lend itself to several readings and will, I think, allow each and everyone to complete or update their knowledge without having to forget everything they have previously known.

On the other hand, each reader, according to their curiosity at a given moment, might regret that such and such a particular point is so little discussed, maybe not even at all. However, the detailed bibliographic notes allow the reader to go back to the specialized works to study these questions at greater length.

In summary, I enjoyed this text and recommend its reading and consultation without any reservations. I wish it great success.

Ecole Polytechnique, Paris
April 1990

Michel Demazure

Acknowledgements

This work is the result of teaching Logic and Computer Science to senior students at the Ecole Nationale des Ponts et Chaussées (ENPC) since 1985, as well as to graduate students in Artificial Intelligence and Pattern Recognition at the Université de Paris VI.

The book was written at the CERMA, the Centre of Applied Mathematics at the ENPC, with the encouragement and friendly advice of Nicolas Bouleau, whose taste for logic led him to create this course, which the author oriented towards programming. Many at the CERMA helped with their advice, even reading the preliminary versions of the book.

The author's interest in using logic to study computer science stems from attending enriching seminars given at the various research centers in Paris: the Laboratoire de Recherche en Informatique at Orsay, the Logic group at the Université de Paris VII, the Formal group at the Institut National de Recherche en Informatique et en Automatique (INRIA) and the Ecole Normale Supérieure.

Jean-Bernard Saint, with whom the author taught this course, shared his taste for logic programming with us and partially contributed to the writing of this book. He was a good friend of both of us and will be missed.

The translator would like to thank Catherine and Cyril for putting up with his nonexistence for many nights and weekends.

Ottawa
March 1992

René Lalement
John Plaice

Chapter 1

Introduction

Logic has at least two roles to play in computer science. First, computer science can be studied from the outside, theoretically, and logic can be used to study its foundations, as for mathematics. Second, the methods used in computer science are themselves logical methods.

In its first role, logic acts as meta-computer science: it offers computability and undecidability results such as Church's, Gödel's and Rice's, and the classification of problems according to their complexity.

The second role has to do with programming being a logical activity. To design a program is to prove a proposition in a constructive manner. To execute a functional program is to normalize a proof. To execute a logic program is to construct a proof and extract information from it. To find an expression's type is to find the proposition of which the expression is a proof. To interpret or optimize a program is to construct a more or less standard model, as if one were evaluating the validity or the consistency of a proposition. Logic's inference systems and semantics are used extensively. Interpreters, compilers and optimizers are formalized using recursive function theory.

This work deals with two computing paradigms where logic plays a useful and essential role. An attempt will be made to show that logic is not only a language in which to state facts or knowledge accurately, even though it did once play this role, when the teaching of mathematics was overly formal, and still does so today, in the current guise of artificial intelligence.

The two computing paradigms to be studied are reduction and resolution, associated, respectively, with functional programming and logic programming.

Computing can be done by *reducing an expression* to a simpler one. Let $(2 + 3) \times (1 + 1)$ be a term to compute. A reduction of this term is

$$(2 + 3) \times (1 + 1) \rightarrow 5 \times (1 + 1) \rightarrow 5 \times 2 \rightarrow 10.$$

It stops as there is no rule which applies to 10. Other reductions are possible, as in

$$(2+3) \times (1+1) \to 2 \times (1+1) + 3 \times (1+1) \to 2 \times 2 + 3 \times 2 \to 4 + 6 \to 10,$$

which ends with the same term. This *confluence* of all the reductions flowing out of the same term is an essential property of this form of computing, as it allows a choice of 'reduction strategies'.

Another approach is to consider computing as the *resolution of a problem.* A typical example is an equation such as

$$x^2 - 4 = 0 \rightsquigarrow x = 2 \text{ or } x = -2.$$

There are two solutions to this equation.

The resolution method must be *complete*, i.e., it must find all the solutions. Since the resolution of a problem often calls for solving other problems, many solutions must be managed and an order must be found to compute, test, accept, reject or combine these partial solutions.

Computer science will be approached in a formal manner, by first considering the need to manipulate *finite* objects. Computer science uses numbers, programs and files. Logic uses terms, formulas and proofs. In both cases, *words*, i.e., character strings, are used to represent these objects. Each of these notations has a concrete syntax, chosen to enhance communication, be it between humans or computers. For example, the notation for formulas, which uses variables, relational and functional symbols, connectives and quantifiers, is quite familiar. The formula

$$(x > 0) \wedge \forall y \forall z \Big(\neg(y.z = x) \vee y = x \vee z = x\Big)$$

is a word over the alphabet formed of the *symbols* (, x, $>$, 0,), $\wedge$, $\forall$, y, z, $\neg$, ., $=$ and $\vee$. The surface representation of programs and formulas is similar, and so is their abstract structure, which is stronger than the structure of words. For the preceding formula, $> 0) \wedge \forall$ is a subword but not a subformula, while $x > 0$ is a subformula.

More precisely, there are three fundamental structures by which these objects can be classified:

- Arithmetic applicative-language expressions, first-order terms, proofs in a Hilbert inference system and Prolog values are *finite terms.*
- Functional language programs, first-order formulas and natural deduction proofs are *lambda* (λ) *terms.*
- Sequential control structures and Prolog–II values are *rational terms.*

It turns out that λ-terms and rational terms can be represented by finite terms, which makes the latter the major structure studied in this book.

To abstract is to simplify: by ignoring the details of concrete syntax, one can study in a general manner these elementary objects which form the basis for the syntactic landscape of logic and computing; this is the purpose of Chapter 1. The same methods, for definition and proof, will be used to reason about these objects, whether they be programs, formulas or proofs. Structural induction is a powerful method which generalizes induction over the integers for finite terms. More flexible techniques are also presented: monotone and well-founded induction.

A second aim of Chapter 1 is to show the reader the difference between what is elementary and what is not, a difference which will be elaborated in Chapter 7. Since its beginnings, it has been clear that computer science deals with finite objects: finite state automata, finite alphabets and finite programs. Mathematics has long escaped from finite domains, as in real analysis. The use of algebra in mathematics at the end of the nineteenth century reintroduced the concept of 'finite', as did the development of computer algebra more recently. Elementary algebraic geometry studies manifolds, such as curves and surfaces, under the form of ideals of polynomials which have a finite number of generators. Algebraic topology studies topological spaces through algebraic structures such as the fundamental group, constructed using generators and relations from a finite description analogous to a graph. The need, still not fully formalized, for finiteness in the description and solving of problems yielded problems which were considered 'difficult': for example, the solving of polynomial equations over the integers, or the untying of a knot. One can imagine the surprise experienced by mathematicians when this need, extended by Hilbert to cover reasoning itself—to ensure, he thought, its foundation—brought forth a new degree of difficulty: 'undecidability'. The meta-mathematical problem of deduction in a theory as simple as that of arithmetic turned out to be undecidable (Church's theorem). Later, it was discovered that some of the most difficult mathematical problems are in fact undecidable. In computer science, where finiteness is necessary, undecidable problems abound: Rice's theorem is a remarkable generator of undecidable problems. These questions will be addressed in the last chapter, which examines computability, i.e., general problems of computation.

Functional programming corresponds to 'computing as reduction'. Two examples of reduction are studied in Chapter 2: the λ-calculus and rewriting systems. The λ-calculus was created as an attempt to transform logic into a mechanically verifiable calculus. It is a formal method with few elements, the symbol λ, an infinite number of variables and parentheses, and one computing rule, β-reduction, which represents the passing of a parameter to a function. It is remarkable that one can encode all data structures and 'run' all computing algorithms with this simple calculus. For example, the Booleans *true* and *false* are encoded by the λ-terms $\mathbf{T} \equiv \lambda x \lambda y\, x$ and $\mathbf{F} \equiv \lambda x \lambda y\, y$, which represent branchings in the sense that the λ-term BMN, which transforms itself by β-reduction to M if B is $\mathbf{T}$ and to N if B is $\mathbf{F}$, behaves as the `if` B `then` M `else` N of usual languages. The logical *and* is coded by $\mathbf{and} \equiv \lambda x \lambda y(xy\mathbf{F})$, so that, for example, $\mathbf{and}\,\mathbf{TF}$ reduces to $\mathbf{F}$. It will be shown how integers, pairs, arithmetic operations and recursive definitions of functions can be encoded. One naturally writes 'higher-order functional programs', which take functions as parameters and return functions as results.

Strangely enough, most programming does not venture beyond first-order functions. In mathematics, higher-order functions have been used with success for more than a century, e.g., the functionals and operators of analysis, to implement general methods. Lisp has contributed to this situation: for a long time it was the only language allowing higher-order programming. Its faults were therefore associated with higher-order programming. But Lisp was not a functional language, despite appearances! There now exist real functional languages whose compiled code is of acceptable efficiency. For non-numeric applications dealing with highly structured objects, such as in logic or software engineering, e.g., languages, programs and proofs, functional programming is an ideal programming paradigm. Functional programming is also widely used in artificial intelligence because of its ability to manipulate symbolic data and to define 'strategies'.

The functional language used in this book is Standard ML (of New Jersey).[1] ML programs are 'typable' λ-terms. For example, the Booleans **T** and **F** can be coded by

```
val T = fn x => (fn y => x)
and F = fn x => (fn y => y) ;
```

where `fn` corresponds to λ. The compiler computes the types of `T` and `F`,

```
↷ val T = fn : 'a -> 'b -> 'a
  val F = fn : 'a -> 'b -> 'b
```

which contain the type variables `'a` and `'b`, which represent arbitrary types.

A logic consists of a language, an inference system and an interpretative system. It it used *to prove and interpret.*

Chapter 3 introduces the language of first-order logic and several inference systems: natural deduction systems, the sequent calculus and Hilbert systems. These systems will first be presented in the restricted form of minimal logic, without negation, then in intuitionistic logic and finally in classical logic. The latter is the most familiar, with its rule of reasoning by contradiction. The connectives follow certain algebraic properties which clearly appear in the sequent calculus and allow one to 'normalize' formulas.

Here is a proof in natural deduction of the formula

$$\varphi \equiv ((A \Rightarrow B) \wedge (A \Rightarrow C)) \Rightarrow (A \Rightarrow (B \wedge C)),$$

where A, B and C are arbitrary formulas. To prove φ, suppose $(A \Rightarrow B) \wedge (A \Rightarrow C)$. From this assumption, deduce $A \Rightarrow B$ and $A \Rightarrow C$. In addition, suppose A. From $A \Rightarrow B$ and A, conclude B. From $A \Rightarrow C$ and A, conclude C. Thus B and C have been deduced. Therefore $B \wedge C$ has been deduced, by supposing $(A \Rightarrow B) \wedge (A \Rightarrow C)$ and A. By discharging the assumption A, $A \Rightarrow (B \wedge C)$ is proven by supposing

[1]The (original) French version of this book used CAML.

only $(A \Rightarrow B) \wedge (A \Rightarrow C)$. By discharging this last assumption, φ is proven with no assumption. It is simpler to represent this proof as a derivation, a formal object which is drawn as a tree where the deduction proceeds from the assumptions at the top to the conclusion at the bottom:

$$\dfrac{\dfrac{\dfrac{\dfrac{\dfrac{[(A \Rightarrow B) \wedge (A \Rightarrow C)]}{A \Rightarrow B} \quad [A]}{B} \qquad \dfrac{\dfrac{[(A \Rightarrow B) \wedge (A \Rightarrow C)]}{A \Rightarrow C} \quad [A]}{C}}{B \wedge C}}{A \Rightarrow (B \wedge C)}}{((A \Rightarrow B) \wedge (A \Rightarrow C)) \Rightarrow (A \Rightarrow (B \wedge C))}$$

The above derivation can be written as a (typed) λ-term. In ML, this becomes

```
fn h => (fn x => (#1 h x, #2 h x)) ;
```

where `#1` returns the first element in a tuple and `#2` returns the second. The type computed by ML is

```
↷ val it = fn : ('a -> 'b) * ('a -> 'c) -> 'a -> 'b * 'c
```

which is precisely the formula φ, where implication $\Rightarrow$ and conjunction $\wedge$ are replaced by the functional arrow `->` and the product `*` of types, and A, B and C by type variables `'a`, `'b` and `'c`.

This functional interpretation of a proof is based on the *Curry–Howard correspondence*, which plays an essential role in the relations between logic, λ-calculus and programming. This relation can be defined by the equations:

$$\begin{array}{rcccl} \textit{proof} & = & \lambda\textit{-term} & = & \textit{program}, \\ \textit{proposition} & = & \textit{type} & = & \textit{specification}. \end{array}$$

A program specification is always a formula in a logical language which states what a program must do. However, proving that a program conforms to a specification requires an *external* logic, typically formed by putting together programs and first-order formulas. Yet, each proof–proposition and term–type pair forms a single entity. The program–specification pair should be understood in as integrated a manner as the other two pairs. In the same way that one never writes a proof without knowing what one is proving, it should not be possible to write programs without knowing what specification a program is implementing. The Curry–Howard correspondence, which deals particularly with functional programming, allows one to consider a program as a proof of its specification, hence programming as looking for a proof. ML's type system, already powerful for a programming language, is still poor as a specification language. All numeric functions have the same type: they take `int` to `int` or `real` to `real`. There are also type systems capable of distinguishing types in a finer manner, and one hopes that they will be the basis for future programming languages. These systems correspond to constructive logics: this is true for intuitionistic logic, but not for classical logic.

Here is a well-known example of a non-constructive proof. One must show that there exist two irrational numbers a and b such that a^b is rational. Consider $\sqrt{3}^{\sqrt{2}}$:

- if $\sqrt{3}^{\sqrt{2}} \in \mathbf{Q}$, then $a = \sqrt{3}$ and $b = \sqrt{2}$;
- if $\sqrt{3}^{\sqrt{2}} \notin \mathbf{Q}$, then $a = \sqrt{3}^{\sqrt{2}}$ and $b = \sqrt{2}$, such that $a^b = 3 \in \mathbf{Q}$.

This proof is correct but it does not yield a value: which of the pairs (a, b) is correct?[2] One of the ideas of logic programming will be to extract a value from a result.

Chapter 4 introduces the other logical approach: *semantics.* A semantics's purpose is to examine the possible meaning of an object. In general, this meaning is the result of an evaluation, as is the value of an expression in programming or the truth value of a logical formula.

The computation of this value reduces the operational content of the object: the two expressions $((1 + 1) + 1) + 1$ and $2 * 2$ have the same value in $\mathbf{N}$, yet they have distinct operational meanings. What is lost by equating objects with the same denotation is gained in terms of freedom of interpretation: the two preceding expressions have value 4 in $\mathbf{N}$, as well as 0 in $\mathbf{Z}/2\mathbf{Z}$.

Take the example of arithmetic expressions, which are first-order terms: if $x, y > 0$ and $z < 0$, what is the sign of xyz? The axioms of arithmetic could be used to construct a derivation of the formula $xyz < 0$. However, the variables can be interpreted in $\{-1, +1\}$, where x and y become $+1$ and z becomes -1, and the expression to be evaluated becomes $(+1)(+1)(-1) = -1$. Even if it is known that $x = 3$, $y = 7$ and $z = -2$, it is unnecessary to compute the value of xyz *in* $\mathbf{Z}$ to determine its sign, the computation in $\{-1, +1\}$ being sufficient. Although $\mathbf{Z}$ is the standard domain of interpretation for arithmetic, there are many 'non-standard' domains, such as $\{-1, +1\}$.

In standard mathematical usage, formulas in predicate logic represent set properties. For example, a set with an operation which satisfies the axioms of group theory is a group. To answer the question, 'can the formula $\forall x \forall y(xy = yx)$ be deduced from the axioms of group theory?', it suffices to know that there exist non-commutative groups. It is much easier—for a mathematician, not a computer—than to prove that it is impossible to construct a proof of this formula.

Tarski's concepts of interpretation and model allow such reasoning. The meaning of logical connectives given by the syntactic construction of derivations is the same as the interpreted meaning in a set. Even though syntactic methods are naturally adapted to computer science, semantic methods appear more intelligent, as they are more flexible. To study the semantics of a language is to consider its possible meanings. According to need, special cases can be treated by choosing a particular meaning with the purpose of refuting a proposition or to obtain partial information about an object.

Hence non-standard interpretations of programming languages allow optimizations, e.g. parallelization of code, without creating the standard interpretation of

[2]The Galfond–Schneider theorem, which solves one of Hilbert's problems, states that an algebraic number raised to an irrational power is transcendental. Hence the correct pair is $a = \sqrt{3}^{\sqrt{2}}$ and $b = \sqrt{2}$.

a program. In artificial intelligence, qualitative interpretations of numeric problems allow an estimation of essential phenomena such as growth or stability of measures: this is called qualitative physics. Normal life is led with one or several familiar models in mind. In mathematics, one might manipulate the consequences of the axioms of Hilbert spaces, but one lives in the space L^2. In computer science, one writes specifications of programs and of data structures, but the physical implementation is in one's mind. Is this a problem of too much generality?

Not at all! Semantics offers an incredibly rich set of possibilities. There are uncountable models of arithmetic and countable models of set theory! These models are unintuitive, but inevitable. Semantics offers an additional means of abstracting when faced with complex problems. Chapter 4 is concerned more with justifying such methods than with showing their use in the above domains. Logic is probably the domain of appplication where the methods have been illustrated most brilliantly, particularly for the proof of independence theorems in set theory.

It is easy to use these deductive or semantics methods to construct a formalism with inference rules or semantics. However, the definitions given for minimal, intuitionistic or classical logic are not arbitrary. They respect some important properties such as cut-elimination and normalization, symmetries such as introduction/elimination, and completeness and soundness results. It is not by presenting a formalism as an inference system that one does 'logic'. There are similar risks in mathematics when one chooses axioms liberally: these have little chance of defining anything of mathematical interest.

Chapter 5 deals with *equational logic*. Equations can be used to state *problems* to solve algebraic, differential and fixpoint equations as well as to state universal *laws* ($x + y = y + x$) or relations ($i^2 = -1$, $E = mc^2$) between values which are *a priori* unequal.

Of course, in predicate logic, this difference in meaning is written using quantifiers. To solve an equation $x^2 + 1 = 0$ proves $\exists x(x^2 + 1 = 0)$, while commutativity is written $\forall xy(x + y = y + x)$. Equational problems can clearly be written using a fragment of first-order logic, but the methods used are not those of the predicate calculus. Furthermore, solving an equation does not entail just proving an existential theorem, since an *explicit* value is sought in a particular *set*,i.e., one does not look for a proof of $\exists x \phi(x)$, rather for a value a such that there exists a proof of $\phi(a)$. For example, an algebraic equation does not have the same solutions in **R** or in **C**. The search for solutions will be restricted to sets of terms, where resolution becomes *unification*.

It turns out that many algebraic structures and many data structures used in algorithms are axiomatized by equations stating such properties as the associativity or commutativity of an operation, and that the semantics of classical logic can be given in Boolean algebras, which can be given an equational axiomatization. Functional programming defines functions in an equational manner, and evaluation of expressions can be seen as a rewriting process which uses the directed rules. The λ-calculus and combinatory logic are equational formalisms. The type inference done by the ML compiler consists of solving equations generated by the constraints of the formation of typed λ-terms. As for logic programming, even though the programs themselves are not equations, the interpreter essentially solves equations, and their solutions are returned as results. All of these applications, through different problems and paradigms, show the importance of equational logic.

Logic programming, the purpose of study of Chapter 6, corresponds to 'computing as resolution'. Unification is already an example of this paradigm, but it is such a simple case that the essential properties of soundness, completeness and switching are hardly visible. To talk of resolution in logic programming is to play with words. A query's solutions are generated by the *resolution* of equations, but this process is also a particular case of Robinson's *resolution principle*, which is a general method of automatic theorem proving. Logic programs and queries can be read as formulas of a particular form, Horn clauses, for which it is possible to generate constructive proofs. Clause $p(M)$ is proven for a first-order term M, and not just $\exists x p(x)$. The declarative semantics of logic programs describes what can be computed, and the operational semantics describes how computations must be done. The history of logic programming is shorter than that of functional programming, yet there are already more interpretations of logic programming: theorem proving, constraint solving, communicating processes and more.

The value of different versions of Prolog will be shown. In particular, Prolog–II and its rational terms, and NU–Prolog and its sound negation and quantifiers and its use of coroutines to eliminate useless non-determinism.

Chapter 2

The syntactic landscape

2.1 Words

Words are the basic syntactic objects of computer science. Typically known as character strings, they are used to write the more structured objects described later on in this chapter.

Let A be a set, called *alphabet*, whose elements are called *symbols*. A *word* over A is a finite sequence of elements of A. The set of words over A, written $A^\star$, is defined by

$$A^\star = \bigcup_{n \geqslant 0} A^n.$$

If $u \in A^n$, $|u| = n$ is the *length* of u. There is a binary operation over $A^\star$, called *concatenation*, written '$\cdot$', or by simple juxtaposition,

$$\begin{array}{rcl} A^p \times A^q & \to & A^{p+q}, \\ (u, v) & \mapsto & u \cdot v \quad \text{or} \quad uv, \end{array}$$

where $u \cdot v = (u_1, \ldots, u_p, v_1, \ldots, v_q)$ if $u = (u_1, \ldots, u_p)$ and $v = (v_1, \ldots, v_q)$. This operation is associative and has as its neutral element the empty word ϵ, the unique element of A^0. Thus $(A^\star, \cdot, \epsilon)$ is a monoid, called the *free monoid generated by A*.

Given these definitions, a useful notation is needed. The single element (a) is written a, and so $u = (u_1, \ldots, u_n) \in A^n$ is the result of the concatenation of the symbols that make it up: $u = u_1 \cdot u_2 \cdot \ldots \cdot u_n$, or simply $u = u_1 \ldots u_n$.

Any subset of $A^\star$ is called a (*formal*) *language* over the alphabet A.

Examples

- If $A = \varnothing$, then $A^\star$ is simply $\{\epsilon\}$.
- If $A = \{1\}$, then the monoid $(A^\star, \cdot, \epsilon)$ is isomorphic to $(\mathbf{N}, +, 0)$. There are as many languages over the alphabet A as there are subsets of $\mathbf{N}$, i.e, as many as there are real numbers—uncountably many.

- If $A = \{0, 1\}$, then $A^\star$ is the set of binary words.
- Programming languages generally use the 'ASCII' set as the alphabet. It has 128 symbols.

The syntactic study of languages is an important branch of computer science. If $A \neq \varnothing$, there exists an uncountably infinite number of languages. Only a countably infinite number of subsets can be obtained through 'finite means', or *recursively*. The other subsets will be of little use to computer science. To define languages, completely operational techniques, such as automata and grammars, must be devised. These techniques must at the very least be able to determine, for a given language L, whether a word belongs to L. This problem is called the recognition problem for L. These questions will be addressed in the final chapter of the book.

The following operations over languages are commonly used:

- union: $L \cup L'$;
- concatenation: $LL' = \{uu'; u \in L, u' \in L'\}$;
- Kleene closure: $L^\star = \{u_1 \dots u_n; n \geqslant 0, u_1, \dots , u_n \in L\}$.

For $a \in A$, $\{a\}L$ will be written aL.

Some order relations will be used over $A^\star$. Let $u, v \in A^\star$. Word u is called a *prefix* of v, written $u \leqslant v$, if there exists a $u' \in A^\star$ such that $v = uu'$. This unique u' is written v/u. The relation $\leqslant$ is called the *prefix order*. Write $u \perp v$ if u and v are incomparable under this order, i.e, if $u \nleqslant v$ and $v \nleqslant u$. The two are then called independent.

Recall a few useful definitions on order relations and variants. A relation is

- a *strict order* if it is irreflexive and transitive;
- a *preorder* if it is reflexive and transitive;
- an *order* if it is reflexive, antisymmetric and transitive.

If $>$ is a strict order, and if $\geqslant$ is defined by $x \geqslant y$ if $x > y$ or $x = y$, then $\geqslant$ is an order. If $\geqslant$ is an order, then $>$, defined by $x > y$ if $x \geqslant y$ and $x \neq y$, is a strict order. Finally, if $\geqslant$ is a preorder, $x \sim y$ if $x \geqslant y$ or $y \geqslant x$ is an equivalence relation, and $x > y$ if $x \geqslant y$ and $y \ngeqslant x$ is a strict order.

2.2 Finite terms

The structure of words, or rather that of the free monoid, is poor: all symbols are of the same kind. The terms presented in this section are given a structure which differentiates symbols by their usage.

2.2.1 Signatures

For arithmetic expressions, there are operators (+, $\times$), variables and constants. The structure of these expressions comes from the functional aspect of the symbols, each of which requires a fixed number of arguments, called its *arity*. In a programming language, one might find

- constant symbols: `true` , `256` , `' '` ;
- unary symbols: `not __` ;
- binary symbols: `__ or __` , `while __ do __` , `__ < __` ;
- ternary symbols: `if __ then __ else __` .

where __ indicates where an argument should be placed. The introduction of these symbols defines the signature of the language.

Definition 2.1 A *signature* is a graded set, i.e, a set Σ with a mapping $\mathrm{ar} : \Sigma \to \mathbf{N}$. If $f \in \Sigma$ and $\mathrm{ar}(f) = n$, f is of *arity* n, or f is n*-ary*. If $n \geqslant 1$, f is a *functional symbol*, and if $n = 0$, it is a *constant symbol*.

T_Σ is the smallest subset E of the free monoid $\Sigma^\star$ such that

1. if $c \in \Sigma$, $\mathrm{ar}(c) = 0$, then $c \in E$;
2. if $f \in \Sigma$, $\mathrm{ar}(f) = n \geqslant 1$, and if $M_1, \ldots, M_n \in E$, then $fM_1 \ldots M_n \in E$.

The elements of T_Σ are called Σ*-finite terms*, first-order terms or just terms.

Example Let $\Sigma = \{0, s, +\}$ with $\mathrm{ar}(0) = 0$, $\mathrm{ar}(s) = 1$, and $\mathrm{ar}(+) = 2$. Then 0, $s0$, $+00$, $+s00 \in \mathrm{T}_\Sigma$, but $+0$, $s00 \notin \mathrm{T}_\Sigma$. □

So, in attempting to construct objects more structured than words, more words were produced! This is not a problem, since the structure is in the definition itself, not in the objects produced.

In addition, the definition of T_Σ is given using a family of subsets already containing T_Σ: it appears to be a circular definition. The following section clarifies these questions.

2.2.2 Inductive definitions

Inductive reasoning on the integers is well known: to prove a property P over the non-negative integers, one proves it for 0 and then proves that if it is true for n, then it is true for $n+1$. Intuitively, one proves P for 0, then for 1, then for 2, etc. If this reasoning works for the integers, it is because the integers were defined—using Peano's definition—so that it would work. The gist of this definition is that a constructive procedure is itself a proof. One can conceive of this procedure, either in a relative manner, if the integers are constructed from already known objects (such as *in* set theory or *in* a 'notation' system), or in an absolute manner, if one starts with nothing.

Inductive definitions and inference systems

The definition given for terms is relative to the free monoid; the result is that they can be written using words. The relative aspect can be seen in the sentence, 'T_Σ is the smallest subset E of the free monoid $\Sigma^\star$ such that ... .' Since T_Σ is now defined, a proof procedure using *induction over terms* can be validated in $\Sigma^\star$.

Proposition 2.1 Let P be a property over words. If P is true for each constant symbol in Σ, and if for each $f \in \Sigma$ of arity $n \geqslant 1$, one can prove that P is true for the word $fM_1 \dots M_n$ if it is true for the words $M_1, \dots, M_n$, then P is true for all terms.

Proof Let E be the subset of $\Sigma^\star$ which contains all the words for which P holds. The set E contains the constants, and contains $fM_1 \dots M_n$ as soon as it contains $M_1, \dots, M_n$. By the definition of T_Σ, $\mathrm{T}_\Sigma \subseteq E$, i.e, P holds for all terms. □

This procedure is called proof by *structural induction.* It is valid for all objects defined inductively and will often be used in this book. It will also be used for definitions. For example, the *height* $h(M)$ of a term M is defined using structural induction:

$$\begin{aligned} h(a) &= 0, \\ h(fM_1 \dots M_n) &= \max(h(M_1), \dots, h(M_n)) + 1, \end{aligned}$$

for every constant symbol a and every symbol f of arity $n \geqslant 1$.

An inductive definition can also be presented as an inference system. An *inference system* is a set of inference rules dealing with *judgments* that are to be proven. The following judgments are of interest: 'M is a term', 'the formula φ is true', 'the type of term M is τ', or 'the value of expression E is v'.

An *inference rule* has the following form:

$$(name) : \frac{j_1 \dots j_n}{j},$$

where $j_1, \dots, j_n$ are judgments, called *premises*, j is a judgment, called *conclusion*, *name* is the name of the rule, and the integer $n \geqslant 0$ is its arity. The rule presents the inference of the conclusion from its premises. If $n = 0$, the rule asserts the conclusion j, which is then called an *axiom* of the inference system. If such is the case, the horizontal bar is often not written.

The simplest example of an inference system is given by the rules of an inductive definition. For terms, what is of interest is the judgment 'term M' stating that M is a term. The signature Σ determines the following rules, with one (n-ary) rule for each (n-ary) symbol:

$$(c) : \frac{}{\text{term } c}$$

$$(f) : \frac{\text{term } M_1 \quad \dots \quad \text{term } M_n}{\text{term } fM_1 \dots M_n}$$

where $c, f \in \Sigma$ and $\mathrm{ar}(c) = 0$, $\mathrm{ar}(f) = n$. These rules are completely deterministic. Their premises are uniquely determined by their conclusions. This property is held by all definitions which use structural induction.

Consider for a moment the possibility that the M in 'term M' might designate any word in $\Sigma^\star$. These rules can be used to prove that a word is a term. The proof is set up in the form of a tree with the root 'at the bottom'. Consider the word $+0s0$. Since 0 is a term, the system can infer that $s0$ is a term, and with this intermediate result, that $+0s0$ is a term. This reasoning is presented using a new syntactic form, a *derivation*:

$$(+) : \frac{(0) : \dfrac{}{\mathrm{term}\ 0} \qquad (s) : \dfrac{(0) : \dfrac{}{\mathrm{term}\ 0}}{\mathrm{term}\ s0}}{\mathrm{term}\ +0s0}.$$

A derivation is a tree in which every node is labeled by a judgment and by an inference rule (in the current case, by a term and by a symbol from the signature). The judgment at the root is the *conclusion* of the derivation. The rules in the leaves are of arity zero (here, the introduction of a constant). If $M \in \Sigma^\star$, then $M \in \mathrm{T}_\Sigma$ if and only if there exists a derivation of this inference system whose conclusion is 'term M'. In general, the symbol $\vdash$ ('turnstile') is used to note, as in $\vdash j$, that there exists a derivation of j. Judgment j is called a *theorem* of the inference system. A new definition for terms is thus obtained:

$$\mathrm{T}_\Sigma = \{M \in \Sigma^\star; \vdash \mathrm{term}\ M\}.$$

It is clear that if $M \in \mathrm{T}_\Sigma$, the judgment 'term M' has a unique derivation whose form is the same as that of M: it is a very particular form of inference system.

Note that if the M is omitted from 'term M', a derivation for the judgment 'term' is obtained. The derivations are of the same form. In fact, the new derivation is another notation, treelike, for that term:

$$(+) : \frac{(0) : \dfrac{}{\mathrm{term}} \qquad (s) : \dfrac{(0) : \dfrac{}{\mathrm{term}}}{\mathrm{term}}}{\mathrm{term}}. \qquad \begin{array}{c} 0 \\ 0 \quad s0 \\ +0s0 \end{array}$$

Derivations of the judgment 'term' (which can be read as 'there exists a term') are all terms (or rather their notation in tree form). This structure of derivations is taken as the abstract syntax of terms. This syntax will be defined in the next section, and used extensively later on. A derivation of 'term' can then be defined to be a term.

Inference systems used from now on will of course be more complex than those associated with a signature.

Well-founded relations

In logic, reasoning must be 'well-founded', in a very definite sense. Write $j \succ j'$ if the judgment j is the conclusion of a rule for which j' is a premise. For terms, term $M \succ$ term M' means that M' is an immediate subterm of M. The *immediate subterms* of $fM_1 \dots M_n$ are $M_1, \dots, M_n$. For example, $s0$ is an immediate subterm of $+0s0$. A proof of judgment j would not be well-founded if there were an infinite number of premises $j \succ j_1 \succ j_2 \succ \dots$: this is the problem of 'primary causes' of philosophy. This situation cannot occur with terms since an immediate subterm of M can only be written with fewer symbols than M, and there does not exist an infinitely decreasing sequence in **N**.

Definition 2.2 A binary relation $\prec$ over a set E is *well-founded*, in short *wf*, if every non-empty subset A of E contains a minimal element a_0 in A, i.e, such that $a \not\prec a_0$ for all $a \in A$.

A wf relation is irreflexive (i.e, $x \not\prec x$ for all $x \in E$; take $A = \{x\}$ in the definition) and, more generally, acyclic (if $a_1 \prec a_2 \prec \dots \prec a_n \prec a_1$, take $A = \{a_0, \dots, a_n\}$). An order relation can be wf only over the empty set. A *wf order* will refer to a *strict* well-founded order. Even if a wf relation is not necessarily transitive, the terminology of order relations such as minimal element and decreasing sequence will be used.

An equivalent description of wf relations is: there does not exist an infinitely decreasing sequence $a_0 \succ a_1 \succ a_2 \succ \dots$. In fact, if there exists an infinitely decreasing sequence, the set A has no minimal element. On the other hand, if there exists a subset A which has no minimal element, take any $a_0 \in A$. Since a_0 is not minimal, there exists $a_1 \in A$ such that $a_1 \prec a_0$ and one can iterate to construct an infinite decreasing sequence.

Examples

- The usual strict order is wf over **N**, but not over **Z** or **Q**.
- The relation over **N**, '$m \succ n$ if m is the successor of n' is wf and is not transitive.
- The relation over T_Σ, '$M \prec N$ if N is an immediate subterm of M' is wf.
- The relation of divisibility over **Z** is a wf preorder.

Wf relations are also called *noetherian* due to an important algebraic property of which the previous example is only a particular case. A commutative ring is noetherian if every infinite sequence of ideals, increasing over the subset relation $\subseteq$, is stationary. In the case of **Z**, which is principal, an increasing sequence of ideals $(p_0) \subseteq (p_1) \subseteq (p_2) \subseteq \dots$ corresponds to a decreasing sequence of their generators: p_0 is divisible by p_1, which is divisible by p_2, etc. Many commonly used rings in algebra and geometry, such as polynomial rings and their quotients, are noetherian. It is a finiteness property: one of Hilbert's theorems asserts that every algebraic set is the set of zeros of a *finite* number of polynomials. It is therefore not surprising that computer science uses this notion to express the finite character of its operations.

Two general constructions of wf orders are useful: products and multisets.

Let $\succ_i$ be a well-founded order over a set E_i, $i = 1, 2$. Define the *lexical order* $\succ_\times$ over the product $E_1 \times E_2$ by $(x_1, x_2) \succ_\times (y_1, y_2)$ if $x_1 \succ_1 y_1$, or $x_1 = y_1$ and $x_2 \succ_2 y_2$. The product $\succ_\times$ is wf if and only if $\succ_1$ and $\succ_2$ are wf.

A *multiset* is a set with possibly multiple occurrences of its elements. More precisely, if E is a set, a multiset over E is a mapping $M : E \to \mathbf{N}$. For $x \in E$, $M(x)$ is the number of occurrences of x in M. The elements of a multiset M are the $x \in E$ such that $M(x) \neq 0$, written $x \Subset M$. A multiset M is finite if it has a finite number of elements. We will use a setlike notation for multisets. For example, $M = \{\!\{a, a, a, c\}\!\}$ instead of $M(a) = 3$, $M(c) = 1$, and $M(x) = 0$ for $x \neq a, c$. The set of finite multisets over E is written $\mathcal{M}(E)$. Multiset operations are defined in an analogous manner to the union, difference and intersection of sets. For example, union $\uplus$ is defined by $(M \uplus N)(x) = M(x) + N(x)$.

Let $\succ$ be a strict order relation over E. Extend it to $\mathcal{M}(E)$ by $M \ggcurly M'$ if M' is obtained from M by replacing an occurrence of an element x of M by a multiset of elements, each smaller than x.

Proposition 2.2 If $\succ$ is a strict order over E, then $(\mathcal{M}(E), \ggcurly)$ is well-founded if and only if $(E, \succ)$ is well-founded.

Well-founded induction This proof method uses induction over a well-founded relation, generalizing induction over the integers and structural induction over terms.

Definition 2.3 Let $\prec$ be a relation over a set E. A subset P of E is called $\prec$-*progressive* (or hereditary) if it contains an element whenever it contains its predecessors:

$$\forall x(\forall y(y \prec x \Rightarrow y \in P) \Rightarrow x \in P).$$

The property $x \in P$ is also called progressive.

Remark A progressive subset P contains, by definition, all the minimal elements of E (P contains the predecessors of a minimal element, since there are none!) □

Proposition 2.3 If $\prec$ is wf over E and if $P \subseteq E$ is $\prec$-progressive, then $P = E$.

Proof If $P \neq E$, then $A = E \setminus P$ is non-empty and so contains a minimal element a_0, which means that $y \prec a_0$ implies $y \notin A$. Since P is progressive, a_0 must be in P, which contradicts the assumption $a_0 \in A$. □

Examples

- Induction over the relation 'm is the successor of n' in $\mathbf{N}$ is exactly the usual induction.
- Induction over the usual order of $\mathbf{N}$ is called *complete induction*; to prove $P(m)$, suppose $P(n)$ for all $n < m$.
- Induction over the relation '$M \prec N$ if N is an immediate subterm of M' is structural induction over terms.

Well-orders and ordinals A *well-order* is a well-founded total order. The underlying set is called *well-ordered*. An equivalent characterization is that every subset has a least element; the set $\mathbf{N}$, with the usual order, is an example. But there are many other ways to well-order the same set: for example, $0 < 2 < 4 < 6 < \ldots < 1 < 3 < 5 < \ldots$ is a well-order for $\mathbf{N}$. The concept of ordinal classifies all well-orders. An *ordinal* is a set (in fact, if it is non-empty, it will be a set of ordinals, hence a set of sets) which is transitive and well-ordered by the relation $\in$. A set T is *transitive* if every element of T is a subset of T. It is easy to construct ordinals. Starting with the empty set, $\varnothing$, $\{\varnothing\}$, $\{\varnothing, \{\varnothing\}\}$ and $\{\varnothing, \{\varnothing\}, \{\varnothing, \{\varnothing\}\}\}$ are all ordinals. By writing them 0, 1, 2 and 3, we obtain $1 = \{0\}$, $2 = \{0, 1\}$ and $3 = \{0, 1, 2\}$, so $0 \in 1 \in 2 \in 3$. All the finite ordinals can be obtained from $0 = \varnothing$ by the successor operation $\alpha \mapsto \alpha + 1 = \alpha \cup \{\alpha\}$. The set of all the finite ordinals is an ordinal written ω—this is how $\mathbf{N}$ is constructed in set theory—which is not a successor. It is called a *limit* ordinal. This construction can be continued: $\omega + 1$ is the set of integers along with a greatest element, and $\omega + \omega$, the set of all $\omega + n$, is the ordinal corresponding to the well-order 'even integers < odd integers' over $\mathbf{N}$.

Monotone induction
There are two ways to define the set of terms; both are formulations of an inductive definition. From above, as in 'the smallest subset T such that ... ,' or from below, using an inference system, which is inherently more constructive. These two approaches can be interpreted using an operator over $\mathcal{P}(\Sigma^\star)$, the set of languages with alphabet Σ. If A is a language, define

$$\Phi_\Sigma(A) = \bigcup_{f \in \Sigma} \{f M_1 \ldots M_{\mathrm{ar}(f)}; M_1, \ldots, M_{\mathrm{ar}(f)} \in A\}.$$

Note that $\Phi_\Sigma(A)$ contains the constant symbols of Σ. The definition of T_Σ from above is given by

$$\mathrm{T}_\Sigma = \bigcap_{\Phi_\Sigma(E) \subseteq E} E.$$

Consider a more general situation. Let X be a set and $\Phi : \mathcal{P}(X) \to \mathcal{P}(X)$ a *monotone* (increasing) operator:

$$\text{if } A \subseteq B, \text{ then } \Phi(A) \subseteq \Phi(B).$$

(Throughout this discussion, increasing will not imply strictly increasing.) We will show that every monotone operator Φ has a least fixpoint $\mu\Phi$, and a greatest fixpoint $\bar{\mu}\Phi$, and that there are several ways to define them, as well as to compute them.

Not all monotone operators are defined through structural induction. For example, the set of words over an alphabet A can be obtained by the following rules: a word is empty, a symbol of A or the concatenation of two non-empty words. Structural induction is not used, for a non-empty word may be defined through concatenation in many different ways. This inductive definition can still be translated into a monotone operator Φ, where $\Phi(E) = \varnothing \cup A \cup \{uv; u, v \in E\}$, for $E \subseteq A^\star$.

These constructions include the usual induction over the integers, since the latter can be obtained from a monotone induction over the operator $\Phi : \mathcal{P}(\mathbf{N}) \to \mathcal{P}(\mathbf{N})$, defined by $\Phi(A) = \{0\} \cup \{n+1; n \in A\}$.

A subset E of X is *closed under* Φ if $\Phi(E) \subseteq E$. Consider the intersection of the subsets closed under Φ,

$$\operatorname{Ind}\Phi = \bigcap_{\Phi(E)\subseteq E} E,$$

called the set inductively defined by Φ. The following result is a special case of one of Tarski's theorems.

Proposition 2.4 If Φ is monotone, $\operatorname{Ind}\Phi$, the set inductively defined by Φ, is the least fixpoint of Φ.

Proof Let $I = \operatorname{Ind}\Phi$ and $\mathcal{E}$ be the set of E such that $\Phi(E) \subseteq E$. Since Φ is monotone, $I \subseteq E$ implies $\Phi(I) \subseteq \Phi(E)$, and so if $E \in \mathcal{E}$, then $\Phi(I) \subseteq E$. Hence $\Phi(I) \subseteq \bigcap_{E\in\mathcal{E}} E = I$. Conversely, $\Phi(I) \subseteq I$ implies $\Phi(\Phi(I)) \subseteq \Phi(I)$, i.e, $\Phi(I) \in \mathcal{E}$. By the definition of I, it follows that $I \subseteq \Phi(I)$: I is therefore a fixpoint of Φ. If I' is another fixpoint of Φ, then in particular $I' \in \mathcal{E}$, hence $I \subseteq I'$, which proves that $\operatorname{Ind}\Phi = \mu\Phi$. □

Let $A \subseteq X$. Since $\varnothing \subseteq A$, the increasing nature of Φ implies $\varnothing \subseteq \Phi(\varnothing) \subseteq \Phi(\Phi(\varnothing)) \subseteq \ldots$. An idea familiar in analysis is to iterate towards this fixpoint. An increasing sequence of subsets of X is computed by

$$\begin{aligned} \Phi^{\uparrow 0} &= \varnothing, \\ \Phi^{\uparrow n+1} &= \Phi(\Phi^{\uparrow n}), \\ \Phi^{\uparrow \omega} &= \bigcup_{n\geqslant 0} \Phi^{\uparrow n}. \end{aligned}$$

The assumption that Φ is monotone does not necessarily imply that $\Phi^{\uparrow\omega}$ is a fixpoint. The process must continue, starting with $\Phi^{\uparrow\omega+1} = \Phi(\Phi^{\uparrow\omega})$. However, there is an important property which allows one to stop at ω.

An operator is called *finitary* if its values are uniquely determined by the values taken when applying the operator to finite sets, i.e., if $x \in \Phi(A)$, there exists a finite $A' \subseteq A$ such that $x \in \Phi(A')$. If Φ is finitary and increasing, $\Phi(A) = \bigcup \Phi(E)$, E consisting of all the finite subsets of A.

If Σ is finite, the Φ_Σ operator is finitary, which means that each inference rule has a finite set of premises. It is easy to check, using induction over n, that $M \in \Phi_\Sigma^{\uparrow n}$ if and only if the judgment 'term M' has a derivation of height n, i.e, the term M is of height n. The Φ_Σ operator can be used to generate the theorems of the inference system starting from $\Phi_\Sigma^{\uparrow 1}$, i.e, from the axioms. This generation is 'stratified' by the height of the derivation. Hence

$$\mathrm{T}_\Sigma = \bigcup_{p \geqslant 0} \Phi_\Sigma^p(\varnothing).$$

Note that $\Phi_\Sigma(\varnothing) = \Sigma_0 = \{c \in \Sigma; \mathrm{ar}(c) = 0\}$.

Definition 2.4 A monotone operator Φ is *continuous* if for every increasing sequence $(E_n)_{n \geqslant 0}$ of subsets of X,

$$\Phi(\bigcup_{n \geqslant 0} E_n) = \bigcup_{n \geqslant 0} \Phi(E_n).$$

Lemma 2.5 A monotone and finitary operator Φ is continuous.

Proof Since Φ is monotone, $\bigcup_{n \geqslant 0} \Phi(E_n) \subseteq \Phi(\bigcup_{n \geqslant 0} E_n)$. Conversely, let $x \in \Phi(\bigcup_{n \geqslant 0} E_n)$. Since Φ is finitary, there exists a finite $A \subseteq \bigcup_{n \geqslant 0} E_n$ such that $x \in \Phi(A)$. As A is finite, there exists an integer p such that $A \subseteq \bigcup_{0 \leqslant n \leqslant p} E_n = E_p$, the sequence E_n being increasing. Hence $x \in \Phi(E_p)$. And so $\Phi(\bigcup_{n \geqslant 0} E_n) \subseteq \bigcup_{n \geqslant 0} \Phi(E_n)$. □

Proposition 2.6 If Φ is continuous, then $\mu\Phi = \mathrm{Ind}\,\Phi = \Phi^{\uparrow\omega}$.

Proof Since $\mathrm{Ind}\,\Phi$ is a fixpoint of Φ, it follows from $\varnothing \subseteq \mathrm{Ind}\,\Phi$ that $\Phi^{\uparrow n} \subseteq \mathrm{Ind}\,\Phi$, hence that $\Phi^{\uparrow\omega} \subseteq \mathrm{Ind}\,\Phi$. It remains to be shown that $\Phi^{\uparrow\omega}$ is a fixpoint of Φ, i.e, that $\Phi^{\uparrow\omega+1} = \Phi^{\uparrow\omega}$. By the continuity hypothesis,

$$\Phi^{\uparrow\omega+1} = \Phi(\bigcup_{n \geqslant 0} \Phi^{\uparrow n}) = \bigcup_{n \geqslant 0} \Phi(\Phi^{\uparrow n}) = \bigcup_{n \geqslant 1} \Phi^{\uparrow n} = \Phi^{\uparrow\omega}.$$

□

In the non-continuous case, the iteration must be continued. $\Phi^{\uparrow\alpha}$ can be defined for all ordinals α,

$$\begin{aligned} \Phi^{\uparrow 0} &= \varnothing, \\ \Phi^{\uparrow\alpha+1} &= \Phi(\Phi^{\uparrow\alpha}), \\ \Phi^{\uparrow\lambda} &= \bigcup_{\mu<\lambda} \Phi^{\uparrow\mu}, \end{aligned}$$

where λ is a limit ordinal, e.g., ω, $\omega + \omega$, ...

What can be shown, by a cardinality argument, because $\Phi^{\uparrow\alpha} \subseteq X$, is that there exists a smallest ordinal α such that $\operatorname{Ind}\Phi = \Phi^{\uparrow\alpha}$. Continuity implies that $\alpha \leqslant \omega$.

This transfinite iteration is not necessary in the case of continuous operators, the typical case for the usual inference systems, but it is necessary in the 'dual' case of iterative approximations from above, to compute the greatest fixpoint:

$$\begin{aligned} \Phi^{\downarrow 0} &= X, \\ \Phi^{\downarrow \alpha+1} &= \Phi(\Phi^{\downarrow \alpha}), \\ \Phi^{\downarrow \lambda} &= \bigcap_{\mu<\lambda} \Phi^{\downarrow \mu}. \end{aligned}$$

Continuity is no longer sufficient to ensure that $\Phi^{\downarrow\omega}$ is the greatest fixpoint. What is needed is an ordinal $\bar{\alpha}$ such that $\Phi^{\downarrow\bar{\alpha}} = \bar{\mu}\Phi$. Concrete examples of this situation can be found in logic programming, related to the proof method of negation as failure.

These results can be formulated in the more general structure of complete lattices. A lattice is an ordered set such that every pair of elements has an upper bound and a lower bound. A lattice is complete if every subset has an upper bound.

When inductive definitions were being presented, it was stated that these can provide proof methods, be it by induction over the integers or structural induction over terms. In the case of a monotone operator Φ, one proves a property P over the set $\operatorname{Ind}\Phi$. If E_P is the set of $x \in X$ satisfying P, it suffices to establish that E_P is closed under Φ to deduce that $\operatorname{Ind}\Phi \subseteq E_P$. One can also reason 'from below' by proving that for all n, $\Phi^{\uparrow n} \subseteq E_P$, i.e,

- $P(x)$ for all $x \in \Phi^{\uparrow 0}$ (vacuously true);
- if $P(y)$ for all $y \in \Phi^{\uparrow n}$, then $P(x)$ for all $x \in \Phi^{\uparrow n+1}$.

The proof has been converted into a proof by induction over the integers. For terms, the proof would be over the height of a term. However, using integers or, more generally, ordinals (*transfinite induction*) is not necessary when structural or monotone induction is possible.

2.2.3 Examples

Terms are the natural syntactic entities to represent the objects described in this book: inference system derivations, data structures, applicative terms, and propositional calculus formulas. For the moment, only the form of these objects will be described: names will be given only cursory meanings. Later on, their semantics, or precise definitions of their meanings, will be given.

Data structures

Terms are the most important objects of symbolic computation, as well as of functional and logic programming: Lisp atoms and pairs, Prolog terms and (first-order) ML concrete types are all terms. Their abstract structure favors a clear and concise 'algebraic' programming style.

A type is declared by introducing a signature Σ. The concrete values of this type are the elements of T_Σ. At least one constant symbol is needed for the type to be non-empty.

Example The signature for integers is given by $\Sigma = \{0, s\}$, with $\mathrm{ar}(0) = 0$ and $\mathrm{ar}(s) = 1$, since $\mathrm{T}_\Sigma = \{0, s0, \dots, s^p0, \dots\} \simeq \mathrm{N}$. The integers are written in a unary form, which, although not convenient for arithmetic computations, does represent the operational character of the integers: consider loop increment (`i++` of C), iteration and recursion. In ML, this type would be declared by

```
datatype nat = zero | s of nat ;
```

and in Prolog by

```
nat(zero).
nat(s(X)) :- nat(X).
```

□

Normally one manipulates several different sorts (or types) of objects, for example integers, strings, atoms and lists. An operation of a given arity cannot be applied to any sort of object. After all, what is the successor of a list? The rules for term formation will be restricted through the use of type declarations.

Example Two *sort* symbols are introduced: 'atom' and 'list', and typed constant and function symbols:

$$\begin{aligned} a, b, c, \dots &: \text{atom}, \\ nil &: \text{list}, \\ cons &: \text{atom}, \text{list} \to \text{list}, \end{aligned}$$

The rules for term formation are:

1. a, b, c, ... are terms of sort 'atom'.
2. *nil* is a term of sort 'list'.
3. If A is a term of sort 'atom', and if L is a term of sort 'list', then $cons(A, L)$ is a term of sort 'list'.

Hence $cons(a,nil)$ is a term, but not $cons(a, b)$. The ML declaration for the type could be

```
datatype atom = a | b
and      list = nil | cons of atom * list ;
```

(one can do better in ML, using type variables; we will return to this subject), and the Prolog one could be

```
atom(a).
atom(b).
list(nil).
list(cons(A,L)) :- atom(A), list(L).
```

□

Definition 2.5 A *many-sorted signature* is a pair (S, Σ) where

- S is a set of *sort* symbols;
- Σ is a set of typed constant and function symbols

$$\begin{aligned} c &: s, \\ f &: s_1, \ldots, s_n \to s, \end{aligned}$$

with $s_1, \ldots, s_n, s \in S$ and $n \geqslant 1$.

The rules for term formation are given by the following inference system, dealing with judgments of the form 'term $M : s$', with one n-ary rule for each n-ary symbol:

$$(c) : \frac{}{\text{term}\, c : s} \qquad\qquad (f) : \frac{\text{term}\, M_1 : s_1 \qquad \ldots \qquad \text{term}\, M_n : s_n}{\text{term}\, f M_1 \ldots M_n : s},$$

where $c, f \in \Sigma$ and $c : s$, $f : s_1 \ldots s_n \to s$. M is a term of sort s if there exists a closed derivation of 'term $M : s$' using this inference system, written $\vdash \text{term}\, M : s$. The set of terms of sort s is written $\mathrm{T}_{\Sigma,s}$, and T_Σ is defined by $\mathrm{T}_\Sigma = \bigcup_{s \in S} \mathrm{T}_{\Sigma,s}$.

The set of these terms can be implemented in ML as a concrete type, in a manner similar to lists. The many-sorted signature (S, Σ) is associated with the type definition:

```
datatype s1 = d1 and s2 = d2 and ...  and sn = dn ;
```

where $S = \{s_1, s_2, \ldots, s_n\}$ and each `si = di` is of the form

```
si = a | b | ... | f of s_p1 * ... * s_pr | ...
```

where a, b, ... are constant symbols of sort `si`, and the `f` are constructors whose type is $\mathtt{s}_{p_1}$ `* ... *` $\mathtt{s}_{p_r}$ `-> s`.

Applicative terms

Undoubtedly one of the most common operations is the application of a 'function' to an 'argument'. The meaning of this operation will be given in the next chapter. From a syntactic point of view, an *applicative* signature will be used. The only symbol of arity $\geqslant 1$ is the lone binary symbol, **App** (also written @): terms of this signature are called *applicative terms*. Application is here noted in an explicit manner: one would write **App** $\cos x$ where standard mathematical usage would be cos(x). A simplified notation will be used most of the time. Write (M_1M_2) instead of $\mathbf{App}\, M_1M_2$ and omit unnecessary parentheses, assuming that application is left-associative:

$$(\ldots((M_1M_2)M_3)\ldots M_n) \qquad \text{is written} \qquad M_1\, M_2\, M_3\, \ldots M_n.$$

External parentheses are removed.

Applicative terms are a sort of universal syntax for terms. In fact, a mapping $M \mapsto M^@$ can be defined from terms constructed over an arbitrary signature Σ into terms over an applicative signature $\Sigma^@$ containing all of the elements of Σ as constants, along with **App**. A constant is translated into itself and a Σ-term $fM_1\ldots M_n$ is translated into the $\Sigma^@$-term $\mathbf{App}\ldots\mathbf{App}\,\mathbf{App}\, fM_1^@M_2^@\ldots M_n^@$ or, using the simplified notation omitting the **App**, into the $\Sigma^@$-term $fM_1\ldots M_n$.

Example The term $f(a, g(a,a), h(a))$ is represented in Figure 2.1 as a Σ-term (to the left) and as a $\Sigma^@$-term (to the right). □

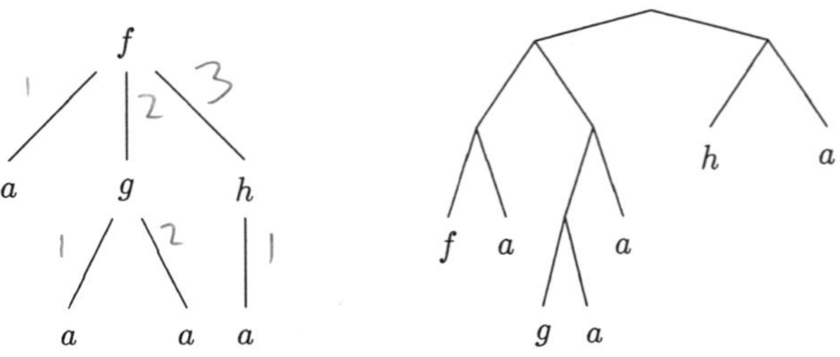

Figure 2.1 Representations of a term by a tree

This mapping is clearly not a bijection, for the $\Sigma^@$-term $\mathbf{App}(a, a)$ cannot be obtained from a Σ-term. The left (respectively right) leaves of $M^@$ are labeled by function (respectively constant) symbols of Σ. The right nodes of $M^@$ correspond to subterms of M. Note that this mapping of a term into an applicative term is different, at the tree level, from the usual transformation in computer science, where a tree is mapped into a binary tree, by linking each node to its 'eldest child' and its 'younger sibling'.

Formulas of the propositional calculus

Formulas are terms constructed with connectives and atoms. Begin by introducing a signature Σ_c formed of *connectives*: here are the common ones. There are binary connectives,

- *implication* ($\Rightarrow$, sometimes written $\supset$ or $\rightarrow$), *equivalence* ($\Longleftrightarrow$), *conjunction* ($\wedge$, 'and'), and *disjunction* ($\vee$, 'or');

a unary connective,

- *negation* ($\neg$, 'not', sometimes written $\sim$);

and constant connectives,

- *false* ($\bot$) and *true* ($\top$).

To these connectives is added a set A of *atomic propositions*, or atoms, or propositional variables, as constant symbols. These are written p, q, r, ... A new signature $\Sigma = \Sigma_c \cup A$ is generated. A term of this signature is called a *proposition*, or formula of the propositional calculus. The set of propositions is written Prop$[A]$.

2.2.4 Terms and trees

Let M be a term represented by a tree. Number the arcs emanating from each node from left to right, starting with 1. Consider as an example the term represented by the tree on the left of Figure 2.1. A particular occurrence of a symbol of M can be designated by the word obtained by concatenating the numbers of the arcs of the path leading from the root to the symbol occurrence: ϵ, 1, 2 and 21, respectively, yield f, a, g and a. These words are actually called *occurrences*. They are elements of $A^\star$, where A is the set of integers $\geqslant 1$. The prefix order will be used between occurrences: $u \leqslant v$ states that u is an ancestor of v. A given occurrence u of M determines a node in a tree, the symbol that labels it and the subterm (subtree) which emanates from it.

For each term M, define

- the set $\mathcal{O}(M)$ of occurrences of M;
- the symbol $M(u)$, for $u \in \mathcal{O}(M)$;
- the subterm M/u, for $u \in \mathcal{O}(M)$;

by

1. if $M = c \in \Sigma$, then $\mathcal{O}(M) = \{\epsilon\}$, $M(\epsilon) = c$, $M/\epsilon = c$;
2. if $M = fM_1 \dots M_n$, then

$$\mathcal{O}(M) = \{\epsilon\} \cup \textstyle\bigcup_{i=1}^{n} i \cdot \mathcal{O}(M_i)$$

$$\begin{array}{rclcrcl} M(\epsilon) & = & f & \quad & M/\epsilon & = & M \\ M(i \cdot u) & = & M_i(u) & \quad & M/i \cdot u & = & M_i/u. \end{array}$$

The following properties are easy to visualize on the tree representation. They are proven using induction over proof terms. See exercise 6.

Proposition 2.7 If M is a finite term, then

1. if $u \leqslant v$ and $v \in \mathcal{O}(M)$, then $u \in \mathcal{O}(M)$;
2. if $1 \leqslant i \leqslant j$ and $u \cdot j \in \mathcal{O}(M)$, then $u \cdot i \in \mathcal{O}(M)$;
3. if $u \in \mathcal{O}(M)$ and $\mathrm{ar}(M(u)) = n$, then $u \cdot 1, \ldots, u \cdot n \in \mathcal{O}(M)$, $u \cdot (n+1) \notin \mathcal{O}(M)$;
4. $\mathcal{O}(M)$ is a finite set.

These properties give a new definition of terms, as trees. The first two properties define trees to be subsets of $A^\star = \{1, 2, 3, \ldots\}^\star$, i.e, languages. The third property restricts how trees can be labeled (the degree of a node = the arity of a symbol). The last property is a finiteness condition, although infinite trees and terms can also exist (Prolog II uses infinite terms).

A subset $\mathcal{O}$ of $A^\star$ is a *tree domain*, or an *occurrence language*, if it satisfies properties (1) and (2) (prefix and 'age' stability). Tree terminology is used:

- ϵ is the *root* of $\mathcal{O}$;
- elements of $\mathcal{O}$ are *nodes*;
- nodes of degree zero are *leaves*.

Note that if $\mathcal{O}$ is a tree domain and that if $u \in \mathcal{O}$, then $\mathcal{O}/u = \{v/u;\, v \in \mathcal{O} \text{ and } v \geqslant u\}$ is also a tree domain.

Example $\{\epsilon\}$, $\{\epsilon, 1, 2\}$, $\{\epsilon, 1, 2, 11\}$ and $\{1, 2\}^\star$ are tree domains. The last one is infinite. □

Terms are well-labeled trees:

Definition 2.6 Let Σ be a signature. A (finite or infinite) Σ-*term* is a mapping $M : \mathcal{O} \to \Sigma$ satisfying properties (1), (2) and (3) of the preceding proposition. The term is a *finite* Σ-term if (4) is also satisfied.

The domain of M, as a mapping, will be written $\mathcal{O}(M)$. For the moment only finite terms will be considered. It is then easy to define the subterm M at occurrence u by

$$\begin{aligned} M/u : \mathcal{O}(M)/u &\to \Sigma, \\ v &\mapsto M(u \cdot v), \end{aligned}$$

and to check that M/u is a term.

The operation complementary to subterm construction is the graft, whose inductive definition is given here. Let $M, N \in \mathrm{T}_\Sigma$ and $u \in \mathcal{O}(M)$. If $M = fM_1 \ldots M_n$, with $n \geqslant 0$, define

$$\begin{aligned} M[\epsilon \leftarrow N] &= N, \\ M[i \cdot u \leftarrow N] &= fM_1 \ldots M_{i-1}M_i[u \leftarrow N]M_{i+1} \ldots M_n. \end{aligned}$$

$M[u \leftarrow N]$ is a term and (see exercise 7)

$$M[u \leftarrow N](v) = \begin{cases} N(w), & \text{if } v = u \cdot w \text{ and } w \in \mathcal{O}(N) \\ M(v), & \text{if } u \not\leqslant v \text{ and } v \in \mathcal{O}(M). \end{cases} \tag{2.1}$$

$M[u \leftarrow N]$ is obtained from *grafting* N *on to* M *at occurrence* u. The following properties, given as an exercise, will be used:

$$\begin{aligned} M[u \leftarrow N]/(uv) &= N/v, \\ M[u \leftarrow N][uv \leftarrow P] &= M[u \leftarrow N[v \leftarrow P]], \\ M[u \leftarrow N]/v &= M/v \quad \text{if } u \perp v, \\ M[u \leftarrow N][v \leftarrow P] &= M[v \leftarrow P][u \leftarrow N] \quad \text{if } u \perp v, \\ M[u \leftarrow N]/v &= (M/v)[u/v \leftarrow N] \quad \text{if } v \leqslant u, \\ M[u \leftarrow N][v \leftarrow P] &= M[v \leftarrow P] \quad \text{if } v \leqslant u. \end{aligned}$$

It is useful to introduce the concept of *context*, which allows one to consider the graft as a binary operation between a context, rather than a term, and an occurrence. A context is a term with a single hole, ready to receive the graft of a term. For example, $f(a, h(a, [\ ], b))$ is a context $\mathcal{C}$ and $\mathcal{C}[N] = f(a, h(a, N, b))$. For a precise definition, see exercise 8.

As an example of the use of occurrences, here is a precise definition of the translation of Σ-terms into applicative $\Sigma^{@}$-terms introduced in §2.2.3. If M is a Σ-term, define the $\Sigma^{@}$-term $M^{@}$ by

$$\begin{aligned} c^{@} &= c & &\text{if } c \in \Sigma_0 \\ M^{@}(1^p) &= \mathbf{App} & &(0 \leqslant p < \mathrm{ar}(M)) \\ M^{@}(1^{\mathrm{ar}(M)}) &= M(\epsilon) & & \\ M^{@}/(1^{\mathrm{ar}(M)-p}\, 2) &= (M/p)^{@} & &(1 \leqslant p \leqslant \mathrm{ar}(M)). \end{aligned}$$

2.2.5 Concrete syntax

It is just as useful to have a useful syntactic representation as it is to know the abstract syntax of terms. An injective mapping from T_Σ to a particular subset of $\Gamma^\star$ must be constructed. Its range must be easily recognizable and the mapping must be easily inverted. Such a mapping is called a *concrete syntax* for terms. A few common examples will be examined.

Prefix notation was used right from the beginning, i.e., this is Łukasiewicz's 'Polish notation':

$$\begin{aligned} \mathrm{pref} : \mathrm{T}_\Sigma &\rightarrow \Sigma^\star, \\ M &\mapsto f \cdot \mathrm{pref}(M/1) \ldots \mathrm{pref}(M/n), \end{aligned}$$

where $f = M(\epsilon)$ and $n = \mathrm{ar}(f)$.

Example Let M be the term represented in Figure 2.1.

$$\begin{aligned}\text{pref}(M) &= f \cdot \text{pref}(M/1) \cdot \text{pref}(M/2) \cdot \text{pref}(M/3) \\ &= f \cdot a \cdot g \cdot \text{pref}(M/21) \cdot \text{pref}(M/22) \cdot h \cdot \text{pref}(M/31) \\ &= fagaaha.\end{aligned}$$

□

The problem of syntactic analysis is: given a word $x \in \Sigma^\star$, is x the prefix notation of a term, and, if so, of what term? The answer to this problem must be a *parsing* algorithm. Using exercise 9, a recursive descent (top–down) parser can be generated. On the other hand, suffix notation is well adapted to bottom–up parsing:

$$\begin{aligned}\text{suff} : \mathrm{T}_\Sigma &\rightarrow \Sigma^\star, \\ M &\mapsto \text{suff}(M/1)\ldots\text{suff}(M/n) \cdot f,\end{aligned}$$

where $f = M(\epsilon)$ and $n = \text{ar}(f)$.

Example (cont.) $\text{suff}(M) = aaagahf$. □

Parenthesized notation inserts parentheses and commas in the middle of infix notation. What is added is called 'syntactic sugar'. It facilitates reading by indicating the arity of symbols and corresponds to mathematical usage:

$$\begin{aligned}\text{paren} : \mathrm{T}_\Sigma &\rightarrow \Sigma^\star, \\ M &\mapsto f\text{`('}\text{paren}(M/1)\text{`,'}\ldots\text{`,'}\text{paren}(M/n)\text{`)'},\end{aligned}$$

where $f = M(\epsilon)$ and $n = \text{ar}(f)$.

Example (cont.) $\text{paren}(M) = f(a, g(a, a), h(a))$. □

Lisp also uses parentheses, in the prefix expressions of the form `(f M1 ... Mn)`. Syntactic sugar is not being used here, as Lisp's functions are not always of fixed arity. The parentheses are therefore needed.

Binary notation is often written using *infix* notation: $(2 + 3)$. Parentheses are necessary to avoid ambiguities, but they can be eliminated at the external level. The concrete syntax of each programming language represents terms in its own manner. Here is an example from C: `(a-- ? a+a : a++)`.

2.2.6 Variables

The definition given for terms is not the classical one, as it omits variables. Yet variables can be found everywhere, in numerous forms, in logic and in programming. From a syntactic point of view, a distinction will be made between a free variable, as in terms, and a bound variable, as in λ-terms. In programming, each kind of variable is associated with a computing paradigm. For cultural reasons, Pascal's *sequential* variable is probably the most familiar. Its behavior is very different from that of the simple *applicative* variable (a name designating a value, with no possibility of the latter being changed), of the *functional* variable of ML and of the *logic* variable of Prolog, all of which will be studied, as well as stream variables (designating sequences of values) of dataflow languages and architectures. In mathematics, variables are used to write down equations, in particular, fixpoint equations of the form $x = \phi(x)$. From a logical point of view, two kinds of variable will be studied: the *assumption* variable, the same as the λ-calculus variable, from the Curry–Howard correspondence; and the *unknown* variable, the basis for logic programming.

The introduction of variables will first be justified for finite terms.

Symbols and metasymbols

Since the beginning of this chapter, the notation has included explicitly declared symbols such as 0, s and nil, as well as *metasymbols* such as M in 'let M be a term'. It would be difficult to write down rules for constructing terms without using such metasymbols. One could not write such things as $+MsN$, which is not a term unless $M, N \in \Sigma$. These are *metaterms*, which have been used constantly without being mentioned explicitly.

Consider the definition of terms as an inference system. Note first that the writing of the inference rules uses metaterms,

$$(+) : \frac{\text{term } M_1 \qquad \text{term } M_2}{\text{term } +M_1M_2},$$

which can be replaced by any term in the construction of derivations. Derivations whose judgments contain metaterms can also be constructed: what meaning should be attributed to these derivations?

The difference between term $+0s0$ and metaterm $+0sM$ can be seen in the derivations:

$$(+) : \frac{(0) : \dfrac{}{\text{term } 0} \qquad (s) : \dfrac{(0) : \dfrac{}{\text{term } 0}}{\text{term } s0}}{\text{term } +0s0} \qquad\qquad (+) : \frac{(0) : \dfrac{}{\text{term } 0} \qquad (s) : \dfrac{\text{term } M}{\text{term } sM}}{\text{term } +0sM}.$$

In the first case, there is a derivation of 'term $+0s0$'. In the second case, there is a derivation of 'term $+0sM$' which depends upon the *assumption* 'term M'. The symbol $\vdash$'s use is expanded, with the assumptions of the derivation written to its left:

$$\text{term } M \vdash \text{term } +0sM.$$

The concept of derivation becomes relative, hence useful, in logic.

Finally, when computing rules are introduced to state, for example, that 0 is the right neutral element for $+$, term metasymbols can be used to avoid writing an infinite number of rules between terms. Instead of writing

$$0+0 \to 0 \qquad 1+0 \to 1 \qquad 2+0 \to 2 \; \dots,$$

write $M+0 \to M$.

The need for metasymbols is not unique to logic. Whatever the case may be, it is important to distinguish metasymbols from the objects they denote when the latter are themselves syntactic objects.

To form *metaobjects*, a set of metasymbols is declared explicitly, along with the initial symbols, those of the signature. Formal language definitions use this method, where the set A of *terminals*, the symbols of the language, is distinct from the set Z of *non-terminals*, or syntactic variables. The latter are the metasymbols: examples from a programming language would be *program*, *expression* and *instruction*. If the words of the language are formed over the alphabet A, *syntactic forms* are words formed over $Z \cup A$.

The same procedure is used for terms. A set X of *variable* symbols is introduced, the arity function ar is extended to $\Sigma \cup X$ by assigning 0 to each element of X and the set $\mathrm{T}_\Sigma[X] = \mathrm{T}_{\Sigma \cup X}$ is constructed. Its elements are called *terms*.

For the moment, nothing distinguishes a variable from a constant. However, the way in which variables are going to be used requires a reformulation of the inductive definition of terms:

Definition 2.7 $\mathrm{T}_\Sigma[X]$ is the smallest subset E of $(\Sigma \cup X)^\star$ such that

1. if $c \in \Sigma$, $\mathrm{ar}(c) = 0$, then $c \in E$;
2. if $f \in \Sigma$, $\mathrm{ar}(f) = n \geqslant 1$, and if $M_1, \dots, M_n \in E$, then $fM_1 \dots M_n \in E$;
3. if $x \in X$, then $x \in E$.

The first and last cases would be treated differently, for example, to define the set of variables of a term M, written as $\mathrm{var}(M)$:

1. $\mathrm{var}(c) = \varnothing$ for $c \in \Sigma$, $\mathrm{ar}(c) = 0$;
2. $\mathrm{var}(fM_1 \dots M_n) = \bigcup_{i=1}^{n} \mathrm{var}(M_i)$ if $f \in \Sigma$, $\mathrm{ar}(f) = n$;
3. $\mathrm{var}(x) = \{x\}$, if $x \in X$.

The set $\mathrm{T}_\Sigma = \mathrm{T}_\Sigma[\varnothing]$ is called the set of *closed* terms.

Substitutions

So that variables can play the role of term metasymbol, a substitution operation must be introduced.

Definition 2.8 A *substitution* is any function $\theta : X \to \mathrm{T}_\Sigma[X]$, which is the identity function almost everywhere, i.e., except for a finite subset of X.

The finite set $\mathrm{dom}(\theta) = \{x \in X \,;\, \theta(x) \neq x\}$ is called the *domain* of θ.

If $\mathrm{dom}(\theta) = \{x_1, \ldots, x_n\}$, x_i distinct, then θ can be represented by the set of variable-term pairs $\{(x_1, \theta(x_1)), \ldots, (x_n, \theta(x_n))\}$, or by the 'array' $\left[\begin{smallmatrix} x_1 & \ldots & x_n \\ \theta(x_1) & \ldots & \theta(x_n) \end{smallmatrix}\right]$. Substitution θ operates on $\mathrm{T}_\Sigma[X]$ by

1. $\theta\, c = c$ for $c \in \Sigma$, $\mathrm{ar}(c) = 0$;
2. $\theta\,(f M_1 \ldots M_n) = f(\theta\, M_1) \ldots (\theta\, M_n)$ if $f \in \Sigma$, $\mathrm{ar}(f) = n$;
3. $\theta\, x = \theta(x)$, if $x \in X$, otherwise x.

This action extends θ to a function from $\mathrm{T}_\Sigma[X]$ to itself, also called θ. The third clause replaces each variable by the corresponding term and the first two clauses propagate this transformation in the term. Note also that $\theta\, M = M[x_1 := \theta(x_1); \ldots ; x_n := \theta(x_n)]$. The interpretation of a substitution on a tree is simple: $\theta\, M$ is obtained by grafting the term $\theta(x_i)$ on to each occurrence of x_i in M (see Figure 2.2).

Figure 2.2 Substitution

In the case of a variable, $M[x := N] = M[u_1 \leftarrow N, \ldots , u_n \leftarrow N]$, where the u_i, $1 \leqslant i \leqslant n$, are the (independent) occurrences of x in M.

The set of substitutions is written $\mathbb{S}$. An object containing metasymbols can now be considered to be a class of terms obtained by substitutions: for example the metaterm $+MssN$ denotes the set of terms $\theta(+xssy)$, with θ covering $\mathbb{S}$.

When variables are introduced into inference systems, derivations can depend on assumptions. When derivations are considered as trees, their leaves can be labeled either by a zero-arity rule and by the judgment that it introduces, or by an arbitrary judgment called an assumption. The set of hypotheses of a derivation is defined in the same way as the set of variables of a term. If a judgment j is an assumption in a derivation, j must be interpreted as an 'undefined' derivation, or a 'variable' derivation of j, which can be substituted with any derivation of j. A derivation without assumptions is called closed.

Other kinds of variable will be introduced later: bound variables, those of the λ-calculus (the functional variable) and of predicate calculus (the logic variable). Furthermore, the variables of programming languages with assignments, such as Pascal, are themselves of a different nature. Terms, generally called expressions, can also be constructed, but the assignment operation prevents substitutions from taking place. For example, one can write $x := x + 1$, but one certainly cannot make the substitution $\left[\begin{smallmatrix} x \\ 2 \end{smallmatrix}\right]$ which would give $2 := 3$. In fact, a clearer syntax would be to write something like $x := !x + 1$, as in ML, but it is difficult to break tradition.

2.2.7 Shared representations and term implementations

To define a signature as an inference system and terms as trees allows for inductive reasoning over terms as well as for programming in a simple algebraic style. But this representation is not optimal. From an implementation point of view, it is correct to see $\mathcal{O}(M)$ as the set of 'addresses' where the symbols of M are stored. The definition of the set $\mathcal{O}(M)$ of occurrences of a term M is too fine. One can graft distinct terms on to each of the occurrences of variable x in $f(x, x)$, even though the substitution operation, much used below, is incapable of distinguishing the two occurrences. If substitutions only are used, it suffices to have one address per variable.

More generally, subterms of a term can be shared. The address–occurrence set has a directed acyclic graph (dag) structure, instead of a tree structure. This dag can be obtained by a quotient of M as a labeled tree structure, by an equivalence relation $\sim_M$ over $\mathcal{O}(M)$ defining the sharing of subterms. This relation must satisfy the property: if $u \sim_M v$ then $M/u = M/v$. In particular,

1. if $u \sim_M v$, then $ui \in \mathcal{O}(M)$ if and only if $vi \in \mathcal{O}(M)$, and then $ui \sim_M vi$;
2. if $u \sim_M v$, then u and v are independent under the prefix order.

A uniform method of sharing is needed, valid for all terms. In particular, a sharing relation $\sim_M$ naturally defines a relation $\sim_{M/w}$ for each subterm M/w, defined by $u \sim_{M/w} v$ if $wu \sim_M wv$. Several degrees of sharing can be used:

- $u = v$: no sharing;
- $M(u) = M(v) \in X$: sharing of variables;
- $M/u = M/v$: maximal sharing, of all the subterms.

In the latter case, a term M will be represented by a *dag* whose nodes are the subterms of M. For example, $h(h(f(a), f(a)), g(a))$ is represented by the graph

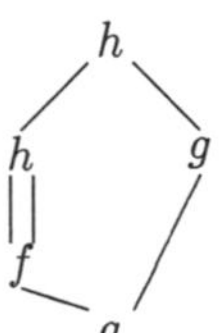

To implement terms, one must be able to represent trees and graphs and implement operations to compute to $M(u)$, M/u and M/ui. 'Linked' structures are used. These can easily be written in a language with pointers, such as C. An occurrence is implemented as a pair formed of a symbol (a pointer to a string of characters) and a pointer to a sequence of occurrences:

```
struct occurrence
  {char *symbol ;
   struct occurrence **arguments ;}
```

With the declaration

```
struct occurrence *M ;
```

the value of $M(\epsilon)$ is `M->symbol` and that of $M/i+1$ is `M->arguments[i]`.

The term $f(a, g(a, x), h(x), h(x))$ would be represented by one of the memory structures of Figure 2.3, depending on the level of sharing: no sharing, sharing of variables or sharing of subterms.

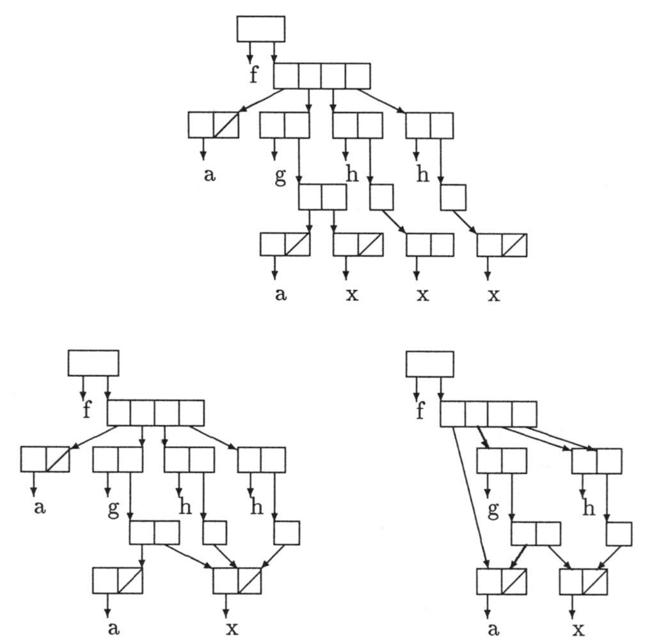

Figure 2.3 Three representations of $f(a, g(a, x), h(x), h(x))$

Another method of representation consists of coding them as applicative terms (§2.2.3); in this case, the constituents of a pair are either a pointer to a pair or a symbol.

2.3 Terms of the λ-calculus

The terms of the λ-calculus, or λ-terms, will be used to define functions. The next chapter will show how to compute using these terms. The notation is familiar in computer science: the function $x \mapsto x + 1$ is written

```
(lambda (x) (+ x 1))
```

in Lisp and

```
fn x => x+1
```

in ML. It would be written $\lambda x\,.\,(+\,x\,1)$ in a λ-calculus extended with the constants $+$ and 1.

2.3.1 The λ-notation

The symbols 'λ', '(' and ')' are introduced, along with a countably infinite set X, whose elements are called *variables*.

Definition 2.9 The rules for forming λ-*calculus terms*, or λ-terms, follow:

1. Each variable is a term.
2. If x is a variable and M is a term, then $(\lambda x\, M)$ is a term.
3. If M and N are terms, then $(M N)$ is a term.

This syntax can be described by the productions:

$$\textit{term} ::= \textit{variable} \,|\, (\lambda \textit{variable term}) \,|\, (\textit{term term}).$$

The term $(\lambda x\, M)$ is an *abstraction* which stands for the 'function $x \mapsto M$'. The term $(M N)$ stands for the application of the 'function M' to the 'argument N'. It is an *application*. The syntactic equality of two terms is written $\equiv$. The symbol $=$ will be reserved for semantic equality, in the same way that $2+2=3+1$ while $2+2 \not\equiv 1+3$, the two expressions being distinct syntactically.

Example x, $(\lambda x\, x)$, $(\lambda x\, y)$, $(x\,(\lambda x\, y))$ and $(\lambda x\,(y\,(\lambda z\,(x\, z))))$ are all terms. □

Note that the same term can be used to the right (as 'argument') or to the left (as 'function') of an application.

Simplified notation To simplify reading and writing, application is considered to be left-associative and abstraction right-associative. Simplify

$$(((M_1\, M_2)\, M_3) \ldots M_n) \quad \text{into} \quad M_1\, M_2\, M_3\, \ldots M_n$$

and

$$(\lambda x_1\,(\lambda x_2\, \ldots (\lambda x_n\, M))) \quad \text{into} \quad \lambda x_1 x_2 \ldots x_n\,.\,M.$$

The outer parentheses are removed.

Curryfication A 'function of several variables' is defined in the λ-calculus by successive abstractions on its variables. Thus one can consider only 'functions of a single variable'. This can only be done using 'higher-order functions', which take functions as arguments or return functions as results. This fact, discovered by Schönfinkel in 1924 ([132]), gives the process known as *curryfication*, named after H. Curry, of functions in functional programming. For example, the function h of two variables is replaced by the function $h^\star$ of one variable, defined by $(h^\star(x))(y) = h(x, y)$.

λ-calculi with constants As for applicative terms, one can define λ-terms constructed from a given set of constants which play the role of predefined values and functions of programming languages. The constants $\mathbf{0}, 1, 2, \ldots$, $+$, $*$, `if`, ... , each with specific rules, can be introduced to write terms similar to Lisp expressions, as in $\lambda x \,.\, +(*\, \mathbf{2}\, x)\mathbf{1}$. These constants will sometimes be used in examples, particularly in functional programming. However, the 'expressive power' of the calculus is in no way changed by the introduction of the constants, for the pure λ-calculus, with its impoverished syntax, can represent all imaginable data structures and algorithms.

Examples The following terms are well known. The symbol $\equiv$ is also used to associate a term with a name:

$\mathbf{I} \equiv \lambda x \,.\, x$	*identity*
$\circ \equiv \lambda fgx \,.\, f(gx)$	*composition*
$\mathbf{0} \equiv \lambda fx \,.\, x$	
$\mathbf{1} \equiv \lambda fx \,.\, fx$	*Church's iterators or numerals*
$\mathbf{2} \equiv \lambda fx \,.\, f(fx)$	

□

2.3.2 Abstract syntax

Now that the concrete syntax has been given, the manipulation of terms and their implementation in a programming language requires the introduction of an abstract syntax. There are two levels of abstract syntax: the first corresponds to a superficial analysis of a term, the second incorporates an interpretation of abstraction.

Shallow level
By analogy with Σ-terms, a λ-term is represented as a tree by considering each λx as a unary symbol, application as a binary symbol written '@' and each variable a symbol of arity zero. Hence $M \equiv \lambda x \,.\, x(\lambda y \,.\, yx)$ is represented by the tree on the left of Figure 2.4. This representation is convenient and allows one to define the domain $\mathcal{O}(M)$ of a term M, occurrences of variables (the leaves of the tree), λ (unary nodes) and application (binary nodes), subterms M/u and grafts $M[u \leftarrow N]$, for $u \in \mathcal{O}(M)$. It is one of the abstract representations of λ-terms.

Unlike the case for finite terms, where it was necessary to manipulate arbitrary signatures, a specialized alphabet will be used here to designate symbol occurrences in a tree: $A = \{\lambda, 1, 2\}$. Let M be a term represented by a tree. Number the arcs emanating from each node: λ for the arc from a λ, 1 and 2 for the left and right arcs from an application @. A particular symbol of M is designated by the word obtained by concatenating the numbers of the arcs of the path leading from the root to this symbol.

The definitions given for finite Σ-terms are easily adapted to λ-terms:

1. variable: $\mathcal{O}(x) = \{\epsilon\}$ and

$$x(\epsilon) = x \qquad x/\epsilon = x;$$

2. abstraction: $\mathcal{O}(\lambda x\, M) = \{\epsilon\} \cup \lambda \cdot \mathcal{O}(M)$ and

$$\begin{aligned} (\lambda x\, M)(\epsilon) &= \lambda x & (\lambda x\, M)/\epsilon &= \lambda x\, M \\ (\lambda x\, M)(\lambda \cdot u) &= M(u) & (\lambda x\, M)/\lambda \cdot u &= M/u; \end{aligned}$$

3. application: $\mathcal{O}(MN) = \{\epsilon\} \cup 1 \cdot \mathcal{O}(M) \cup 2 \cdot \mathcal{O}(N)$ and

$$\begin{aligned} (MN)(\epsilon) &= @ & (MN)/\epsilon &= MN \\ (MN)(1 \cdot u) &= M(u) & (MN)/1 \cdot u &= M/u \\ (MN)(2 \cdot u) &= N(u) & (MN)/2 \cdot u &= N/u. \end{aligned}$$

Example If $M \equiv \lambda x \,.\, x(\lambda y \,.\, yx)$, then

$$\begin{aligned} O(M) &= \{\epsilon, \lambda, \lambda 1, \lambda 2, \lambda 2 \lambda, \lambda 2 \lambda 1, \lambda 2 \lambda 2\} \\ M(\lambda 2) &= \lambda y \\ M/\lambda 2 &= \lambda y \,.\, (yx). \end{aligned}$$

□

Binders and binding trees

Although convenient, this representation is far from satisfactory as it does not take into account the role of λ, which is to bind a variable. There are many other *binders* in mathematics and programming:

- $\int_0^1 t\, dt$
- $\{x \in R;\, x^3 - x + 1 = 0\}$
- $\sup_{i \in I}(a_i)$
- $\forall x\, (x \geqslant 0)$
- `procedure A(x) ; begin ... x ... end ;`

In each case, the binder introduces a variable and binds it. This mechanism is distinguished by

- the possibility of renaming bound variables, as is typical in mathematics,

$$\int_0^1 t\,dt = \int_0^1 u\,du,$$

with the risk of incorrect renaming (clash)

$$\int_0^1 (t+u)\,dt \neq \int_0^1 (u+u)\,du;$$

- the concept of scope, well known since the introduction of Algol 60 and its block structure.

Here are definitions for the λ-calculus. Let M be a term represented by a tree and $u \in \mathcal{O}(M)$ be an occurrence of the variable x. Trace back in the tree from a leaf u to the root ϵ. There are two cases:

- If λx is not encountered on this path, u is a *free* occurrence of variable x in M.
- Otherwise, u is a *bound* occurrence of x. The first encountered occurrence of λx on this path is the binder of u. The number of λy, where $y \neq x$, which must be crossed before reaching the binder of u is the *binding height* of u.

Example The binding height is written as an exponent on the bound variables of the term:

$$M \equiv \lambda x\,.\,x^0(\lambda y\,.\,y^0x^1).$$

□

An an arc to the tree representing M between each bound occurrence and its binder, then remove the names of the bound occurrences. The result is a *binding tree* (see Figure 2.4): it is a tree to which arcs attaching a leaf u with a unary node $v < u$ have been added. This kind of representation is the second level of abstract syntax for λ-terms. A variant simpler to program will be given by replacing the bindings by integers.

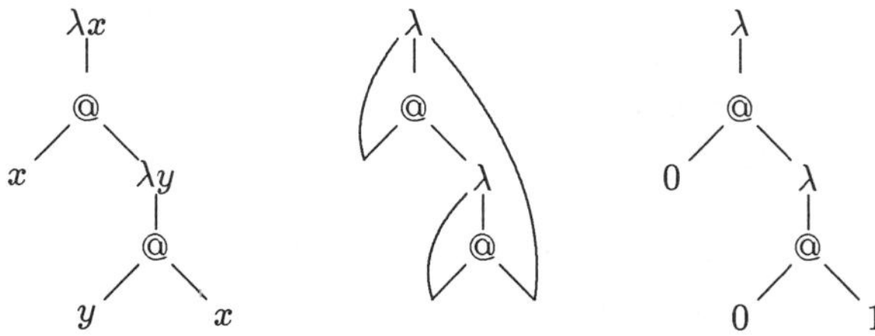

Figure 2.4 Concrete and abstract syntaxes of λ-terms

De Bruijn Terms
Since bindings can be constructed from the binding heights, a purely arborescent representation is obtained by noting its binding height at each occurrence.

Example $\lambda x\,.\,x$ and $\lambda xy\,.x$ are represented by $\lambda 0$ and $\lambda(\lambda 1)$. □

This form, invented by de Bruijn in 1972 ([35]), can easily implement λ-terms in any programming language which can be used to manipulate trees. It was introduced by de Bruijn to ease the implementation of Automath, a language, or, rather, a family of languages, designed for the writing and proving of mathematical proofs, and based on the typed λ-calculi. Note that the implementation of languages with lexical scoping also uses bindings, represented by integer-valued fields, or lexical addresses, to reach variables which are not local to a procedure. The same holds true for rational control structures (loops with exit).

The free occurrences of a variable could be simply represented by its name, as in $x(\lambda 0)$ for $x(\lambda y\, y)$. However, to have a simple notation which uses only integers, all free variables are numbered, as in

$$X = \{v_0, v_1, v_2, \dots\},$$

and the free occurrence u of variable v_i is represented by the integer $i + p$, where p is the *abstraction depth* of u, i.e., the number of λ encountered in the path leading from u to the root ϵ, ends included.

Example The term $\lambda x\,.\,(\lambda y\,.\,yv_4)(v_4 x)$ is represented by

$$\lambda(\lambda(0\,6)(5\,0)) \qquad \text{instead of} \qquad \lambda(\lambda(0\,v_4)(v_4\,0)),$$

the first occurrence of v_4 being at abstraction depth 2 ($6 = 4 + 2$), and the second occurrence at depth 1 ($5 = 4 + 1$): it is the left-hand term in Figure 2.5 on page 38.
□

An occurrence of integer i corresponds to a free variable if i is greater than or equal to its abstraction depth.

λ-terms are therefore represented as finite terms, but over an infinite signature, formed of a binary application symbol, @, a unary abstraction symbol, λ, and of all the integers as constant symbols. These finite terms are called *de Bruijn terms* (or trees). Their set is written Λ. Here is a concrete syntax, analogous to the one previously given for λ-terms:

$$\mathit{term} ::= \mathit{integer} \mid \lambda(\mathit{term}) \mid (\mathit{term}\ \mathit{term}).$$

This representation is a simple and correct implementation of λ-terms. It also defines the abstract syntax of terms through a translation of terms from a concrete notation, the elements of $(X \cup \{\lambda, (,)\})^\star$, to de Bruijn terms. Two terms are *α-convertible* if they have the same representation as de Bruijn terms. This relation is an equivalence relation. So, from now on, terms will be treated up to α-equivalence, i.e., as elements of Λ, but they will be written in the standard manner, e.g. $\lambda x\,.\,x \equiv \lambda y\,.\,y \in \Lambda$.

In fact, the relation of α-convertability is the first step towards the semantics of the λ-calculus. It gives the precise meaning of the bound variables. That this can be done purely at the syntactic level is impressive.

It remains to be shown how a concrete representation of a term can be produced from a de Bruijn term. A variable is *free* in a term M if it has at least one free occurrence in M (it can also have bound occurrences). The set $\text{var}(M)$ of free variables of M can be defined by induction over the terms:

$$\begin{aligned} \text{var}(x) &= \{x\}, \\ \text{var}(\lambda x M) &= \text{var}(M) \setminus \{x\}, \\ \text{var}(MN) &= \text{var}(M) \cup \text{var}(N). \end{aligned}$$

A term M is *closed* if $\text{var}(M) = \varnothing$. The term M is then called a *combinator*.

A term using concrete syntax can be produced from a binding tree by passing names from the root to the leaves:

- The same variable name is associated with all the leaves linked to the same occurrence of λ.
- If a subterm which is an abstraction contains free variables, the head-λ should be associated with a different variable name.

2.3.3 Graft and substitution

The main operations over λ-terms are grafting and substitution. Unlike those over Σ-terms, each of these operations is defined for an appropriate abstract representation.

Graft

The *graft* $M[u \leftarrow P]$ of a term P on to an occurrence u of M is defined for the tree representations of M and P, as if they were Σ-terms. Define

$$\begin{aligned} M[\epsilon \leftarrow P] &= P, \\ (MN)[1 \cdot u \leftarrow P] &= M[u \leftarrow P]\,N, \\ (MN)[2 \cdot u \leftarrow P] &= M\,N[u \leftarrow P], \\ (\lambda x\, M)[\lambda \cdot u \leftarrow P] &= \lambda x\,(M[u \leftarrow P]). \end{aligned}$$

This operation can imply the 'capture' of a free variable of P by a λ of M.

Example Let $P \equiv \mathbf{K}x$, the 'constant function of value x', and $M \equiv \lambda x\, N$, the 'function $x \mapsto N$'. Then $M[\lambda \leftarrow P] \equiv \lambda x\,(\mathbf{K}x)$, the 'function $x \mapsto \mathbf{K}x$, the constant function of value x'. The free variable x was bound by the graft, while it was free in P. □

In programming, the graft of code is called a *macro*: this is what is done by the C preprocessor when handling **#define**'s. As for Σ-terms, grafting can also be defined using the concept of context.

Substitution

On the other hand, *substitution* of a term N for the free occurrences of variable x in M is done at the level of binding trees representing M and N. The idea is to graft the binding tree of N to each of the free occurrences of x in the binding tree of M; the result is a new binding tree. If one starts with concrete representations of M and N, constructing their binding trees essentially means renaming variables to avoid capture.

For de Bruijn terms M and N, let $u_1, \ldots, u_k$ be the occurrences in M representing the same free variable, i.e., if $M(u_i) = n_i$ and if p_i is the abstraction depth of u_i, then $n_i - p_i = m \geqslant 0$ is independent of i and its occurrences represent the free variable v_m. Consider the terms N^{+p_i} obtained by incrementing the occurrences of free variables in N by p_i. Then

$$M[v_m := N] \equiv M[u_1 \leftarrow N^{+p_1}] \ldots [u_k \leftarrow N^{+p_k}].$$

Example To the two occurrences of free variable v_4 are grafted the terms $12 = 10^{+2}$ and $11 = 10^{+1}$, as seen in Figure 2.5.

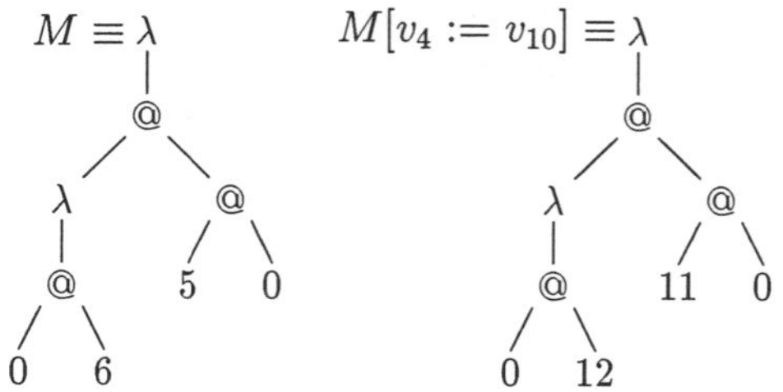

Figure 2.5 Substitution over de Bruijn terms

□

While grafts can be considered to be symbolic manipulations of code, it is the mechanism of substitution which is basic to (functional or not) programming to implement parameter passing during a function or procedure call. This mechanism, defined by a relation between terms, is called β-reduction and will be studied in the next chapter. Its basis is the contraction of redexes.

2.3.4 Redexes

Two kinds of terms have already been presented: abstractions $(\lambda x\, M)$ and applications (MN). A term of the form $(\lambda x\, M)N$ is called a *redex*: it is the combination of an abstraction and an application. The role of these terms is important enough for functional languages to have their own notation for them. For example, in Standard ML, two notations are possible:

```
(fn x => M) N
let val x = N in M end
```

The last notation, due to Landin, is quite natural and means that λ can often be avoided.

The basic mechanism of the λ-calculus and of functional programming is the contraction of redexes. The transformation of the redex $(\lambda x\, M)N$ into $M[x := N]$ is called a *β-contraction.* The term $M[x := N]$ is the result of contracting the redex $(\lambda x\, M)N$. This transformation describes the passing of an argument to a function.

Example With $\mathbf{I} \equiv \lambda x \,.\, x$, the redex $\mathbf{I}M$ contracts to M. □

For nested redexes, it will be convenient to do simultaneous contractions (equivalent to successive contractions, as there can be no capture of variables):

$$(\lambda x_1 \ldots x_p \,.\, M)N_1 \ldots N_p \qquad \text{contracts to} \qquad M[x_1 := N_1] \ldots [x_p := N_p].$$

For example, with $\mathbf{2} \equiv \lambda f x \,.\, f(fx)$, $\mathbf{2}PQ$ contracts to $P(PQ)$

A term for which no subterm is a redex is called *normal.* The set of occurrences of subterms of M which are redexes is written $\mathcal{O}_{red}(M)$. Then M is normal if and only if $\mathcal{O}_{red}(M) = \varnothing$.

2.4 Formal terms

Infinite terms appear in computer science as naturally as do finite terms. In programming, they appear as data structures: streams in functional programming, trees of Prolog–II. There are certain problems whose solutions, in general, require infinite terms. One can sometimes arrange to consider only finite terms. The situation would be analogous to that in algebra where, to compute the real roots of a second-order polynomial, one would insist that its discriminant be positive, and insist that the other equations are 'impossible'. Why restrict oneself artificially when there is a simple, more general method?

Recall the definition of terms. To avoid confusion, these will be called formal terms, be they finite or infinite, alluding to the formal series of algebra. Tree domains are subsets of the free monoid $A^\star$ stable under prefixes and age, with $A = \{n \in \mathrm{N}; n \geqslant 1\}$. The monoid $A^\star$ is itself an example of an infinite tree domain.

Definition 2.10 A *formal Σ-term* is a mapping $M : \mathcal{O} \to \Sigma$ such that $\mathcal{O}$ is a tree domain, and if $u \in \mathcal{O}$ and $\mathrm{ar}(M(u)) = n$, then $u{\cdot}1, \ldots, u{\cdot}n \in \mathcal{O}$ and $u{\cdot}(n{+}1) \notin \mathcal{O}$. Their set is written T_Σ^∞.

As T_Σ^∞ is generally uncountable, this set appears to be of no interest to computer science, since it can deal only with 'finite' objects. However, all programming languages use 'real' numbers without any problem! There are therefore two solutions: one is to work with 'floating point numbers' which are truncated reals and designate intervals; the other is to work with rationals (these are often only decimal fractions, but there are languages such as Lisp and CAML which do exact rational arithmetic). Both of these solutions are used to give a finite representation for an infinite term, 'exact' for rational terms or 'approximated' in general.

In this book, the most important application of rational terms will be the resolution of equations between terms, the problem of *unification*. Another application in computer science concerns process algebras. 'Irrational' infinite terms appear when expanding arbitrary, recursive set of functions: it suffices to approximate them enough to compute the values of the functions.

2.4.1 Approximations

An infinite term can be approximated by finite terms by truncating it. Start by adding a new constant symbol Ω to signature Σ and construct a 'truncating to depth n' operator τ_n. If M is a formal $\Sigma \cup \{\Omega\}$-term, then $\tau_n(M)$ is a finite $\Sigma \cup \{\Omega\}$-term of domain $\{u \in \mathcal{O}(M); |u| \leqslant n\}$, defined by

$$(\tau_n(M))(u) = \begin{cases} M(u) & \text{if } u \in \mathcal{O}(M) \text{ and } |u| < n \\ \Omega & \text{if } u \in \mathcal{O}(M) \text{ and } |u| = n. \end{cases}$$

These $\Sigma \cup \{\Omega\}$-terms are called *partial terms* by analogy with partial functions. The meaning of symbol Ω reflects a *semantics* of the computation: it designates an undefined value, or, rather, a value which has yet to be defined, which could later be substituted with a term at a later stage of the computation, to obtain a better approximation of an infinite term. This is a case where a purely syntactic construction allows one to describe a semantic behavior.

The simplest method of introducing approximation consists of defining a distance, actually an ultrametric, between partial formal terms by

$$d(M, N) = \begin{cases} 0 & \text{if } M = N \\ 2^{-p} & \text{if } p = \inf\{k; \tau_k(M) \neq \tau_k(N)\}. \end{cases}$$

In particular, d is a distance over the set T_Σ (and over $\mathrm{T}_{\Sigma\cup\{\Omega\}}$), thereby allowing one to reconstruct T_Σ^∞ (and $\mathrm{T}_{\Sigma\cup\{\Omega\}}^\infty$). One shows in fact that $\mathrm{T}_{\Sigma\cup\{\Omega\}}^\infty$ is the completion of the metric space $(\mathrm{T}_{\Sigma\cup\{\Omega\}}, d)$.

Another concept of approximation, less artificial, arises from the order relation $\leqslant_\Omega$, from which one can also construct $\mathrm{T}^\infty_{\Sigma\cup\{\Omega\}}$, making it into a 'complete' ordered set. Here is its inductive definition over finite terms:

$$\begin{array}{rcll} \Omega & \leqslant_\Omega & M & \text{for all } M \\ fM_1\dots M_r & \leqslant_\Omega & fN_1\dots N_r & \text{and } M_i \leqslant_\Omega N_i \text{ for all } i. \end{array}$$

Since the constant Ω is only an intermediary technique, the infinite terms of interest are the maximal infinite elements of $\mathrm{T}^\infty_{\Sigma\cup\{\Omega\}}$, i.e., those having no occurrence of Ω.

2.4.2 Rational terms

Rational terms are described similarly to rational numbers: in the same way that $2/3$ is the solution of $3x = 2$, the rational term $f^\omega = fff\dots$ is the solution of $f(x) = x$. As process trees, they represent sequential programs. They are the solutions of equations, as programs with terminal recursivity. They can be represented by finite terms over a signature extended with a unary iteration symbol (`loop`) and an infinite number of 0-ary symbols for leaving loops (`exit n`).

Consider the set of equations over terms

$$(E) : x_i = t_i, \qquad t_i \in \mathrm{T}_\Sigma[x_1, \dots , x_d], \quad (1 \leqslant i \leqslant d).$$

It is supposed that no t_i is a variable and that the variables x_i are all distinct. In Chapter 5, where equations of the form $s = t$ will be studied, such a system will be referred to as reduced. The solution is an n-tuple of infinite terms, all elements of T^∞_Σ. By definition, these solutions are *rational* terms. The existence of a solution is easy to prove. In fact, associate with the system of equations an operator Φ over $(\mathrm{T}^\infty_\Sigma)^d$ (or over $(\mathrm{T}^\infty_{\Sigma\cup\{\Omega\}})^d$) defined by

$$(M_1, \dots , M_d) \mapsto (t_1[x_i := M_i], \dots , t_d[x_i := M_i]).$$

This operator has a least fixpoint, whose existence comes from general theorems—the fixpoint of a contracting operator in a complete metric space, here $(\mathrm{T}^\infty_\Sigma)^d$, or else the fixpoint of a continuous operator in a complete ordered set, here $(\mathrm{T}^\infty_{\Sigma\cup\{\Omega\}})^d$. In the last case, the fixpoint is the least upper bound of the $\Phi^n(\Omega)$. Chapter 4 will elaborate on the properties of complete orders.

Solutions can easily be drawn by hand: draw the d terms M_i as trees and connect the leaves labeled by variables with the corresponding terms: the result is a graph, a perfectly finite object, but which no longer has the structure of a term. By choosing one of the variables and by making the appropriate substitutions for the graph pointed to by a variable, an infinite term is produced:

A rational term can also be considered to be a formal term whose set of subterms is finite, each subterm being able to appear an infinite number of times: if a term

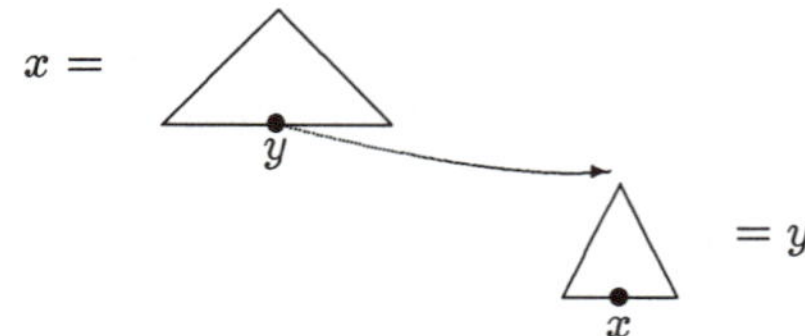

has a finite number of subterms, a system for which it is a solution can be obtained by associating a variable with each subterm.

To represent the finite graph or the infinite term by a finite term, the signature is extended by defining

$$\Sigma^\star = \Sigma \cup \mathbf{N} \cup \{\star\}.$$

This countable signature is similar to that defining de Bruijn terms in λ-calculus. For the latter, a constant n represents a variable bound by a λ of height n, if n is less than the abstraction depth. In $\Sigma^\star$, the unary symbol $\star$ plays the same role as λ with integer constants, called *branching indices*. The free/bound distinction here becomes exit/iteration, thanks to the following definitions.

Let $M \in \mathrm{T}_{\Sigma^\star}[X]$ and $u \in \mathcal{O}(M)$. The *$\star$-depth* of u is the number of occurrences v of M such that $M(v) = \star$ and $v < u$. If p is the $\star$-depth of u and $M(u) = n \in \mathbf{N}$, then the difference $n-p$ is the escape value of u. If $n - p < 0$, u is an iteration leaf and n is its iteration height, i.e., the number of $\star$ between u and the root ϵ to reach the $\star$ to which u is bound. Otherwise it is an exit leaf. Constant occurrences other than indices and variable occurrences are also exit leaves. A $\Sigma^\star$-term is *$\star$-closed* if all the indices are iteration indices.

Example Figure 2.6 represents a rational term in graph form (to the left) and as a finite $\Sigma^\star$-term (to the right). □

Figure 2.6 Two forms of a rational term

The result of exercise 16 can easily be extended to define a mapping $M \mapsto M^\star$ of $\mathrm{T}^\infty_\Sigma[x_1, \dots, x_n, y]$ to $\mathrm{T}^\infty_\Sigma[x_1, \dots, x_n]$, variables $x_1, \dots, x_n$ being considered as constants: $M^\star$ is the limit of sequence M, $M; M$, $(M; M); M$, ... , where $M; N = M[y := N]$.

For each $\star$-closed finite $\Sigma^\star$-term M, an infinite Σ-term $\delta(M)$ can be defined by induction on the height of M. When $M(\epsilon) \in \Sigma$, define

$$\delta(fM_1 \dots M_n) = f\delta(M_1)\dots\delta(M_n), \qquad \text{for } f \in \Sigma.$$

Let $M = \star N \in \mathrm{T}_{\Sigma^\star}[x_1, \dots, x_n]$ be a $\star$-closed finite term: the indices of M are all iteration indices, by definition. Let y be a new variable and N' be the term obtained by grafting y on to each of the occurrences of an exit index of N (these are the indices p at $\star$-depth p in N): N' is a $\star$-closed of height inferior to that of M. Let $\delta(N')$ be constructed by induction, and

$$\delta(M) = \delta(N')^\star,$$

using the operation $M \mapsto M^\star$ defined above (see exercise 16). The formal term $\delta(M)$ is the solution to $y = N'$.

Solving systems of equations whose right-hand sides are in $\mathrm{T}_\Sigma[X]$ produces systems of equations whose right-hand sides are in $\mathrm{T}_{\Sigma^\star}[X]$. Solving them will therefore be done in this more general setting, where $M_i \in \mathrm{T}_{\Sigma^\star}[x_1, \dots, x_d]$.

Fix a principal variable x_1, e.g., in C, the `main` function, and eliminate the other variables using x_1. The same technical problems are met as in the λ-calculus when a substitution must be made over de Bruijn terms: an index incrementing operator, written $M \mapsto M^+$, is used: it replaces each variable x (a leaf) by x^+ and increments each exit index by 1.

An equation $x = M$ is solved by $x = \star M'$, where M' is obtained from M by replacing each occurrence of x by 0, each other variable $y \neq x$ by y^+ and by incrementing each of the exit indices.

Example Consider the set of equations (s is a constant):

$$x = p(a(x), b(y)), \qquad y = q(c(x), d(s)).$$

The second equation is substituted into the first one:

$$x = p(a(x), b(q(c(x), d(s)))),$$

which is solved by

$$x = \star p(a(0), b(q(c(0), d(s)))).$$

By solving first for x,

$$x = \star p(a(0), b(y^+)),$$

is obtained. When substituted into y's definition, it gives

$$y = q(c(\star p(a(0), b(y^+))), d(s)),$$

which is solved by (since $0^+ = 1$)

$$y = \star q(c(\star p(a(0), b(1))), d(s)).$$

□

Note that the solution to a system whose right-hand side is in $\mathrm{T}_\Sigma[X]$ is a $\star$-closed $\Sigma^\star$-term.

2.4.3 Application to sequential processes

Consider a signature formed with binary symbols p, q, ... , with unary symbols a, b, ... and a zero-arity symbol `stop`. The binary symbols are tests $p(x, y)$ meaning 'if p then x else y': a test is binary because it has two continuations. The unary symbols are atomic actions, which take their continuation as argument. The `stop` symbol designates the end (no continuation). A sequential program defines an infinite term over this signature, each branch designating a possible behavior. For example, the program which computes the greatest common divisor (gcd),

```
while x<>y do
  if x>y then x := x-y else y := y-x end if
end do ;
r := x
```

corresponds to the infinite term

where p and q denote the tests `x<>y` and `x>y`, and a, b and c the actions `x := x-y`, `y := y-x` and `r := x`. The same infinite term would have been obtained for the program

```
1 : if x<>y then
              if x>y then x := x-y ; goto 1
                      else y := y-x ; goto 1
              end if
    else r := x
    end if
```

and it is the same computation process (the same 'algorithm') which is executed. These terms describe the computations independently of the form of the programs. The general problem for structures denoting sequential processes is to obtain a finite notation for the infinite terms. The situation is similar for formal languages, which generally denote infinite sets, but which must be defined with finite means, for example with grammars or automata. This is possible if one restricts oneself to certain classes of languages, for example rational and algebraic languages. This is also possible with process terms, which are rational terms.

Adding variables to the signature formed over tests, actions and `stop` allows one to have variables in the leaves instead of just `stop`. One can then substitute a variable with any other process, and the latter will be seen as a branch by other processes. To these process variables correspond, in a programming language, labels, or, equivalently, the names of procedures with no parameters. Their use allows one to write programs as systems of equations: they are recursive definitions of procedures with terminal recursivity. For example, the above gcd program can be written abstractly as follows:

$$x = p(q(a(x), b(x)), c(\texttt{stop})).$$

The solution to this equation is an infinite term, whose most natural finite notation is to indicate looping by a link from its leaves x to the root, which could be written as follows:

$$x = \texttt{loop}\ p(q(a(\texttt{continue}), b(\texttt{continue})), c(\texttt{stop})),$$

by adding symbols `loop` and `continue` to the signature.

To have finite terms denote these processes, the signature has to be extended: $\star$ is an interaction, `while true` in Pascal or `loop` in Ada. Placed inside p iterations, a branching index $n < p$ means looping on the $(n+1)$-th $\star$ above n, i.e., escaping n loops containing this index. If $n \geqslant p$, the exit is equivalent to `stop`. This is why 0 is equated with `stop` and that `exit(n)` usually denotes the branching index n (loops can be named in Ada, which allows one to write `exit` *identifier*).

Very few languages use these general structures. The `while` loop is the most popular iteration structure. It was chosen for reasons of readability (the condition appears at the front of the loop) and ease of compilation. It has even been associated with a model of 'structured programming'. Yet, it is not possible, in general, to transform a program with branching into a program with `while` without introducing supplementary Boolean variables. It is therefore only a means of structuring control if one is willing to modify the environment (new variables) and the process (new tests, new assignments). It is a global modification of the *algorithm*, which requires new memory and computation resources, and not just a simple transformation of the program.

The formalization of processes just presented, although simplified, isolates the 'control' and distinguishes programs and algorithms. It is quite remarkable that one can avoid sequencing (;) as one of the instructions of our language, even though it is the characteristic operation of sequential programming!

Exercises[1]

1. Let $\leqslant$ be the prefix order over $A^\star$. Does it have a smallest element? a greatest element? What is the set of lower bounds of a word? Does each pair of words have a least upper bound? a greatest lower bound?

2.★ If $\leqslant_A$ is a total order over alphabet A, the total (lexical) order $\leqslant^\star$ over $A^\star$ is defined by: $u \leqslant^\star v$ if u is a prefix of v, or if $u = wiu'$ and $v = wjv'$ with

[1]Hints to the exercises labeled with a ★ can be found at the end of the book.

$i, j \in A$ and $i \leqslant j$ (w is the maximal prefix common to u and v). Find a sufficient and necessary condition over A so that the set of lower bounds of a word is finite. Does every word have a successor? Is every word the successor of another one?

3. Show that $K(\Phi) = \bigcup_{E \subseteq \Phi(E)} E$ is the greatest fixpoint of Φ, and that if one defines $\Phi^\circ(E) = X \setminus \Phi(X \setminus E)$, then $K(\Phi) = X \setminus \text{Ind}\,\Phi^\circ$. Does the continuity of Φ ensure that $K(\Phi) = \text{Ind}\,\Phi$?

4. Let $X = E \times E$ and $R \subseteq X$ be a binary relation over E. Define Φ such that $\text{Ind}\,\Phi$ is the transitive closure of R. Show that if E is a finite set, $\text{Ind}\,\Phi = \Phi^{\uparrow n}$ for some integer n.

5. Define binary trees with leaves labeled by atoms (the directed pairs of Lisp) and binary trees with nodes labeled by atoms.

6.★ Show that $\mathcal{O}(M)$ satisfies the four properties of proposition 2.7.

7. Show that $M[u \leftarrow N]$ satisfies the four properties of proposition 2.7.

8. Since a context $\mathcal{C}$ determines a mapping $M \mapsto \mathcal{C}[M]$, give an inductive definition of contexts as mappings from $\mathrm{T}_\Sigma[X]$ to itself. Show also that every context can be designated by a term over a certain signature Σ_c (conversely?). To each of $M \in \mathrm{T}_\Sigma[X]$ and $u \in \mathcal{O}(M)$, associate a context $M_{u\leftarrow}$. Show how to graft a context on to another context.

9. Let $x = x_1 \ldots x_n \in \Sigma^\star$. Define integers r_i, $0 \leqslant i \leqslant n$, in the following way: after having read the word $x_1 \ldots x_i$, the integer r_i is the number of constants which must be concatenated to the right of $x_1 \ldots x_i$ to obtain the prefix notation of a term; in particular $r_0 = 1$. For the word $fagaaha$, with $\text{ar}(f) = 3$, $\text{ar}(g) = 2$ and $\text{ar}(h) = 1$, $(r_i)_{0 \leqslant i \leqslant n} = (1, 3, 2, 3, 2, 1, 1, 0)$. Show that $r_i = r_{i-1} + \text{ar}(x_i) - 1$ for $1 \leqslant i \leqslant n$, and that x is the notation for a term if and only if $r_i > 0$ for $1 \leqslant i < n$ and $r_n = 0$.

10. Write in C the following functions manipulating terms: creation, head, subterms and equality.

11.★ Show that every λ-term M is in one of the following forms:

$$\begin{array}{rl} \lambda x_1 \ldots x_n \,.\, x M_1 \ldots M_m & n, m \geqslant 0, \\ \lambda x_1 \ldots x_n \,.\, (\lambda x\, M_0) M_1 \ldots M_m & n \geqslant 0, \quad m \geqslant 1. \end{array}$$

12.★ Define the set $\text{var}(M) \subseteq \{v_0, v_1, \ldots\}$ when M is a de Bruijn term.

13. Find T_Σ and T_Σ^∞ for the following subsignatures of arithmetic: $\Sigma_1 = \{0\}$, $\Sigma_2 = \{s\}$, $\Sigma_3 = \{0, s\}$ and $\Sigma_4 = \{+\}$.

14. Show that T_Σ^∞ is uncountable if and only if Σ contains at least two symbols whose sum of arities is $\geqslant 2$. Find similar conditions for T_Σ^∞ to be countable, finite, empty or equal to T_Σ.

15. Show that $M \leqslant_\Omega N$ if and only if for every occurrence u of M, $M(u) = N(u)$ or $M(u) = \Omega$. Deduce that $M \leqslant_\Omega N$ is equivalent to the existence of $u_1, \ldots, u_k \in \mathcal{O}(M)$ and of terms $P_1, \ldots, P_k$ such that $M(u_i) = \Omega$ and $N = M[u_1 \leftarrow P_1, \ldots, u_k \leftarrow P_k]$.

16. Show that the set $T_\Sigma^\infty[y]$, along with the operation ';' defined by $M; N = M[y := N]$, is a monoid. Let $M^0 = y$ and $M^{n+1} = M^n; M$. Show that the sequence $(M^n)_{n\geqslant 0}$ is a Cauchy sequence. Let $M^\star = \lim_{n\to\infty} M^n$. Show that $M^\star \in T_\Sigma^\infty$.

Bibliographic notes

Formal languages are studied in [65]. For a deep study of infinite terms, see B. Courcelle's article [27]. The representation of rational terms by finite terms is due to G. Cousineau[28], who saw their usefulness for studying control structures in sequential programming.

Chapter 3

Reduction

3.1 The λ-calculus

The concept of function used in contemporary mathematics, as a particular case of a relation—a set of pairs—is inadequate for computer science. Apart from the special case of a function defined over a finite set, for which a finite representation of its extension is possible, it is necessary to represent functions by their intensions, i.e., by computation rules. A return is thus made to the concepts of the nineteenth century. Based on this idea, the λ-calculus was introduced by A. Church in 1933 [14], and developed by B. Rosser and S Kleene.

The λ-calculus has today become the paradigm of functional programming, after having been, first, an attempt to mechanize logic, then a general theory of functions, and finally, in the hands of computer scientists, a low-level programming language.

It is also used regularly to experiment with new computing paradigms, where certain properties, such as termination and confluence, play a crucial role comparable to that of the fundamental principles of physics.

3.1.1 β-reduction

Recall that a λ-calculus term is either a variable, an application or an abstraction. A redex is an application whose left term is an abstraction. The result of contracting redex $(\lambda x\, M)N$ is the term $M[x := N]$.

To explain the functional character of λ-calculus terms, a new relation, called β-reduction, based on the contraction of redexes, is introduced. In fact, several binary relations will be introduced: immediate β-reduction ($\longrightarrow_\beta$), β-reduction ($\stackrel{*}{\longrightarrow}_\beta$) and β-conversion ($=_\beta$). Each of these definitions is given using an inference system containing the same judgments $M \rhd N$, where M and N are terms and $\rhd$ is a new symbol. This judgment means 'M reduces to N': for $\longrightarrow_\beta$, it is a reduction in one step, for $\xrightarrow{\star}_\beta$, it is a reduction in several steps, and for $=_\beta$ it is an inter-reduction.

The inference system for immediate β-reduction contains the contraction rule, with no premises, and three unary rules for context passing.

Definition 3.1 Immediate β–reduction is a binary relation, written $\longrightarrow_\beta$, or simply $\longrightarrow$: $M \longrightarrow_\beta N$ if and only if there exists a closed derivation, i.e., without assumption, of the judgment $M \rhd N$ by the inference system:

$$\textbf{(red)}: \frac{}{(\lambda x P)Q \rhd P[x := Q]}$$

$$(\lambda): \frac{M \rhd M'}{\lambda x M \rhd \lambda x M'} \qquad (1): \frac{M \rhd M'}{MN \rhd M'N} \qquad (2): \frac{M \rhd M'}{NM \rhd NM'}$$

If $M \longrightarrow_\beta N$, M *reduces immediately* to N.

Example Let R be a redex which contracts to C, and M be an arbitrary term. Here is a derivation of $\lambda x\,.\,RM \rhd \lambda x\,.\,CM$ (recall that $\lambda x\,.\,RM$ is a simplified notation for $(\lambda x\,(RM))$):

$$(\lambda): \cfrac{(1): \cfrac{\textbf{(red)}: \cfrac{}{R \rhd C}}{RM \rhd CM}}{\lambda x\,.\,RM \rhd \lambda x\,.\,CM}.$$

□

Recall that a term M is *normal* (§2.3.4) if it has no redex subterms, i.e., if $\mathcal{O}_{red}(M) = \varnothing$.

It is not a coincidence that the context-passing rules are designated by the names λ, 1 and 2, which form the alphabet of occurrences for λ-terms (see §2.3.2). A ternary relation is defined between two terms M, N and an occurrence u by $M \xrightarrow{u} N$ if $u \in \mathcal{O}_{red}(M)$ and $M/u \equiv (\lambda x\, P)Q$, $C \equiv P[x := Q]$ is the result of contracting M/u, and $N \equiv M[u \leftarrow C]$.

Proposition 3.1 $M \longrightarrow_\beta N$ if and only if there exists $u \in \mathcal{O}_{red}(M)$ such that $M \xrightarrow{u} N$.

Proof Reason by induction over the derivation of $M \rhd N$. Recall that all the rules are of arity 0 or 1.

If M is a redex and N the result of contracting M, define $u = \epsilon$.

Otherwise, $M \rhd N$ is inferred by one of the unary rules. Suppose that it is the first one, (λ), with $M \equiv \lambda x\, M_1$, $N = \lambda x\, N_1$ and $M_1 \longrightarrow_\beta N_1$. By the inductive hypothesis, there exists $u_1 \in \mathcal{O}_{red}(M_1)$ such that $M_1 \xrightarrow{u_1} N_1$. Deduce that $u = \lambda \cdot u_1 \in \mathcal{O}_{red}(M)$ and that $M \xrightarrow{u} N$. The second and third cases are treated similarly.

Conversely, if $M \xrightarrow{u} N$, construct a derivation of $M \rhd N$ by starting with the contraction rule of M/u and successively applying the unary rules by reading u from right to left. In the preceding example, the redex R of $\lambda x \,.\, RM$ is at occurrence $u = \lambda 1$, and rules (**red**), (1) and (λ)) are applied. □

Context passing can also be written concisely, using the single rule,

$$(\mathbf{cont}) : \frac{P \rhd Q}{\mathcal{C}[P] \rhd \mathcal{C}[Q]},$$

$\mathcal{C}$ being an arbitrary term context. Proposition 3.1 shows that it is sufficient to apply this rule when P is a redex.

An attempt will be made to reduce a term by successive immediate β-reductions, hoping that a normal form will be reached. To do this means to compute the *reflexive, transitive closure* of immediate β-reduction, called β-reduction.

Definition 3.2 β–reduction is a binary relation, written $\xrightarrow{\star}_\beta$, or simply $\xrightarrow{\star}$: $M \xrightarrow{\star} N$ if and only if there exists a closed derivation of judgment $M \rhd N$ by the following inference system:

$$(\beta) : \frac{}{M \rhd M'} \quad \text{if} \quad M \rightarrow_\beta M'$$

$$(\mathbf{R}) : \frac{}{M \rhd M}$$

$$(\mathbf{T}) : \frac{M \rhd M' \qquad M' \rhd M''}{M \rhd M''}$$

Proposition 3.2 $M \xrightarrow{\star} M'$ if and only if there exist terms $M_0, M_1, \ldots, M_n, n > 0$, such that $M_0 \equiv M$, $M_n \equiv M'$, and $M_i \rightarrow_\beta M_{i+1}$ or $M_i \equiv M_{i+1}$, for $0 \leqslant i < n$.

Proof By induction over the derivation of $M \rhd M'$ proving $M \xrightarrow{\star} M'$.

- If $M \rhd M'$ is derived from (β) or from (**R**), then $M_0 \equiv M$ and $M_1 \equiv M'$.
- If $M \rhd M''$ is derived by applying (**T**), by the inductive hypothesis, there are two sequences,

  $$M \equiv M_0,\; M_1, \ldots, M_n \equiv M'$$

  for $M \rhd M'$ and

  $$M' \equiv M'_0,\; M'_1, \ldots, M'_{n'} \equiv M''$$

  for $M' \rhd M''$. The required sequence is

  $$M \equiv M_0,\; M_1, \ldots, M_n,\; M'_1, \ldots, M'_{n'} \equiv M''.$$

Conversely, one can easily construct a derivation of $M \rhd M'$ from a sequence of immediate β-reductions. □

The $\xrightarrow{\star}$ relation expresses the concept of computation as reduction. Also of interest is the *transitive* closure of $\rightarrow_\beta$, written $\xrightarrow{+}$ and defined analogously to $\xrightarrow{\star}$, but without rule (**R**). Finally, the equivalence relation induced by $\rightarrow_\beta$ is written $=_\beta$, or simply $=$, or even $\xleftrightarrow{\star}_\beta$: it is called *β–conversion*. In programming, it corresponds to equivalence between programs. As the reflexive, symmetric and transitive closure of $\rightarrow_\beta$, it is defined by adding the symmetry rule

$$(\mathbf{S}) : \frac{M \rhd N}{N \rhd M}$$

to the inference system defining $\xrightarrow{\star}$.

3.1.2 Fundamental properties

Note that β-reduction does not always reduce the *size* of terms, and that it is not always possible to reach a normal form.

Definition 3.3 A term M is *normalizable* if there exists a β-reduction $M \xrightarrow{\star} N$, where N is normal.

Examples Normal terms are normalizable. The term $\Omega \equiv \Delta\Delta$, where $\Delta \equiv (\lambda x\,.\,xx)$, is not normalizable, since the only possible immediate β-reduction of Ω is $\Omega \xrightarrow{\epsilon} \Omega$, so the only β-reduction is $\Omega \rightarrow_\beta \Omega \rightarrow_\beta \Omega \rightarrow_\beta \ldots$. Its *reduction graph* has one vertex, Ω, and a loop. □

The following result, due to Church, indicates the level of difficulty of the λ-calculus; it will be proven in Chapter 7, where decidability will be defined:

Theorem 3.3 The normalization problem is undecidable.

The fundamental property of β-reduction is confluence. To define it, the transitive, reflexive closure of a binary relation $\rightarrow$ is written $\xrightarrow{\star}$.

Definition 3.4 Let $x \in E$. The binary relation $\rightarrow$ over E is

- *confluent* in x, if for every x_1, x_2, if $x \xrightarrow{\star} x_1$ and $x \xrightarrow{\star} x_2$, there exists x' such that $x_1 \xrightarrow{\star} x'$ and $x_2 \xrightarrow{\star} x'$;
- *locally confluent* in x, if for every x_1, x_2, if $x \rightarrow x_1$ and $x \rightarrow x_2$, there exists x' such that $x_1 \xrightarrow{\star} x'$ and $x_2 \xrightarrow{\star} x'$;
- *strongly confluent* in x, if for every x_1, x_2, if $x \rightarrow x_1$ and $x \rightarrow x_2$, there exists x' such that $x_1 \rightarrow x'$ and $x_2 \rightarrow x'$.

These properties can be expressed by the following diagrams, where a complete line indicates a hypothesis ('if ... ') and a dotted line indicates a conclusion ('there exists ... '):

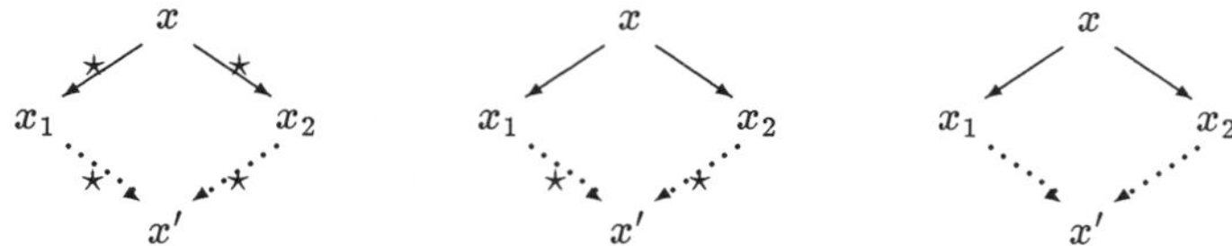

A binary relation $\longrightarrow$ over a set E is confluent (respectively locally confluent, strongly confluent) if it is in every element of E.

Confluence is a fundamental property of every model using reduction for computation: it means the independence of the result of a computation from the reduction taken. The conversion relation turns out to mean reducing to the same term: this is called the *Church–Rosser property*.

Proposition 3.4 If relation $\longrightarrow$ is confluent, then $x \overset{\star}{\longleftrightarrow} y$ is equivalent to the existence of z such that $x \overset{\star}{\longrightarrow} z$ and $y \overset{\star}{\longrightarrow} z$.

Proof By induction on the derivation of $x \overset{\star}{\longleftrightarrow} y$. If $x \overset{\star}{\longleftrightarrow} y$ results from

- $x \longrightarrow y$, take $z \equiv y$;
- the reflexivity rule (with $x \equiv y$), take $z \equiv x$;
- $y \overset{\star}{\longleftrightarrow} x$, by the symmetry rule, the inductive hypothesis applied to $y \overset{\star}{\longleftrightarrow} x$ directly yields z;
- $x \overset{\star}{\longleftrightarrow} t$ and $t \overset{\star}{\longleftrightarrow} y$ by the transitivity rule, the inductive hypothesis applied to $x \overset{\star}{\longleftrightarrow} t$ and to $t \overset{\star}{\longleftrightarrow} y$, respectively, provides x_1 and y_1 such that $x \overset{\star}{\longrightarrow} x_1$, $t \overset{\star}{\longrightarrow} x_1$, $t \overset{\star}{\longrightarrow} y_1$ and $y \overset{\star}{\longrightarrow} y_1$. By applying the confluence property to the two reductions emanating from t, z is obtained such that $x_1 \overset{\star}{\longrightarrow} z$ and $y_1 \overset{\star}{\longrightarrow} z$.

□

This proposition states that confluence implies the *consistency* of a reduction relation, i.e., that it is impossible to prove $M \overset{\star}{\longleftrightarrow} N$ for all terms. It suffices to take two normal terms M and N which are syntactically different, for example two distinct variables: if $M \overset{\star}{\longleftrightarrow} N$, then there exists P such that $M \overset{\star}{\longrightarrow} P$ and $N \overset{\star}{\longrightarrow} P$, hence $M \equiv P$ and $N \equiv P$, since M and N are normal, which contradicts $M \not\equiv N$.

Theorem 3.5 (Church–Rosser) Immediate β-reduction $\longrightarrow_\beta$ is confluent.

This theorem will be proven when the reduction mechanism is more finely presented, using the concept of residue.

Corollary 3.6 If M is normalizable, there exists a unique normal term, written $\overline{M}$, such that $M \overset{\star}{\longrightarrow} \overline{M}$.

Proof If M is normalizable, there exists a normal M_1 such that $M \xrightarrow{\star} M_1$. If $M \xrightarrow{\star} M_2$, M_2 normal, then by confluence there exists M_3 such that $M_1 \xrightarrow{\star} M_3$ and $M_2 \xrightarrow{\star} M_3$. If M_1 and M_2 are normal, then $M_1 \equiv M_3$ and $M_2 \equiv M_3$, hence $M_1 \equiv M_2$. □

$\overline{M}$ is called the *normal form* of M. This uniqueness allows one to look for the normal form in a non-deterministic manner.

One must not conclude that any reduction of a normalizable term leads to a normal term: the successive choice of redexes to be contracted plays a crucial role. For example, $M \equiv (\lambda x\, y)\Omega$ is normalizable. Its normal form is obtained by contracting M. If Ω were always contracted, the infinite reduction $M \longrightarrow M \longrightarrow M \longrightarrow \ldots$ would be obtained.

Definition 3.5 A term M is *strongly normalizable* if it does not have an infinite reduction $M \longrightarrow M_1 \longrightarrow M_2 \longrightarrow \ldots$.

There are normalizable terms which are not strongly normalizable. This is the case for $(\lambda x\, y)\Omega$. This example also shows that a term can be normalizable even if not all of its subterms are, and that precautions must be taken when reducing.

If the λ-calculus is to be useful, as a method of computing the normal form of a term, a *reduction strategy* must be fixed: for each term, the redex to be contracted must be chosen. For example, the *leftmost strategy* consists of contracting the leftmost redex (whose λ is leftmost). By making the search for the normal form deterministic, the choice of strategy determines for each M a unique reduction which is either finite, with $M \xrightarrow{\star} M_n$ and M_n normal, or infinite. A strategy is *normalizing* if the reduction that it determines for every normalizable term is finite. The following theorem is proven in [8, theorem 13.2.2].

Theorem 3.7 The leftmost reduction strategy is normalizing.

Corollary 3.8 If the leftmost reduction of a term is infinite, then the term is not normalizable.

Corollary 3.9 The normalization problem is semi-decidable.

Proof If M is normalizable, the leftmost reduction gives the normal form of M 'in a finite time'. If M is not normalizable, one can always wait … . □

3.1.3 Programming in λ-calculus

Although it is easy to write interpreters for the λ-calculus that apply a reduction strategy, it is very difficult to program in it, i.e., to write terms defining the usual objects and algorithms of computer science. In general, it is the operational aspect of the objects that yields their λ-calculus representation. For example, integers are 'function iterators' and Booleans are 'branchings'.

In my system you can ignore the internals and just give something a name. This is how the seperate elements compute → there is no global knowledge.

Booleans The combinators **T** and **F** representing the Booleans *true* and *false* must serve as branchings: $\mathbf{T}\,x\,y \xrightarrow{\star} x$ and $\mathbf{F}\,x\,y \xrightarrow{\star} y$. Thus

$$\mathbf{T} \equiv \lambda xy\,.\,x, \qquad \mathbf{F} \equiv \lambda xy\,.\,y.$$

A conditional expression, analogous to `(b ? x : y)` in C, is simply represented by the application of the Boolean condition to its two branches:

$$\mathbf{if} \equiv \lambda bxy\,.\,bxy,$$

so that $\mathbf{if}\,\mathbf{T}xy \xrightarrow{\star} x$ and $\mathbf{if}\,\mathbf{F}xy \xrightarrow{\star} y$. The Boolean operations can be defined by the combinators

$$\begin{aligned} \mathbf{and} &\equiv \lambda xy\,.\,xy\mathbf{F}, \\ \mathbf{or} &\equiv \lambda xy\,.\,x\mathbf{T}y, \\ \mathbf{not} &\equiv \lambda bxy\,.\,byx. \end{aligned}$$

So, $\mathbf{and}\,\mathbf{T}y = \mathbf{T}y\mathbf{F} = y$ and $\mathbf{and}\,\mathbf{F}y = \mathbf{F}y\mathbf{F} = \mathbf{F}$ (note, however, that $\mathbf{and}\,x\mathbf{F} = x\mathbf{F}\mathbf{F}$ does not reduce to **F**; the two arguments do not play the same role), and $\mathbf{or}\,\mathbf{T}y = \mathbf{T}$ and $\mathbf{or}\,\mathbf{F}y = y$.

Pairs Booleans allow one to define pairs as *objects* responding to the messages **T** and **F**, in the object-oriented programming sense, by giving the appropriate component: $(\mathbf{cons}\,xy)\mathbf{T} \xrightarrow{\star} x$ and $(\mathbf{cons}\,xy)\mathbf{F} \xrightarrow{\star} y$. Hence the projections:

$$\pi^1 \equiv \lambda c\,.\,c\mathbf{T}, \qquad \pi^2 \equiv \lambda c\,.\,c\mathbf{F},$$

and the pairing constructor:

$$\mathbf{cons} \equiv \lambda xym\,.\,mxy.$$

A purely functional language, i.e., based on the λ-calculus, is therefore adapted to the 'message-passing' style of programming. The necessarily functional nature of objects allows them to respond to messages: it suffices to pass those messages as arguments.

Union The structure dual to pairs is union, constructed using injections: $\imath^1 M$ receiving messages f and g must apply f to M while $\imath^2 M$ applies g to M: $(\imath^1 x)fg \xrightarrow{\star} fx$ and $(\imath^2 x)fg \xrightarrow{\star} gx$. This gives the two constructors

$$\imath^1 \equiv \lambda xfg\,.\,fx, \qquad \imath^2 \equiv \lambda xfg\,.\,gx.$$

Integers A natural method, due to Church, is to represent the integers as iterators of functions: the integer n is coded by the combinator (Church numeral) $\mathbf{n} \equiv \lambda fx\,.\,f^n x$. In particular,

$$\begin{aligned} \mathbf{0} &\equiv \lambda fx\,.\,x \equiv \mathbf{F}, \\ \mathbf{1} &\equiv \lambda fx\,.\,fx. \end{aligned}$$

The two operations essential for them to be considered as integers are the successor function and the zero equality test:

$$\begin{aligned} \textbf{succ} &\equiv \lambda nfx\,.\,f((nf)x),\\ \textbf{zero} &\equiv \lambda nxy\,.\,n(\mathbf{T}x)y, \end{aligned}$$

where $K \equiv \mathbf{T} \equiv \lambda xy\,.\,x$. So, $\textbf{succ}\,\mathbf{n} \xrightarrow{\star} \mathbf{p}$ if $p = n+1$, and

$$\textbf{zero}\,\mathbf{n}\,xy = \begin{cases} x & \text{if } n \neq 0\\ y & \text{if } n = 0. \end{cases}$$

The other arithmetic operations are easy to program by successive iterations of successor:

$$\begin{aligned} \textbf{add} &\equiv \lambda mngz\,.\,(mg)((ng)z),\\ \textbf{mult} &\equiv \lambda mng\,.\,m(ng),\\ \textbf{exp} &\equiv \lambda mn\,.\,nm. \end{aligned}$$

Fixpoint combinators A closed term Y is called a *fixpoint combinator* if for every term M, $YM =_\beta M(YM)$. Here are two, due to Church (Y_C) and to Turing (Y_T):

$$\begin{aligned} Y_C \equiv \lambda f\,.\,VV &\quad \text{where} \quad V \equiv \lambda x\,.\,f(xx),\\ Y_T \equiv ZZ &\quad \text{where} \quad Z \equiv \lambda zf\,.\,f(zzf). \end{aligned}$$

For each term M,

$$\begin{aligned} Y_T M &\equiv (\lambda zf\,.\,f(zzf))ZM\\ &\xrightarrow{\star} f(zzf)[f := M, z := Z]\\ &\equiv M(ZZM)\\ &\equiv M(Y_T M). \end{aligned}$$

For Y_C, there is only a β-conversion, not a *beta*-reduction.

These combinators allow one to solve *equations*. Let M be an arbitrary term and $n \geqslant 0$. To solve for x in equation

$$xy_1 \ldots y_n = M$$

is to find a term F such that $Fy_1 \ldots y_n =_\beta M[x := F]$, where x, $y_1, \ldots, y_n$ are all variables of the λ-calculus. There is always a solution, which is

$$F \equiv Y(\lambda xy_1 \ldots y_n\,.\,M),$$

where Y is a fixpoint combinator.

In programming, one can write recursive programs with the help of fixpoint combinators. For example, suppose the factorial function must be programmed. It is the solution of the equation

$$\mathit{fact}(x) = \text{if } (x = 0) \text{ then } 1 \text{ else } x \cdot \mathit{fact}(\mathit{pred}(x)),$$

so *fact* is a fixpoint of the 'functional'

$$\lambda f x \,.\, \text{if } (x = 0) \text{ then } 1 \text{ else } x \cdot f(\mathit{pred}(x)).$$

One can therefore define

$$\mathit{fact} \equiv Y(\lambda f x \,.\, \text{if } (x = 0) \text{ then } 1 \text{ else } x \cdot f(\mathit{pred}(x))),$$

where Y is a fixpoint combinator. To make this term explicit, define

$$\begin{aligned} \mathbf{K} &\equiv \lambda x y \,.\, x, \\ \mathbf{pred} &\equiv \lambda x \,.\, (n(\lambda p u \,.\, u(\mathbf{succ}(p\,\mathbf{T}))(p\,\mathbf{T}))(\lambda u \,.\, u\,\mathbf{0}\,\mathbf{0}))\mathbf{F}. \end{aligned}$$

The combinator **pred** represents the predecessor, where if **n** is a Church numeral,

$$\mathbf{pred}\,\mathbf{n} = \begin{cases} \mathbf{m} & \text{if } n = m + 1 \\ \mathbf{0} & \text{if } n = 0. \end{cases}$$

See exercise 4.

A λ-term for the factorial is computed as follows:

$$\mathit{fact} \equiv Y(\lambda f x \,.\, \mathbf{zero}\, x\, \mathbf{1}(\mathbf{mult}\, x(f(\mathbf{pred}\, x)))).$$

The use of abbreviations (giving names to certain combinators) and of syntactic sugar (`let ... in ... end`) already greatly simplifies terms, but it is mostly the introduction of a sufficiently expressive system of types which allows for real programming. The pure λ-calculus should be considered to be a low-level language, a kind of 'machine language'. This assumption allows one to study language mechanisms and semantics uniformly. This path was first followed in the 1960s by Landin [93].

3.1.4 Typed λ-calculus

In mathematics, a mapping's definition includes the domain and range (co-domain). This aspect was not treated by λ-calculus terms. So the set-theoretic meaning of terms such as $\lambda x \,.\, xx$ is unclear.

Typed terms

The first solution consists of introducing simple types.

Definition 3.6 Given a set A of *atomic type* symbols, the closed terms over a signature $A \cup \{\rightarrow\}$, $\mathrm{ar}(\rightarrow) = 2$ and $\mathrm{ar}(\mathbf{a}) = 0$, $\mathbf{a} \in A$, are called *simple types*.

To do arithmetic, one could define $A = \{\mathbf{nat}\}$. Objects of type **nat** are then interpreted as integers, objects of type $\mathbf{nat} \rightarrow \mathbf{nat}$ as functions from $\mathbf{N}$ into $\mathbf{N}$, those of type (**nat** $\to$ **nat**) $\to$ (**nat** $\to$ **nat**) transform functions into functions, and so on. To program, **bool**, **string** and others would also have to be included among the atomic types.

Write types with σ, τ and abbreviate $\tau_1 \to (\tau_2 \to (\ldots \to (\tau_{n-1} \to \tau_n)\ldots))$ by $\tau_1 \to \tau_2 \to \ldots \to \tau_{n-1} \to \tau_n$, or even $\tau_1, \tau_2, \ldots, \tau_{n-1} \to \tau_n$ by noticing that the type can be interpreted by 'decurryfying' as that of functions of $n-1$ variables, respectively, in $\tau_1, \ldots, \tau_{n-1}$, and values in τ_n.

Definition 3.7 Let X^σ be a countable infinite set of *type variables*, which are written $x^\sigma, y^\sigma, \ldots$ for each type σ. *Typed terms* are defined by:

1. Each variable x^σ is a typed term of type σ.
2. If $M^{\sigma\to\tau}$ and N^σ are typed terms of respective types $\sigma \to \tau$ and σ, then $(M^{\sigma\to\tau}N^\sigma)^\tau$ is a typed term of type τ.
3. If x^σ is a variable of type σ, and if M^τ is a typed term of type τ, then $(\lambda x^\sigma.M^\tau)^{\sigma\to\tau}$ is a typed term of type $\sigma \to \tau$.

The marking of the type as superscript on a term and its subterms is indispensable (even though it will often be omitted for convenience). Hence, as for finite terms and pure λ-terms, the very writing of a term is a proof that it is formed correctly. By omitting the type annotations, a pure term is obtained: the formation rules for typed terms thus appear to be a restriction on the formation rules for pure terms.

For each type τ there is an identity $(\lambda x^\tau . x^\tau)^{\tau\to\tau}$, but there is no typed term corresponding to $\lambda x \,.\, xx$.

As for the pure λ-calculus, this set of terms can be extended by *typed* constants, for example 0, 1, 2, ... of type **nat**, *succ* of type **nat** $\to$ **nat**, + of type **nat** $\to$ **nat** $\to$ **nat**; a *higher-order signature* Σ is produced, formed from the set of its constants, along with an arity function from Σ into the set of types, and the following clause must be added to the preceding definition:

4. If $f \in \Sigma$ and $\mathrm{ar}(f) = \tau$, then f is a typed term of type τ.

For it to be useful, this extension supposes that specific computing rules (the so-called δ rules) be given for the constants, for example $1 + 1 \to_\delta 2$. From a practical point of view, in a functional programming language these rules consist of calling external procedures to handle arithmetic efficiently. Furthermore, other computing mechanisms, which the type system cannot allow, such as recursion, can be added in the same manner. However, the addition of constants is theoretically unnecessary to increase the computational power of pure λ-calculus. These δ rules complicate the study of reduction considerably: the confluence of β-reduction can be destroyed by adding such rules. Therefore, no δ rule will be considered here, and, in general, only typed terms without constants will be considered.

Simple types are stratified according to their *order*. A function ord is defined, associating an integer to each atomic type, for example $\mathrm{ord}(\mathbf{nat}) = 1$. This function is extended by:

$$\mathrm{ord}(\tau_1 \to \tau_2) \quad = \quad \max(\mathrm{ord}(\tau_1) + 1, \mathrm{ord}(\tau_2)).$$

Hence, **nat** is of order 1, **nat** $\to$ **nat** and **nat** $\to$ (**nat** $\to$ **nat**) are of order 2, and (**nat** $\to$ **nat**) $\to$ **nat** and (**nat** $\to$ **nat**) $\to$ (**nat** $\to$ **nat**) are of order 3. Note the difference between the *false* higher-order type **nat** $\to$ (**nat** $\to$ **nat**), which is actually the type of 'curryfied functions of two variables', and (**nat** $\to$ **nat**) $\to$ **nat**, which is the type of 'functionals'.

The order of a variable is that of its type. A typed term is of order $\leqslant n$ if all of its variables (free or bound) are of order $\leqslant n$: $x^{\mathbf{nat}}$ and $\lambda x^{\mathbf{nat}}.\, x^{\mathbf{nat}}$ are of order 1, and $\lambda f^{\mathbf{nat} \to \mathbf{nat}} \lambda x^{\mathbf{nat}}.\, f(f(x))$ is of order 2. When a term contains constants, it is of order $\leqslant n$ if the types of its constants are also of order $\leqslant n+1$: $0^{\mathbf{nat}}$ is of order 0, $1+1$ and $x^{\mathbf{nat}}+1$ are of order 1. Hence, a 'first-order' Σ-term is of order $\leqslant 1$: there is an atomic type of order 1 (or several, in the case of many-sorted signatures), its variables are of order 1, and the types of symbols in Σ are of order 1 for the constants, and 2 for the function symbols. A Σ-term of order 0 is a constant symbol.

This notion of order allows one to classify computer languages. A language allowing definitions of functions over numeric values is of order 1. Pascal is of order 2, as function definitions can have functional parameters of order 1. Functional languages such as ML are of order ω, since they allow the use of variables of arbitrarily high order.

The main definitions of the untyped λ-calculus translate directly to the typed one (graft, substitution, redex, β-reduction, residues, ...). For substitution, a variable must be substituted with a term of the same type. The Church–Rosser theorem is therefore valid, as are its corollaries. There are new phenomena: the weak and strong normalization properties.

Theorem 3.10 All typed terms are normalizable.

Theorem 3.11 All typed terms are strongly normalizable.

Hence the normal form of a typed term is obtained by any reduction. Strong normalization is an important result which is difficult to prove. The β-convertibility of two terms is therefore decidable: it suffices to compute the normal forms of two terms and to compare them syntactically to decide if they are β-convertible. This follows from the strong normalization theorem and the proof of the weak normalization theorem, which can be adapted into an effective reduction strategy. The proof of the weak normalization theorem will be given in the last section of the chapter, which deals with the problem of termination.

Typable terms

Typed terms were introduced so that terms would better suit the set definition of a function. However, this approach is not really natural. There are an infinite number of terms $\mathbf{I}^\tau$ for the identity function, but they all do the same thing, regardless of the argument. Furthermore, even in mathematics, it is common to define a function—using an algorithm, for example—before determining the domain over which it is defined, e.g., the set of values for which the algorithm terminates. This suggests that it is the pure terms which are essential. Typing them just adds more information. To assign a type to a pure term is already giving it a kind of semantics. It is an *abstract interpretation*, which does not compute the value of a term, but, rather, the set of possible values: a pure term of type **nat** $\rightarrow$ **nat** can be *interpreted* as a function over the integers with integer values.

The problem is therefore to type pure terms. A first difficulty comes from the variables. Variables must be assigned a type, as in programming, but if one wants polymorphic types, as for the identity, what type should they be given? The solution is to introduce type variables, written $\xi, \eta, \ldots$. Simple types were terms closed over a signature $A \cup \{\rightarrow\}$. They are now terms, with variables, over this signature.

Since every term is obtained from its variables, it suffices to type its variables, then to infer its type. For example, if x is declared to be of type ξ, it can be inferred that $\mathbf{I} \equiv \lambda x\, x$ is of type $\xi \rightarrow \xi$. One can also declare x to be of type **nat** $\rightarrow$ **nat** and then deduce that $\mathbf{I}$ is of type (**nat** $\rightarrow$ **nat**) $\rightarrow$ (**nat** $\rightarrow$ **nat**). An 'assumption' is made about the types of the variables. A finite set of variable type declarations forms a *typing context*, of the form $x_1 : \tau_1, \ldots, x_n : \tau_n$, where $x_1, \ldots, x_n$ are variables of the λ-calculus and $\tau_1, \ldots, \tau_n$ are types.

Here is an inference system. Its judgments are of the form '$C \vdash M : \tau$', stating that in the typing context C, the term M is assigned type τ. When C is empty, write $\vdash M : \tau$ instead of $\varnothing \vdash M : \tau$:

$$\frac{}{x_1 : \tau_1, \ldots, x_n : \tau_n \vdash x_i : \tau_i}$$

$$\frac{C \vdash M : \sigma \rightarrow \tau \qquad C \vdash N : \sigma}{C \vdash MN : \tau}$$

$$\frac{\{x : \sigma\} \cup C \vdash M : \tau}{C \vdash \lambda x\, M : \sigma \rightarrow \tau}$$

A closed term M is *typable* if there exists a type τ such that the judgment $\vdash M : \tau$ can be derived.

Example Recall that $\mathbf{K} \equiv \lambda xy\,.\,x$. To prove that $\vdash \mathbf{K} : \xi \rightarrow \eta \rightarrow \xi$, start with $x : \xi$ and $y : \eta \vdash x : \xi$. Abstracting over y yields $x : \xi \vdash \lambda y\, x : \eta \rightarrow \xi$; abstracting over x yields $\vdash \lambda x(\lambda y\, x) : \xi \rightarrow (\eta \rightarrow \xi)$. □

Note that in the third rule (for $\lambda x\, M$), the assumption $x : \sigma$ appearing in the context of the premise does not appear in the context of the conclusion (unless it is also in C): the assumption $x : \sigma$ is *discharged* by the inference rule. On the other hand, in the second rule (for MN), it seems superfluous to mention context C since it is not modified. One can then write these rules simply by indicating the discharged hypotheses, usually written between square brackets ([...]) or slashed $(\cdot/\!\!\cdot)$.

$$\frac{M : \sigma \to \tau \qquad N : \sigma}{MN : \tau}, \qquad \frac{\begin{array}{c}[x : \sigma] \\ M : \tau\end{array}}{\lambda x\, M : \sigma \to \tau}.$$

This formulation will be used when discussing natural deduction, where the problem will be different: the type τ will be known, and terms of type τ will have to be found.

A type τ such that $C \vdash M : \tau$ is not necessarily unique. Consider the case of **I**. There is only one rule for which the conclusion is an abstraction; one can only have inferred $\lambda x\, x : \sigma \to \tau$ from $x : \sigma \vdash x : \tau$ using the first rule, and with $\sigma = \tau$. Hence, all judgments derivable for **I** are the $\vdash \mathbf{I} : \tau \to \tau$, with τ arbitrary. They are therefore all instances of $\xi \to \xi$, where ξ is an arbitrary variable. The type $\xi \to \xi$ is called a principal type of **I**. This principal type is not even unique: $\eta \to \eta$ is another one, obtained by renaming. This question will be cleared up when the subsumption preorder is defined over types.

Typed programming languages Each type language has a 'type system' which contains atomic types, such as **nat** or **bool**, type formation rules and expression typing rules. The formation rules introduce type constructors. In functional languages, the $\to$ constructor is essential, but the majority of languages also have constructors for the product or the union of types. Typing rules, including typed constant declarations, e.g., **true** : **bool** and **succ** : **nat** $\to$ **nat**, allowing the inference of judgments $C \vdash M : \tau$, can be used:

- if C, M and τ are known, to check that M is correctly typed: this is *type-checking*, used for typed languages such as Pascal where all the variables, functions, etc., must be declared along with their type;
- if only C and M are known, to infer the (principal) type τ of M: this is *type inference*, used by ML and most other contemporary functional languages.

3.2 Rewriting systems

The second general construction forming the basis for functional languages has a geometric flavor. It is the subsumption preordering over first-order terms. When combined with the λ-calculus, this relation creates an extension of β-reduction, call by pattern matching, found in most functional languages. A more general form, call by unification, is the basis for logic programming. Pattern matching allows one to write computation rules with more flexibility than does the bare λ-calculus: these rules are rewrite rules, whose study is now an integral part of computer science.

3.2.1 Pattern matching

Recall the notation of Chapter 1: Σ is a signature, X, $\mathrm{T}_\Sigma[X]$ and $\mathbb{S}$ are, respectively, the set of variables, terms and substitutions. Instead of writing $\mathrm{T}_\Sigma[X]$, $\mathrm{T}[X]$ will be used. The definition of substitutions allows one to distinguish variables from constants, by internalizing the concept of metasymbol. It allows one to define a 'topology' over $\mathrm{T}[X]$, and, more precisely, to define a preorder.

The definition of a term $M \in \mathrm{T}[X]$ determines the set $\mathbb{S}M = \{\theta\, M\,;\, \theta \in \mathbb{S}\}$. Its elements are the *instances* of M, and M is its *pattern*. If x is a variable, then $\mathbb{S}x = \mathrm{T}[X]$, and if M is a closed term, then $\mathbb{S}M = \{M\}$. Hence, the 'size' of $\mathbb{S}M$ measures the genericity of M. As $\mathbb{S}M$ becomes smaller, term M becomes more defined, i.e., gives information. This level of information defines an order relation. These ideas are very general and can of course be applied to many objects other than terms, such as words, graphs and 'forms'.

Definition 3.8 Let $M, M' \in \mathrm{T}[X]$. The term M' is an *instance* of M, or M is a *generalization* of M', written $M \leqslant M'$, if there exists a substitution $\theta \in \mathbb{S}$ such that $\theta M \equiv M'$. The restriction of θ to $\mathrm{var}(M)$, which is uniquely determined by M and M', is called the *matcher* of M towards M'. The relation $\leqslant$ is the *subsumption preorder* over terms.

It is clear that $M \leqslant M'$ if and only if $\mathbb{S}M \supseteq \mathbb{S}M'$. Hence, $\leqslant$ is a preorder relation, i.e. it is reflexive and transitive. But it is not antisymmetric, so it is not an order relation: for example, if x and y are distinct variables, $x \leqslant y$ and $y \leqslant x$, and $x \neq y$. This defect comes from the nature of variables, which is to carry minimal information. Consider the equivalence relation generated by the preorder.

Definition 3.9 Let $M, M' \in \mathrm{T}[X]$. The term M is a *variant* of M', written $M \sim M'$, if $M \leqslant M'$ and $M' \leqslant M$. Define $\tilde{T} = T/\sim$ and use $\leqslant$ for the order relation induced by $\leqslant$ over $\tilde{\mathrm{T}}[X]$.

Proposition 3.12 $M \sim M'$ if and only if there exists a renaming, i.e., a bijective substitution ρ, taking variables to variables, such that $M' = \rho M$.

Proof If $M' = \rho M$, where ρ is a renaming, then $M = \rho^{-1} M'$, hence $M \leqslant M'$ and $M' \leqslant M$, so $M \sim M'$.

Conversely, if $M \sim M'$, let θ and θ' be matchers from M towards M' and from M' towards M respectively. Then $\theta'\theta M = M$ and $\theta\theta' M' = M'$. In particular, if $x \in \mathrm{var}(M)$, $\theta'\theta(x) = x$, which shows that the restriction of θ to $\mathrm{var}(M)$ is injective. Since $\theta(x)$ is a variable as well as a subterm of M', $\theta(x) \in \mathrm{var}(M')$. Similarly, $\theta'(x) \in \mathrm{var}(M)$ if $x \in \mathrm{var}(M')$. Hence, $\theta\restriction_{\mathrm{var}(M)} : \mathrm{var}(M) \to \mathrm{var}(M')$ and $\theta'\restriction_{\mathrm{var}(M')} : \mathrm{var}(M') \to \mathrm{var}(M)$ are injections, which proves that $\mathrm{var}(M)$ and $\mathrm{var}(M')$ are equipotent. Define the substitution ρ by

$$\rho(x) = \begin{cases} \theta(x), & \text{if } x \in \mathrm{var}(M) \\ \phi(x), & \text{if } x \in \mathrm{var}(M') \setminus \mathrm{var}(M) \\ x & \text{otherwise,} \end{cases}$$

where ϕ is a bijection from $\text{var}(M') \setminus \text{var}(M)$ into $\text{var}(M) \setminus \text{var}(M')$. □

The equivalence $\sim$ equates terms which are equal up to the renaming of variables. In particular, all variables are equivalent and their class constitutes the smallest element of $\tilde{\mathrm{T}}[X]$, written $\perp$. The maximal elements of $\tilde{\mathrm{T}}[X]$ are the closed terms. The class of term M is written $[M]$.

Example If $\Sigma = \{0, s\}$, then $\tilde{\mathrm{T}}[X]$ includes the terms $[s^p 0]$ and $[s^p x]$, with relations $[s^p x] \leqslant [s^{p+1} x]$ and $[s^p x] \leqslant [s^p 0]$ for all $p \in \mathbf{N}$. □

The ordered set $(\tilde{T}, \leqslant)$ has remarkable properties. In what follows, the strict order $>$ is used: $t > t'$ if $t \geqslant t'$ and $t \neq t'$, for $t, t' \in \tilde{\mathrm{T}}[X]$, and in $\mathrm{T}[X]$, $M > M'$ if $M \geqslant M'$ and $M \nleqslant M'$.

Definition 3.10 An ordered set $(E, \leqslant)$ is a *well-founded inf-semilattice* if

1. $<$ is a well-founded strict order over E, i.e, there does not exist any sequence $(e_i)_{i \geqslant 0}$ such that $e_i > e_{i+1}$ for all i;
2. every pair e, e' of elements of E has a greatest lower bound, written $e \wedge e'$.

Proposition 3.13 The order $\leqslant$ over $\tilde{\mathrm{T}}[X]$ is well-founded.

Proof Define the *size* $|M|$ of a term M by

1. $|x| = 0$, if $x \in X$;
2. $|f M_1 \ldots M_n| = 1 + \sum_{i=1}^{n} |M_i|$, if $\text{ar}(f) = n \geqslant 0$;

and let $v(M) = \text{card var}(M)$. Note that $|\ |$, and v are defined over $\tilde{\mathrm{T}}[X]$, i.e., are invariant under renaming of variables.

If $M' > M$, then $M' = \sigma M$, where σ is not a permutation, and $|M'| \geqslant |M|$.

Let $(M_i)_{i \geqslant 0}$ be a strictly decreasing sequence of terms. Then the sequence $(|M_i|)_{i \geqslant 0}$ is decreasing in $\mathbf{N}$. Since $\mathbf{N}$ is well-founded, this sequence becomes stationary at some point: there exists $p \geqslant 0$ such that $|M_i| = |M_p|$ for all $i \geqslant p$.

If $|\sigma M| = |M|$, then σ does not introduce any new symbol. By assumption σ is not a renaming, i.e., is not injective, hence $v(\sigma M) < v(M)$.

Therefore, $v(M_i) < v(M_{i+1})$ for every $i \geqslant p$. Yet, since the terms M_i all have the same height, the sequence $(v(M_i))_{i \geqslant p}$ is bounded by the number of occurrences of variables in M_p.

The existence of a strictly decreasing sequence of terms is therefore contradictory, which proves that $\tilde{\mathrm{T}}[X]$ is well-founded. □

This property of the order over $\tilde{\mathrm{T}}[X]$ can be made stronger. It is not just that a class of terms cannot have a strictly decreasing infinite sequence of lower bounds, but that the number of lower bounds of each class is finite. This property is proven by examining the *covering* relation associated with the order. In a preordered set $(E, \leqslant)$, *e covers e'* if $e > e'$ and there does not exist an element e'' such that $e > e'' > e'$. By interpreting the subsumption preorder in terms of information, the covering relation determines the 'quanta' of information. In general, it is not true that, given an order relation λ, the order generated by the covering relation of λ is still λ. This is what actually happens with terms. By using this relation, from exercises 9 and 10 (p. 108), one can deduce that the set of lower bounds of a term is finite.

Generalization The problem of *generalization* is, given a set A of terms, to find the largest term M_0, with $M \geqslant M_0$ for all $M \in A$. This situation is typical of artificial intelligence, where, starting from a set of objects such as forms or frames, an attempt is made to find an object of which these objects are all particular cases. The theorem below means that $\tilde{\mathrm{T}}[X]$ is a good domain for artificial intelligence.

According to a general result on well-founded inf-semilattices, every non-empty subset admits a least upper bound (the proof is left as an exercise). There is no analogous result for upper bounds. However, each set with an upper bound has a least upper bound, since the non-empty set of upper bounds has a greatest lower bound, which is the least upper bound.

To solve the generalization problem, construct the greatest lower bound of two terms, called the *greatest generalization* of those terms. This lower bound is not unique in $\mathrm{T}[X]$, but it is in $\tilde{\mathrm{T}}[X]$. It is, however, easier to work in $\mathrm{T}[X]$.

For $M, N \in \mathrm{T}[X]$, one constructs a term $M \wedge N$ such that $[M \wedge N]$ is the greatest lower bound of $[M]$ and $[N]$.

If M and N have the same head symbol, the inf is defined recursively:

$$fM_1 \ldots M_p \wedge fN_1 \ldots N_p = f(M_1 \wedge N_1) \ldots (M_p \wedge N_p).$$

If they do not, $M \wedge N$ is an arbitrary variable whose class is $\perp$ in $\tilde{\mathrm{T}}[X]$. To apply the recursive case, pairs of terms (M_i, N_i) must be distinguished by different variables.

Example Let $M = f(g(a,b),a,b)$ and $N = f(g(b,c),b,h(c))$. Then $M \wedge N = f(g(x,y),x,z)$, variable x being the inf of pair (a,b), which has two occurrences in (M,N). □

The easiest way to distinguish pairs of terms is to use an arbitrary bijection Φ from T^2 to X (the sets X and Σ are supposed to be countable). Write the converse bijection $\pi : X \to T^2$. Its component functions are π^1 and π^2 from X to $\mathrm{T}[X]$. Let

$$M \wedge N = \Phi(M, N)$$

if M and N do not have the same head symbol.

It is easy to prove that $M \wedge N$ is a lower bound for M and N (exercise 11). It is more difficult to prove that it is the greatest lower bound, since the order is not compatible with the term structure.

Lemma 3.14 If $M_i' = \sigma' M_i$ and $M_i' = \sigma'' M_i$, $1 \leqslant i \leqslant n$, then there exists a substitution σ such that $M_i' \wedge M_i'' = \sigma M_i$ for all i.

Proof By induction over $l = \sum_{i=1}^{n} |M_i|$.

If $l = 0$, then each M_i is a variable x_i, and $\sigma(x_i) = M_i' \wedge M_i''$ defines the wanted substitution.

If $l > 0$, then there exist indices $k_1, \ldots, k_p$ such that $|M_{k_i}| > 0$. Define $M_{k_i} = f_i(M_{k_i}{}^j)$. Then $\sum_i \sum_j |M_{k_i}{}^j| = \sum_i(|M_{k_i}| - 1) < \sum_i |M_{k_i}| \leqslant l$, and $M_{k_i}' = \sigma' M_{k_i} = f_i(\sigma' M_{k_i}{}^j)$, $M_{k_i}'' = \sigma' M_{k_i} = f_i(\sigma'' M_{k_i}{}^j)$.

By the inductive hypothesis, there exists σ such that $M_{k_i}'{}^j \wedge M_{k_i}''{}^j = \sigma M_{k_i}{}^j$. Deduce that $M_{k_i}' \wedge M_{k_i}'' = \sigma M_{k_i}$. All that is now needed is to extend σ to the M_i which are variables, as in the case $l = 0$: a σ is thus obtained such that $M_i' \wedge M_i'' = \sigma M_i$ for all i. □

It suffices to consider the lemma for $n = 1$ to determine that $M' \wedge M''$ is a greatest lower bound for M' and M'', i.e., that if $M \leqslant M'$ and $M \leqslant M''$, then $M \leqslant M' \wedge M''$. However, if an attempt had been made to prove the property by induction over terms directly, the inductive hypothesis would have been too weak and would not have sufficed to prove the inductive step. It was necessary to take $M_i' = \sigma' M_i$ instead of $M_i' \geqslant M_i$, for all i. This is a typical example of reinforcing or generalizing a property before proving it by induction: one does not always generalize for the fun of it!

Pattern matching The problem is to determine, for $M, M' \in \mathrm{T}[X]$, if $M \leqslant M'$. If so, the matcher which transforms M to M' must be computed. To write a pattern-matching program, it is useful to consider a substitution as an environment, i.e., as a sequence of variable-term bindings: in this manner, for a fixed M, all instances θM share the same structure, that of M.

A matcher from M to N in the environment θ is needed. Initially, θ is the identity. There are two failure cases (halt):

- if M and N have distinct head symbols;
- if M is a variable bound by θ to a distinct term of N;

one success case (halt):

- if M is a variable not bound by θ or already bound to N, the matcher is θ, extended by (M, N);

one iteration case:

- if M and N have the same head symbol, iterate over the children of M and N.

The last case means that if $M = f M_1 \ldots M_p$ and $N = f N_1 \ldots N_p$, the matching continues in parallel with M_1 to N_1, …, M_p to N_p. The matching will not be done in parallel, but, rather, sequentially for each pair (M_i, N_i), and by propagating the matcher obtained in the environment. This procedure is standard in functional programming.

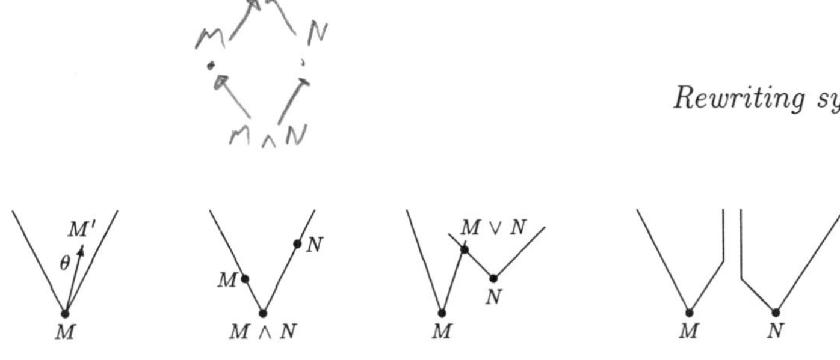

Figure 3.1 The term lattice

The 'cone' properties of the instances of a term are illustrated by the four sketches of Figure 3.1: matcher θ of M to M', greatest generalization of M and N, least upper bound of two terms, when it exists, and disjoint cones of terms otherwise.

3.2.2 Rewrite rules

The subsumption preorder allows one to define computing rules over first-order terms, analogous to β-reduction, but more flexible.

Definition 3.11 A *rewriting system* consists of a signature Σ and a set $\mathcal{R}$ of pairs of terms $(P, Q) \in \mathrm{T}_\Sigma[X] \times \mathrm{T}_\Sigma[X]$ such that $\mathrm{var}(Q) \subseteq \mathrm{var}(P)$ and $P \notin X$.

Each pair (P, Q), called a *rule*, also written $P \to Q$, defines a generic computing rule: P is a *generic redex* and Q is the result of contracting P. An instance σP is called a *redex*; the result of contracting σP is σQ.

The definition of a reduction relation from a rewriting system is analogous to β-reduction. A construction can be made using inference rules or by directly using grafts.

Definition 3.12 A term M rewrites itself into a term N by applying the rule $(P, Q) \in \mathcal{R}$ to the occurrence $u \in \mathcal{O}(M)$ if $M/u \geqslant P$ and $N = M[u \leftarrow \sigma Q]$, where σ is the matcher of P to M/u. This relation is written $M \xrightarrow{P \to Q, u} N$ or simply $M \xrightarrow{u} N$.

The system $(\Sigma, \mathcal{R})$ determines an inference system over judgments $M \rhd M'$, M and M' being terms. For each $(P, Q) \in \mathcal{R}$, there is a family of redex contraction rules. There are also context-passing rules.

$(\text{red}) : \dfrac{}{\sigma P \rhd \sigma Q}$	$(P, Q) \in \mathcal{R}, \quad \sigma \in \mathbb{S}$
$(f, i) : \dfrac{M_i \rhd M_i'}{f M_1 \dots M_i \dots M_n \rhd f M_1 \dots M_i' \dots M_n}$	$f \in \Sigma, \quad 1 \leqslant i \leqslant \mathrm{ar}(f)$

This inference system defines a binary relation over $\mathrm{T}[X]$, written $\to_\mathcal{R}$: $M \to_\mathcal{R} M'$ if and only if there exists a closed derivation of $M \rhd M'$. As for the λ-calculus, one proves that $M \longrightarrow_{\mathcal{R}} M'$ is equivalent to the existence of $u \in \mathcal{O}(M)$ and of $(P, Q) \in \mathcal{R}$ such that $M \xrightarrow{P \to Q, u} M'$. Passing the context is also written using the single rule

$$(\mathbf{cont}) : \frac{P \rhd Q}{\mathcal{C}[P] \rhd \mathcal{C}[Q]},$$

$\mathcal{C}$ being a term context, which states that reduction is a local process, independent of the context, as in context-free grammars.

3.2.3 Properties of rewriting systems

Unlike the λ-calculus, which uses only one rule, β-reduction rewriting systems use a number of forms. As there are no general theorems, each case must be examined individually.

A rewriting system is not necessarily confluent. For example, the rules $x \,\mathbf{or}\, y \longrightarrow x$ and $x \,\mathbf{or}\, y \longrightarrow y$ which express non-determinism are not confluent. A technique allowing, in certain cases, the proof of confluence will be presented later on. In general, the confluence problem is undecidable.

A system $\mathcal{R}$ is *noetherian* if all its terms are strongly normalizable by $\longrightarrow_{\mathcal{R}}$, i.e. if every reduction terminates. A system containing a permutative rule of the form $x + y \longrightarrow y + x$ cannot be noetherian. But there are more subtle cases (see exercise 12). The proof of noetherianity of a system is usually difficult, and linked to the concept of well-order.

Theorem 3.15 The noetherianity of systems formed of at least two rules is undecidable.

3.2.4 Examples

Rewriting systems are used in many areas.

An example is for non-deterministic functional programs, where function calls are made by *pattern matching*. Here is the definition of addition

$$\begin{aligned} x + 0 &\longrightarrow x, \\ x + s(y) &\longrightarrow s(x + y), \end{aligned}$$

and of Ackermann's function

$$\begin{aligned} A(x+1, 0) &\longrightarrow A(x, 1), \\ A(0, x) &\longrightarrow x + 1, \\ A(x+1, y+1) &\longrightarrow A(x, A(x+1, y)). \end{aligned}$$

These systems are confluent and noetherian. The proof of noetherianity, easy for the first, is a termination proof for the program.

The calculus also uses rules implemented in computer algebra systems. Here is a program which differentiates with respect to ξ:

$$
\begin{array}{rcl}
\partial\xi & \to & 1, \\
\partial a & \to & 0, \\
\partial(u+v) & \to & \partial u + \partial v, \\
\partial(uv) & \to & (\partial u)v + u(\partial v), \\
\partial(-u) & \to & -(\partial u), \\
\partial(u-v) & \to & \partial u - \partial v, \\
\partial(u/v) & \to & ((\partial u)v - u(\partial v))/v^2, \\
\partial(\ln u) & \to & (\partial u)/u, \\
\partial(u^v) & \to & vu^{v-1}\partial u + u^v(\ln u)\partial v,
\end{array}
$$

where ξ is a constant, the second rule being defined for all constants $a \neq \xi$.

3.3 Functional programming

The λ-calculus is the paradigm of functional programming. High-level functional languages exist, often with a syntax inspired by the λ-calculus, along similar evaluation mechanisms. The best known are Haskell, Hope, Lisp (particularly Scheme), Miranda and ML.

3.3.1 Evaluation rules in λ-calculus

Weak β-reduction
No language, functional or not, fully implements β-reduction, for procedure bodies are never evaluated, except in a partial manner, for optimization. In the case of the λ-calculus, restrictions are made on β-reduction; the resulting *weak β-reduction* has only the following context rules:

$$(1) : \frac{M \rhd M'}{MN \rhd M'N} \qquad (2) : \frac{N \rhd N'}{MN \rhd MN'}.$$

For this reduction, an abstraction $\lambda x\, M$ is irreducible, even if M contains a redex. Weakly normal terms are therefore the variables, abstractions and applications MN, where M and N are weakly normal, M not being an abstraction. In particular, weakly-normal closed terms are the closed abstractions.

The purpose of evaluation is to obtain a value from a program. Only closed terms correspond to programs, entities which can run by themselves. In pure λ-calculus, a value is a closed abstraction. In an extended λ-calculus, the set of values also contains constants such as numbers, Booleans and arithmetic operators, and partial applications, such as the term *plus* M, for all M, which needs a second argument to be evaluated.

However, weak β-reduction does not give a deterministic reduction order: given an application, should the left or the right subterm be reduced first, and to what point?

Weak reduction strategies

There are two weak reduction strategies for terms: call by value and call by name.

Call by value is a new restriction for β-reduction, with the rules

$$(\mathbf{red}_v) : \frac{}{(\lambda x\, M)V \rhd M[x := V]}$$

$$(1v) : \frac{M \rhd M'}{MV \rhd M'V} \qquad (2) : \frac{N \rhd N'}{MN \rhd MN'}$$

if V is a value

This strategy is easy to implement: most functional languages use it. Its major fault is its 'eagerness': when presented with an application MN, it evaluates both M and N, even if the value of N is not needed, or if computing it is harmful (if N is not normalizable).

To avoid this problem, it suffices to delay the evaluation of N by passing it directly as an argument to an abstraction: this is called *call by name*, another restriction on β-reduction, with only one context passing rule:

$$(\mathbf{red}) : \frac{}{(\lambda x\, M)N \rhd M[x := N]}$$

$$(1) : \frac{M \rhd M'}{MN \rhd M'N}$$

Call by value is easier to implement in a machine with environments. It is the solution used by the majority of languages, from C to Lisp, ML and FP. Call by name is normally implemented on graph reduction machines; this is done for Lazy ML, Haskell and Miranda. It produces better-defined programs.

Evaluation rules

The judgment $M \rhd M'$ says only that M reduces to M'. To evaluate M, one must successively reduce $M \rhd M'$, then $M' \rhd M''$, ... until a value is produced. It is preferable to construct a judgment $M \longrightarrow V$ which directly expresses that V is the value of M.

Furthermore, the substitution which appears in the contraction of a redex is not a free operation. It is replaced by a transformation of the environment. Instead of reducing a term, it will be evaluated, and a non-closed term can only be evaluated in an environment where its free variables are bound.

Since Landin [92], the following definitions are standard in programming.

- A *binding* is a pair (x, v), where x is a variable and v a value.
- An *environment* is a list of bindings.
- A *closure* is a pair (M, ρ), where M is a term, and ρ an environment containing a binding for each free variable of M. A closed term is a closure (in any environment) and a closure generates a closed term through repeated substitutions.
- A *value* is a closure (M, ρ), where M is normal for weak β-reduction. In particular, a closed abstraction is a value.

Reduction rules are applied to λ-terms, but evaluation rules treat terms 'in an environment'. The judgment '$\rho \vdash M \longrightarrow v$' is introduced to express that, in the environment ρ, the term M has value v.

Only the evaluation rule for an application is given. For call by value,

$$(\mathbf{App}_v) : \frac{\rho \vdash M \longrightarrow (\lambda x\, M', \rho') \qquad \rho \vdash N \longrightarrow v \qquad (x, v); \rho' \vdash M' \longrightarrow v'}{\rho \vdash MN \longrightarrow v'}$$

For call by name, the notion of value must be extended to arbitrary closures: environments are therefore formed of bindings of the form (x, F), where F is a closure:

$$(\mathbf{App}_n) : \frac{\rho \vdash M \longrightarrow (\lambda x\, M', \rho') \qquad (x, (N, \rho)); \rho' \vdash M' \longrightarrow v'}{\rho \vdash MN \longrightarrow v'}$$

Note that, in both cases, the evaluation of M', the body of the abstraction, is done in the environment ρ', extended by a new binding because of parameter passing, and not in ρ. This is characteristic of *static* or *lexical* binding. If ρ were used instead of ρ', the result would be *dynamic* binding. Lisp1.5, like many interpreted languages, used dynamic binding. Today, new Lisps, such as Scheme and Common Lisp, use static binding.

These evaluation rules are perfectly executable. They can easily be translated into a program. They constitute the dynamic semantics, expressed in the formalism of *structural operational semantics*, due to G.D. Plotkin, and developed under the name of *natural semantics* at the French INRIA.

Evaluation of data structures

Pairs are introduced in the λ-calculus by the rules:

1. if M and N are two terms, then (M, N) is a term;
2. if M is a term then $\pi^1 M$ and $\pi^2 M$ are terms.

Pairs allow one to construct all sorts of data structures, such as binary trees and lists, just as in Lisp.

From the point of view of reductions, the introduction of pairs creates a new kind of redex, terms of the form $\pi^i(M, N)$, which contract naturally:

$$\pi^1(M, N) \longrightarrow M, \qquad \pi^2(M, N) \longrightarrow N.$$

When this notion of contraction is done within a context, four new reduction rules are introduced (two for pair formation, one for π^1, one for π^2), along with the β-reduction rules. The notion of value must be extended to pairs: there are at least two possibilities, which define strict pairs and lazy pairs. In each case, contraction and how evaluation is done in context must be restricted.

For strict pairs, a pair (M,N) is a value if M and N are values, and $\pi^i M$ is a value if M is a value which is not a pair. This choice is taken by most programming languages (see Table 3.1), which implement call by value.

$$
\begin{array}{ll}
(\mathbf{pair}_s) : \dfrac{}{\pi^1(U,V) \rhd U} & (\mathbf{pair}_s) : \dfrac{}{\pi^2(U,V) \rhd V} \\[2ex]
(1,\_) : \dfrac{M \rhd M'}{(M,N) \rhd (M',N)} & (v,2) : \dfrac{N \rhd N'}{(U,N) \rhd (U,N')} \\[2ex]
(\pi^{1s}) : \dfrac{M \rhd M'}{\pi^1 M \rhd \pi^1 M'} & (\pi^{2s}) : \dfrac{M \rhd M'}{\pi^2 M \rhd \pi^2 M'} \\[2ex]
\multicolumn{2}{c}{\text{if } U \text{ and } V \text{ are values}}
\end{array}
$$

Table 3.1 Reduction of strict pairs

For *lazy* pairs, or *non-strict* pairs, every pair (M,N) is a value. These lazy pairs are similar to abstractions for weak β-reduction: they are values, whatever their contents. They can be implemented by placing strict pairs within an abstraction. Their reduction rules are similar to the rules used by programming languages that implement call by name.

$$
\begin{array}{ll}
(\mathbf{pair}) : \dfrac{}{\pi^1(M,N) \rhd M} & (\mathbf{pair}) : \dfrac{}{\pi^2(M,N) \rhd N} \\[2ex]
(\pi^{1p}) : \dfrac{M \rhd M'}{\pi^1 M \rhd \pi^1 M'} & (\pi^{2p}) : \dfrac{M \rhd M'}{\pi^2 M \rhd \pi^2 M'} \\
\multicolumn{2}{c}{\text{if } M \text{ is not a pair}}
\end{array}
$$

Table 3.2 Reduction of lazy pairs

Lazy pairs (see Table 3.2) are interesting as they allow for the incremental evaluation of objects, which might normally be infinite. As an example program, the Sieve of Erastothenes, which generates all the primes, will be given in the form of a rewriting system.

Lazy reduction of first-order terms
Consider the following rewriting system:

$$\begin{array}{rcl} \mathit{primes} & \to & \mathit{sieve}(\mathit{integers}(2)), \\ \mathit{integers}(x) & \to & \mathit{cons}(x, \mathit{integers}(s(x)), \\ \mathit{sieve}(\mathit{cons}(x,l)) & \to & \mathit{cons}(x, \mathit{sieve}(\mathit{filter}(x,l))), \\ \mathit{filter}(m, \mathit{cons}(x,l)) & \to & \text{if } m \text{ divides } x \text{ then } \mathit{filter}(m,l) \\ & & \text{else } \mathit{cons}(x, \mathit{filter}(m,l)). \end{array}$$

These rules state that $\mathit{filter}(m,l)$ removes from list l the multiples of the integer m; $\mathit{integers}(x)$ is the list of integers $\geqslant x$; *primes* is the list of prime numbers: $2, 3, 5, 7, 11, \ldots$. It is supposed that appropriate rules have been given for *divide* and for *if then else*.

Consider the reduction of *primes*. If the innermost redex is contracted at each step, only the first two rules are used and the following infinite reduction is produced:

$$\begin{array}{rcl} \mathit{primes} & \to & \mathit{sieve}(\mathit{integers}(2)) \\ & \to & \mathit{sieve}(\mathit{cons}(2, \mathit{integers}(3))) \\ & \to & \mathit{sieve}(\mathit{cons}(2, \mathit{cons}(3, \mathit{integers}(4)))) \\ & \to & \ldots \\ & \stackrel{\infty}{\to} & \mathit{sieve}(2, 3, 4, 5, 6, \ldots). \end{array}$$

The reduction is infinite, yet its 'limit', written informally using the symbol $\stackrel{\infty}{\to}$, has not even begun computing *sieve*. The other rules are never used, even though the third rule can be applied at every step, starting from $\mathit{sieve}(\mathit{cons}(2, \mathit{integers}(3)))$. This reduction is called unfair.

Definition 3.13 A reduction is *fair* if it is finite or if any rule which can be applied an infinite number of times is actually applied an infinite number of times.

If the leftmost–outermost redex is always contracted, another reduction is obtained. It is still infinite, but fair:

$$\begin{array}{rl} & \mathit{sieve}(\mathit{cons}(2, \mathit{integers}(3))) \\ \to & \mathit{cons}(2, \mathit{sieve}(\mathit{filter}(2, \mathit{integers}(3)))) \\ \to & \mathit{cons}(2, \mathit{sieve}(\mathit{cons}(3, \mathit{filter}(2, \mathit{integers}(3))))) \\ \to & \mathit{cons}(2, \mathit{cons}(3, \mathit{sieve}(\mathit{filter}(3, \mathit{filter}(2, \mathit{integers}(4)))))) \\ \to & \ldots . \end{array}$$

Even though this reduction is infinite, and *primes* is not normalizable, the computation is done in an incremental manner,

$$\mathit{primes} \stackrel{\infty}{\to} 2, 3, 5, 7, 11, \ldots ,$$

in the sense that, for all n, there exists a p such that the n first primes appear at the head of the term obtained at the end of p steps.

What has been used is the analogue to call by name in the λ-calculus. To avoid an infinite computation, this strategy must be restricted by making *cons* a lazy constructor, i.e., by allowing terms $cons(M, L)$ as values, no matter what the terms M and L are. Lazy reduction consists of contracting the outermost redexes not contained in a *cons*.

The reduction of *primes* stops at $cons(2, sieve(filter(2, integers(3))))$. To obtain the successive elements of *primes*, the reduction inside the *cons* must be explicitly *forced*, with the rules

$$hd(cons(M, L)) \to M, \qquad tl(cons(M, L)) \to L.$$

The second prime number is obtained by:

$$\begin{aligned} hd(tl(primes)) &\xrightarrow{\star} hd(tl(sieve(cons(2, integers(3))))) \\ &\xrightarrow{\star} hd(sieve(filter(2, integers(3)))) \\ &\to hd(cons(2, sieve(cons(3, filter(2, integers(3)))))) \\ &\to 3. \end{aligned}$$

3.3.2 Call by pattern matching

β-reduction models a procedure call in programming. However, there are more intelligent mechanisms, based on *pattern matching*, which are used in functional programming languages. A rewriting system can be interpreted as a non-deterministic program, each term determining, according to its form, the rule which must be used or fired.

This mechanism can be introduced in the framework of the λ-calculus. Take a signature Σ (single-sorted to simplify) and consider the $\lambda\Sigma$-terms formed from variables and constants, and using symbols from Σ, as well as application and abstraction:

- if M and N are $\lambda\Sigma$-terms, then (MN) is a $\lambda\Sigma$-term;
- if m is a Σ-term (a *pattern*) and P a $\lambda\Sigma$-term, then $\lambda m\, P$ is a $\lambda\Sigma$-term;
- if $f \in \Sigma$ and $\mathrm{ar}(f) = r$, if $M_1, \ldots, M_r$ are $\lambda\Sigma$-terms, then $fM_1 \ldots M_r$ is a $\lambda\Sigma$-term;
- every variable and every constant of Σ is a $\lambda\Sigma$-term.

For example, if Σ contains a binary pair forming symbol $(\_, \_)$, terms such as $\lambda(x, (y, \mathbf{0})) \,.\, (x, x(yx))$ can be formed.

The contraction of a redex is done by pattern matching,

$$\boxed{(\lambda m\, P)Q \to \theta P}$$

if $m \leqslant Q$ and θ is the matcher of m to Q.

Hence

$$[\lambda(x,(y,\mathbf{0})) \,.\, (x, x(yx))](R,(S,\mathbf{0}))$$

is a redex yielding $(R, R(SR))$, but

$$[\lambda(x,(y,\mathbf{0})) \,.\, (x, x(yx))]\mathbf{I}$$

is not a redex, since $(x,(y,\mathbf{0})) \not\leqslant \mathbf{I}$.

For reasons of efficiency and decidability, patterns m must be *linear*, i.e, a variable name can only occur once. Hence, terms such as $(\lambda(x,x) \,.\, M)(P,Q)$, whose contraction would require an equality test on P and Q, are not allowed. A programming language can only reasonably be expected to test textual equality ($P \equiv Q$) or equality of representations in memory (equality of two pointers of the same type). These two kinds of equality are not satisfactory from a programming point of view. The equality between values of P and Q ($P =_\beta Q$) is undecidable in general. In particular, the extensional equality between two functions is undecidable.

3.3.3 The principal type scheme of a λ-term

A type inference system was presented for λ-calculus terms and simple types, which are terms with variables over a signature $A \cup \{\rightarrow\}$. Write T for the set of these types. It can be shown that every typable closed term has a most general type, unique up to renaming of its type variables, i.e., it is an element of $\tilde{T}$. Instead of associating a type with a term, it is better to associate a set of types. For example, I has as types all of the instantiations of the form $\xi \rightarrow \xi$, a set which can be written $(\xi) \,.\, (\xi \rightarrow \xi)$. The typed terms remain monomorphic, so this choice remains a convenience.

Despite what was stated previously, the term `let val` x = Q `in` P `end` is not exactly an abbreviation for the redex $(\lambda x P)Q$. The difference is that in a redex, all occurrences of x in P must have the same type, while in a `let`, they can have distinct types. This allows one to write terms such as: `let val` i = $\lambda x\, x$ `in` $(i\,\mathbf{true}, i\,\mathbf{2})$ `end` It is the `let` form which introduces *polymorphism* in the system of simple types. The inference system must be modified to take this into account, by introducing type schemes.

If τ is a type, and if $\{\xi_1, \ldots, \xi_n\}$ is a set of variables of type ($n \geqslant 0$), then

$$(\xi_1 \ldots \xi_n) \,.\, \tau$$

is a *type scheme*, representing the set of types $\theta(\tau)$, for all substitutions θ whose domains are included in $\{\xi_1, \ldots, \xi_n\}$. The following properties hold, among others:

$$\begin{aligned} () \,.\, \tau &= \{\tau\}, \\ (\xi) \,.\, \tau &= \{\tau\} \quad \text{if } \xi \notin \operatorname{var} \tau, \\ (\xi) \,.\, \xi &= T. \end{aligned}$$

Note that by equating the type τ with the type scheme $().\tau$, types can be considered to be particular cases of type schemes. These schemes are also more general than type classes for the $\sim$ of renaming. The class in $\tilde{T}$ of a type $\tau \in T$ is the type scheme $[\tau] = (\xi_1 \dots \xi_n).\tau$, where $\{\xi_1, \dots, \xi_n\} = \mathrm{var}(\tau)$. On the other hand, a type scheme is not necessarily closed by renaming type variables, since every type is a type scheme according to the above equality. Finally, if $\sigma' \subseteq \sigma$, σ' is a *generic instance* of σ. For example, $(\eta).(\eta \to b) \to a$ is a generic instance of $(\xi).\xi \to a$, where a and b are type constants.

The λ-terms continue to be typed by types, but in a typing context, a list of declarations $x : \sigma$ is allowed, where x is a λ-calculus variable and σ is a type scheme. Table 3.3 gives the Damas–Milner system, where $\mathrm{gen}_C(\tau)$ is the type scheme $(\vec{\xi}).\tau$ and $\vec{\xi}$ is formed of the variables of τ which have no free occurrence in the schemes figuring in the typing context C.

The (var) rule allows a bound variable i to have two distinct types. For example:

$$i : (\xi).\xi \to \xi \vdash i : \mathbf{bool} \to \mathbf{bool}$$

from which the type for `let val` $i = \lambda x\, x$ `in` $(i\,\mathbf{true}, i\,\mathbf{2})$end is computed:

$$(\mathtt{let}) : \frac{\vdash \lambda x\, x : \xi \to \xi \qquad i : (\xi).\xi \to \xi \vdash (i\,\mathbf{true}, i\,\mathbf{2}) : \mathbf{bool} * \mathbf{nat}}{\vdash \mathtt{let}\ i = \lambda x\, x\ \mathtt{in}\ (i\,\mathbf{true}, i\,\mathbf{2})\mathrm{end} : \mathbf{bool} * \mathbf{nat}}$$

$$(\mathrm{var}) : \frac{}{x_1 : \sigma_1, \dots, x_n : \sigma_n \vdash x_i : \tau_i} \quad \text{if } \tau_i \in \sigma_i$$

$$(\mathtt{App}) : \frac{C \vdash M : \tau \to \tau' \qquad C \vdash N : \tau}{C \vdash MN : \tau'}$$

$$(\lambda) : \frac{\{x : \tau\} \cup C \vdash M : \tau'}{C \vdash \lambda x\, M : \tau \to \tau'}$$

$$(\mathtt{let}) : \frac{C \vdash Q : \tau \qquad \{x : \mathrm{gen}_C(\tau)\} \cup C \vdash P : \tau'}{C \vdash \mathtt{let}\ x = Q\ \mathtt{in}\ P\ \mathtt{end} : \tau'}$$

Table 3.3 The Damas–Milner type inference system

3.3.4 ML, a typed functional language

ML started out as the metalanguage for the proof system LCF (*Logic for Computable Functions*) due to D. Scott and developed by R. Milner at Edinburgh in 1978 [110]. It was designed to define proof strategies, by manipulating objects of type 'theorem', 'rule', ... The ideas essentially go back to Landin, whose ISWIM (*If You See What I Mean*) language can be considered to be the ancestor of functional languages.

ML has since become the prototype of functional languages. There are several versions: Standard ML (Cambridge, UK, and Bell Labs, USA), LML (Gothenburg, Sweden) and CAML (Paris, France). Lazy ML is a *lazy* version, in which programs are compiled to a (virtual) supercombinator graph reduction machine. The name CAML comes from its compilation to the CAM (*Categorical Abstract Machine*), developed by P.-L. Curien and G. Cousineau. The dialect used here is Standard ML of New Jersey, written by A. Appel and D. MacQueen.

The applicative language
ML is, first of all, an applicative language. A program is a sequence of definitions, each introducing a name along with its value, and expressions:

```
(* all the commands, definitions and expressions to evaluate *)
(* end with a semicolon *)
let val a = 1 and b = 2 and c = 3
in b*b - 4*a*c
end;
```

```
↷ val it = ~8 : int        (* value and type returned by SML *)
```

The execution of a program consists of constructing an environment, then of evaluating expressions in the new environment. The evaluation mechanism has no side effects on the environment. Only declarations, local to an expression (preceding example), or global, as in

```
val pi = 3.1415926535 ;
```

```
↷ val pi = 3.1415926535 : real
```

can change the environment. The latter consists of name–value bindings, unlike many languages where names are bound to locations which can contain values.

Numbers, Booleans, strings and lists The power of the language comes from the expressions which can be constructed. The usual expressions of programming languages are available. Here are some examples:
Numbers:

```
5 ;
```

```
↷ val it = 5 : int          (* the value of 5 is 5, of type int *)
```

```
3.5 ;
```

```
↷ val it = 3.5 : real       (* a floating point number *)
```

The syntax of arithmetic expressions is infix, as is usual:

```
((2 + 7)*5) + (2*(3 div (1+1))) ;
```

Booleans:

```
(false orelse true) andalso true ;
```

Tuples:

```
(1,2) ;
```

```
↷ val it = (1,2) : int * int
```

```
#1(1,2) ;
```

```
↷ 1 : int
```

```
#2(1,2) ;
```

```
↷ 2 : int
```

```
(1,2,3) ;
```

```
↷ val it = (1,2,3) : int * int * int
```

Strings:

```
"here " ^ "is very ordinary Standard ML" ;
                              (* concatenation, infix operation *)
```

```
↷ val it = "here is very ordinary Standard ML"  : string
```

Lists:

```
[1,2,3,4] ;
```

```
↷ val it = [1,2,3,4] : int list
```

```
hd [1,2,3,4] ;           (* hd is the equivalent of Lisp's car *)
```

```
↷ val it = 1 : int
```

```
tl [1,2,3,4] ;           (* tl is the equivalent of Lisp's cdr *)
```

```
↷ val it = [2,3,4] : int list
```

```
1 :: [2,3,4] ;           (* infix list constructor *)
```

```
↷ val it = [1,2,3,4] : int list
```

```
[1,2,3] @ [4,5] ;        (* list concatenation *)
```

```
↷ val it = [1,2,3,4,5] : int list
```

```
[] ;                     (* the empty list, generic constant *)
```

```
↷ val it = [] : 'a list
```

Functions ML is also, and particularly, a functional language, in the sense that it allows the construction of 'functions', by abstraction of an expression with respect to a variable, as in λ-calculus. The syntax of functional constructions in ML is that of the λ-calculus, with its simplifications (left-associativity of applications, right-associativity of abstractions). One writes `fn` instead of 'λ', and `=>` instead of '.'.

```
val discriminant = fn (a,b,c) => b*b - 4*a*c ;
```

```
↷ val discriminant = fn : int * int * int -> int
```

or, more simply,

```
fun discriminant (a,b,c) = b*b - 4*a*c ;
```

The value of a function, which cannot be printed, is represented by a `fn` in front of its type. Local declarations are in fact redexes, for which the `let` form is an alternate notation, introduced by Landin in the ISWIM language:

```
(fn x => x+1) 4 ;
let val x = 4 in x+1 end ;
```

CAML also allows the form

```
x+1 where x = 4 ;;
```

it is, too, a notation used in ISWIM.

As for any functional language, binding is static or lexical, i.e., the value associated with a name is determined by the environment in which the name is introduced, and not by the environment in which it is used. The use of this form of binding means that when a function is defined, the current environment must be packaged in a *closure*, along with the code implementing the function:

```
let val x = 0
in let fun f y = x+y
   in let val x = 2
      in f(0)
      end
   end
end;
```

```
↷ val it = 0 : int
```

Argument passing is done *by value*, and is guided by *pattern matching*: the redex

```
let val (x,y) = (2,3) in x+y end ;
```

```
↷ val it = 5 : int
```

is equivalent to

```
(fn (x,y) => x+y) (2,3) ;
```

Recursion The type system does not allow the definition of fixpoint combinators as does Lisp. There is therefore a primitive construct in the language for recursive definitions, called `val rec`:

```
val rec f = fn x => if (x=0) then 1 else x * f(x-1) ;
```

A derived form of the same definition would be

```
fun f x = if (x=0) then 1 else x * f(x-1) ;
```

Since general recursion has been introduced, ML programs do not necessarily terminate.

An example of mutual recursion is:

```
fun even x = if x=0 then true  else odd(x-1)
and odd  x = if x=0 then false else even(x-1) ;
```

The type system
ML is a strongly typed language, i.e., every name is typed (in a functional Lisp, such as Scheme, *objects* are typed, but not the *names* which denote them). However, it is not necessary to declare the type of a name when it is introduced, as must be done in Pascal when a variable's name is being declared: the expression defining the name's value can be used to infer the type. This inference is done at compilation time: it is a *static* process, based on the analysis of the text of the program (it is the expression which is used, not its value, which can only be known at execution time). The first phase of the semantic analysis of a program is to infer a type. If the compiler can not infer a type, the definition must be incorrect. However, the error messages are not always helpful:

```
fun f x = x x ;

↷ std_in:2.11-2-13 Error: operator is not a function
    operator: 'Z
    in expression:
     x x
```

Similarly, the fixpoint combinators Y_C and Y_T cannot be typed.

Predefined constructors The type system contains binary type constructors (`->`, `*`), unary ones (`list`), as well as type constants, such as `bool` (along with the two constants of that type, `true` and `false`), `int`, `real`, `string` (character strings, written between '"'), and `unit` (with `()` as the constant and only value of that type).

First-order terms over this signature are types; in particular, because of types such as `int -> (int -> int)`, a 'higher-order' functional language is achieved.

The unary type constructor `list` is predefined, with a generic constant `[]` (the empty list), a constructor `::` ('cons') and two selectors `hd` and `tl` analogous to `car` and `cdr` in Lisp. For example, `int list` is the type of lists of numbers.

The n-products of types are written using `*`; it is an n-ary type constructor. As for n-tuples, they are generated by putting n i expressions in parentheses with surrounding parentheses. `("Hello", "world", 3)` is a triple of type `string * string * int`. For each n, there is a $\#n$ operator to obtain the n-th field. N-tuples are special cases of records.

Type abbreviations can be declared as follows:

```
type relation = int * int -> bool ;
```

```
↷ type relation = int * int -> bool
```

Furthermore, types can contain type variables, written `'a, 'b, ...` , which allows the definition of generic objects:

```
type 'a relation = 'a * 'a -> bool ;
```

Concrete types Finally, new types, called concrete types, can be defined. The declaration

```
datatype direction = south | east | north | west ;
```

```
↷ datatype  direction
  con east : direction
  con north : direction
  con south : direction
  con west : direction
```

introduces the type `direction`, with four new values which are type constants. Call by pattern can be extended from predefined types to concrete types:

```
fun mirror south = north
|   mirror east = west
|   mirror north = south
|   mirror west = east ;
```

Recursive types can be defined by defining their constructors. Here are the natural numbers:

```
type natural = Zero | Succ of natural ;
```

and generic lists, which will be redefined using the name `List` (`list` being predefined):

```
datatype 'a List = Nil | Cons of 'a * 'a List ;
```

```
↷ datatype 'a List
  con Cons: 'a * 'a List -> 'a List
  con Nil : 'a List
```

Concrete types allow an algebraic style of programming with control structures particular to a datatype, to traverse or iterate, among other things.

Higher-order functions

A functional language must allow the writing of 'higher-order' functional programs, as in λ-calculus. There are several reasons:

- writing combinators, either general (composition, curryfication, ...) or particular to a given datatype (iteration over a list);
- conceiving general reusable programs, which is one of the objectives of software engineering (for example a sorting algorithm using an order relation as a parameter; this can be done using a functional parameter in Pascal or a `generic` procedure in Ada);
- showing the higher-order nature of a computation even if its result is of type `int -> int`.

Combinators Here is a higher-order function, composition, whose definition shows the utility of infix notation:

```
infix o
fun f o g = fn x => f(g(x)) ;
```

```
↷ val o = fn : ('a -> 'b) * ('c -> 'a) -> 'c -> 'b
```

The operation `o` can be used in infix form (i.e, `f o g`) or in prefix form, as for any other function, if it is preceded by `op`: `op o (f,g)`. The following combinators are well known in λ-calculus:

```
fun I x = x ;         (* the identity  I ≡ λxx *)
```

```
↷ val I = fn : 'a -> 'a
```

```
fun K x y = x ;                    (* K ≡ λxy.x *)
```

```
↷ val K = fn : 'a -> 'b -> 'a
```

The `K` combinator is a higher-order generic functional value. It is not a function of two variables, rather of one variable, and `K 0`, for example, is a function (the constant function of value 0): the parenthesization of `K x y` is `(K x) y`. It is generic, in the sense that one can evaluate `K 0` or `K []` just as easily as `K ("qwerty",[1;2;3])`.

One can define an operator which takes an uncurryfied function and produces a curryfied function, as well as the reverse:

```
fun curry f x y = f(x,y) ;
fun uncurry g (x,y) = g x y ;
```

Here is the successor function:

```
val succ = (curry (op +)) 1 ;
```

Concrete types and iterators Iteration combinators linked to concrete types will now be examined, first with the natural integers:

```
datatype natural = Zero | Succ of natural ;
```

A definition using patterns over the integers uses the iterator:

```
fun natural_iter z s = let fun f Zero = z
                             |   f (Succ n) = s (f n)
                           in f end ;
```

```
↷ val natural_iter = fn : 'a -> ('a -> 'a) -> natural -> 'a
```

This iterator allows one to translate the elements of the concrete type `natural` to predefined integers:

```
val natural_int = natural_iter 0 succ ;
```

```
↷ val natural_int = fn : natural -> int
```

```
natural_int (Succ(Succ(Succ Zero))) ;
```

```
↷ 3 : int
```

Programming an iterator corresponds exactly to reasoning by induction; this will be shown for lists. A proof by induction of a property P over lists goes as follows:

- prove $P(\texttt{Nil})$;
- prove that for each a of type `'a` and for each list l of type `'a List`, if $P(l)$, then $P(\texttt{Cons}(a, l))$.

Similarly, to define a function f over lists, one needs:

- a value n for $f(\texttt{Nil})$;
- for each a of type `'a` and each list l of type `'a List`, knowing $f(l)$, compute $f(\texttt{Cons}(a, l))$ with the help of function c: $f(\texttt{Cons}(a, l)) = c(a, f(l))$.

The function definition is transcribed using pattern matching:

```
datatype 'a List = Nil | Cons of 'a * 'a List
fun List_iter n c = let fun f Nil = n
                          |   f (Cons(a,l)) = c(a,f(l))
                        in f end ;
```

```
↷ val List_iter = fn : 'a -> ('b * 'a -> 'a) -> 'b List -> 'a
```

Note that the 'signature' of type `'b List` is of the form `n :'t` and `c :'b * 't -> 't` with two sorts of objects `'b` and `'t`. Building the iterator means just 'interpreting' `Nil` and `Cons` by `n` and `c`. From a functional point of view, a list operates over functions of this type: that is how lists are defined in the higher-order typed λ-calculi such as Girard's system F. The above iterator is often better known as `reduce`, or `fold-right`.

Here are some simple applications of the iterator over lists: the first computes the sum of the elements of a list of integers, the second (better known as map) applies a function of one variable to all the elements of a list and forms a list of the results:

```
val List_sum = List_iter 0 (op +) ;
```

```
↷ val List_sum = fn : int List -> int
```

```
fun List_map f  = List_iter Nil (fn (a,l) => Cons(f(a),l)) ;
```

```
↷ val List_map = fn : ('a -> 'b) -> 'a List -> 'b List
```

An implementation of first-order terms When one is working with a given signature Σ, one can construct a concrete type using Σ, as was done for lists. The values of the introduced concrete type are then exactly the Σ-terms. In order to be able to work with an arbitrary signature, a general type Term is defined. A term is either a variable, associated with an integer, or the combination of a symbol and a list of terms:

```
type symbol = string
datatype Term = Var of int | Comb of symbol * Term List ;
```

```
↷ type  symbol = string
  datatype Term
  con Comb : symbol * Term List -> Term
  con Var : int -> Term
```

To this definition, which introduces the constructors Var and Comb, is associated an iterator. The signature is not quite first-order, due to the use of List:

```
fun Term_iter v c n cs =
    let fun g (Var n) = v n
          |  g (Comb(f,l)) =
               c(f, List_iter n (uncurry ((curry cs) o g)) l)
    in g end ;
```

```
↷ val Term_iter = fn : (int -> 'a) ->
                       (symbol * 'b -> 'a) ->
                       'b ->
                       ('a * 'b -> 'b) ->
                       Term -> 'a
```

Here is a simpler iterator, less general as it does not 'interpret' the list constructors (the variable of type 'b and the variables n and cs of Term_iter are replaced by the type 'a List along with its constructors Nil and Cons):

```
fun Term_iter' v c = let fun g (Var n) = v n
                           |  g (Comb(f,l)) = c(f, List_map g l)
                     in g end ;
```

```
↝ val Term_iter' = fn : (int -> 'a) ->
                        (symbol * 'a List -> 'a) ->
                        Term -> 'a
```

Examples of higher-order computation From now on, the predefined integers will be used instead of the concrete type `natural`. The curryfied form of the Ackerman function can be written simply by using the following recursive definition:

```
fun A 0 n = n+1
|   A m 0 = A (m-1) 1
|   A m n = A (m-1) (A m (n-1)) ;
```

Although the Ackerman function is of type `int -> int -> int`, hence first-order, the algorithm which computes it is not first-order. A generic iteration operator `It` over the integers is introduced, analogous to the `natural_iter` defined above: given a type variable `'a`, it takes as arguments `z : 'a` and `s : 'a -> 'a` to produce a function `F : int -> 'a` which satisfies `F n =` s^n `z`:

```
fun It z s = let fun F 0 = z
                 |   F n = s(F(n-1))
             in F end ;
```

```
↝ val It = fn : 'a -> ('a -> 'a) -> int -> 'a
```

Reconsider Ackerman's function. It can be proved by double induction that $A(m+1)n = (A(m))^{n+1}1$. One can then use the following, concise definition of `A`, which emphasizes the double iteration:

```
val A = It succ (fn f => It (f 1) f) ;
```

Note the polymorphism of `It` in this definition: the first occurrence is of (fourth-order) type

```
(int -> int) ->
((int -> int) -> (int -> int)) ->
int -> (int ->int),
```

and the second is of (third-order) type

```
int -> (int -> int)  -> int -> int
```

A *recursor*, or recursion operator, can also be introduced. It is the **R** of Gödel's system T:

```
fun R z g  = let fun F 0 = z
                 |   F n = g (F(n-1)) (n-1)
             in F end ;
```

```
↝ val R = fn : 'a -> ('a -> int -> 'a) -> int -> 'a
```

The recursion operator is generic, with a type variable `'a`, and is applied to `z :'a` and `g :'a -> int -> 'a`. When `'a` is `int`, the classical primitive recursion operator is produced. It is first-order, and here is its decurryfied version:

$$
\begin{aligned}
F(0) &= z, \\
F(n+1) &= g(F(n), n).
\end{aligned}
$$

The usual definition contains a parameter, often a tuple of integers, which corresponds to the case where one adds a variable `p` to `z` (which becomes a function), `R` and `g`.

A higher-order type The preceding examples showed higher-order functions, but the concrete types used, lists, terms and trees, are only first-order: they are first-order terms over a many-sorted signature. Here is an example of a higher-order concrete type that resembles the description of the fundamental sequences of countable ordinals. This analogy will be used to explain the construction of this type and the 'arithmetic' operations.

An ordinal is zero, the successor of an ordinal or the limit of a sequence of ordinals:

```
datatype ord = Zero | Succ of ord | Lim of (int -> ord) ;
```

```
↷ datatype ord
    con Lim : (int -> ord) -> ord
    con Succ : ord -> ord
    con Zero : ord
```

Finite ordinals correspond to natural numbers, by iteration of `Succ` over `Zero`:

```
fun nat n = if n=0 then Zero else Succ (nat (n-1)) ;
```

```
↷ val nat = fn : int -> ord
```

```
nat 4 ;
```

```
↷ (Succ (Succ (Succ (Succ Zero)))) : ord
```

`nat` could have been defined using the iterator `It` over the integers:

```
val nat = It Zero Succ ;
```

In set theory, ω designates the smallest infinite ordinal: it is the set of all the finite ordinals, i.e., the integers:

```
val omega = Lim nat ;
```

```
↷ val omega = Lim fn : ord
```

An iterator is introduced over `ord` by basing its definition on the proof by induction scheme over the ordinals:

```
fun ord_iter z s l  = let fun F Zero = z
                          |   F (Succ x) = s(F(x))
                          |   F (Lim f) = l(F o f)
                      in F end ;
```

```
↷ val ord_iter = fn : 'a ->
                      ('a -> 'a) ->
                      ((int -> 'a) -> 'a) ->
                      ord -> 'a
```

This iteration operator allows one to define an arithmetic over the ordinals in an analogous manner to integer arithmetic: addition is obtained by iterating the successor, multiplication by iterating addition and exponentiation by iterating multiplication. The curryfied versions of these functions are given:

```
fun ord_add x y = ord_iter x Succ Lim y ;
```

```
↷ val ord_add = fn : ord -> ord -> ord
```

```
fun ord_mult x y = ord_iter Zero
                             (fn u => ord_add u x)
                             Lim  y ;
fun ord_exp x y = ord_iter (Succ Zero)
                            (fn u => ord_mult u x)
                            Lim y ;
```

By using type ord, one can define a hierarchy of *numeric* functions, by repeated iterations and diagonalizations, generalizing the construction of Ackerman's function. In fact, the *fast hierarchy* is reproduced: the family, indexed by the ordinals, consisting of more and more complex functions, is given by:

$$
\begin{aligned}
f_0(n) &= n+1, \\
f_{\alpha+1}(n) &= f_\alpha^{n+1}(n), \\
f_{\lim \alpha_k}(n) &= f_{\alpha_n}(n).
\end{aligned}
$$

The first function is the successor function succ over the integers. To compute $f_{\alpha+1}(n)$, an iterated iterator It2 is used over the integers:

```
fun It2 f n = It (f n) f n ;
```

```
↷ val It2 = fn : (int -> int) -> int -> int
```

To compute $f_{\lim \alpha_k}(n)$, the diagonal combinator is used:

```
fun diag f n = f n n ;
```

```
↷ val diag = fn : ('a -> 'a -> 'b) -> 'a -> 'b
```

```
val hierarchy = ord_iter succ It2 diag ;
```

```
↷ val hierarchy = fn : ord -> int -> int
```

Ackermann's function is very low in the hierarchy: ω is the first infinite ordinal. There are many others. The ordinal ϵ_0 became very famous in 1936 when Gentzen proved the consistency of Peano's arithmetic, by transfinite induction up to ϵ_0 (see [144]). This proof cannot be formulated in Peano's arithmetic (Gödel's second incompleteness theorem). Besides, Peano's induction axiom allows induction only over the integers, i.e., up to ω. By definition, ϵ_0 is the limit of the sequence $\omega_0 = 1$, $\omega_1 = \omega$, $\omega_2 = \omega^\omega$, $\omega_{n+1} = \omega^{\omega_n}$, ... Although infinite, these ordinals correspond to perfectly ordinary types in functional programming. The sequence is generated by the local function `f` in the following definition:

```
val epsilon_0 =
    let val f = ord_iter (Succ Zero) (ord_exp omega) Lim
    in f omega end ;
```

This ordinal is the limit of the usual ordinals. It is the smallest ordinal x which is a solution to the equation $x = \omega^x$.

In the fast hierarchy, $f_{\epsilon_0}(n) = f_{\omega_n}(n)$ (an iteration over the ordinals was done *up to* ϵ_0); it is a (total) function, but the fact that it is defined for all the integers is not provable in arithmetic. From a practical point of view in arithmetic, all of these functions are easily programmed. They can be *defined* by a program, but it would be unthinkable to actually *compute* their values, except for a few at the beginning. For these functions, the *algorithmic* complexity of a program computing its values is related to the *logical* complexity of the termination proof of its program.

Non-functional aspects

Functional programming, in fact, the λ-calculus, has a serious limitation for the formulation of algorithms: expressions must always produce a value. The λ-calculus lacks a means of controlling the sequentiality of the calculus. For example, it should be possible to pop out of a recursion with a final result immediately without being forced to pass the result back through preceding recursive invocations. As in logic programming, where a goal can fail, it should be possible for a function to return no result, and to raise an exception signaling its failure. This 'exception' should then be treatable. This requirement is mandatory for error recuperation. Such mechanisms exist in most languages, e.g., Lisp's exits and Ada's exceptions.

The most general escaping mechanism, found in Scheme, allows direct access to the *continuation* of an expression, a low-level theoretical concept from denotational semantics, akin to the execution stack. The `call-with-current-continuation` construction which allows this access must be handled carefully; it motivated the study of extensions of the λ-calculus, such as the $\lambda_v C$-calculus of M. Felleisen[42].

Standard ML now offers continuations. However, it also offers a solution of medium power: exceptions are declared by `exception` ... , triggered by `raise` ... , and handled selectively by pattern matching, using ... `handle` ... . Exceptions are declared in a similar manner to concrete types, by introducing constant exceptions and constructors:

```
exception error of string and stop ;
```

```
↷ exception error
  exception stop
```

A raised exception is propagated so long as it has not been `handled`:

```
if false then ""
else raise error "bang"
handle (error x) -> x ^ "!!!"
|      stop -> "finished" ;
```

```
↷ "bang!!!" : string
```

Note also that Standard ML has a unary type constructor **ref** such that an object of type t **ref** is a reference to an object of type t. Objects with a mutable local state can be created, and purely sequential programming becomes possible by the use of assignment `:=`, the dereferencing operator `!` and the sequencing operator `;`

```
val place = ref 3 ;
```

```
↷ val place = ref 3 : int ref
```

```
!place ;
```

```
↷ val it = 3 : int
```

```
(place := 10 ; place := !place + 1 ; 1 ; 2)
```

```
↷ val it = () : unit
  val it = () : unit
  val it = 1 : int
  val it = 2 : int
```

It is possible to construct mutable data structures. Here is a model of Lisp objects:

```
datatype lisp = Nil | Pair of lisp ref * lisp ref ;
```

```
↷ datatype lisp
  con Nil : lisp
  con Pair : lisp ref * lisp ref -> lisp
```

```
exception NilPair
fun set_car Nil _ = raise NilPair
|   set_car (d as Pair(car,cdr)) v = (car := v ; d) ;
```

```
↷ val set_car = fn : lisp -> lisp -> lisp
```

Finally, one can program in an 'object-oriented' style by constructing objects which are closures with mutable local state, i.e., closures whose environment contains name–reference bindings, the reference value being modifiable. Here is a creator of counters which responds to the messages `0!`, `++`, `--` and `?`

```
fun counter () =
  let val value = ref 0 in
      let fun F "0!"= (value := 0 ; !value)
          |   F "++"= (value := !value + 1; !value)
```

```
               |   F "--"= (if !value > 0
                           then value := !value - 1
                           else raise error "decrement: null value";
                           !value)
               |   F "?" = !value
               |   F _   = raise error "unknown command";
          in F end
      end
↷ val counter = fn : unit -> string -> int
```

```
val c = counter () ;
c "++" ; c "++" ; c "--" ; c "?" ;
```

```
↷ val it = 1 : int
  val it = 2 : int
  val it = 1 : int
  val it = 1 : int
```

```
(c "a" handle error x => (print x; print "\n"; 1))
```

```
↷ unknown command
  val it = 1 : int
```

Closures and delayed evaluation
The evaluation mechanism of Standard ML reproduces weak β-reduction with call by value. Hence, the body of functions is never evaluated. An expression can be frozen by making it an abstraction; this is not possible in non-functional languages.

In Standard ML, an expression such as

```
fun Suc_r () = x+1 ;
```

has as value a closure containing the value of `x` and the code necessary to increment `x`. It is only when `Suc_r()` is called that `x+1` will be computed.

With this technique, call by name can be simulated in Standard ML.

3.3.5 Scheme, an untyped functional language

Scheme is a dialect of Lisp with static and monovalued binding. The same space is used for all names, be they 'values' or 'functions'. As for other Lisps, environments define name–memory location bindings rather than name–value bindings. Recursion and redefinition of global variables become easy. First conceived at MIT as an interpreter for the λ-calculus, Scheme has had considerable influence on the evolution of Common Lisp towards static binding, and is at the center of the European project of Lisp normalization.

Since the language is not typed, it is possible, as in the λ-calculus, to define a fixpoint combinator, which is not possible in ML, except by using some recursive concrete types:

```
(define Y
  (lambda(f)
    (let ([g (lambda(h) (lambda(x) ((f (h h)) x)))])
         (g g)
)))
```

The factorial can be defined by applying `Y` to the 'functional' `(lambda(f) ...)`:

```
(define fact
  (Y (lambda(f)
       (lambda(x) (if (zero? x) 1 (* x (f (- x 1)))))
)))
```

The qualities of Scheme can be appreciated when comparing it to Common Lisp, which is 'bivalued', which means that one must state explicitly which of the two name spaces (ordinary values or functions) should be used to look for the 'value' bound to a name:

```
(defun Y (f)
  (let ((g #'(lambda(h)
               #'(lambda(x)
                   (funcall (funcall f (funcall h h)) x)))))
       (funcall g g)
))

(defun fact (n)
  (funcall (Y #'(lambda(f)
                  #'(lambda(x)
                     (if (zerop x) 1 (* x (funcall f (- x 1)))))))
           n))
```

3.3.6 Compilation of λ-terms into applicative terms

What purpose do bound variables play in the λ-calculus? They are only introduced, by abstraction, to be eliminated, by application. Can one make do without them? There exists a formalism, called *combinatory logic*, which shows effectively that one can construct a calculus of the same expressive power as the λ-calculus, without bound variables. It is a variant of the latter, but was developed independently by Schönfinkel in 1924 [132], and later by Curry, with the intention of creating a 'logic' which could be manipulated in a combinatorial manner.

The debate has been taken up in the area of programming languages: is abstraction, made popular by Lisp's `lambda`, useful and necessary? Backus, upon his reception of the Turing Award in 1978 [7], said no: abstraction is not necessary and reduces the clarity of programs. Backus's presentation was particularly designed to attack languages such as Fortran which are based on the von Neumann architecture, but he was also taking a position against Lisp-like languages.

From an implementation point of view, bound variables require an environment access mechanism which makes compilation quite complex. It would be simpler if this step could be avoided.

Applicative terms and concrete syntax were introduced in Chapter 1. They are first-order terms formed of constant symbols and a single binary function symbol, called application, written using a notation similar to that of the λ-calculus. In programming, an *applicative language* is a language based on the evaluation of expressions (first-order terms), just as arithmetic expressions are evaluated in calculators. For these expressions, the available constants are the integers and the elementary operations ($+$, $*$, ...), and there are rules to compute with these. Of course, other constants are needed to produce more complex calculi. In fact, the λ-calculus example shows that arithmetic operations can be represented as soon as a sufficiently general functional calculus is defined. Combinators which permute or duplicate the arguments of a function, or which solve a given equation, can easily be defined.

The compilation of several functional languages, such as SASL and Miranda, is based on the translation of programs which are terms of a λ-calculus extended with applicative terms constructed using $\boldsymbol{S}$, $\boldsymbol{K}$ and some other combinators. The resulting term, represented as a tree, and even by a graph if subterms are shared, is then reduced by the rewrite rules of combinatory logic. Since the rule associated with each combinator can be seen as an elementary machine instruction, there have been many attempts to construct a 'graph reduction machine'. Use of such a virtual machine is also a standard implementation technique.

A mapping of λ-terms to applicative terms, $M \mapsto M_{@}$, will be defined; the mapping is used for compiling λ-terms. A reverse mapping, $A \mapsto A_\lambda$, is also defined. For the moment, the constants with which applicative terms are formed have not been made explicit. This will be done by associating a closed λ-term c_λ with each constant c, along with a rewrite rule.

The problem is to translate abstractions into applicative terms. The first step is to simulate abstraction in combinatory logic. For each variable x, define an abstraction operator $A \mapsto [x].A$, associating the applicative term A with the applicative term $[x].A$ representing the 'function $x \mapsto A$'. By structural induction over A, the cases of a constant c, of an application PQ, and of a variable y must be treated; in the latter case, $y \equiv x$ and $y \not\equiv x$ must be distinguished.

So that the simulation of abstraction accurately reflects the meaning in λ, at least one renaming property must be verified:

$$[x].A \;\equiv\; [y].(A[x := y])$$

if $y \notin \operatorname{var} A$. In particular, $[x].x$ must be independent of x: a constant symbol $\boldsymbol{I}$ is used for this value and is defined by $\boldsymbol{I}_\lambda \equiv \lambda x\,.\,x \equiv \mathbf{I}$, with the rule $\boldsymbol{I}x \longrightarrow x$, obviously derived from $\mathbf{I}x \longrightarrow_\beta x$.

Furthermore, if A is a distinct variable from x, if A is a constant symbol (for example $\mathbf{I}$), or more generally, if x does not appear in A, the 'function $x \mapsto A$' is a constant value function A. A second constant symbol is introduced: $\boldsymbol{K}$, with $\boldsymbol{K}_\lambda \equiv \lambda xy \,.\, x$ and the rule $\boldsymbol{K}xy \longrightarrow x$ to define the case $[x].A$ by $\boldsymbol{K}A$.

Finally, consider an application AB in the case where $x \in \mathrm{var}(AB)$. A new constant symbol $\boldsymbol{S}$ is introduced to define $[x].AB \equiv \boldsymbol{S}([x].A)([x].B)$. Using the standard notation of functions, the combinator applied to two functions f and g produces the function $\boldsymbol{S}(f,g) : x \mapsto f(x, g(x))$. Hence its translation $\boldsymbol{S}_\lambda \equiv \lambda fgx \,.\, fx(gx) \equiv \mathbf{S}$ and the rule $\boldsymbol{S}fgx \longrightarrow fx(gx)$.

Recall the definition of $[x].A$:

$$\begin{array}{rcll} [x].x & \equiv & \boldsymbol{I} & \\ [x].A & \equiv & \boldsymbol{K}A & \text{if } x \notin \mathrm{var}\, A \\ [x].AB & \equiv & \boldsymbol{S}([x].A)([x].B) & \text{if } x \in \mathrm{var}\, AB \end{array}$$

the definition of the translation of λ-terms into applicative terms:

$$\begin{array}{rcl} (x)_@ & \equiv & x \\ (PQ)_@ & \equiv & (P)_@(Q)_@ \\ (\lambda x \,.\, M)_@ & \equiv & [x].(M)_@ \end{array}$$

and the rewrite rules:

$$\begin{array}{rcl} \boldsymbol{I}x & \longrightarrow & x \\ \boldsymbol{K}xy & \longrightarrow & x \\ \boldsymbol{S}xyz & \longrightarrow & xz(yz) \end{array}$$

These rules define *weak reduction* of combinatory logic, written $\longrightarrow_@$.

The reverse translation $A \mapsto A_\lambda$ is perfectly natural (if there were other constants, one would add their translation to those of $\mathbf{K}$ and $\mathbf{S}$):

$$\begin{array}{ccc} & x_\lambda \equiv x & \\ \boldsymbol{I}_\lambda \equiv \mathbf{I} & \boldsymbol{K}_\lambda \equiv \mathbf{K} & \boldsymbol{S}_\lambda \equiv \mathbf{S} \\ & AB_\lambda \equiv A_\lambda B_\lambda & \end{array}$$

Proposition 3.16

- if $A \stackrel{\star}{\longrightarrow}_@ B$, then $A_\lambda \stackrel{\star}{\longrightarrow}_\beta B_\lambda$;
- $(M_@)_\lambda \stackrel{\star}{\longrightarrow}_\beta M$.

The proof is trivial. Conversely, it could be that $M \longrightarrow_\beta N$ without $M_@ \longrightarrow_@ N_@$.

Example $\mathbf{SK} \stackrel{\star}{\longrightarrow}_\beta \mathbf{0}$, but $\boldsymbol{SK}$ is irreducible for $\longrightarrow_@$. □

This fact is due to the definition of the rewrite rules $\longrightarrow_@$ which have no context passing rule similar to

$$\frac{M \rhd M'}{\lambda x M \rhd \lambda x M'}.$$

Furthermore, it is not necessarily true that $(A_\lambda)_@ =_@ A$.

Example $(\boldsymbol{K}_\lambda)_@ \equiv \boldsymbol{S}(\boldsymbol{K}\boldsymbol{K})\boldsymbol{I}$, which does not reduce to $\boldsymbol{K}$. However, it is true that $\boldsymbol{S}(\boldsymbol{K}\boldsymbol{K})\boldsymbol{I}xy \xrightarrow{\star} x \leftarrow \boldsymbol{K}xy$. □

It follows that the two theories, although related, are not equivalent.

One can immediately check that $(\boldsymbol{SKK})x \xrightarrow{\star}_@ x$, hence that the term $\boldsymbol{SKK}$ plays the same role as the constant $\boldsymbol{I}$. It therefore suffices to add the constants $\boldsymbol{S}$ and $\boldsymbol{K}$ and to define $[x].x = \boldsymbol{SKK}$. The set $\{\boldsymbol{S}, \boldsymbol{K}\}$ (similarly $\{\boldsymbol{S}, \boldsymbol{K}, \boldsymbol{I}\}$) forms a *combinatory basis* for the λ-calculus. There exist other bases, some with a single element. However, from a practical point of view the presented compilation rules are unusable because of the combinatorial explosion of the resulting terms.

Example Consider the complexity of the translation of the recursive definition of the factorial function:

$$f \equiv \lambda n\,.\,zero\,n\,1(mult\,n\,(f(pred\,n))).$$

No attempt will be made to solve this equation using a fixpoint combinator. Rather, the right hand side will be translated using applicative terms. Here are a few application steps of the translation rules:

$$\begin{array}{ll} & [n]\,.\,zero\,n1(mult\,n(f(pred\,n))) \\ \equiv & \boldsymbol{S}([n].zero\,n1)([n].mult\,n(f(pred\,n))) \\ \equiv & \boldsymbol{S}(\boldsymbol{S}([n].zero\,n)([n].1))(\boldsymbol{S}([n].mult\,n)([n].f(pred\,n))) \\ \equiv & \boldsymbol{S}(\boldsymbol{S}(\boldsymbol{S}([n].zero)([n].n))(([n].1))) \\ & \quad (\boldsymbol{S}(\boldsymbol{S}([n].mult)([n].n))(\boldsymbol{S}([n].f)(\boldsymbol{S}([n].pred)([n].n)))) \\ \equiv & \boldsymbol{S}(\boldsymbol{S}(\boldsymbol{S}(\boldsymbol{K}\,zero)\boldsymbol{I})(\boldsymbol{K}1))(\boldsymbol{S}(\boldsymbol{S}(\boldsymbol{K}\,mult)\boldsymbol{I})(\boldsymbol{S}(\boldsymbol{K}f)(\boldsymbol{S}(\boldsymbol{K}\,pred)I))). \end{array}$$

□

To avoid this proliferation of incomprehensible applicative terms, it is possible to modify the definition of $[x].A$ by adding combinators less universal than $\boldsymbol{S}$, $\boldsymbol{K}$, $\boldsymbol{I}$. For example, $\boldsymbol{B}$ (respectively $\boldsymbol{C}$) is a specialized version of $\boldsymbol{S}$ for applications whose right-hand (respectively left-hand) subterm contains the variable which is being abstracted. Their rules are:

$$\begin{array}{ll} \boldsymbol{B}fgx & \rightarrow f(gx), \\ \boldsymbol{C}fax & \rightarrow (fx)a. \end{array}$$

These new combinators allow significant optimizations. The following rules are used:

$$\begin{array}{rcl} \boldsymbol{S}(\boldsymbol{K}M)\boldsymbol{I} & =_{opt} & M, \\ \boldsymbol{S}(\boldsymbol{K}M)(\boldsymbol{K}N) & =_{opt} & \boldsymbol{K}(MN), \\ \boldsymbol{S}(\boldsymbol{K}M)N & =_{opt} & \boldsymbol{B}MN, \\ \boldsymbol{S}M(\boldsymbol{K}N) & =_{opt} & \boldsymbol{C}MN. \end{array}$$

They can be justified by adding an extensionality rule to combinatory logic:

$$\frac{Ax = Bx}{A = B} \qquad \text{if } x \notin \text{var}(AB).$$

See exercise 20.

Example (cont.) Here is the optimized form of the preceding recursive definition:

$$\begin{array}{rl} f & \equiv \quad \boldsymbol{S}(\boldsymbol{S}(\boldsymbol{S}(\boldsymbol{K}\,zero)\boldsymbol{I})(\boldsymbol{K}1))(\boldsymbol{S}(\boldsymbol{S}(\boldsymbol{K}\,mult)\boldsymbol{I})(\boldsymbol{S}(\boldsymbol{K}f)(\boldsymbol{S}(\boldsymbol{K}\,pred)I))) \\ & =_{opt} \quad \boldsymbol{S}(\boldsymbol{C}\,zero\,1)(\boldsymbol{S}\,mult\,(\boldsymbol{B}f\,pred)), \end{array}$$

which is much simpler than the preceding one. □

3.3.7 Supercombinators

A more flexible translation of λ-terms into applicative terms, proposed by Hughes in 1982 [72], uses *supercombinators.*

Definition 3.14 Let E be a set of terms. The set comb(E) of *(applicative) combinations* of E is defined by induction:

1. $E \subseteq \text{comb}(E)$;
2. if $M, N \in \text{comb}(E)$, then $MN \in \text{comb}(E)$.

A closed term M is a *supercombinator* of arity r if it has form $M \equiv \lambda x_1 \ldots x_r \,.\, C$, where C is a combination of $x_1, \ldots, x_r$ and of supercombinators; $x_1, \ldots, x_r$ are the parameters and C is the body of M.

Note that a supercombinator is a combinator, and that in the base case the body is simply a combination of the parameters.

Examples

- $(M_2M_3)(M_1(M_3M_1)) \in \text{comb}(\{M_1, M_2, M_3, M_4\})$;
- $\mathbf{S} \equiv \lambda xyz \,.\, xz(yz)$ is a supercombinator of arity 3, with body $xz(yz) \in \text{comb}(x, y, z)$;
- $\lambda y \,.\, xy$ is not a supercombinator because it is not a combinator;
- $\lambda x \,.\, x(\lambda y \,.\, xy)$ is a combinator but not a supercombinator since $(\lambda y \,.\, xy)$ is not one;
- $\lambda x \,.\, \mathbf{K}x$ is a supercombinator.

Proposition 3.17 Any term M can be transformed into a supercombinator M^* such that $M^*x_1 \ldots x_r \xrightarrow{\star} M$, if $\{x_1, \ldots, x_r\} = \text{var}(M)$. In particular, if M is a combinator, then $M^* \xrightarrow{\star} M$.

Proof If $M \equiv x$, let $M^* \equiv \mathbf{I}$.

Let $M \equiv \lambda\vec{y}\,.\,C$, where C is not an abstraction, so that all the leading λ are in $\lambda\vec{y}$, and $\{\vec{x},\vec{y}\} = \text{var}(C)$. The term C is a combination of $\vec{x},\vec{y}$ and abstractions $C_1,\ldots,C_p$ which have their free variables in $\{\vec{x},\vec{y}\}$. By induction, there exist supercombinators C_i^* such that $C_i^*\vec{u}^{(i)} \xrightarrow{\star} C_i$, where $\vec{u}^{(i)}$ is formed with free variables of C_i. The term C' is constructed by replacing C_i by $C_i^*\vec{u}^{(i)}$ in C for $i = 1,\ldots,p$: C' is therefore a combination of $\{\vec{x},\vec{y}\}$ and of supercombinators C_i^*. Let $M^* \equiv \lambda\vec{x}\vec{y}\,.\,C'$: it is a supercombinator. From $C_i^*\vec{u}^{(i)} \xrightarrow{\star} C_i$, deduce $C' \xrightarrow{\star} C$; hence, $M^*\vec{x} \xrightarrow{\star} M$. □

A supercombinator M can be associated with an applicative rewriting system $(\Sigma,\mathcal{R})$, containing a principal rule $fx_1\ldots x_r \rightarrow \ldots$, f being a new constant symbol corresponding to M.

Let $M \equiv \lambda\vec{x}\,.\,C$ be a supercombinator of arity r. Write its body C as a combination of its parameters $x_1,\ldots,x_r$, and of supercombinators $S_1,\ldots,S_p$, respectively of arity $r_1,\ldots,r_p$. Suppose that one has already associated a symbol f_i and a rewriting system $(\Sigma_i,\mathcal{R}_i)$ with the supercombinator S_i. Let

$$\Sigma = \Sigma_1 \cup \ldots \cup \Sigma_p \cup \{f\},$$

where the signatures Σ_i are disjoint and f is a new constant symbol, and

$$\mathcal{R} = \mathcal{R}_1 \cup \ldots \cup \mathcal{R}_p \cup \{r\},$$

where r is the rule $fx_1\ldots x_r \rightarrow C_@$, the applicative term $C_@$ being obtained by replacing S_i by f_i in C.

Example Let $M \equiv \lambda x\,.\,(\lambda y\,.\,y(\lambda zt\,.\,xt(zy)))(\lambda yt\,.\,x(yt))$. The term Mx is a combination of two applications C_1 and C_2 having x as free variable. The term C_1xy is a combination yC_3, and $C_2xyt = x(yt)$ and $C_3xyzt = xt(zy)$. Let

$$\begin{aligned} C_3^* &\equiv \lambda xyzt\,.\,xt(zy),\\ C_2^* &\equiv \lambda xyt\,.\,x(yt),\\ C_1^* &\equiv \lambda xy\,.\,y(C_3^*xy),\\ M^* &\equiv \lambda x\,.\,(C_1^*x)(C_2^*x). \end{aligned}$$

Associate with the supercombinators M^*, C_1^*, C_2^* and C_3^* the constant symbols f, f_1, f_2 and f_3. The following rewriting system is obtained:

$$\begin{aligned} f^*x &\rightarrow (f_1x)(f_2x),\\ f_1^*xy &\rightarrow y(f_3xy),\\ f_2^*xyt &\rightarrow x(yt),\\ f_3^*xyzt &\rightarrow xt(zy). \end{aligned}$$

□

It would be unthinkable to ask a programmer to write all programs using a finite base of combinators such as $\{\boldsymbol{S}, \boldsymbol{K}, \boldsymbol{I}, \boldsymbol{B}, \boldsymbol{C}\}$. However, it would be possible to write definitions using only supercombinators, leading to a descending method of programming by refinement. The result would be a programming style intermediate between the pure applicative programming (no variables) and general functional programming (arbitrary λ-terms). In fact, the criticisms made by Backus about 'λ-languages' do not touch these programs. The result could be considered as a form of 'structured functional programming'.

In fact, supercombinators are of interest for the implementer, as an intermediate language between the source and machine language. It is easy to conceive the first phase of a compiler which transforms an arbitrary term into a supercombinator. Do note that this compilation is more complex when recursion must be treated and when optimizations such as maximal sharing are done. The result is then an applicative rewriting system, bound variables having been eliminated. To write an interpreter for this intermediate language means designing a program to evaluate rewriting systems. By choosing the data structures aiding this evaluation correctly, the program can run using the operations of an abstract machine. The best actual implementation based on this method is Lazy ML, a lazy dialect of ML designed at Chalmers University of Technology in Gothenburg (Sweden).

3.4 Termination and confluence

If a computation is seen as a process producing a finite result, the problem of termination is crucial. This will not be the case for all processes, in particular concurrent programs, since, for some, stopping can mean deadlock or failure! Furthermore, as was shown with the lazy reduction of first-order terms, non-terminating processes may produce results by the 'increasing' evaluation of some data structure. Other properties must then be studied, such as liveness or fairness. In the 'general' case, it will be seen that the termination problem is undecidable. This in no way prevents the study of special cases. A reduction relation satisfies the (weak) normalization property if all terms are normalizable. It satisfies the strong normalization property if there are no infinite reductions.

Confluence is an even more important property. If the writing of programs which do not necessarily end is allowed, one would expect at least that, no matter what the execution strategy, the eventual result be the same. To ensure this, β-reduction will be shown to be confluent.

3.4.1 Termination and well-orders

A rewrite rule can reduce the size visibly, be it measured in the number of symbols, the height or some other measure, of its terms. In the case of rule $-(-x) \rightarrow x$, a function $h : \mathrm{T}[X] \longrightarrow \mathbf{N}$ (for example $h(M) = |M|$) is used to check that $h(P) > h(Q)$ for any rule (P, Q). To deduce that $h(M) > h(M')$ if $M \longrightarrow M'$, it suffices to show that h satisfies the following conditions, which follow from the definition of rewriting by contraction and context passing:

1. $h(\sigma P) > h(\sigma Q)$ for every substitution σ and for every rule (P, Q);
2. $h(M_i) > h(M_i')$ implies $h(fM_1 \dots M_i \dots M_n) > h(fM_1 \dots M_i' \dots M_n)$.

However, a usual rule such as associativity

$$(x + y) + z \longrightarrow x + (y + z),$$

cannot be treated by simply reasoning about the size of terms. The solution is to assign a heavier weight to the left subterm of a sum. A function h is defined by structural induction:

$$\begin{aligned} h(a) &= 1, \\ h(M + N) &= 2h(M) + h(N), \end{aligned}$$

where a is a constant or a variable. This inductive definition consists of giving $\mathbf{N}$ a Σ-algebra structure (see Chapter 4):

$$\begin{aligned} a^{\mathbf{N}} &= 1, \\ +^{\mathbf{N}}(x, y) &= 2x + y. \end{aligned}$$

It is easy to see that function h satisfies conditions (1) and (2) above.

Finally, the size of terms can quite easily increase by rewriting, as the following example shows, where f and g are unary symbols:

$$ffx \longrightarrow gfgx.$$

In this case, $h(M)$ could be defined by the number of occurrences of f in M.

So, it suffices to have a function h satisfying properties (1) and (2) to prove the noetherianity of a rewriting system. It is not necessary that h has its values in $\mathbf{N}$. The only property of the integers that is used is that the order $>$ is well-founded: there is no infinite, strictly decreasing sequence of natural integers. A function h can be sought whose range is in a well-ordered set $(E, >)$. Referring to E is a notational convenience, since one can reason directly in $\mathrm{T}[X]$ with the order,

$$M \succ_h N \quad \text{if} \quad h(M) > h(N).$$

Hence a well-founded order $\succ$ is sought over the set of terms such that it satisfies properties (1) and (2): decreasing the instances of rules and context passing. This condition is also necessary because if $\longrightarrow$ is noetherian, the $\xrightarrow{+}$ itself is a well-order, by the definition of noetherianity. This necessary and sufficient condition constitutes the Manna–Ness theorem [97].

Theorem 3.18 A rewriting system $\mathcal{R}$ is noetherian if and only if there exists a well-order $\succ$ over the set of terms such that

1. $\sigma P \succ \sigma Q$ for every rule $(P, Q) \in \mathcal{R}$ and every substitution σ;
2. $M_i \succ M_i'$ implies $fM_1 \ldots M_i \ldots M_n \succ fM_1 \ldots M_i' \ldots M_n$ ($\succ$ is *compatible with* Σ).

In practice, the external form is more easily used, i.e., with a function h into a well-ordered set E.

3.4.2 Simplification orders

The Manna–Ness theorem is very strict: it requires a strict order $\succ$ to be well-founded over the set of terms when in fact it would suffice if it were well-founded 'along the reductions'. That there be an infinitely decreasing sequence of terms has no importance if the sequence is not a reduction.

In applications of the Manna–Ness theorem, one works with a known set $(E, \succ)$, for example $\mathbf{N}$ or $\mathcal{M}(\mathbf{N})$, and one knows once and for all that the order $\succ$ is well-founded. It is much more difficult to prove that an order is well-founded along a chain of reductions: this proof would normally depend on the rewriting system being studied. There does exist, however, a class of orders which are necessarily well-founded along reduction chains. They are the simplification orders, introduced by N. Dershowitz.

Definition 3.15 A strict order relation $\succ$ over $\mathrm{T}[X]$ is a *simplification* order if for every symbol f of arity n,

1. $M_i \succ M_i'$ implies $fM_1 \ldots M_i \ldots M_n \succ fM_1 \ldots M_i' \ldots M_n$;
2. $fM_1 \ldots M_i \ldots M_n \succ M_i$ for every i.

The *embedding* relation is defined by

$$M = fM_1 \ldots M_p \sqsubseteq gM_1' \ldots M_q' = M'$$

if one of the following three conditions is satisfied:

1. $M = M'$;
2. $f = g$ and $M_i \sqsubseteq M_i'$ for every i, $1 \leqslant i \leqslant p$;
3. there exists i, $1 \leqslant i \leqslant p$ such that $M \sqsubseteq M_i'$.

Hence, $M \sqsubseteq M'$ if M can be obtained from M' by 'pruning' M'. It is a (partial) order relation. The slightly more general case of preorders is considered, for a stronger concept than well-order will be useful.

Definition 3.16 A preorder $\sqsubseteq$ is a *well-preorder* if for every infinite sequence of elements $(x_n)_{n \geqslant 0}$, there exists i, j such that $i < j$ and $x_i \sqsubseteq x_j$.

Lemma 3.19 A preorder $\sqsubseteq$ is a well preorder if and only if for every infinite sequence of elements $(x_n)_{n \geqslant 0}$, there exists an increasing extracted sequence $(x_{k_i})_{i \geqslant 0}$.

Proof There can only be a finite set of elements x_i which are not bounded by any x_j, $j > i$: if there were infinitely many, the definition of $\sqsubseteq$ would imply that two of the x_i would be comparable, and the smallest would be bounded by the other. An increasing sequence is extracted by starting with an arbitrary element of the sequence, beyond that finite set, by choosing an upper bound in the sequence and by iterating the procedure. □

There are other useful characterizations. For example, a preorder is a well-preorder if the associated strict order is well-founded and if every set of incomparable elements is finite.

Theorem 3.20 (Kruskal) If Σ is a finite signature, $\sqsubseteq$ is a well-preorder over T_Σ.

Proof Suppose that a counterexample exists, i.e., there exists a sequence $(M_n)_{n\geqslant 0}$ not satisfying the well-preorder property, which is minimal, i.e., chosen in the following manner: if $M_0, \ldots, M_p$ are already constructed, choose for M_{p+1} a term C_{p+1} of minimal size among all the counter-examples $(C_n)_{n\geqslant 0}$ such that $C_0 = M_0$, $\ldots$, $C_p = M_p$.

Since Σ is finite, one can extract from $(M_n)_{n\geqslant 0}$ a sequence of terms whose elements have the same head symbol, $(f(N^1_m, \ldots, N^p_m))_{m\geqslant 0}$.

By the minimality assumption, the sequence $(N^1_m)_{m\geqslant 0}$ cannot be a counterexample: if it were, the sequence $M_0, \ldots, M_{s-1}, N^1_0, \ldots, N^1_m, \ldots$, where s is such that $f(N^1_0, \ldots, N^p_0) = M_s$, would be a counterexample strictly 'smaller than' $(M_n)_{n\geqslant 0}$, which is impossible according to the construction.

According to the lemma, a sequence of terms can be extracted, each embedded in the next: $N^1_{m_1} \sqsubseteq N^1_{m_2} \ldots$ is the result. Consider, then, the sequence $(f(N^1_{m_i}, \ldots, N^p_{m_i}))_{i\geqslant 0}$ extracted from $(M_n)_{n\geqslant 0}$. The minimality assumption also applies to $(N^2_{m_i})_{i\geqslant 0}$. By iterating this extraction procedure for each of the arguments of f, the sequence $(f({M'_k}^1, \ldots, {M'_k}^p))_{k\geqslant 0}$ is extracted from $(M_n)_{n\geqslant 0}$ such that for every r $(1 \leqslant r \leqslant p)$, and every $k \geqslant 0$, ${M'_k}^r \sqsubseteq {M'_{k+1}}^r$.

By the definition of the embedding relation, deduce for every k that

$$f({M'_k}^1, \ldots, {M'_k}^p) \sqsubseteq f({M'_{k+1}}^1, \ldots, {M'_{k+1}}^p),$$

hence that the sequence $(M_n)_{n\geqslant 0}$ has an increasing subsequence for $\sqsubseteq$. The initial assumption over $(M_n)_{n\geqslant 0}$ is contradicted. □

One proves by induction over M' that

Lemma 3.21 If a term M is embedded in a term M', then $M' \succcurlyeq M$ for every simplification order $\succ$.

Theorem 3.22 (Dershowitz) R is noetherian if there exists a simplification order $\succ$ such that $\sigma P \succ \sigma Q$ for every rule $(P, Q) \in \mathcal{R}$ and every substitution σ.

Proof If there exists an infinite reduction $M_1 \longrightarrow M_2 \longrightarrow \ldots$, Kruskal's theorem ensures the existence of i and $j > i$ such that $M_i \sqsubseteq M_j$. According to the preceding lemma, it follows that $M_j \succcurlyeq M_i$. Yet, by definition, $\succ$ verifies conditions (1) and (2) of the Manna–Ness theorem. Therefore, $M \longrightarrow M'$ implies $M \succ M'$ and, in particular, $M_i \xrightarrow{+} M_j$ implies $M_i \succ M_j$. But $\succ$ being a strict order, $M_j \succcurlyeq M_i$ and $M_i \succ M_j$ are incompatible. □

A necessary condition can be given for non-noetherianity: if a rewriting system is not noetherian then there exist terms M and M' such that $M \sqsubseteq M'$ and $M \xrightarrow{+} M'$. Note that this necessary condition is not sufficient, as is shown by the example of rule $ffx \longrightarrow gfgx$. The noetherianity of this rule cannot be proven using a simplification order.

There are systematic methods for constructing simplification orders. They are used in rewriting systems such as REVE.

Let $\succ$ be a 'precedence' order over the signature Σ. Using $\succ$, define the *recursive path ordering* $\succ_{\mathrm{rpo}}$ over T_Σ and the associated order $\ggcurly_{\mathrm{rpo}}$ over $\mathcal{M}(\mathrm{T}_\Sigma)$, the set of multisets of terms. Write $\asymp$ for the equivalence under permutation of subterms: $M = fM_1 \ldots M_p \asymp fM'_1 \ldots M'_p = M'$ if $M_i \asymp M'_{\pi(i)}$ for a permutation π of indices $\{1, \ldots, n\}$, and $M \succsim_{\mathrm{rpo}} M'$ if $M \succ_{\mathrm{rpo}} M'$ or $M \asymp M'$.

$$M = fM_1 \ldots M_p \succ_{\mathrm{rpo}} gM'_1 \ldots M'_q = M'$$

if one of the three following conditions is satisfied:

RPO1 $f = g$ (hence $p = q$) and $\{\!\{M_1, \ldots, M_p\}\!\} \ggcurly_{\mathrm{rpo}} \{\!\{M'_1, \ldots, M'_q\}\!\}$;
RPO2 $f \succ g$ and for every j, $1 \leqslant j \leqslant q$, $M \succ_{\mathrm{rpo}} M'_j$;
RPO3 $f \not\succcurlyeq g$ and there exists i, $1 \leqslant i \leqslant p$ such that $M_i \succsim_{\mathrm{rpo}} M'$.

Examples Suppose

$$x \wedge (y \vee z) \succ_{\mathrm{rpo}} (x \wedge y) \vee (x \wedge z)$$

is to be proven. Choose $\wedge \succ \vee$, which corresponds to case RPO2: $x \wedge (y \vee z) \succ_{\mathrm{rpo}} x \wedge y$ and $x \wedge (y \vee z) \succ_{\mathrm{rpo}} x \wedge z$ must be proven. Then $x \wedge (y \vee z) \succ_{\mathrm{rpo}} x \wedge y$ corresponds to case RPO1 (same head symbol): $\{\!\{x, y \vee z\}\!\} \ggcurly_{\mathrm{rpo}} \{\!\{x, y\}\!\}$ must be proven. By the definition of the order over multisets, this follows from $y \vee z \succ_{\mathrm{rpo}} y$, which is true according to RPO3 ($y =$ hence $\asymp$ a subterm of $y \vee z$).

In the following example, a decreasing sequence is proven using *rpo* even though the size increases:

$$\neg(x \vee y) \succ_{\mathrm{rpo}} \neg x \wedge \neg y.$$

Suppose $\neg \succ \wedge$. One must prove (case RPO2) that $\neg(x \vee y) \succ_{\mathrm{rpo}} \neg x$, and $\neg y$, then (case RPO1) that $\{\!\{x \vee y\}\!\} \ggcurly_{\mathrm{rpo}} \{\!\{x\}\!\}$ and $\{\!\{x \vee y\}\!\} \ggcurly_{\mathrm{rpo}} \{\!\{y\}\!\}$. These results follow from $x \vee y \succ_{\mathrm{rpo}} x$ and $x \vee y \succ_{\mathrm{rpo}} y$, respectively, which are true by RPO3. □

Certain rules are not suitable for an *rpo*:

Example For the associativity rule, $(xy)z \rightarrow x(yz)$, one would have to prove $\{\!\{xy, z\}\!\} \not\succ_{\text{rpo}} \{\!\{x, yz\}\!\}$, which does not work. However, $(xy)z \succ_{\text{rpo}} x$ and yz, and by replacing the multisets by n-tuples, $(xy, z) \succ (x, yz)$ for the lexical order, since $xy \succ x$ by RPO3. □

This leads to another case for defining the *rpo*, which applies when a functional symbol has a lexical *status*:

RPO1LEX $f = g$ (hence $p = q$), and $(M_1, \ldots, M_p) \succ^{\star}_{\text{rpo}} (M'_1, \ldots, M'_q)$, where $\succ^{\star}_{\text{rpo}}$ is the lexical extension of $\succ_{\text{rpo}}$ to n-tuples of terms.

3.4.3 The Church–Rosser theorem

The Church–Rosser theorem, stating that β-reduction is confluent, is proven by introducing the concept of reduction relative to a set of 'visible' redexes.

Residues
Recall the choice of redex which is contracted in a term and the consequences of that choice on subsequent reductions. Note, first, certain phenomena:

- Let R be a redex, C the result of contracting R, and $M \equiv (\lambda x \,.\, xx)R$. M has two redexes, which are the subterms $M/\epsilon \equiv M$ and $M/2 \equiv R$. If R is contracted first, only one contraction of the obtained term, $(\lambda x \,.\, xx)C$, yields CC. If M is contracted to RR, two contractions of R are necessary to reach CC.
- Let $M \equiv (\lambda x\, y)N$. It is unnecessary to try to reduce N, since the contraction of M immediately yields y.
- Contracting can generate new redexes, e.g., $(\lambda x \,.\, x\mathbf{I})(\lambda y\, N) \longrightarrow (\lambda y\, N)\mathbf{I}$.

To study these phenomena, the concept of *residue* is introduced. Let $M \xrightarrow{u} M'$, $M/u \equiv (\lambda x\, P)Q$ and $v \in \mathcal{O}_{red}(M)$. According to the relative position of u and v under the prefix order, we examine the effect of contracting M/u at redex M/v, which yields a set of redex subterms of M', residues of M/v:

- If $u \perp v$, the contraction of M/u does not modify M/v; the residue of M/v is M'/v (*conservation*).
- If $u > v$, M/v is reduced, as M, by contraction of M/u; the residue of M/v is M'/v (*reduction*).
- If $u = v$, M/v is contracted and does not have a residue (*vanishing*).
- If $u < v$, then M/v is a redex of $P \equiv M/u1\lambda$, or of $Q \equiv M/u2$:
  - if M/v is a redex of P, it is transformed in the same way that P is by substituting Q for x; the residue of M/v is the subterm of M' corresponding to $(M/v)[x := Q]$ (*substitution*);

– if M/v is a redex of Q, it is grafted on to each occurrence of x in P; these grafted subterms are the residues of M/v in M' (*multiplication*); the occurrences of the residues are independent in $\mathcal{O}(M')$.

If this definition appears to be complicated, the practical determination of residues is simple. It suffices to mark the λ of the redex M/v and to 'trace' its reduction in M. The symbol $\tilde{}$ is used for this marking. With an ML-like syntax, a marked redex $(\tilde{\lambda}xP)Q$ could be written `let val` $x = Q$ `in` P `end`.

Example

$$(\lambda x\,.\,(\lambda z\,.\,zx)x)((\tilde{\lambda}yP)Q) \longrightarrow \lambda z\,.\,z((\tilde{\lambda}yP)Q))((\tilde{\lambda}yP)Q).$$

□

The definition of residues can be extended:

- The residues of a redex by a β-reduction are easily obtained by tracking the movements of the $\tilde{\lambda}$ of the redex.
- The residues of a set of redexes are obtained by marking the λ of each of the considered redexes.

Internal developments
It is convenient to name reductions. One can write $u : M \longrightarrow M'$ for the immediate β-reduction $M \xrightarrow{u} M'$, and $U : M \xrightarrow{\star} M'$ for the parallel reduction of a set U of *independent* occurrences of redexes. If $\rho : M \xrightarrow{\star} M'$ and $\rho' : M' \xrightarrow{\star} M''$ are two reductions, $\rho;\rho'$ denotes the composed reduction $M \xrightarrow{\star} M' \xrightarrow{\star} M''$. The composition has as neutral element $0 : M \longrightarrow^0 M$.

If u and v are two redex occurrences, write $u\backslash v$ for the set of residues of u by contraction of v, and extend this notation to $U\backslash v$, $u\backslash\rho$ and $U\backslash\rho$, when U is a set of redexes and ρ a reduction, by

$$\begin{aligned} U\backslash v &= \bigcup_{u\in U} u\backslash v, \\ U\backslash 0 &= U, \\ U\backslash(\rho;\rho') &= (U\backslash\rho)\backslash\rho'. \end{aligned}$$

If $U \subset \mathcal{O}_{red}(M)$, the pair (M, U) denotes the term M, with the set U of marked redexes.

Definition 3.17 Let M be a term and $U \subset \mathcal{O}_{red}(M)$. A *reduction* of M *relative* to U, or reduction of (M, U), is any β-reduction of M, such that at each step the contracted redex is either an element of U or a residue of an element of U by the preceding steps.

By contracting only marked redexes, new redexes, invisible in M, cannot be contracted. More precisely, a relative reduction is a reduction $u_1; \dots ; u_n$, where for every i, $1 \leqslant i \leqslant n$, $u_i \in U \backslash (u_1; \dots ; u_{i-1})$. This reduction is written $(M, U) \xrightarrow{\star} (M', U')$, where U' is the set of residues of the redexes of U by this β-reduction of M. If $U' = \varnothing$, this reduction is called a *development* of (M, U), or of M, if $U = \mathcal{O}_{red}(M)$.

Which redex M/u, $u \in U$, should be contracted first? For independent occurrences, the initial choice has no influence on subsequent choices.

Let M be a term and u_1 and u_2 be two redex occurrences in M, with $M/u_i \equiv (\lambda x_i P_i)Q_i$, $i = 1, 2$. Consider the two immediate β-reductions u_1 and u_2 of M, $u_i : M \xrightarrow{u_i} M_i$. The sets of residues $u_1 \backslash u_2$ and $u_2 \backslash u_1$ are formed of independent occurrences which can therefore be contracted in any order. Form the reductions $M_1 \xrightarrow{u_2 \backslash u_1} M'$ and $M_2 \xrightarrow{u_1 \backslash u_2} M'$: then $M' \equiv M''$. This is a *local* confluence result, but it does not suffice to prove the Church–Rosser theorem.

In the case of an arbitrary set of redex occurrences, to avoid the multiplication of residues, one must begin with the innermost occurrences, i.e., maximal under the prefix order. Let $u \in U$ be a maximal occurrence in U. Then, by contracting M/u, each $v \in U \setminus \{u\}$ has a unique residue (the case $v > u$, which could multiply v, is excluded by the assumption about u). If U has n elements, the set of residues of U by contraction of u has $(n - 1)$ elements. By subsequently contracting a maximal redex in the residues of U, a sequence of n immediate β-reductions $M \to M_1 \to \dots \to M_n$ is obtained, and the set of residues of U by this reduction is empty. Write $M^\imath$ for the term M_n when U is the set of marked redexes in M, which is uniquely determined by M and U, and can be computed by

$$\begin{aligned}
(x)^\imath &\equiv x, \\
(M_1 M_2)^\imath &\equiv (M_1)^\imath (M_2)^\imath \quad \text{if } M_1 M_2 \text{ is not a marked redex }, \\
(\lambda x M)^\imath &\equiv \lambda x M^\imath, \\
((\tilde{\lambda} x M_1) M_2)^\imath &\equiv (M_1)^\imath [x := (M_2)^\imath].
\end{aligned}$$

Hence, relative reduction has the (weak) normalization property:

Proposition 3.23 Every term has a finite development.

One should not think that $M^\imath$ is normal: this is due to the creation of new redexes which are not residues of redexes.

The parallelogram property

To continue with the proof of the Church–Rosser theorem, the *parallelogram property* is intermediate between confluence and local confluence: it is satisfied by a relation $\to$ if for all x, x_1 and y, if $x \to x_1$ and $x \xrightarrow{\star} y$, then there exists z such that $x_1 \xrightarrow{\star} z$ and $y \xrightarrow{\star} z$. See Figure 3.2

Lemma 3.24 (Parallelogram) Confluence follows from the parallelogram property.

Proof By induction on the length of the reduction $x \longrightarrow^p x_p$. Let $x \xrightarrow{\star} y$ and $x \longrightarrow x_1 \longrightarrow \ldots \longrightarrow x_p$. If $p = 0$, let $z \equiv y$.

If $p > 0$, the parallelogram property implies the existence of z_1 such that $x_1 \xrightarrow{\star} z_1$ and $y \xrightarrow{\star} z_1$. By the inductive hypothesis, there exists z such that $x_p \xrightarrow{\star} z$ and $z_1 \xrightarrow{\star} z$, from which, $x_p \xrightarrow{\star} z$ and $y \xrightarrow{\star} z$. □

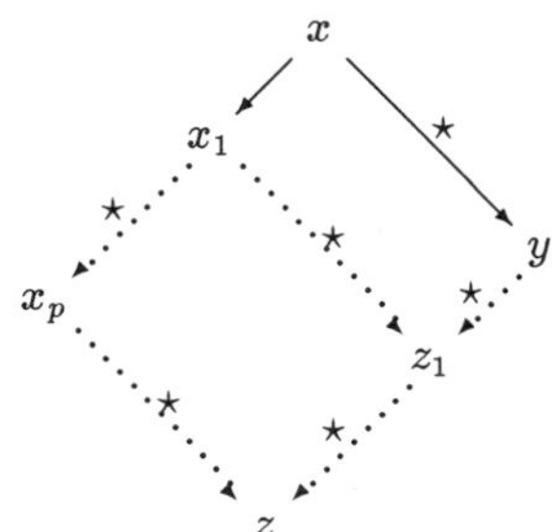

Figure 3.2 Proof of the parallelogram property

All that is needed now is to prove the parallelogram property for β-reduction. The proof is constructive. Let $M \longrightarrow M_1$ by contraction of the redex u, and $\rho : M \xrightarrow{\star} N$. Compute the set $u\backslash\rho$ of the residues of u under ρ, by tracing u in ρ. It suffices to contract the elements of $u\backslash\rho$ in N to obtain P such that $M_1 \xrightarrow{\star} P$. To be sure of the termination of this reduction, an internal development of $(N, u\backslash\rho)$ is used. Define $P \equiv N^\imath$. Now $M_1 \xrightarrow{\star} P$ must still be proven, using the following lemma:

Lemma 3.25 Let M, N be marked terms. If $M \xrightarrow{\star} N$ then $M^\imath \xrightarrow{\star} N^\imath$.

Proof It suffices to treat the case of an immediate β-reduction $M \longrightarrow_\beta N$, by induction on the length of $M \xrightarrow{\star} N$. The only case to be studied is that of redex contraction. Since it has already been shown that

$$\begin{aligned} ((\lambda x P)Q)^\imath &\equiv (\lambda x P^\imath)Q^\imath \longrightarrow_\beta P^\imath[x := Q^\imath], \\ ((\tilde{\lambda} x P)Q)^\imath &\equiv P^\imath[x := Q^\imath], \end{aligned}$$

it suffices to prove

$$(P[x := Q])^\imath \equiv P^\imath[x := Q^\imath]$$

by induction over P. Consider the case where

$$P \equiv (\tilde{\lambda} y P_1)P_2.$$

Then

$$\begin{aligned} (P[x := Q])^\imath &\equiv ((\tilde{\lambda} y P_1[x := Q])\, P_2[x := Q])^\imath \\ &\equiv (P_1[x := Q])^\imath[y := (P_2[x := Q])^\imath)], \end{aligned}$$

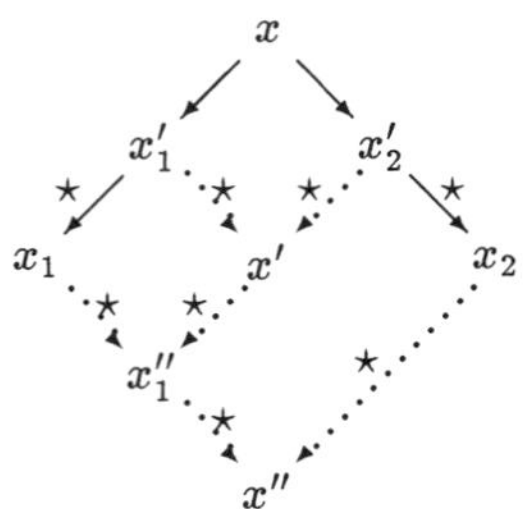

Figure 3.3 Proof of Newman's lemma

from which, by the inductive hypothesis,

$$(P[x := Q])^\imath \equiv (P_1^\imath[x := Q^\imath])[y := (P_2^\imath[x := Q^\imath])];$$

by substitution,

$$\begin{aligned} &\equiv \; P_1^\imath[y := P_2^\imath][x := Q^\imath] \\ &\equiv \; ((\tilde{\lambda} y P_1)P_2)^\imath[x := Q^\imath]. \end{aligned}$$

□

With this lemma, the Church–Rosser theorem is proven.

3.4.4 Newman's lemma

Local confluence is a property much easier to prove than confluence.

Lemma 3.26 (Newman) A noetherian relation is confluent if and only if it is locally confluent (see Figure 3.3).

Proof By induction over the well-founded relation $\xrightarrow{+}$, by showing that the set P of x where $\longrightarrow$ is confluent is progressive. Let x be such that $y \in P$ for every y such that $x \xrightarrow{+} y$. One must prove that $x \in P$.

Let $x \xrightarrow{+} x_1$ and $x \xrightarrow{+} x_2$. Split the two relations into $x \longrightarrow x'_1 \xrightarrow{\star} x_1$ and $x \longrightarrow x'_2 \xrightarrow{\star} x_2$. By local confluence, there exists x' such that $x'_1 \xrightarrow{\star} x'$ and $x'_2 \xrightarrow{\star} x'$. By the inductive hypothesis, $x'_1 \in P$ and $x'_2 \in P$, which means that there exist, respectively, x''_1 such that $x_1 \xrightarrow{\star} x''_1$ and $x' \xrightarrow{\star} x''_1$, and x'' such that $x''_1 \xrightarrow{\star} x''$ and $x_2 \xrightarrow{\star} x''$.

To prove that $x \in P$, it remains to check the trivial cases $x \equiv x_1$ or $x \equiv x_2$ where there is confluence, without using induction. □

3.4.5 Normalization in λ-calculus

β-reduction has no normalization property. However, if the term formation rules are restricted, for example using types, or the use of β-reduction is restricted, by marking redexes, then there are normalization results.

Finiteness of developments
Let (M, U) be a marked term, U being a set of redex occurrences in M. It has already been shown that relative reduction satisfies the weak normalization property by successively contracting the innermost redexes. Here it will be shown that relative reduction also satisfies the strong normalization property, i.e., it is noetherian.

Define a function h over terms with values of $\mathbf{N}$ which is strictly decreasing when a marked redex is contracted. The principle is to associate an integer weight > 0 with each variable occurrence, and to represent it by annotation: for example $M \equiv (\tilde{\lambda}x \,.\, x^{11}x^8)(y^2(\lambda x \,.\, x^5))$. Let $h(M)$ be the sum of the weights: for the preceding term $h(M) = 11 + 8 + 2 + 5 = 26$. The weight of each occurrence is carried by reduction, exactly like the variable it annotates. Hence $M \longrightarrow (y^2(\lambda x \,.\, x^5))(y^2(\lambda x \,.\, x^5))$. Clearly it should be the case that $h(M) > h(M')$ if $(M, U) \longrightarrow (M', U')$. To ensure this decreasing property, then, for every marked redex $(\tilde{\lambda}xP)Q$ and for every occurrence of x in P, $h(x) > h(Q)$: it is easy to prove that this condition suffices. A decreasing weighting is constructed by numbering the variable occurrence from right to left starting from 0 and by giving to the n-th the weight 2^n: the above term M would be annotated as $M \equiv (\tilde{\lambda}x \,.\, x^8x^4)(y^2(\lambda x \,.\, x^1))$. For the reasoning to work, the decreasing nature of the weighting must be conserved by reduction. This property can be checked by distinguishing the cases of the relative positions of the redexes.

See Barendregt [8, section 11.2] for the details. Note that this result is constructive: if the initial term contains n variable occurrences, there are at most 2^n reduction steps.

Proposition 3.27

1. Every reduction of (M, U) can be extended to a development.
2. The set of terms obtained by any relative reduction of M is finite.

There is also a confluence result. The Church–Rosser theorem cannot be applied directly, for one would have to check that the β-reductions showing confluence contract only marked redexes. The proof of the local confluence of β-reduction only used residues of initial redexes: the reduction of marked redexes is therefore locally confluent. From Newman's lemma, confluence follows from local confluence and strong normalization:

Theorem 3.28 Relative reduction is noetherian and confluent.

Because of this theorem, development is as basic as immediate β-reduction. In the same manner that N is uniquely determined by M and by the contracted redex in the relation $M \rightarrow_\beta N$, the final term N of a development (M, U) is uniquely determined by M and U, written $M \overset{U}{\Rrightarrow} N$. The term N cannot be computed, for example, by an internal reduction of M. If the `let` form of functional languages is interpreted as the marking of a redex, the preceding theorem allows this form to be considered as a 'syntactic extension' of the λ-calculus, definable as a redex, with the rule:

$$\texttt{let val } x = Q \texttt{ in } P \texttt{ end} \rhd P[x := Q]$$

The application of this rule a finite number of times produces a term with no `let`.

Normalization in typed λ-calculus
The weak normalization theorem is proven for the typed λ-calculus, which guarantees the existence of a normal form.

A few useful definitions are given. The *degree* of a redex is

$$\deg((\lambda x^\sigma M^\tau)N^\sigma) = \mathrm{ord}(\sigma \rightarrow \tau).$$

Note that $\mathrm{ord}(\sigma) < \mathrm{ord}(\sigma \rightarrow \tau)$ and that $\deg(R^\tau) > \mathrm{ord}(\tau)$. The *total degree* $\mathrm{Deg}(M)$ of a term M is the lub of the degrees of its redexes. A normal term has total degree 0.

Lemma 3.29 Let R be a redex of M, $M \longrightarrow M'$ by contraction of R and R' be a redex of M':

- if R' is the residue $R_0 \backslash R$ of a redex R_0 of M, then $\deg(R') = \deg(R_0)$;
- otherwise, R' is a new redex, and $\deg(R') < \deg(R)$.

Proof M' is obtained by contracting the redex $R \equiv M/u \equiv (\lambda x^\sigma P^\tau)^{\sigma \rightarrow \tau} Q^\sigma$ of M, to $C \equiv M'/u \equiv P[x := Q]$. Let $R' \equiv M'/v$ be a redex of M'. There are three cases depending on the position of u with respect to v.

1. If $v \perp u$, R' is the residue of a redex R_0 of M which was conserved in M'.
2. If $v < u$, R' is the residue of a redex R_0 of M which was reduced by contraction of R.
3. If $v \geqslant u$, R' is a redex of C. It is the residue of a redex R_0 of M in Q, multiplied for each occurrence of x in P, the residue of a redex R_0 of M in P substituted by $[x := Q]$, or a redex which was created during the contraction of R.

For the cases of conservation, reduction, multiplication or substitution of a redex, the degree of the residue does not change: $\deg(R') = \deg(R_0)$.

A new redex is created when Q is an abstraction $\lambda y^{\sigma'} Q'^{\tau'}$, for each occurrence of x^σ in P^τ which is the left-hand side of an application, written $x^\sigma P'^{\sigma'}$. The type constraints force $\sigma' \to \tau' = \sigma$. The contraction of R does the substitutions $(x^\sigma P'^{\sigma'})[x := \lambda y Q']$ which create new redexes $(\lambda y^{\sigma'} Q'^{\tau'})^{\sigma' \to \tau'} P'^{\sigma'}[x := \lambda y Q']$. Their degree is $\mathrm{ord}(\sigma' \to \tau') = \mathrm{ord}(\sigma) < \mathrm{ord}(\sigma \to \tau) = \deg(R)$. □

To conclude, a redex of M' has a degree strictly bounded by that of R if it is new or it has the degree of the redex of which it is a residue. This shows in particular that $\mathrm{Deg}(M) \geqslant \mathrm{Deg}(M')$. This lemma is used to prove the weak normalization theorem of the typed λ-calculus.

Proof Let $h(M) = (d, n)$, where $d = \mathrm{Deg}(M)$ and n is the number of redexes of M of degree d. If M is normal, $h(M) = (0, 0)$. The term M is reduced in order to decrease $h(M)$ in N^2 for the lexical order (i.e. $(d, n) > (d', n')$ if $d > d'$ or else $d = d'$ and $n > n'$). If $h(M) = (d, n)$, with $d > 0$, choose a redex R of M of degree d, itself not containing any redex of degree d, and let $M \to M'$ by contraction of R.

By choosing a redex R of maximum degree d of which every redex subterm is of degree strictly less than itself, the multiplication of redexes of degree $\geqslant d$ is avoided. By contracting R, a redex of degree d disappears, and the created redexes are of strictly inferior degree. This proves that $h(M) > h(M')$. □

Exercises[1]

1. Construct the graph of immediate β-reduction of the terms WWW and $NNNN$, where $W \equiv \lambda xy \,.\, xyy$ and $N \equiv \lambda xyz \,.\, zyxx$.

2.★ Construct the graph of immediate β-reduction of $\Delta(W\mathbf{I})$, W as above.

3.★ Let $M \equiv \lambda pq \,.\, pF(\mathbf{K}a)(qF(\mathbf{K}b))$, where a, b are variables, and $F \equiv \lambda fg \,.\, gf$. What is the normal form of $M\mathbf{pq}$ if $\mathbf{p}$ and $\mathbf{q}$ are Church numerals. What is the length of the reduction?

4. Check that the λ-term **zero** represents an arithmetic if-then-else, and that **pred** represents the predecessor.

5. Show that $\lambda x \,.\, xx$ is not typable but that **II** is typable. Give a principal type.

6.★ Give a principal type for the combinators **S**, **B** and **C**. Their definition in ML is:

```
fun S f g x = f x (g x)
and B f g x = f (g x)
and C f x y = f y x ;
```

[1]Hints to the exercises labeled with a ★ can be found at the end of the book.

7. Find the covering relations defined by the usual order over **N** in **R**, the prefix order in a free monoid $A^\star$ and the lexical order $\leqslant^\star$ over $A^\star$, for A totally ordered (see Chapter 1, exercise 2).

8. Let M, $N \in \mathrm{T}[X]$; show that M covers N if and only if $M = \sigma N$, where σ is a substitution equating two distinct variables of N or associating a term $fy_1 \ldots y_n$ (with new variables y_i) with one of the variables of N.

9. Using the result of the preceding exercise, compute the number of classes in $\tilde{\mathrm{T}}[X]$ of terms covering a term N, as a function of the number of variables of N. Show that the number of classes of terms covered by N is finite.

10. Show that the order relation generated by the covering relation is the subsumption preorder.

11.★ Show that $M \wedge N$ is a lower bound of M and N.

12. Is the system formed by the rule $-x + y \longrightarrow (- - x + y) + y$ noetherian?

13. Define a generic data type for binary trees (with leaves of type `'a`) and an iterator over these trees. Use the iterator to compute the height, the number of leaves and the list of leaves of a tree.

14.★ Here is a recursive definition of trees and forests:

```
datatype 'a Tree = Union of 'a * 'a Forest
and    'a Forest = Empty | Cons of 'a Tree * 'a Forest ;
```

Define two iterators `Tree_iter` and `Forest_iter`.

15. Let

```
type signature = symbol -> int ;
```

Define a function `correct` of type `signature * term -> bool` which indicates if a term is correctly formed over a signature. Implement the usual operations over terms and occurrences, computing $\mathcal{O}(M)$, $M(u)$, M/u and $M[u \leftarrow N]$.

16. Use the iterator `It` to define addition, multiplication and exponentiation over integers.

17. Define the iterator `It` over the integers using the recursor `R`, and inversely.

18. What functions do `hierarchy Zero` and `hierarchy (Succ Zero)` compute? Prove that `hierarchy omega` is the diagonal $\lambda n A(n, n)$ of the Ackermann function.

19. Derive the rewrite rules for the applicative terms B and C, where $B \equiv S(KS)K$ and $C \equiv S(BBS)(KK)$.

20. Prove the optimization rules of combinatory logic using the extensionality rule.

21. Prove that the lexical order $\succ_\times$ over a product of well-founded orders is also well-founded.

22. Using an *rpo* order, prove the termination of the rewriting system computing Ackerman's function A, giving A a lexical status.

23. Show that the redexes which contract to a redex are of one of three forms: $(\lambda x \,.\, xM)A$, $(\lambda x\, R)M$ or $\mathbf{I}R$, where M is an arbitrary term, A is an abstraction and R is a redex.

Bibliographic notes

The principal reference for the pure λ-calculus and combinatory logic is Barendregt's book [8]. The typed λ-calculi and their relation to logic are treated in [54, 64, 90]. Rewriting systems have been the subject of surveys by Huet and Oppen [71], and Dershowitz and Jouannaud [39].

Landin's articles [92, 94] are the origins of functional programming. Another point of view was presented by Backus [7] as he received the Turing Award. Structural operational semantics, due to G.D. Plotkin [122], are developed under the name of *natural semantics* by G. Kahn and T. Despeyroux at the French INRIA [18]. The Damas–Milner type system, which is implemented in ML, is described in [110]. The variant presented here is found in [19].

Standard ML is described in [111, 159]; the first is a good introduction to functional programming. The book [1] by Abelson, Sussman and Sussman is highly recommended for programming in general. Scheme is described in [20], Ada, for example, in [9] and Common Lisp in [140]. CAML and its implementation are described in [29, 158]. Miranda is a trademark of Research Software Ltd.

The idea of compilation of functional languages by translation into applicative terms is due to Turner [152]. Supercombinators were introduced by Hughes [72]. The implementation of functional languages (interpretation and compilation) is treated in [43, 119, 120].

Kruskal's theorem, proved in [91], is a fundamental result in combinatorics. The given proof is Dershowitz's [38] (according to Nash–Williams), where he introduces simplification orders. Termination proof techniques have been particularly studied by Manna and Dershowitz. The REVE rewriting system is a joint project of MIT and the Centre for Research in Computer Science of Nancy, France.

Term M of exercise 3 is due to B. Maurey, as given in [89].

Chapter 4

First-order logic

4.1 First-order formulas

First-order languages were first introduced by Frege in 1879, who used a very complicated notation. Today, their use is widespread in mathematics.

Example In the expression

$$(x > 0) \wedge \forall y \forall z \big(\neg(y * z = x) \vee y = x \vee z = x\big)$$

there are

- one constant (or 0-ary function) symbol 0;
- one binary (or 2-ary) function symbol $*$;
- two binary relation symbols $=$ and $>$;
- variables x, y and z; y and z are quantified, x is not;
- terms x, 0, $y * z$, y and z;
- connectives $\wedge$, $\neg$ and $\vee$; and
- a quantifier $\forall$.

The whole expression is a formula stating a property about x. It will be shown later on that this formula *can* mean that x is prime in **N**. □

4.1.1 Signature of a first-order language

The signature of a first-order language is the set Σ of function and relation symbols which are found in its formulas. In the above example, 0, $*$, $=$ and $>$ are members of the signature.

A *predicate calculus signature* is a pair of disjoint signatures Σ_f and Σ_r, respectively called *functional* and *relational* signatures (see § 2.2.1). Write $\Sigma = \Sigma_f \cup \Sigma_r$.

The elements of Σ_f (respectively Σ_r) are the function (respectively relation) symbols.

Examples

- The signature of the language for arithmetic is $\{0, s, +, \times, <, =\}$, where 0, s, $+$ and $\times$ are the functional symbols and $=$ and $<$ are the binary functional symbols.
- The signature of the language for groups is $\{e, *, ^{-1}, =\}$.

4.1.2 Formulas

Let a *countably infinite* set X of *variables* be given. Then $\mathrm{T}_{\Sigma_f}[X]$ is the set of finite terms over the signature Σ_f, also written $\mathrm{T}_\Sigma[X]$ or simply $\mathrm{T}[X]$. The construction of $\mathrm{T}[X]$ gives a functional meaning to the elements of Σ_f. A relational meaning must still be given to the elements of Σ_r.

The *atomic formulas* (or *atoms*) are the $Rt_1 \dots t_n$ with $R \in \Sigma_r$ of arity $n \geqslant 0$ and $t_1, \dots, t_n \in \mathrm{T}[X]$: they express a relation between terms. When they exist, standard notations for relations will be used. For example, $t_1 = t_2$ will be used instead of $= t_1 t_2$.

Example $x = x$ and $x > (y * 0)$ are atoms. □

The set of variables of an atom $Rt_1 \dots t_n$ is:

$$\mathrm{var}(Rt_1 \dots t_n) = \bigcup_{i=1}^{n} \mathrm{var}(t_i).$$

The set of atoms is written $\mathrm{At}_\Sigma[X]$ or $\mathrm{At}[X]$ and the set of *closed* atoms (no variable) is written At_Σ.

Formulas are constructed from atoms using connectives and quantifiers. The connectives can be chosen quite freely from the propositional calculus (§2.2.3). To begin, the *connectives* $\wedge$ ('and'), $\vee$ ('or'), $\Rightarrow$ ('implies') and $\perp$ ('false') will be used, as will the *universal* ($\forall$) and *existential* ($\exists$) *quantifiers*.

Definition 4.1 The set $\mathrm{L}_\Sigma[X]$ of *formulas of the first-order predicate calculus* over the signature Σ is the smallest set E of the free monoid $(\Sigma \cup X \cup \{\wedge, \vee, \Rightarrow, \perp, \forall, \exists\})^\star$ such that:

1. $\mathrm{At}_\Sigma[X] \subseteq E$;
2. $\perp \in E$, and if $\varphi, \psi \in E$ then $\varphi \wedge \psi$, $\varphi \vee \psi$, $\varphi \Rightarrow \psi \in E$;
3. if $x \in X$, $\varphi \in E$, then $\forall x \varphi$ and $\exists x \varphi \in E$.

Example $\forall x\,(x = x)$, $(x > y*y) \wedge (x = z)$ and $\forall x\,(y > z)$ are formulas, but $x = \perp$ and $\forall(x * x) > 0$ are not. □

The construction of L[X] only gave a functional meaning to connectives and quantifiers: $\wedge, \vee, \Rightarrow$ are binary, $\forall x, \exists x$ are unary, and $\perp$ is 0-ary Their logical meaning has yet to be given. The quantifier $\forall$ could have been presented as a unary operator applied to 'functional–formulas', in which case one would write $\forall(\lambda x\,.\,(x = x))$ instead of $\forall x\,(x = x)$.

Higher-order languages L[X] is a first-order language because variables only range over terms. One cannot quantify over functional or relational symbols, which are 'constants' of the language. In a *second-order* language, there are also an infinite number of relational variables of each arity, all quantifiable. The power of expression is much greater: for example, equality is defined by Leibniz's formula,

$$(x = y) \equiv \forall P(P(x) \Rightarrow P(y)),$$

where P is a unary relation symbol, and conjunction by

$$\varphi \wedge \psi \equiv \forall P((\varphi \Rightarrow (\psi \Rightarrow P)) \Rightarrow P).$$

Higher-order logic is starting to be studied seriously in computer science.

Occurrences As for λ-terms, there are several levels of abstract syntax. The occurrences of a formula , its subformulas, and the logical symbol at an occurrence are defined as if the formula consisted of terms over the signature formed of connectives, $\forall x$ and $\exists x$. For example, for $\varphi \equiv \forall x(A(x) \Rightarrow \exists y B(x, y))$, $\mathcal{O}(\varphi) = \{\epsilon, 1, 11, 12, 121\}$, $\varphi/12 \equiv \exists y B(x, y)$ and $\varphi(12) = \exists y$.

Sign of a subformula A formula occurrence is *positive* (respectively *negative*) if it is in the scope of an even (respectively odd) number of negations (explicitly, or implicitly, in the left-hand side of an implication).

More precisely, the sign of occurrences of φ is defined by a mapping sg : $\mathcal{O}(\varphi) \rightarrow \{+, -\}$, setting $+' = -$ and $-' = +$:

- $\text{sg}(\epsilon) = +$;
- if $\varphi/u = \chi \wedge \psi$ or $\chi \vee \psi$, $\text{sg}(u1) = \text{sg}(u2) = \text{sg}(u)$;
- if $\varphi/u = \chi \Rightarrow \psi$, $\text{sg}(u1) = \text{sg}(u)'$ and $\text{sg}(u2) = \text{sg}(u)$;
- if $\varphi/u = \forall x\psi$ or $\exists x\psi$, $\text{sg}(u1) = \text{sg}(u)$.

The sign of u is, by definition, that of the subformula φ/u and the logical symbol $\varphi(u)$. Note that it is not a deductive definition over formulas: in terms of attribute grammars, the sign of a subformula is not *synthesized* starting from the signs of its components, but, rather, *inherited* from its context.

Free and bound variables The set $\mathrm{var}(\varphi)$ of *free variables* of a formula φ is defined by induction by:

$$\begin{aligned}\mathrm{var}(Rt_1 \dots t_n) &= \bigcup_{i=1}^{n} \mathrm{var}(t_i),\\ \mathrm{var}(\bot) &= \varnothing,\\ \mathrm{var}(\varphi \wedge \psi) = \mathrm{var}(\varphi \vee \psi) = \mathrm{var}(\varphi \Rightarrow \psi) &= \mathrm{var}(\varphi) \cup \mathrm{var}(\psi),\\ \mathrm{var}(\forall x \varphi) = \mathrm{var}(\exists x \varphi) &= \mathrm{var}(\varphi) \setminus \{x\}.\end{aligned}$$

Every occurrence of x in $\forall x\varphi$ and $\exists x\varphi$ is *bound.* A quantifier binds a variable, like the λ does in the λ-calculus. Recall that the name of a bound variable does not really have any meaning: $\forall x(x > 0)$ and $\forall y(y > 0)$ have the same meaning, but not $\forall x(x > y)$ and $\forall y(y > y)$. Since a written formula contains only a finite number of variables, defining X to be infinite allows renaming (implicit α-conversion): replacing a bound variable in φ by a new variable not appearing in φ yields a new formula equal to φ.

Examples

- In $\forall x\exists y(x * y = z)$, y can be renamed to u, but not to x or z.
- $(x = 0) \vee \forall x(x \geqslant 0)$ is the same formula as $(x = 0) \vee \forall y(y \geqslant 0)$, since only the bound occurrence of x was renamed.

The elements of the set $\mathrm{var}(\varphi)$ defined above are called *free variables* of φ. One can check they are not modified by renaming bound variables of φ. Henceforth, any formula φ will be treated up to renaming of bound variables (implicit equivalence relation) and its bound variables will be written so they are not in $\mathrm{var}(\varphi)$, or not in a given set containing $\mathrm{var}(\varphi)$. The syntactic equality of two terms, just as for the equality of two formulas up to renaming of bound variables, is written $\equiv$, when it is necessary, or simply $=$.

$\mathrm{L}[x_1, \dots, x_n]$ is the set of formulas $\varphi \in \mathrm{L}[X]$ such that $\mathrm{var}(\varphi) \subseteq \{x_1, \dots, x_n\}$, $x_1, \dots, x_n \in X$. One often writes $\varphi(x_1, \dots, x_n)$ to indicate that $\varphi \in \mathrm{L}[x_1, \dots, x_n]$. *Closed formulas* or *propositions* are the φ such that $\mathrm{var}(\varphi) = \varnothing$. Their set is written L_Σ instead of $\mathrm{L}_\Sigma[\varnothing]$. Formulas containing no quantifier are called *open.* Their set, written $\mathrm{L}^o[X]$, can be equated with $\mathrm{Prop}[\mathrm{At}[X]]$, the set of propositional formulas formed over atoms. If $\varphi \in \mathrm{L}[x_1, \dots, x_n]$ is an arbitrary formula, its *universal* and *existential closures* are, respectively,

$$\begin{aligned}\forall(\varphi) \text{ or } \forall\vec{x}\varphi &\equiv \forall x_1 \dots \forall x_n\varphi \in \mathrm{L}_\Sigma \quad \text{and}\\ \exists(\varphi) \text{ or } \exists\vec{x}\varphi &\equiv \exists x_1 \dots \exists x_n\varphi \in \mathrm{L}_\Sigma.\end{aligned}$$

Substitutions Grafting a term on to a term occurrence and substituting a term for a variable were presented in §2.2.4 and §2.2.6.

Example $(x + yz)[y := x + y] \equiv x + (x + y)z$. □

The definition of substitution in a formula is analogous to the λ-calculus case (§2.3.3), due to the bound variables. Here is a simple definition. Let $\varphi \in \mathrm{L}[X]$ and a term $t \in \mathrm{T}[X]$ such that the bound variables of φ are renamed outside of $\mathrm{var}(t)$ (always possible): $\varphi[x := t]$ is the formula obtained from φ by replacing each occurrence of x (necessarily free) in φ by t. If $x \notin \mathrm{var}(\varphi)$, then $\varphi[x := t]$ is φ.

Example Let φ be the formula $\exists y(x > y)$. To replace $x \in \mathrm{var}(\varphi)$ by the term $y+u$, one must first rename y in φ by rewriting φ as $\exists z(x > z)$ and then $\varphi[x := y + u]$ becomes $\exists z(y + u > z)$. It is certainly not $\exists y(y + u > y)$! □

4.2 Natural deduction

In this section, reasoning as it has been done since the ancient Greeks will be formalized. The standard meaning will also be given to the logical symbols. Doing this requires a new syntactic object, the *derivation*. Its inductive definition is made up of inference rules (§2.2.2).

There are several kinds of inference systems. The natural deduction systems that will be presented in detail allow the derivation of formulas modulo assumptions which are introduced and then discharged during a proof: the structure of derivations is that of λ-terms. In Hilbert systems, formulas are derived from axioms: derivations have a first-order term structure, where constants correspond to axioms. Finally, derivations of the sequent calculus also have a term structure, but the judgments that they contain are *sequents*, not formulas.

Gentzen proposed several inference systems in 1934, including the natural deduction systems NK (Natural Classic) for classical logic, NJ for intuitionistic logic and NM for minimal logic (see [144]). An 'elementary' formulation (in the sense of subatomic particles in physics) of the meaning of a logical formula is given using the meanings of its components: it is an operational interpretation of the connectives and quantifiers. For each of the logical symbols, there is an introduction (I) rule and an elimination (E) rule. The natural deduction system will be presented in four steps: the propositional minimal calculus, formed of rules for implication, conjunction and disjunction; the intuitionistic calculus, which adds an elimination rule for $\bot$ and the quantifiers; and the classical system, with the law of the excluded middle.

4.2.1 The minimal system

There are two possible presentations for these rules. The first presentation derives formulas in a context of assumptions, i.e, judgments of the form $\Gamma \vdash \psi$, where $\Gamma = \{\varphi_1, \dots, \varphi_n\}$ is a finite set of formulas and $\vdash$ is the deduction symbol. This judgment is generally written $\varphi_1, \ldots, \varphi_n \vdash \psi$ and states that ψ can be deduced from $\varphi_1, \ldots, \varphi_n$. Write Γ, φ instead of $\Gamma \cup \{\varphi\}$ and Γ, Δ instead of $\Gamma \cup \Delta$.

The rules for implication are

$$(I_{\Rightarrow}) : \frac{\Gamma, \varphi \vdash \psi}{\Gamma \vdash \varphi \Rightarrow \psi} \qquad\qquad (E_{\Rightarrow}) : \frac{\Gamma \vdash \varphi \Rightarrow \psi \qquad \Delta \vdash \varphi}{\Gamma, \Delta \vdash \psi}.$$

Consider the management of the assumptions, i.e., the expressions to the left of the $\vdash$, in these rules. In general, the assumptions of the conclusion of a rule are the union of the assumptions of the premises, except when an assumption is 'discharged'. In the case of $(I_{\Rightarrow})$, the formula φ is an assumption of the premise $\Gamma, \varphi \vdash \psi$, but no longer occurs as an assumption in the conclusion $\Gamma \vdash \varphi \Rightarrow \psi$— except if $\varphi \in \Gamma$: that occurrence of φ has been *discharged* by applying $(E_{\Rightarrow})$. Applying $(E_{\Rightarrow})$ can discharge a bunch of assumptions, i.e., a possibly empty set of occurrences of φ appearing to the left of $\vdash$ in the premise.

To avoid having to carry around the assumptions, the formulas themselves will be used as judgments, and only the assumptions to be discharged will be shown. The latter will be placed above the judgments for which they are assumptions, in square brackets or slashed. This second presentation for natural deduction rules will be used henceforth.

The rules for the minimal propositional system are given in Table 4.1, where φ, χ and ψ are formulas.

$$(I_{\Rightarrow}) : \frac{\overset{[\varphi]}{\psi}}{\varphi \Rightarrow \psi} \qquad\qquad (E_{\Rightarrow}) : \frac{\varphi \Rightarrow \psi \qquad \varphi}{\psi}$$

$$(I_{\wedge}) : \frac{\varphi \qquad \psi}{\varphi \wedge \psi} \qquad\qquad (E^1_{\wedge}) : \frac{\varphi \wedge \psi}{\varphi} \qquad\qquad (E^2_{\wedge}) : \frac{\varphi \wedge \psi}{\psi}$$

$$(I^1_{\vee}) : \frac{\varphi}{\varphi \vee \psi} \qquad\qquad (I^2_{\vee}) : \frac{\psi}{\varphi \vee \psi} \qquad\qquad (E_{\vee}) : \frac{\varphi \vee \psi \qquad \overset{[\varphi]}{\chi} \qquad \overset{[\psi]}{\chi}}{\chi}$$

Table 4.1 The NM propositional system

- $I_{\wedge}$, $E_{\wedge}$, $I_{\vee}$ and $E_{\Rightarrow}$ (*modus ponens*) are clear.
- $I_{\Rightarrow}$ and $E_{\vee}$ contain discharged assumptions, in square brackets. For example, $I_{\Rightarrow}$ (rule often used in mathematical reasoning) means: 'if ψ can be deduced from φ then deduce $\varphi \Rightarrow \psi$ without supposing φ'. Rule $E_{\vee}$ reasons by case: if $\varphi \vee \psi$ has been proven, then if χ follows from φ and if χ follows from ψ, then deduce χ without supposing φ or ψ.

An inference rule acts as a rewriting rule, by pattern matching. The variables φ, ψ and χ appearing in such rules are formula metasymbols.

4.2.2 Derivations

Inference rules are used to construct the set $D_\Sigma[X]$ of derivations inductively. These have the form of a binding tree whose nodes are labeled by a rule and by the judgment that is inferred. A binding links a node (the application of a rule) and one or more assumptions which were discharged by applying the rule. The root of a derivation tree is its conclusion. Each inference rule gives a rule for constructing derivations, which defines how assumptions are to be managed during an inference.

(D_O) For each formula φ, the tree with one node (labeled by) φ is a derivation, of root φ and assumption φ.

$(DI_\wedge)$ If d is a derivation with root φ, d' is a derivation with root ψ, then

$$(I_\wedge) : \frac{\begin{array}{cc} d & d' \\ \varphi & \psi \end{array}}{\varphi \wedge \psi}$$

is a derivation of root $\varphi \wedge \psi$.

$(DI_\Rightarrow)$ If d is a derivation with root ψ, then

$$(I_\Rightarrow) : \frac{\begin{array}{c} d \\ \psi \end{array}}{\varphi \Rightarrow \psi},$$

with an *arbitrary* number ($\geqslant 0$) of links between the root and the leaves labeled by φ not already linked, is a derivation of root $\varphi \Rightarrow \psi$.

The other rules are written in an analogous manner. The set of leaves of a derivation is the set of its assumptions and the root is its conclusion. The application of either of the rules $E_\vee$ and $I_\Rightarrow$ discharges the slashed assumptions appearing in the rule: each occurrence of the assumptions slashed in the rule can also be (but is not forced to) slashed in the derivation. In a derivation, an assumption is either *active*, or *discharged*.

Example The two occurrences of assumption $\varphi \wedge \psi$ are discharged by $(I_\Rightarrow)$:

$$(I_\Rightarrow^{[1]}) : \frac{(I_\wedge) : \dfrac{(E_\wedge^2) : \dfrac{[\varphi \wedge \psi]^1}{\psi} \qquad (E_\wedge^1) : \dfrac{[\varphi \wedge \psi]^1}{\varphi}}{\psi \wedge \varphi}}{\varphi \wedge \psi \Rightarrow \psi \wedge \varphi}.$$

The discharge is indicated by the superscript ^{1} of the assumptions and the $(I_\Rightarrow)$.
□

To represent a derivation of φ under assumptions Γ, it is not always convenient to use a tree format. A sequential presentation can also be used: this is a sequence of formulas $\varphi_1, \ldots, \varphi_n$, possibly in brackets, along with their justification, such that $\varphi_n = \varphi$ and for all k $(1 \leqslant k \leqslant n)$, where either $\varphi_k \in \Gamma$, either φ_k results in the application of a rule whose premises appear among the φ_l, with $l < k$, or $[\varphi_k]$ is a discharged assumption introduced during the proof.

Here is a presentation of the derivation of $(\varphi \wedge \psi) \Rightarrow (\psi \wedge \varphi)$ given above:

1	$[\varphi \wedge \psi]$	assumption
2	ψ	$E_\wedge^2$ on 1
3	φ	$E_\wedge^1$ on 1
4	$\psi \wedge \varphi$	$I_\wedge$ on 2 and 3
5	$(\varphi \wedge \psi) \Rightarrow (\psi \wedge \varphi)$	$I_\Rightarrow$ on 4, discharging 1

This representation, often used, masks the real (abstract) nature of derivations.

Furthermore, a derivation dag is often used instead of a tree, which allows certain formulas, such as assumptions and intermediate conclusions, to be shared. In the preceding example, assumption 1 would be shared.

The difficult part is to *create* the derivations. In some cases, the structure of a formula φ to be proved guides the proof: the last rule to be applied is typically the introduction of the principal connector of φ. If it is an implication, an assumption must be put 'above' the tree, and it should be taken advantage of. There are good reasons for applying eliminations to assumptions to decompose them, before applying introductions to construct the conclusion progressively. These remarks can be formalized.

4.2.3 The deduction relation

Building of derivations is usually thought of as an intermediate step for defining the deduction relation.

Definition 4.2 Deduction is the relation $\vdash$ defined over $\mathcal{P}(\mathrm{L}[X]) \times \mathrm{L}[X]$ by: $\Gamma \vdash \varphi$ if there exists a derivation φ whose active assumptions belong to Γ.

The relation is written $\vdash_M$ for minimal logic, $\vdash_I$ for intuitionistic logic and $\vdash_C$ for classical logic. If $\Gamma = \{\varphi_1, \ldots, \varphi_n\}$ is finite, write $\varphi_1, \ldots, \varphi_n \vdash \varphi$ instead of $\Gamma \vdash \varphi$. If $\Gamma = \varnothing$, write $\vdash \varphi$, meaning that φ is a (deductive) *tautology*.

This relation satisfies the following property, which immediately follows from the rules $I_\Rightarrow$ and $E_\Rightarrow$:

$$\varphi \vdash \psi \qquad \text{if and only if} \qquad \vdash \varphi \Rightarrow \psi;$$

more generally, also using $I_\wedge$ and $E_\wedge$,

$$\varphi_1, \ldots, \varphi_n \vdash \psi \qquad \text{if and only if} \qquad \vdash \varphi_1 \wedge \ldots \wedge \varphi_n \Rightarrow \psi.$$

This property, called the 'deduction theorem' is not as easy to prove when other inference systems are used, such as Hilbert systems. A consequence of this property is that it suffices to know how to prove tautologies: a tautology is no easier to prove than a 'difficult' theorem.

Note that much information is lost by only keeping (for $\vdash$) the leaves and the root of a derivation tree. The problem $d \in D_\Sigma[X]$? is decidable, while $\Gamma \vdash \varphi$? is generally not.

Proven equivalence A new binary connective, $\Longleftrightarrow$, short for *equivalence*, is introduced: write $\varphi \Longleftrightarrow \psi$ instead of $(\varphi \Rightarrow \psi) \wedge (\psi \Rightarrow \varphi)$. This connective is often used to state tautologies, as in

$$\vdash (\varphi \wedge \psi) \Longleftrightarrow (\psi \wedge \varphi),$$

where the first part ($\Rightarrow$) was proven above.

It is easy to show that the binary relation between φ and ψ, $\vdash \varphi \Longleftrightarrow \psi$, is an equivalence relation over the set of formulas: it is the *proven equivalence*, which should not be confused with $\Longleftrightarrow$, which is not a relation but a functional symbol just like the other connectives. This equivalence relation is compatible with the formation of formulas. For example, it can easily be shown that:

$$\text{If } \begin{cases} \Gamma \vdash \varphi_1 \Longleftrightarrow \psi_1 \\ \Gamma \vdash \varphi_2 \Longleftrightarrow \psi_2 \end{cases} \text{ then } \Gamma \vdash (\varphi_1 \wedge \varphi_2) \Longleftrightarrow (\psi_1 \wedge \psi_2).$$

There are five other analogous properties (for $\vee$, $\Rightarrow$, $\neg$, $\forall x$ and $\exists x$) which allow the simplification of derivations by replacing, in each formula, a subformula φ by ψ where $\vdash \varphi \Longleftrightarrow \psi$. More generally, if φ is a tautology (i.e, $\vdash \varphi$), then φ can be introduced as an assumption and then immediately discharged, since by the definition of $\vdash$, there exists a derivation of φ whose assumptions are discharged. Hence, a tautology can be used in a manner analogous to an axiom.

The connectives satisfy a number of 'algebraic' properties which can be written using proven equivalence. For example, conjunction is idempotent *modulo* equivalence: $\varphi \wedge \varphi$ and φ are distinct formulas, but provably equivalent. In logic, only the class of a formula is important for the equivalence. This will not be true in logic *programming*, where, from an algorithmic point of view, it can take twice as long to prove $\varphi \wedge \varphi$ than to prove φ.

Conjunction and disjunction are 'associative', 'commutative', 'idempotent' and mutually 'distributive':

$$\begin{gathered} \vdash_M \varphi \wedge \psi \Longleftrightarrow \psi \wedge \varphi, \\ \vdash_M (\varphi \wedge \chi) \wedge \psi \Longleftrightarrow \varphi \wedge (\chi \wedge \psi), \\ \vdash_M (\varphi \wedge \chi) \vee \psi \Longleftrightarrow (\varphi \vee \psi) \wedge (\chi \vee \psi), \\ \vdash_M \varphi \wedge \varphi \Longleftrightarrow \varphi, \end{gathered}$$

and the relations which are yielded by exchanging $\wedge$ and $\vee$.

Example These properties allow the simplification of formulas:

$$\vdash (\varphi \wedge (\psi \wedge \varphi)) \wedge (\chi \wedge \psi) \Longleftrightarrow (\varphi \wedge \chi \wedge \psi)$$

using the commutativity, associativity and idempotence of $\wedge$. □

Derived rules The rules of natural deduction are elementary; they form a 'machine language' for logic. From a practical point of view, other (higher-level) rules can be introduced, allowing much shorter derivations. However, these rules cannot be added in an arbitrary manner, for the consistency of the system can be destroyed. It is best to add a rule *derived* from the initial inference system, i.e., whose conclusion allows a derivation under the assumptions formed by the premises in an inference system.

The use of a derived rule can then be replaced *in situ* by a derivation from the initial system.

Example The transitivity rule of implication

$$(\Rightarrow\Rightarrow) : \frac{\varphi \Rightarrow \chi \quad \chi \Rightarrow \psi}{\varphi \Rightarrow \psi}$$

is a rule derived from the derivation:

$$(I^{[1]}_{\Rightarrow}) : \cfrac{(E_{\Rightarrow}) : \cfrac{\chi \Rightarrow \psi \qquad (E_{\Rightarrow}) : \cfrac{\varphi \Rightarrow \chi \qquad [\varphi]^1}{\chi}}{\psi}}{\varphi \Rightarrow \psi}.$$

□

4.2.4 Abstract syntax of derivations

Derivations have a binding tree structure which resembles that of λ-calculus terms. However, there can be several bindings (0, 1, or more) attached to the same node. To represent a derivation by a λ-term, the set of discharged assumptions must be handled by the application of a single rule: a variable of the λ-calculus is associated with each active assumption and a (bound) variable to each *bunch* of discharged assumptions. The result is a typed λ-calculus whose types are formulas. Each derivation can be written as a term typed by a formula. This notation is based on the following correspondence in Table 4.2.

The typing rules for the typed λ-calculus have already been given. Judgment

$$x_1 : \tau_1, \ldots, x_n : \tau_n \vdash_M M : \tau$$

should be reread as: under assumptions $\tau_1, \ldots, \tau_n$, the term M is a derivation of τ, where each x_i is a unknown derivation of τ_i.

Typed λ-calculus	*Natural deduction*
λ-term	derivation
type	formula
type of term	conclusion of derivation
free variable	active assumption
bound variable	bunch of discharged assumptions
abstraction	⇒-introduction
application	⇒-elimination

Table 4.2 The Curry–Howard correspondence

Examples

- The derivation with one node φ is represented by the variable x^φ.
- The term $x^{\varphi\Rightarrow\psi}y^\varphi$ represents the derivation

$$(E \Rightarrow) : \frac{\varphi \Rightarrow \psi \qquad \varphi}{\psi}.$$

- The term $\lambda x^\varphi \, . \, x$ represents the derivation

$$(I_\Rightarrow) : \frac{[\varphi]}{\varphi \Rightarrow \varphi}.$$

- The term $\lambda f^{\varphi\Rightarrow\varphi}x^\varphi \, . \, f(fx)$ represents the derivation

$$(I^{[2]}_\Rightarrow) : \cfrac{(I^{[1]}_\Rightarrow) : \cfrac{(E_\Rightarrow) : \cfrac{[\varphi \Rightarrow \varphi]^2 \qquad (E_\Rightarrow) : \cfrac{[\varphi \Rightarrow \varphi]^2 \qquad [\varphi]^1}{\varphi}}{\varphi}}{\varphi \Rightarrow \varphi}}{(\varphi \Rightarrow \varphi) \Rightarrow (\varphi \Rightarrow \varphi)}.$$

- The term $\lambda x^{\varphi\Rightarrow(\chi\Rightarrow\psi)}y^{\varphi\Rightarrow\chi}z^\varphi \, . \, xz(yz)$ represents the derivation

$$(I^{[1]}_\Rightarrow) : \cfrac{(I^{[2]}_\Rightarrow) : \cfrac{(I^{[3]}_\Rightarrow) : \cfrac{(E_\Rightarrow) : \cfrac{(E_\Rightarrow) : \cfrac{[\varphi \Rightarrow (\chi \Rightarrow \psi)]^1 \qquad [\varphi]^3}{\chi \Rightarrow \psi} \qquad (E_\Rightarrow) : \cfrac{[\varphi \Rightarrow \chi]^2 \qquad [\varphi]^3}{\chi}}{\psi}}{\varphi \Rightarrow \psi}}{(\varphi \Rightarrow \chi) \Rightarrow (\varphi \Rightarrow \psi)}}{(\varphi \Rightarrow (\chi \Rightarrow \psi)) \Rightarrow ((\varphi \Rightarrow \chi) \Rightarrow (\varphi \Rightarrow \psi))}.$$

4.2.5 Functional interpretation

Only a syntactic justification was given for the term/derivation correspondence. But it is not just a notation system: the correspondence is also semantic.

The basic idea is that a formula's meaning is the set of its proofs, where the proofs are the objects of an effective calculus. This idea, called BHK[1] interpretation, forms the basis for the constructive approach to logic. By refusing the static notion of truthfulness (*being* true, *being* false), intuitionists looked for a constructive, dynamic meaning for the concept of provable formula: what must one *do* to prove a formula? The interest of such an interpretation for programming is manifest in the following double slogan:

$$\begin{array}{rcccl} \textit{proof} & = & \textit{term} & = & \textit{program} \\ \textit{proposition} & = & \textit{type} & = & \textit{specification} \end{array}$$

In its syntactic form, this idea leads to a typed λ-calculus whose types are the formulas and the terms of type φ are the derivations of φ. The logical connectives become type constructors with very standard meanings, corresponding to well-known programming combinators. It is the Curry–Howard correspondence, used in combinatory logic by H. Curry, and made explicit in 1969 by W. Howard [66], which supports several formalisms which link computer science and logic.

Implication is the functional type constructor $\rightarrow$, conjunction acts as the product $\times$ and disjunction acts as the union (or sum) of types.

A proof of $\varphi \Rightarrow \psi$ is a construction associating a proof of ψ with each proof of φ. The introduction of $\Rightarrow$ is the abstraction $\lambda x^{\varphi}\, d$ of a derivation d of ψ over a derivation variable of φ. The elimination of $\Rightarrow$ is the application of a derivation of $\varphi \Rightarrow \psi$ to a derivation of φ to obtain a derivation of ψ. These are the basic operations of the typed λ-calculus. More precisely, the variable x^{φ} bound in $\lambda x^{\varphi}\, d$ corresponds exactly to the bunch of assumptions discharged by applying $I_{\Rightarrow}$. In fact, the result is exactly the above, proposed, simple notation system for derivations.

A derivation of $\varphi \wedge \psi$ is a pair (d, d') formed by a derivation d of φ and a derivation d' of ψ, where the introduction of $\wedge$ meaning pair formation; the eliminations are the projections π^1 and π^2. Conjunction then behaves as the *product* of types.

Disjunction corresponds to a *union* (or sum) of types. A derivation of $\varphi \vee \psi$ is either a derivation of φ or a derivation of ψ. Each introduction is an injection, written ι^1 or ι^2, in a sum. The elimination of $\vee$ is called **case** in programming. If d is a derivation of $\varphi \vee \psi$, d' (respectively d'') a derivation of χ under the assumption φ (respectively ψ), then $\mathbf{case}(d, [x^{\varphi}].d', [y^{\psi}].d'')$ is a derivation of χ. Variables x^{φ} and y^{ψ} are, respectively, bound in $[x^{\varphi}].d'$ and $[y^{\psi}].d''$. This correspondence can be seen in Table 4.3.

Examples

- $\lambda x^{\varphi \wedge \psi} . (\pi^2(x), \pi^1(x))$ is a derivation of $\varphi \wedge \psi \Rightarrow \psi \wedge \varphi$.
- $\lambda z^{\varphi \vee \varphi} . \mathbf{case}(z, [x^{\varphi}].x, [x^{\varphi}].x)$ is a derivation of $(\varphi \vee \varphi) \Rightarrow \varphi$.

[1] for Brouwer, Heyting and Kolmogorov.

Typed λ-calculus	*Natural deduction*
pair	conjunction
pair formation	∧-introduction
projections	∧-eliminations
union	disjunction
injections	∨-introductions
case	∨-elimination

Table 4.3 The Curry–Howard correspondence: the connectives

In the λ-calculus, a redex corresponds to the combined use of an introduction and an elimination. Redex contraction defines reductions over derivations, for which there are strong normalization and confluence results, due to D. Prawitz [125].

For example, the derivation

$$(E_\Rightarrow): \dfrac{(I_\Rightarrow): \dfrac{\begin{matrix}[\varphi] \\ \vdots \\ \psi\end{matrix}}{\varphi \Rightarrow \psi} \qquad \begin{matrix}\vdots \\ \varphi\end{matrix}}{\psi} \qquad \text{is contracted to} \qquad \begin{matrix}\vdots \\ \varphi \\ \vdots \\ \psi\end{matrix}$$

Example

$$(E_\Rightarrow): \dfrac{(I_\Rightarrow): \dfrac{(E_\wedge): \dfrac{\varphi \wedge \psi}{\varphi}}{\varphi \wedge \psi \Rightarrow \varphi} \qquad \varphi \wedge \psi}{\varphi} \quad \to \quad (E_\wedge): \dfrac{\varphi \wedge \psi}{\varphi}\ .$$

□

There are contraction rules for the other connectives ∧ and ∨, corresponding to the combination of introduction and elimination. The weak normalization theorem of the typed λ-calculus guarantees the existence of normal proofs in natural deduction (at least for the fragment corresponding to the type system used). Prawitz obtained not only a strong normalization theorem, but also a description of the form of normal derivations. Some remarkable properties of logic follow from this description. For example, the disjunctive property: if $\vdash_M \varphi \vee \psi$, then $\vdash_M \varphi$ or $\vdash_M \psi$. In fact, consider a derivation of $\varphi \vee \psi$ in normal form. The last rule used is ∨-introduction and its premise is either φ or ψ.

The Curry–Howard correspondence was presented in the framework of minimal propositional logic. It can easily be extended to intuitionistic logic, but does not hold for classical logic. Going to first-order logic consists of adding quantified types, which is not difficult, but is not very interesting for programming languages. However, second-order quantification (over type variables) is a way of introducing generic types in languages, more general than those of ML: this is what is done in J.-Y. Girard's system F (1972 [51]), which, although strongly normalizable, has great expressive power, allowing recursive definitions of functions and of data structures. For example, the type for integers is defined by $\forall p \,.\, (p \to p) \to (p \to p)$, where p is a type variable.

These ideas have led to the design of completely new kinds of programming languages, These languages, completely integrated with logic, allow the simultaneous specification, synthesis and verification of programs. P. Martin-Löf's *Intuitionistic Type Theory* [103] is both a programming language and a formal system for constructive mathematics, where constructive proofs *are* the algorithms. J.-L. Krivine's *Second Order Functional Arithmetic* [90] and T. Coquand and G. Huet's *Calculus of Constructions* [25] are two other recent formalisms based on these ideas.

4.2.6 The intuitionistic system

Deductions in minimal logic contain only implications, conjunctions and disjunctions. Negation will be introduced by giving meaning to the constant connective $\bot$ representing 'false'. There is no introduction rule for $\bot$, only an elimination rule:

$$(E_\bot) : \frac{\bot}{\psi}$$

which means that from nothing, one can infer anything—in Latin, *ex falso quodlibet sequitur*. The Curry–Howard correspondence interprets $\bot$ as the 'empty type', so $\bot$ needs no proof. For each ψ, rule $E_\bot$ defines the unique 'function' from $\bot$ to ψ.

The inference system obtained, NJ, by adding this rule to those of minimal logic is the one of propositional intuitionistic logic. Write $\vdash_I$ for the deduction relation defined by this system, as well as the one obtained by adding the rules for quantifiers.

Negation A new connective is introduced: *negation* ($\neg$, sometimes written $\sim$). $\neg\varphi$ is an abbreviation of $\varphi \Rightarrow \bot$.

It is surprising that many tautologies containing negation can be derived in minimal logic (without using the rule $E_\bot$).

Example One of the *contraposition* laws is $(\varphi \Rightarrow \psi) \Rightarrow (\neg\psi \Rightarrow \neg\varphi)$. It is nothing but a particular case, with $\omega = \bot$, of $(\varphi \Rightarrow \psi) \Rightarrow ((\psi \Rightarrow \omega) \Rightarrow (\varphi \Rightarrow \omega))$, which is a minimal tautology. □

Other theorems must use $E_\bot$, for example $(\neg\varphi \vee \psi) \Rightarrow (\varphi \Rightarrow \psi)$, whose converse will be proven in classical logic. To reason directly with negation, it will be more convenient to use appropriate inference rules, which are written without $\bot$:

$$(I_\neg) : \frac{\begin{array}{cc}[\varphi] & [\varphi] \\ \neg\psi & \psi\end{array}}{\neg\varphi}, \qquad (E_\neg) : \frac{\neg\varphi \quad \varphi}{\psi}.$$

As negation is a particular case of implication, $I_\neg$ and $E_\neg$ are obtained from $I_\Rightarrow$ and $E_\Rightarrow$. The elimination of $\Rightarrow$, applied to φ and $\varphi \Rightarrow \bot$, allows the introduction of $\bot$. Hence $E_\neg$ follows from $E_\Rightarrow$ and $E_\bot$, and $I_\neg$ is a particular case of $I_\Rightarrow$:

$$(I_\Rightarrow) : \frac{\begin{array}{c}[\varphi] \\ \bot\end{array}}{\varphi \Rightarrow \bot} \qquad (E_\bot) : \frac{(E_\Rightarrow) : \dfrac{\varphi \Rightarrow \bot \quad \varphi}{\bot}}{\psi}.$$

Note that only $E_\neg$ uses $E_\bot$, which means that $I_\neg$ is a rule derivable in minimal logic.

4.2.7 Quantifiers and the logic variable

The NM and NJ systems are extended to the predicate calculus by giving rules to reason over quantifiers:

$$(I_\forall) : \frac{\varphi(a)}{\forall x \varphi(x)} \qquad (E_\forall) : \frac{\forall x \varphi}{\varphi[x := t]}$$

$$(I_\exists) : \frac{\varphi[x := t]}{\exists x \varphi} \qquad (E_\exists) : \frac{\exists x \varphi \quad \begin{array}{c}[\varphi(a)] \\ \chi\end{array}}{\chi}$$

where φ, χ, $\psi \in \mathrm{L}[X]$, x, $a \in X$ and $t \in \mathrm{T}[X]$. It is supposed that the bound variables have been renamed outside the name space of the free variables and of the substituted terms (a and t), and that x is free in $\varphi(x)$.

By transforming these inference rules into derivation construction rules, certain restrictions must be placed on the presence of free variables in assumptions.

($DI_\forall$) If d is a derivation of $\varphi(a)$, and if all the assumptions of d in which a is free are discharged, then

$$(I_\forall) : \frac{\begin{array}{c}d \\ \varphi(a)\end{array}}{\forall x \varphi(x)}$$

is a derivation of $\forall x \varphi(x)$.

($DE_\exists$) If d is a derivation of $\exists x \varphi(x)$ and d' a derivation of ψ, and if all the assumptions of d and those of d', other than $\varphi(a)$, in which a is free, are discharged, then

$$(E_\exists): \frac{\begin{matrix} d & & d' \\ \exists x\varphi(x) & & \psi \end{matrix}}{\psi},$$

with an arbitrary number of links between the root and the leaves of d' labeled by $\varphi(a)$ and not already linked, is a derivation of ψ.

The rules $(DI_\exists)$ and $(DE_\forall)$ are elementary: no assumption discharge, no restriction. If the restrictions over $(DI_\forall)$ and $(DE_\exists)$ were not applied, the following unexpected derivations would be obtained:

$$(E_\exists): \frac{\exists x(x=0) \qquad (E_\neg): \dfrac{[a=0] \qquad \neg(a=0)}{\psi}}{\psi}$$

(for arbitrary ψ), and

$$\frac{a=0}{\forall x(x=0)}.$$

Here, on the other hand, is a good derivation. Note that at the moment when the elimination of the $\exists$ is applied, x is bound in the active assumptions $\forall x(\neg\varphi)$ and $\exists x\varphi$, other than φ.

$$(I^1_\neg 2): \frac{(E^1_\exists): \dfrac{[\exists x\varphi]^2 \qquad (E_\neg): \dfrac{[\varphi]^1 \qquad (E_\forall): \dfrac{\forall x\neg\varphi}{\neg\varphi}}{\perp}}{\perp}}{\neg(\exists x\varphi)}$$

By discharging the last hypothesis by $I \Rightarrow$, $\vdash_I (\forall x\neg\varphi) \Rightarrow \neg(\exists x\varphi)$ would be obtained.

Quantifiers also satisfy a form of commutativity and distributivity modulo proven equivalence:

$$\vdash_M \forall x\forall y\varphi \iff \forall y\forall x\varphi$$
$$\vdash_M \forall x\varphi \iff \varphi, \qquad \text{if } x \notin \text{var}(\varphi)$$
$$\vdash_M \forall x(\varphi \wedge \psi) \iff \forall x\varphi \wedge \forall x\psi$$
$$\vdash_M \forall x(\varphi(x) \vee \psi) \iff \forall x\varphi(x) \vee \psi, \qquad \text{if } x \notin \text{var}(\psi)$$

Replacing $\forall$, $\wedge$ and $\vee$ by $\exists$, $\vee$ and $\wedge$ would give more proven equivalences.

The logic variable In any comparison between functional and logic programming languages, attention must be paid to the nature of the variables.

A functional variable is unidirectional: passing an argument to a function always occurs in the same direction, and the process is formalized by β-reduction. A logic variable is bidirectional: it can also receive a value from the invocation of a predicate in which it appears.

This phenomenon must necessarily be explained by the inference rules for quantifiers: they are the only ones with variables. Consider the combination of an elimination and of an introduction.

Let $\varphi \in \mathrm{L}_\Sigma[x_1, \ldots x_m]$ and $t_1, \ldots, t_m \in \mathrm{T}_\Sigma[y_1, \ldots, y_n]$:

$$(I_{\vec{\forall}}) : \frac{(E_{\vec{\forall}}) : \dfrac{\forall \vec{x} \varphi}{\varphi[x_1 := t_1, \ldots, x_n := t_m]}}{\forall \vec{y}(\varphi[x_1 := t_1, \ldots, x_n := t_m])},$$

where, first, the m eliminations of $\forall x_i$, $1 \leqslant i \leqslant m$, are applied, then the n introductions of $\forall y_i$, $1 \leqslant i \leqslant n$. The latter are allowed since the y_i are not free in the assumption, which is closed.

A derived rule is thus obtained, φ being an arbitrary formula and σ a substitution:

$$(\forall\forall) : \frac{\forall(\varphi)}{\forall(\sigma\varphi)}.$$

Note that the inference of $\sigma\varphi$ from φ is not normally allowed. Quantifiers here serve as 'modal' operators in the context from which one can apply a substitution to a formula, yielding an instance.

In an instantiation, the (bound) variables of φ become more defined under the subsumption preorder over terms. Such an inference gives them a value, taking the environment from $\begin{bmatrix} x \\ x \end{bmatrix}$ to $\begin{bmatrix} x \\ \sigma(x) \end{bmatrix}$: the formula is therefore weakened as an environment is created. The typical problem in proofs is: 'prove $\forall(\varphi)$'. But one can be interested by a more general problem: 'find a σ such that $\forall(\sigma\varphi)$ is derivable'. In the case where $\forall(\varphi)$ is derivable, all substitutions σ are solutions, according to the rule $(\forall\forall)$, but the most interesting, because it is the most general, is the identity substitution. In the general case where the identity is not a solution, the most general substitutions, i.e., minimal under a particular preorder, are sought. For example, if the 'goal' is the formula $\forall xy\, R(x, y)$, solutions such as substitutions $\begin{bmatrix} x \\ 0 \end{bmatrix}$ and $\begin{bmatrix} y \\ 1 \end{bmatrix}$ could be obtained, with derivations of $\forall y\, R(0, y)$ and $\forall x\, R(x, 1)$. This illustrates the behavior of the 'logic variable': it can be instantiated during a derivation, and several solutions are possible.

4.2.8 The classical system

Adding another inference method, the excluded middle, to intuitionistic logic yields classical logic. The excluded middle, which states that a property is either true or false, can be defined in several ways: by a nullary inference rule explicitly stating the principle, by a unary inference rule for proof by contradiction or by a unary inference rule stating the involutivity of negation.

$$(EM): \frac{}{\varphi \vee \neg\varphi}, \qquad (C): \frac{\begin{array}{c}[\neg\varphi]\\ \bot\end{array}}{\varphi}, \qquad (\neg\neg): \frac{\neg\neg\varphi}{\varphi}.$$

These rules are equivalent, i.e., derivable from each other. For example, the excluded middle is derivable in NJ extended by the rule (C) of reasoning with contradiction.

$$(C^{[2]}): \dfrac{(E_\neg): \dfrac{[\neg(\varphi \vee \neg\varphi)]^2 \qquad (I_\vee): \dfrac{(C^{[1]}): \dfrac{(E_\neg): \dfrac{[\neg(\varphi \vee \neg\varphi)]^2 \qquad (I_\vee): \dfrac{[\neg\varphi]^1}{\varphi \vee \neg\varphi}}{\bot}}{\varphi}}{\varphi \vee \neg\varphi}}{\bot}}{\varphi \vee \neg\varphi}.$$

The NK system of natural deduction in classical logic is given by Table 4.4. Write $\vdash_C$ or simply $\vdash$ for the deduction relation defined by this system.

Proofs using the excluded middle are often more difficult than proofs in minimal logic. The role of negation will probably be clearer in the sequent calculus. Note, however, that there are formulas derivable in NK and not in NJ which do not contain the negation symbol. One might even be tempted to try to prove them in minimal logic. However, their derivation in NK necessarily uses rule (C), and so introduces negation (see exercise 16). In the sequent calculus, it will not be necessary to introduce negation—but sequents must be derived, not formulas.

The connectives, like the quantifiers, are no longer independent. Morgan's axioms state the relationship between $\vee$, $\wedge$ and $\neg$:

$$\vdash_C \varphi \vee \psi \Longleftrightarrow \neg(\neg\varphi \vee \neg\psi),$$
$$\vdash_C \varphi \wedge \psi \Longleftrightarrow \neg(\neg\varphi \wedge \neg\psi).$$

Implication is definable from $\vee$ and $\neg$:

$$\vdash_C (\varphi \Rightarrow \psi) \Longleftrightarrow (\neg\varphi \vee \psi).$$

The two quantifiers are exchanged by negation:

$$\vdash_C \forall x\varphi \Longleftrightarrow \neg\exists x(\neg\varphi).$$

Proposition 4.1 For any formula φ, there exists φ' formed of logical symbols $\vee$, $\neg$ and $\exists$ such that $\vdash_C \varphi \Longleftrightarrow \varphi'$.

$$(I_\wedge): \frac{\varphi \quad \psi}{\varphi \wedge \psi} \qquad (E^1_\wedge): \frac{\varphi \wedge \psi}{\varphi} \qquad (E^2_\wedge): \frac{\varphi \wedge \psi}{\psi}$$

$$(I^1_\vee): \frac{\varphi}{\varphi \vee \psi} \qquad (I^2_\vee): \frac{\psi}{\varphi \vee \psi} \qquad (E_\vee): \frac{\varphi \vee \psi \quad \begin{matrix}[\varphi]\\ \chi\end{matrix} \quad \begin{matrix}[\psi]\\ \chi\end{matrix}}{\chi}$$

$$(I_\Rightarrow): \frac{\begin{matrix}[\varphi]\\ \psi\end{matrix}}{\varphi \Rightarrow \psi} \qquad (E_\Rightarrow): \frac{\varphi \Rightarrow \psi \quad \varphi}{\psi}$$

$$(I_\forall): \frac{\varphi(a)}{\forall x \varphi(x)} \qquad (E_\forall): \frac{\forall x \varphi}{\varphi[x := t]}$$

$$(I_\exists): \frac{\varphi[x := t]}{\exists x \varphi} \qquad (E_\exists): \frac{\exists x \varphi \quad \begin{matrix}[\varphi(a)]\\ \chi\end{matrix}}{\chi}$$

$$(I_\neg): \frac{\begin{matrix}[\varphi]\\ \psi\end{matrix} \quad \begin{matrix}[\varphi]\\ \neg\psi\end{matrix}}{\neg\varphi} \qquad (E_\neg): \frac{\neg\varphi \quad \varphi}{\psi}$$

$$(C): \frac{\begin{matrix}[\neg\varphi]\\ \psi\end{matrix} \quad \begin{matrix}[\neg\varphi]\\ \neg\psi\end{matrix}}{\varphi}$$

Table 4.4 The NK system

Proof By induction over formulas.

If φ is atomic, let $\varphi' = \varphi$.

If $\varphi = \varphi_1 \vee \varphi_2$, by the inductive hypothesis $\varphi_i \Longleftrightarrow \varphi'_i$, $i = 1, 2$, let $\varphi' = \varphi'_1 \vee \varphi'_2$.

If $\varphi = \varphi_1 \wedge \varphi_2$, by the inductive hypothesis $\varphi_i \Longleftrightarrow \varphi'_i$; since $\vdash \varphi \Longleftrightarrow \neg(\neg\varphi_1 \vee \neg\varphi_2)$, let $\varphi' = \neg(\neg\varphi'_1 \vee \neg\varphi'_2)$.

If $\varphi = \varphi_1 \Rightarrow \varphi_2$, let $\varphi' = \neg\varphi'_1 \vee \varphi'_2$.

Finally, if $\varphi = \exists x \varphi_1$, let $\varphi' = \exists x \varphi'_1$, and if $\varphi = \forall x \varphi_1$, let $\varphi' = \neg\exists x(\neg\varphi'_1)$. □

This reduction to three logical symbols will be useful in proofs by induction over formulas, since it is easier to treat three cases than six. Other symbol 'bases' can be useful, such as the dual basis of the former, consisting of $\wedge$, $\neg$ and $\forall$.

Classical logic, intuitionistic logic and computer science Adding rule (C) to NJ allows the proof of many more tautologies. However, classical logic is not more 'powerful' than intuitionistic logic: it is simply less strict, having lost the constructive character visible through the Curry–Howard correspondence. Gödel discovered a translation explaining the relations between NK and NJ: if φ is a formula, an intuitionistic formula φ^g classically equivalent to φ can be constructed, intuitionistically weaker yet derivable in NJ if φ is in NK:

$$\begin{array}{rcll} \bot^g & = & \bot, & \\ \varphi^g & = & \neg\neg, \varphi & \text{if } \varphi \text{ is atomic, and } \neq \bot, \\ (\varphi \wedge \psi)^g & = & \varphi^g \wedge \psi^g, & \\ (\varphi \vee \psi)^g & = & \neg(\neg\varphi^g \wedge \neg\psi^g), & \\ (\varphi \Rightarrow \psi)^g & = & \varphi^g \Rightarrow \psi^g, & \\ (\forall x\varphi)^g & = & \forall x\varphi^g, & \\ (\exists x\varphi)^g & = & \neg\forall x\neg\varphi^g. & \end{array}$$

Example

$$\begin{array}{rcl} \varphi & \equiv & \forall x R(x) \vee \exists x \neg R(x), \\ \varphi^g & \equiv & \neg(\neg\forall x\neg\neg R(x) \wedge \neg\neg\forall x\neg\neg\neg\neg R(x)). \end{array}$$

□

Note that φ^g has no disjunctions of existential quantifiers, and that every atom appears negated at least once. This transformation allows the elimination of the two logical symbols whose intuitionistic and classical behaviors are very different.

Lemma 4.2 If φ is a formula constructed using $\Rightarrow$, $\wedge$ and $\forall$ from negations of atoms, then $\vdash_M \varphi \Longleftrightarrow \neg\neg\varphi$.

Proof By induction on the formation of this class of formulas. There are four cases: $\neg A$, $\psi \Rightarrow \chi$, $\psi \wedge \chi$ and $\forall x\psi$.

Consider the case $\varphi \equiv \psi \wedge \chi$. Then $\vdash_M \neg\neg(\psi \wedge \chi) \Rightarrow \neg\neg\psi \wedge \neg\neg\chi$. By the inductive hypothesis, $\vdash_M \neg\neg\psi \Rightarrow \psi$ and $\vdash_M \neg\neg\chi \Rightarrow \chi$, hence $\vdash_M \neg\neg(\psi \wedge \chi) \Rightarrow (\psi \wedge \chi)$.

The other cases use the following minimal tautologies:

$$\vdash_M \neg\neg\neg\psi \Longleftrightarrow \neg\psi$$
$$\vdash_M \neg\neg(\psi \Rightarrow \chi) \Rightarrow (\psi \Rightarrow \neg\neg\chi)$$
$$\vdash_M \neg\neg\forall x\psi \Rightarrow \forall x\neg\neg\psi$$

The proof of each of these tautologies is left as an exercise. □

Theorem 4.3

1. $\vdash_C \varphi \Longleftrightarrow \varphi^g$;
2. $\Gamma \vdash_C \varphi$ if and only if $\Gamma^g \vdash_M \varphi^g$.

Proof It is easy to prove (1). For (2), suppose $\Gamma^g \vdash_M \varphi^g$. Then $\vdash_M (\Gamma \Rightarrow \varphi)^g$, and so $\vdash_C (\Gamma \Rightarrow \varphi)^g$. By (1), $\vdash_C (\Gamma \Rightarrow \varphi)$, i.e, $\Gamma \vdash_C \varphi$.

The converse is a proof by induction over derivations of φ in NK. The cases where φ is derived by introduction or elimination of $\Rightarrow$, $\wedge$ or $\forall$ are immediate.

Consider the contradiction rule case. Let d be a derivation of $\bot$ from which φ is derived by discharging the assumption $\neg\varphi$ of d. By the inductive hypothesis, there exists a derivation d^g of $\bot$ under the assumption $\neg\varphi^g$. Furthermore, $\neg\neg\varphi^g \Rightarrow \varphi^g$ is a minimal tautology (the previous lemma). So, the minimal derivation is

$$(E_{\Rightarrow}) : \cfrac{(I_{\Rightarrow}^{[1]}) : \cfrac{\begin{array}{c}[\neg\varphi^g]^1 \\ d^g \\ \bot\end{array}}{\neg\neg\varphi^g} \qquad \neg\neg\varphi^g \Rightarrow \varphi^g}{\varphi^g}.$$

The cases of the logical symbols $\vee$ and $\exists$ do not have to be considered, since in classical logic, they can be defined in terms of $\Rightarrow$, $\wedge$, $\bot$ and $\forall$. □

It is a good idea to be able to recognize those formulas derivable in intuitionistic logic and those which are only classical. The functional interpretation given by minimal logic is useful. Consider the intuitionistic deduction

$$(\varphi \Rightarrow \psi) \vee (\varphi \Rightarrow \chi) \vdash_I \varphi \Rightarrow (\psi \vee \chi).$$

To the left of the $\vdash_I$, there is either a function which generates a proof of ψ from a proof of φ, or a function which generates a proof of χ from a proof of φ. One can then create a function which generates a proof of ψ or a proof of χ from a proof of φ. The inverse deduction, true in NK, is false in NJ, and even incomprehensible from a constructive point of view: how, starting from a function f of A into $B \cup C$, does one define a function $g : A \to B$ or a function $h : A \to C$? One would have to choose an element $b \in B$ to define $g(a) = b$ when $f(a) \notin B$, a choice which is clearly not constructive.

From the classical point of view, limiting the number of connectives does not reduce the expressivity. However, logic is not just used to state properties. The connectives all have an operational aspect, clearly visible in NJ, which is lost in Gödel's translation. From the constructive point of view of computer science, the meaning of $\exists x A(x)$, the existence of an element, is stronger than $\neg(\forall x \neg A(x))$, the non-existence of a derivation.

Intuitionistic logic is today considered to be the logic corresponding to sequential computation. It is not coincidental that logic programming (pure Prolog), normally defined using classical logic, fits perfectly in the framework of intuitionistic logic.

4.2.9 The sequent calculus

The NK system is not as well adapted to classical logic as the NJ system is to intuitionistic logic. The I/E symmetry is not respected by the contradiction proof rule. Furthermore, the classical algebraic properties of connectives and quantifiers are not evident: the $\wedge/\vee$ and $\forall/\exists$ dualities and the involutivity of $\neg$. The excluded middle gives disjunction an opaque character: $\vdash_C \varphi \vee \psi$ no longer means that φ is provable *or* that ψ is provable.

For classical logic, the most appropriate system is Gentzen's *sequent calculus* LK, which reestablishes the symmetry broken by rule (C), but at the cost of a certain redundancy of proofs and by the use of 'undefined' judgments. Furthermore, the intuitionistic version LJ is naturally adapted to logic programming. These two facts justify a short introduction to the sequent calculus.

Judgments are introduced as *sequents*, which are lists of symbols of the form

$$\varphi_1, \dots, \varphi_n : \psi_1, \dots, \psi_m,$$

stating that from φ_1 and φ_2 and ... and φ_n, one can deduce ψ_1 or ψ_2 or ... or ψ_n. As particular cases,

$$: \psi \qquad \text{and} \qquad \varphi :,$$

respectively, state that ψ can be proven and that φ is refutable.

The intuitionistic case arises when m is 0 or 1: this is the LJ system. Until now, one changed logic by adding or removing rules. Here, the form of judgments is restricted, and the rules are retained. The behavior of Prolog fits naturally in this system: Horn clauses are intuitionistic sequents over atomic formulas.

The rules of the sequent calculus respect a left/right symmetry. There are:

- structural rules,
- logical rules,
- the identity axiom, and
- the cut rule.

Since a sequent is a list—not a set—of symbols, their order and repetitions must be taken into account: this role is fulfilled by the *structural rules* of weakening, contraction and permutation. The permutation rule is generally applied implicitly (see Table 4.5).

$$(L_A) : \frac{\Gamma : \Delta}{\Gamma, \varphi : \Delta} \qquad (R_A) : \frac{\Gamma : \Delta}{\Gamma : \Delta, \varphi}$$

$$(L_C) : \frac{\Gamma, \varphi, \varphi : \Delta}{\Gamma, \varphi : \Delta} \qquad (R_C) : \frac{\Gamma : \varphi, \varphi, \Delta}{\Gamma : \varphi, \Delta}$$

$$(L_P) : \frac{\Gamma, \varphi, \psi, \Gamma' : \Delta}{\Gamma, \psi, \varphi, \Gamma' : \Delta} \qquad (R_P) : \frac{\Gamma : \Delta, \varphi, \psi, \Delta'}{\Gamma : \Delta, \psi, \varphi, \Delta'}$$

Table 4.5 The LK system: structural rules

For each connective or quantifier, there is a left- and a right-introduction rule: these are the *logical rules*. Negation ($\neg$) is used directly, instead of $\perp$. There is a restriction in $L_\exists$ and $R_\forall$: x must not occur free in Γ, Δ (see Table 4.6).

$$(L^1_\wedge) : \frac{\Gamma, \varphi : \Delta}{\Gamma, \varphi \wedge \psi : \Delta} \quad (L^2_\wedge) : \frac{\Gamma, \psi : \Delta}{\Gamma, \varphi \wedge \psi : \Delta} \qquad (R_\wedge) : \frac{\Gamma : \varphi, \Delta \quad \Gamma' : \psi, \Delta'}{\Gamma, \Gamma' : \varphi \wedge \psi, \Delta, \Delta'}$$

$$(L_\vee) : \frac{\Gamma, \varphi : \Delta \quad \Gamma', \psi : \Delta'}{\Gamma, \Gamma', \varphi \vee \psi : \Delta, \Delta'} \qquad (R^1_\vee) : \frac{\Gamma : \varphi, \Delta}{\Gamma : \varphi \vee \psi, \Delta} \quad (R^2_\vee) : \frac{\Gamma : \psi, \Delta}{\Gamma : \varphi \vee \psi, \Delta}$$

$$(L_\Rightarrow) : \frac{\Gamma : \varphi, \Delta \quad \Gamma', \psi : \Delta'}{\Gamma, \Gamma', \varphi \Rightarrow \psi : \Delta, \Delta'} \qquad (R_\Rightarrow) : \frac{\Gamma, \varphi : \psi, \Delta}{\Gamma : \varphi \Rightarrow \psi, \Delta}$$

$$(L_\neg) : \frac{\Gamma : \varphi, \Delta}{\Gamma, \neg\varphi : \Delta} \qquad (R_\neg) : \frac{\Gamma, \varphi : \Delta}{\Gamma : \neg\varphi, \Delta}$$

$$(L_\forall) : \frac{\Gamma, \varphi[x := t] : \Delta}{\Gamma, \forall x \varphi : \Delta} \qquad (R_\forall) : \frac{\Gamma : \varphi, \Delta}{\Gamma : \forall x \varphi, \Delta}$$

$$(L_\exists) : \frac{\Gamma, \varphi : \Delta}{\Gamma, \exists x \varphi : \Delta} \qquad (R_\exists) : \frac{\Gamma : \varphi[x := t], \Delta}{\Gamma : \exists x \varphi, \Delta}$$

Table 4.6 The LK system: logical rules

A zero arity rule (axiom) is necessary: the *identity*

$$\boxed{\varphi : \varphi}$$

corresponds to the derivation with node φ in natural deduction. The cut rule is presented below. The system described up to now is the *cut-free* LK system.

Derivations in LK have a tree structure, whose leaves are applications of the identity rule and whose root is the conclusion. There is no reasoning 'modulo assumptions', as in natural deduction.

Example A derivation of the sequent $: ((\varphi \Rightarrow \psi) \Rightarrow \varphi) \Rightarrow \varphi$, which is Peirce's axiom, a classical tautology not derivable in intuitionistic logic. Note that some of the sequents constructed during the derivation have several formulas to the right and that the right-contraction rule must be used.

$$(R_\Rightarrow) : \cfrac{(R_C) : \cfrac{(L_\Rightarrow) : \cfrac{(R_\Rightarrow) : \cfrac{(R_A) : \cfrac{\varphi : \varphi}{\varphi : \psi, \varphi}}{: \varphi \Rightarrow \psi, \varphi} \qquad \varphi : \varphi}{(\varphi \Rightarrow \psi) \Rightarrow \varphi : \varphi, \varphi}}{(\varphi \Rightarrow \psi) \Rightarrow \varphi : \varphi}}{: ((\varphi \Rightarrow \psi) \Rightarrow \varphi) \Rightarrow \varphi}.$$

□

Treating negation in LK is much simpler than in NK: to negate is to pass to the other side of the ':'.

Example A Morgan axiom of classical logic:

$$(R_\Rightarrow) : \cfrac{(R_\vee) : \cfrac{(L_\neg) : \cfrac{(R_\wedge) : \cfrac{(R_\neg) : \cfrac{\varphi : \varphi}{: \varphi, \neg\varphi} \qquad (R_\neg) : \cfrac{\psi : \psi}{: \psi, \neg\psi}}{: \varphi \wedge \psi, \neg\varphi, \neg\psi}}{\neg(\varphi \wedge \psi) : \neg\varphi, \neg\psi}}{\neg(\varphi \wedge \psi) : \neg\varphi \vee \neg\psi}}{: \neg(\varphi \wedge \psi) \Rightarrow \neg\varphi \vee \neg\psi}.$$

□

Consider the relations between natural deduction and the sequent calculus in the intuitionistic case. For each derivation of sequent $\Gamma : \varphi$ in LJ, there is a derivation of formula φ in NJ under the assumptions Γ. In fact, they are bunches of assumptions which correspond to formulas of Γ. This correspondence is more than a simple translation of LJ derivations into NJ derivations. It must be viewed operationally. A derivation in LJ is a 'program' to construct a derivation in NJ. The right rules correspond exactly to the introduction rules in natural deduction, and work on the conclusion. The left rules correspond to eliminations, but work at the level of assumptions. For example, the left introduction of conjunction

$$(L^1_\wedge) : \frac{\Gamma, \varphi : \chi}{\Gamma, \varphi \wedge \psi : \chi}$$

allows one to construct, from a derivation of χ under the assumptions Γ and φ, the following derivation of χ under the assumptions Γ and $\varphi \wedge \psi$:

$$\begin{array}{ccc} \Gamma & & \dfrac{\varphi \wedge \psi}{\varphi} \\ & \vdots & \\ & \chi & \end{array}$$

Conversely, a derivation in LJ can be described in a non-unique manner by a derivation of LJ constructed using structural and logical rules and the identity axiom—hence without using cut.

The cut-free sequent calculus has important formal properties. A simple glance at the rules will show that each formula appearing as a conclusion also already appears as a premise. To be more precise, the subformulas of a formula φ must be redefined: for connectives, it is a standard concept, but for quantifiers, substitutions must be added.

The set of *Gentzen subformulas* of a formula φ is composed of φ and

- if φ is $\varphi_1 \wedge \varphi_2$, $\varphi_1 \vee \varphi_2$ or $\varphi_1 \Rightarrow \varphi_2$, the subformulas of φ_1 and φ_2,

- if φ is $\neg\varphi_1$, the subformulas of φ_1, and
- if φ is $\forall x\varphi_1$ or $\exists x\varphi_1$ the subformulas of $\varphi[x := t]$ for every term t.

Finally, the subformulas of a sequent are the subformulas of its component formulas. The following proposition is immediate, by simple inspection of the rules.

Proposition 4.4 The cut-free LK system satisfies the subformula property: formulas appearing in a sequent derivation are the subformulas of the sequent.

This proposition has significant consequences in automatic theorem-proving. If one is trying to construct a derivation of a sequent, the search space is restricted to the subformulas of the goal. In the case of the propositional calculus, this space is finite, and the calculus is therefore decidable. With quantifiers, there are an infinite number of subformulas.

The subformula property does not hold for all inference systems. In particular, the *cut* rule, presented here,

$$(\text{cut}) : \frac{\Gamma : \varphi, \Delta \qquad \Gamma', \varphi : \Delta'}{\Gamma, \Gamma' : \Delta, \Delta'}$$

does not respect it, since the formula φ is destroyed by inference: it is not possible to reconstitute φ from the conclusion of a cut. On the other hand, the cut allows much more intelligent proofs than can be done using only logical rules. In natural deduction, it substitutes a derivation for φ for the assumption φ in a derivation. It therefore justifies, at the level of LJ, the interpretation of an assumption as an undetermined derivation, just like the free variable of the λ-calculus. The φ which is cut off appears as a lemma, a general result which is picked up during the proof.

Despite the increased flexibility, the cut rule does not increase the expressivity of LK. Better still, if a derivation uses cuts, it can be effectively transformed into a cut-free derivation. This is Gentzen's famous Hauptsatz (cut-elimination theorem, 1934), a result similar to Prawitz's normalization theorem in natural deduction. Its proof uses the L/R symmetry of the rules and the fact that the only axiom (at the leaves of the derivation tree) is the identity. For an axiom, the cut disappears:

$$(\text{cut}) : \frac{\Gamma : \varphi, \Delta \qquad \varphi : \varphi}{\Gamma : \Delta, \varphi} \qquad \longrightarrow \qquad (R_P) : \frac{\Gamma : \varphi, \Delta}{\Gamma : \Delta, \varphi}.$$

In the other cases, the cut is pushed down to the leaves of the derivation, and the formula in which there is a cut becomes smaller. For example, in the following case, the cut in $\neg\varphi$ is replaced by a cut in φ:

$$(\text{cut}) : \frac{(R_\neg) : \dfrac{\Gamma, \varphi : \Delta}{\Gamma : \neg\varphi, \Delta} \qquad (L_\neg) : \dfrac{\Gamma' : \varphi, \Delta'}{\Gamma', \neg\varphi : \Delta'}}{\Gamma, \Gamma' : \Delta, \Delta'} \longrightarrow \frac{(\text{cut}) : \dfrac{\Gamma' : \varphi, \Delta' \qquad \Gamma, \varphi : \Delta}{\Gamma', \Gamma : \Delta', \Delta}}{\Gamma, \Gamma' : \Delta, \Delta'}.$$

This theorem does not generalize if other axioms are used (for example, in a theory; see below). In that case a proof with cuts is truly more powerful, and the rule must be added to LK, which has major consequences. However, cut-elimination is not a confluent reduction, and therefore cannot claim to represent computation in a manner analogous to the Curry–Howard correspondence.

4.2.10 Hilbert systems

The third method, the first historically, is the Hilbert system. Derivations of formulas are made as in natural deduction, but without discharging deductions. Instead of being λ-terms, the derivations are first-order terms, like the derivations used in Chapter 1, to prove that a term is well formed. These derivations, viewed as terms, are formed over a signature containing rules as n-ary functional symbols, the $n = 0$ case corresponding to axioms.

It was shown in Chapter 2 how to translate λ-terms into applicative terms by choosing $\boldsymbol{S}$ and $\boldsymbol{K}$ as constants. From the Curry–Howard correspondence, one can deduce that a natural-deduction derivation, which has a typed λ-term structure, can be translated into a derivation using only modus ponens and formed from axioms. The derivation-term correspondence is described in Table 4.7.

Combinatory logic	*Hilbert system*
type	formula
applicative term	Hilbert derivation
type of term	conclusion of derivation
application (**App**)	modus ponens ($E_{\Rightarrow}$)
constant symbol($\boldsymbol{S}$, $\boldsymbol{K}$)	axiom name
types of the constants	axioms
variable	assumption name
type of variable	assumption
combinatory basis	axiom system

Table 4.7 The derivation-term correspondence

The types of $\boldsymbol{S}$ and $\boldsymbol{K}$ are well known and give the first axioms of Hilbert systems (they are polymorphic constants: they should really be written $\boldsymbol{S}^{\varphi,\chi,\psi}$, and $\boldsymbol{K}^{\varphi,\psi}$, and then typed combinatory logic would be used):

$$\begin{array}{lcl} \boldsymbol{S} & : & (\varphi \Rightarrow (\chi \Rightarrow \psi)) \Rightarrow ((\varphi \Rightarrow \chi) \Rightarrow (\varphi \Rightarrow \psi)) \\ \boldsymbol{K} & : & \varphi \Rightarrow (\psi \Rightarrow \varphi) \end{array}$$

The Hilbert system for the (minimal) logic of implication is formed from the two axioms $\boldsymbol{S}$ and $\boldsymbol{K}$ and the single rule, modus ponens, or implication elimination, which corresponds to implication.

$$\text{(App)} : \frac{\varphi \Rightarrow \psi \qquad \varphi}{\psi}$$

There is no introduction rule: modus ponens and the axioms can simulate $(I_{\Rightarrow})$ in the same manner in which abstraction in the λ-calculus is simulated using the constants $\boldsymbol{S}$ and $\boldsymbol{K}$ in combinatory logic. Write $\vdash_H$ for the deduction relation defined by this inference system: $\varphi_1, \ldots, \varphi_n \vdash_H \varphi$ if there exists a derivation of φ whose assumptions are included in $\varphi_1, \ldots, \varphi_n$. Such a derivation is understood as a typed applicative term constructed over the constants $\boldsymbol{S}$ and $\boldsymbol{K}$ and the variables whose types are included in $\varphi_1, \ldots, \varphi_n$.

The identity λxx can be defined using the applicative rule $\boldsymbol{SKK}$. It corresponds to the following term, a derivation of $\varphi \Rightarrow \varphi$ in a Hilbert system,

$$\text{(App)}: \frac{\text{(App)}: \dfrac{(\varphi\Rightarrow(\varphi\Rightarrow\varphi)\Rightarrow\varphi)\Rightarrow(\varphi\Rightarrow(\varphi\Rightarrow\varphi))\Rightarrow\varphi\Rightarrow\varphi \qquad \varphi\Rightarrow(\varphi\Rightarrow\varphi)\Rightarrow\varphi}{(\varphi\Rightarrow(\varphi\Rightarrow\varphi))\Rightarrow\varphi\Rightarrow\varphi} \qquad \varphi\Rightarrow\varphi\Rightarrow\varphi}{\varphi\Rightarrow\varphi},$$

which can be written as the applicative term

$$(\boldsymbol{S}^{\varphi,\varphi\Rightarrow\varphi,\varphi}\boldsymbol{K}^{\varphi,\varphi\Rightarrow\varphi})\boldsymbol{K}^{\varphi,\varphi}.$$

As can be seen, simple tautologies can have Hilbert derivations much more complex that their natural deduction counterparts. It is not by adding axioms (for example $\varphi \Rightarrow \varphi$) that one can hope to reduce this complexity, similarly to what was observed when λ-terms were compiled into applicative terms. Gentzen systems yield much more natural proofs than do Hilbert systems, at the cost of more important formal difficulties, such as assumption management in natural deduction, or sequent manipulation. The same problem exists in functional programming where one has the choice of languages based on the λ-calculus, with their concept of free and bound variables, and applicative languages with no variable, where a good number of 'functionals' are necessary (see Backus's FP language).

There is a deduction theorem, but unlike that for natural deduction, it is not immediate: $\varphi \vdash_H \psi$ if and only if $\vdash_H \varphi \Rightarrow \psi$. The problem is to transform a derivation of ψ containing φ as assumption into a closed derivation of $\varphi \Rightarrow \psi$. This is done by simulating abstraction in applicative terms: if M is a derivation of ψ and X is a variable of type φ (possibly appearing in M), then $[x]M$ must be computed. The proof of a deduction theorem therefore consists of the abstraction algorithm presented in Chapter 2.

The deduction theorem means that implication 'internalizes' deduction: $\vdash_H$ is replaced by $\Rightarrow$. This will justify the axioms associated with the other connectives. An axiom is associated with each natural deduction rule, by replacing the bar with implication, and by curryfying the multiple premises for $(I_\wedge)$ and $(E_\vee)$. The result is shown in Table 4.8

$$
\begin{array}{|rcl|}
\hline
\pi^1 & : & \varphi \wedge \psi \Rightarrow \varphi \\
\pi^2 & : & \varphi \wedge \psi \Rightarrow \psi \\
(\,,\,) & : & \varphi \Rightarrow (\psi \Rightarrow (\varphi \wedge \psi)) \\
\imath_1 & : & \varphi \Rightarrow \varphi \vee \psi \\
\imath_2 & : & \psi \Rightarrow \varphi \vee \psi \\
case & : & \varphi \vee \psi \Rightarrow ((\varphi \Rightarrow \chi) \Rightarrow ((\psi \Rightarrow \chi) \Rightarrow \chi)) \\
\bot & : & \bot \Rightarrow \varphi \\
\hline
\end{array}
$$

Table 4.8 A Hilbert system: propositional axioms

The transformation is more delicate for quantifiers. Rules $(I_\exists)$ and $(E_\forall)$ directly yield substitution axioms:

$$
\begin{array}{|lcl|}
\hline
(\sigma_\exists) & : & \varphi[x := t] \Rightarrow \exists x\varphi \\
(\sigma_\forall) & : & \forall x\varphi \Rightarrow \varphi[x := t] \\
\hline
\end{array}
$$

However, because of the restrictions on free variables, one cannot transform the rules $(I_\forall)$ and $(E_\exists)$ into axioms. In fact, an axiom $\varphi(a) \Rightarrow \forall x\varphi$ corresponding to $(I_\forall)$ would systematically introduce the free variable a, which is illegal in the derivation of NK. The Hilbert system therefore uses two generalization rules with the same application restrictions:

$$
\boxed{(Gen_\exists) : \frac{\varphi(a) \Rightarrow \chi}{\exists x\varphi \Rightarrow \chi} \qquad (Gen_\forall) : \frac{\chi \Rightarrow \varphi(a)}{\chi \Rightarrow \forall x\varphi}}
$$

Variable a must not be free in χ.

Finally, for classical logic, the excluded-middle axiom is added:

$$
\boxed{(EM) : \varphi \vee \neg\varphi}
$$

Hilbert systems tend to be favored by mathematicians, who are used to working with axioms, while natural deduction and the sequent calculus, much studied by logicians in proof theory, have deep and promising links with computer science.

4.3 Normal forms

In classical logic, the tautologies which were presented show the syntactic redundancy of the first-order languages: the connectives used are not independent and satisfy certain algebraic relations such as commutativity and distributivity. The formulas can then be put into 'normal form'. It is standard to represent a formula by a tree whose leaves are atoms and whose internal nodes are connectives and quantifiers. The usual normal forms consist of pushing negations towards the leaves and quantifiers towards the root. The $\wedge/\vee$ case requires choosing which distributivity to use.

4.3.1 Prenex formulas

Definition 4.3 A formula of the form $Q_1x_1 \ldots Q_nx_n\varphi^o$, with $n \geqslant 0$, $Q_i \in \{\forall, \exists\}$, $\varphi^o \in \mathrm{L}^o[X]$, i.e, φ^o quantifier-free, and $x_1, \ldots, x_n \in \operatorname{var}\varphi^o$, is called *prenex.*

Write $Q\vec{x}$ instead of $Q_1x_1 \ldots Q_nx_n$ and $Q'\vec{x}$ the prefix obtained by replacing each $\forall$ by $\exists$ and conversely.

Proposition 4.5 For each formula φ, there exists a prenex formula ψ such that $\vdash \varphi \Longleftrightarrow \psi$ and $\operatorname{var}(\varphi) = \operatorname{var}(\psi)$.

The following tautologies are used:

$$\begin{array}{rcll} \vdash Qx\varphi(x) \vee \psi & \Longleftrightarrow & Qx(\varphi(x) \vee \psi), & \text{if } x \notin \operatorname{var}(\psi) \\ \vdash Qx\varphi(x) \wedge \psi & \Longleftrightarrow & Qx(\varphi(x) \wedge \psi), & \text{if } x \notin \operatorname{var}(\psi) \\ \vdash \neg(Qx\varphi) & \Longleftrightarrow & Q'x(\neg\varphi). & \end{array}$$

Proof By induction over formulas. It suffices to distinguish four cases: atoms, $\vee$, $\neg$ and $\exists$.

- If φ is atomic, φ is prenex.
- If φ is $\varphi_1 \vee \varphi_2$, with $\vdash \varphi_1 \Longleftrightarrow \psi_1$ and $\vdash \varphi_2 \Longleftrightarrow \psi_2$, and $\psi_1 = Q\vec{x}\,{\psi_1}^o$ and $\psi_2 = Q\vec{y}\,{\psi_2}^o$ are prenex formulas. Suppose that the bound variables are renamed so that they are distinct and do not intersect with $\operatorname{var}(\varphi)$; then

$$\begin{array}{rl} \vdash \varphi \Longleftrightarrow \psi_1 \vee \psi_2 & \text{(property of } \Longleftrightarrow) \\ \vdash \psi_1 \vee \psi_2 \Longleftrightarrow Q\vec{x}\vec{y}\,(\psi_1^o \vee \psi_2^o) & \text{(tautologies)} \end{array}$$

- If φ is $\neg\chi$ with $\vdash \chi \Longleftrightarrow \psi$ and $\psi = Q\vec{x}\psi^o$, then $\vdash \varphi \Longleftrightarrow Q'\vec{x}(\neg\psi^o)$.
- If φ is $\exists x\chi$ with $\vdash \chi \Longleftrightarrow \psi$ and $\psi = Q\vec{y}\psi^o$, then $\vdash \varphi \Longleftrightarrow \exists xQ\vec{y}\psi^o$.

□

Remark It is easy to transform the proof of the theorem into a procedure computing ψ from φ. However, according to the order of use of the prenex transformations, distinct prenex formulas ψ will be obtained. What is important is that they be equivalent to each other as well as with the initial formula. For example, $\forall x\, \exists y\, \forall z(A \wedge B)$ and $\forall x\, \forall z\, \exists y(A \wedge B)$ are two prenex formulas equivalent to $\forall x(\exists y\, A \wedge \forall z\, B)$, where A and B are atomic, $y \notin \operatorname{var}(B)$ and $z \notin \operatorname{var}(A)$. □

Computing an equivalent prenex formula can be seen as rewriting using the system obtained by replacing the $\Longleftrightarrow$ in the above tautologies by $\longrightarrow$.

4.3.2 Disjunctive and conjunctive formulas

Disjunctive and conjunctive normal forms are useful for the formulas of the propositional calculus, as well as for the quantifier-free formulas of first-order languages.

A *literal* is an atom or the negation of an atom. If l is a literal, define the conjugate literal l' of l by

$$l' = \begin{cases} \neg\chi & \text{if } l \text{ is an atom } \chi, \\ \chi & \text{if } l \text{ is the negation of an atom } \chi. \end{cases}$$

Definition 4.4 Let φ be a formula of the propositional calculus: if

$$\varphi = \bigwedge_{i=1}^{m} \bigvee_{j=1}^{p_i} \varphi_i^j \qquad (\text{respectively } \bigvee_{i=1}^{m} \bigwedge_{j=1}^{p_i} \varphi_i^j \ ),$$

where the φ_i^j are literals and φ is called a *conjunctive* (respectively *disjunctive*) formula.

Proposition 4.6 For each formula φ of the propositional calculus, there exists a conjunctive formula φ^c and a disjunctive formula φ^d such that

$$\vdash \varphi \Longleftrightarrow \varphi^c \qquad \text{and} \qquad \vdash \varphi \Longleftrightarrow \varphi^d.$$

Proof Begin by replacing the connectives $\Rightarrow$ and $\Longleftrightarrow$ appearing in φ with $\neg$, $\vee$ and $\wedge$, then reason by induction over formulas:

- If φ is atomic, $\varphi^c = \varphi^d = \varphi$.
- If $\varphi = \chi \wedge \psi$, let $\varphi^c = \chi^c \wedge \psi^c$; furthermore, if by induction $\vdash \chi \Longleftrightarrow \chi^d = \bigvee_i \chi_i$, and $\vdash \psi \Longleftrightarrow \psi^d = \bigvee_j \psi_j$, let $\varphi^d = \bigvee_{i,j}(\chi_i \wedge \psi_j)$.
- If $\varphi = \chi \vee \psi$, proceed in a dual (in ${}^c/{}^d$) manner.
- If $\varphi = \neg\psi$, by induction $\psi^c = \bigwedge_{i=1}^m \bigvee_{j=1}^{p_i} \psi_i^j$. Let $\varphi^d = \bigvee_{i=1}^m \bigwedge_{j=1}^{p_i} \psi_i^{j'}$; constructing φ^c is analogous, starting from ψ^d.

These transformations are equivalences for $\vdash$. □

Example

$$(A \wedge \neg(B \vee \neg C)) \Rightarrow (\neg B \Rightarrow (A \wedge B)).$$

Eliminate the $\Rightarrow$:

$$\neg(A \wedge \neg(B \vee \neg C)) \vee (\neg\neg B \vee (A \wedge B)).$$

Push the $\neg$ towards the leaves and eliminate the $\neg\neg$:

$$(\neg A \vee (B \vee \neg C)) \vee (B \vee (A \wedge B)).$$

Collect the $\vee$:

$$(\neg A \vee B \vee \neg C \vee B \vee A) \wedge (\neg A \vee B \vee \neg C \vee B \vee B),$$

which is a conjunctive formula equivalent to the initial formula. □

These transformations can also be written using noetherian but non-confluent rewriting systems. The one computing the conjunctive forms is given by Table 4.9. The proof of termination uses an *rpo* simplification order (see Chapter 2).

$\varphi \Leftrightarrow \psi$	$\rightarrow$	$(\varphi \Rightarrow \psi) \wedge (\psi \Rightarrow \varphi)$
$\varphi \Rightarrow \psi$	$\rightarrow$	$\neg\varphi \vee \psi$
$\neg\neg\varphi$	$\rightarrow$	φ
$\neg(\varphi \wedge \psi)$	$\rightarrow$	$\neg\varphi \vee \neg\psi$
$\neg(\varphi \vee \psi)$	$\rightarrow$	$\neg\varphi \wedge \neg\psi$
$\varphi \vee (\psi \wedge \chi)$	$\rightarrow$	$(\varphi \vee \psi) \wedge (\varphi \vee \chi)$
$(\varphi \wedge \psi) \vee \chi$	$\rightarrow$	$(\varphi \vee \chi) \wedge (\psi \vee \chi)$

Table 4.9 Computing conjunctive forms

The normal forms of this system are conjunctive formulas, but a formula can be reduced to several conjunctive formulas.

4.4 Theories

In this section, we will work with first-order languages. Be it in mathematics, programming or artificial intelligence, one is always working inside a *theory* which defines 'common knowledge'. The deduction relation $\vdash$ is that of classical logic.

4.4.1 Definitions and examples

A theory is a set of first-order language formulas $\mathrm{L}[X]$. For the moment, unless one is going to enumerate the formulas, a good way to define a theory is to generate it using the deduction relation $\vdash$.

Definition 4.5 Given a subset Γ of $\mathrm{L}[X]$, whose elements are called *axioms* (of the theory), the *theory* generated by Γ is the set of formulas φ of $\mathrm{L}[X]$, called *theorems*, such that $\Gamma \vdash \varphi$, i.e., closed under $\vdash$.

The predicate calculus
It is the theory without axioms, the set of $\varphi \in \mathrm{L}[X]$ such that $\vdash \varphi$. Its theorems are the *tautologies*, some of which have already been seen.

Equality theories

Suppose that Σ contains a binary relation symbol E (or $=$, if there is no problem of confusion with other uses of $=$, such as for β-conversion). The set Ax_E of equality axioms is:

$$
\begin{array}{ll}
(Ax_E 1) & \forall x\, Exx, \\
(Ax_E 2) & \forall xy\, Exy \Rightarrow Eyx, \\
(Ax_E 3) & \forall xyz\, Exy \wedge Eyz \Rightarrow Exz, \\
(Ax_E f) & \forall x_1 \ldots x_n y_1 \ldots y_n\, Ex_1y_1 \wedge \ldots \wedge Ex_ny_n \Rightarrow Efx_1 \ldots x_n\, fy_1 \ldots y_n, \\
(Ax_E R) & \forall x_1 \ldots x_n y_1 \ldots y_n\, Ex_1y_1 \wedge \ldots \wedge Ex_ny_n \Rightarrow (Rx_1 \ldots x_n \Rightarrow Ry_1 \ldots y_n),
\end{array}
$$

for each functional symbol f and relational symbol R in Σ.

By convention, for any theory in a language containing an equality symbol E, the axioms Ax_E will always be implicit in the statement of the (other) axioms. The signature, the language and the theory are said to be with equality.

Since equality has a very special role compared to the other relational symbols, its properties can be incorporated into the inference system, by changing the axioms of the inference rules. For example, $Ax_E 2$ would be transformed into

$$\frac{t = t'}{t' = t}.$$

Note that this rule would allow an infinite derivation that would exchange terms on either side of an equality.

Group theory

Over the usual signature $\Sigma = \{e, ., ^{-1}, =\}$, there are three axioms,

$$
\begin{gathered}
\forall x\, (x.e = x), \\
\forall x\, (x.\, x^{-1} = e), \text{ and} \\
\forall xyz\, ((x.\, y).\, z = x.\, (y.\, z)).
\end{gathered}
$$

It is of course assumed here that the system includes the five $Ax_=$ axioms, listed above.

Many other algebraic structures, such as rings, fields and lattices, have an axiomatic definition in a first-order language.

Peano arithmetic

This is one of the most important theories studied by logicians. It is the standard definition for arithmetic and is used by mathematicians as well as computer scientists. The signature is $\Sigma = \{0, s, +, ., <, =\}$, where s is a unary functional symbol for the successor function.

The axioms (AP) are:

$$\begin{gathered}\forall x\ \neg(sx = 0),\\ \forall xy\ sx = sy \Rightarrow x = y,\\ \forall x\ x + 0 = x,\\ \forall xy\ x + sy = s(x + y),\\ \forall x\ x.0 = 0,\\ \forall xy\ x.sy = x.y + x,\\ \forall x\ \neg(x < 0),\\ \forall xy\ x < sy \Leftrightarrow x < y \vee x = y,\\ \varphi(0) \wedge \forall x(\varphi(x) \Rightarrow \varphi(sx)) \Rightarrow \forall x \varphi(x).\end{gathered}$$

The last line is the induction axiom scheme: there are a countably infinite number of such axioms for each formula φ and each variable x not bound in φ. It allows proofs by induction.

A weaker theory, i.e., having fewer axioms, than Peano arithmetic is often introduced by eliminating the induction axiom scheme and by adding the axiom $\forall xy(x < y \vee x = y \vee y < x)$ which can no longer be proved from the other axioms (see exercise 27). This theory is called *elementary arithmetic*, having the advantage of being finitely axiomatizable, which is not the case for Peano arithmetic, according to the Ryll–Nardzewski theorem. Elementary arithmetic was introduced by Mostowski, Tarski and R.M. Robinson to prove undecidability properties.

4.4.2 General properties

Consistency is the first property to demand of any theory: a subset Γ of L_Σ (or the theory T which it generates) is consistent if $\Gamma \nvdash \bot$, i.e, if there does not exist $\varphi \in L[X]$ such that $\Gamma \vdash \varphi$ and $\Gamma \vdash \neg\varphi$, or equivalently, by rule $E_\neg$, if $T \neq L[X]$. A theory is inconsistent if it is not consistent.

Inconsistency is directly related to the deduction relation.

Proposition 4.7

1. $\Gamma \vdash \varphi$ if and only if $\Gamma \cup \{\neg\varphi\}$ is inconsistent;
2. $\Gamma \vdash \neg\varphi$ if and only if $\Gamma \cup \{\varphi\}$ is inconsistent.

Proof

1. If $\Gamma \vdash \varphi$, then $\Gamma, \neg\varphi \vdash \varphi$ and $\Gamma, \neg\varphi \vdash \neg\varphi$ as well: $\Gamma \cup \{\neg\varphi\}$ is inconsistent. Conversely, by contradiction:

$$(C) : \frac{\dfrac{\Gamma \quad [\neg\varphi]}{\bot}}{\varphi}.$$

2. Idem, with rule $(I\neg)$ instead of (C).

□

Completeness is an important, strong property, which is not satisfied by all interesting theories. A theory of axioms Γ is *complete* if for every closed formula φ, $\Gamma \vdash \varphi$ or $\Gamma \vdash \neg\varphi$.

Examples Field theory is not complete: one can neither prove $1 + 1 = 0$ nor $1 + 1 \neq 0$. But the theory of algebraically closed fields of characteristic 0 is complete. Peano arithmetic is not complete: see Gödel's incompleteness theorem in Chapter 7. □

Decidability is a fundamental property, related to the concept of computability and at the heart of theoretical problems in automatic theorem-proving: a theory with axioms Γ is *decidable* if 'there exists an algorithm which, given a formula φ, determines if $\Gamma \vdash \varphi$ or $\Gamma \nvdash \varphi$'.

A precise definition of decidability will be given in Chapter 7. A theory which is not decidable is called undecidable.

Examples Arithmetic, group theory and the predicate calculus (with a symbol of arity > 1) are undecidable. Presburger arithmetic, commutative group theory and algebraically closed fields of characteristic 0 are decidable. □

4.4.3 Operations over theories

Theories are objects which can be manipulated. Four operations will be considered: extension, relativization, implementation and skolemization.

Extensions

Generally, a theory T' is an extension of a theory T if $T \subseteq \mathrm{L}_\Sigma[X]$, $T' \subseteq \mathrm{L}_{\Sigma'}[X]$, $\Sigma \subseteq \Sigma'$ and $T \subseteq T'$. There are two ways to extend a theory: add more symbols to the signature, or add more axioms.

An important mechanism is that of definitional extension. The idea is to introduce a new symbol with an axiom defining it. There are two cases, according to the functional or relational nature of the symbol. Let $T \subseteq \mathrm{L}_\Sigma$.

Relational extension Let $\chi \in \mathrm{L}_\Sigma[x_1, \dots, x_n]$. Form $\Sigma' = \Sigma \cup \{R\}$, where R is a new n-ary relational symbol and $T' = T \cup \{\mathrm{def}(R)\}$ where $\mathrm{def}(R)$ is the axiom defining R:

$$\forall x_1 \dots x_n \, (Rx_1 \dots x_n \Longleftrightarrow \chi(x_1, \dots, x_n)).$$

Note that R does not appear in χ: it is an explicit definition, not 'recursive'.

Example Add to the signature of arithmetic a unary symbol '*even*' with the axiom def(*even*):

$$\forall x(even(x) \iff \exists y(x = y + y)).$$

□

Functional extension Let $\chi \in \mathrm{L}_\Sigma[x_1, \ldots, x_n, y]$. To state the functional characteristic of χ, it is supposed that

$$T \vdash \forall x_1 \ldots x_n \, \exists y \, [\chi \wedge \forall y'(\chi[y := y'] \Rightarrow y = y')],$$

which can be abbreviated by

$$T \vdash \forall \vec{x} \, \exists! y \, \chi.$$

The $\exists!$ is read as 'there exists a unique'. In this case, let $\Sigma' = \Sigma \cup \{f\}$, f a new n-ary functional symbol, and $T' = T \cup \{\mathrm{def}(f)\}$, where $\mathrm{def}(f)$ is the axiom defining f:

$$\forall x_1 \ldots x_n y \, (y = f x_1 \ldots x_n \iff \chi(x_1, \ldots, x_n, y)).$$

Example In arithmetic, '*max*' is defined by the axiom:

$$\forall x_1 x_2 y \, (y = max(x_1, x_2) \iff (y = x_1 \vee y = x_2) \wedge x_1 \leqslant y \wedge x_2 \leqslant y).$$

□

In each case, there is a translation $\delta : \mathrm{L}_{\Sigma'}[X] \to \mathrm{L}_\Sigma[X]$. For the relational case, $\delta(\varphi)$ is obtained by replacing $Rt_1 \ldots t_n$ with $\chi(t_1, \ldots, t_n)$ in φ. For the functional case, it is not as simple since variables and existential quantifiers must be added to the t_i. The properties of this translation, with $\varphi \in \mathrm{L}_{\Sigma'}[X]$, are

$$T' \vdash \varphi \iff \delta(\varphi) \quad \text{and} \tag{4.1}$$

$$T' \vdash \varphi \text{ if and only if } T \vdash \delta(\varphi). \tag{4.2}$$

Evidently, $\delta(\varphi) = \varphi$ if $\varphi \in \mathrm{L}_\Sigma[X]$.

A theory $T^\star$ is a *definitional extension* of T if $T^\star$ is obtained from T by a finite number of relational or functional extensions. This type of extension is common place. 'Number theory' is a definitional extension of Peano arithmetic with all the useful functions and relations. Mathematics are, from a truly formalist point of view, a gigantic definitional extension of set theory. In programming, for example in Lisp, new function symbols are added as needed, along with their definitions.

According to the above properties, if $\varphi \in \mathrm{L}_\Sigma$, $T^\star \vdash \varphi$ if and only if $T \vdash \varphi$: new theorems are not added over the initial signature. The theory $T^\star$ is a *conservative* extension of T. This property (weaker than a definitional extension) is important. In particular, if the extension $T^\star$ of T is conservative, then $T^\star$ is consistent if T is. In computer science, one talks rather in this case of 'hierarchical consistency'.

Set theory Even though axiomatic set theory is of little interest to computer science, there are good reasons for familiarity with it. First, from a technical point of view, it illustrates better than any other theory the definitional extension of a theory. Second, it was in this context that the various paradoxes of the naive theory were eliminated and that some of the most famous problems of logic, such as the independence of the axiom of choice, were solved.

The language of Zermelo–Fraenkel (ZF) set theory includes equality and uses a single binary relation symbol, $\in$. All variables are of the same nature and designate sets, unlike those in the naïve theory which distinguishes set and element. However, the atomic formula $x \in y$ can be read as 'x is an element of set y'.

Due to the poverty of the language (comparable to that of the λ-calculus), it is necessary to extend it by adding new symbols, along with their definitions. Even the definition of the axioms will be facilitated by the successive introduction of new symbols. For example, the relational symbol $\subseteq$ is defined by $x \subseteq y \iff \forall u(u \in x \Rightarrow u \in y)$. To introduce a functional symbol always requires a previous proof.

Let $\varphi \in \mathrm{L}_{ZF}[\vec{x}, y]$, where the 'parameters' $\vec{x}$ and the variable y are distinct. The formula

$$\exists! z \forall y(y \in z \iff \varphi(\vec{x}, y)),$$

stating that 'the y such that $\varphi(\vec{x}, y)$ form a set', is abbreviated by $\mathrm{Set}\{y; \varphi(\vec{x}, y)\}$. If $ZF \vdash \mathrm{Set}\{y; \varphi(\vec{x}, y)\}$ and $\vec{x} = x_1 \dots x_n$, introduce an n-ary functional symbol F_φ along with its definition:

$$y \in F_\varphi(\vec{x}) \iff \varphi(\vec{x}, y).$$

The expression $\{y; \varphi(\vec{x}, y)\}$ can be used as a term of the language, instead of $F_\varphi(\vec{x})$. However, writing $\{y; \varphi(\vec{x}, y)\}$ has no meaning so long as $\mathrm{Set}\{y; \varphi(\vec{x}, y)\}$ has not been proven. It is, however, a good way to introduce 'paradoxes' of which the simplest ones are $\{x; x \notin x\}$ and $\{x; x = x\}$. This extension is allowed by the axiom of extensionality which follows.

Extensionality 'Two sets are equal if they have the same elements:'

$$\forall z(z \in x \iff z \in y) \Rightarrow x = y.$$

Union 'Joining a family x of sets is a set $\bigcup(x)$:'

$$\mathrm{Set}\{y; \exists u(u \in x \wedge y \in u)\}$$

Power set 'The set of the subsets of a set x is a set $\mathcal{P}(x)$:'

$$\mathrm{Set}\{y; y \subseteq x\}.$$

Infinite 'There exists a set x having an element z which is the empty set, and for which, for any element m, there exists an element n whose elements are those of m along with m itself.'

Replacement scheme 'The image of a set by an operation over sets is a set:'

$$\mathrm{Set}\{y; \exists x(x \in u \wedge y = F_\varphi(x))\},$$

for each formula φ such that $\mathrm{Set}\{y; \varphi(\vec{x}, y)\}$.

Foundation 'Every non-empty set x has an element y distinct from x.'

A few words about the other axioms. The axiom of choice, which states that a family of non-empty sets is non-empty, is crucial to the development of ordinary mathematics: it allows the proof of the facts that every vector space has a basis, that the product of compact spaces is compact and that the union of a countable number of countable sets is countable, among others. The continuum hypothesis states that there exist as many ways of well-ordering **N** as there are real numbers. It is little used, except in 'fine' real analysis. In 1928 Gödel showed the consistency of the axiom of choice and of the continuum hypothesis relative to ZF [56], but it was only in 1963 that Cohen showed that if ZF is consistent, these axioms cannot be theorems of ZF [21].

Relativization

Let D be a unary relational symbol. The relative quantifiers $\forall^D$ and $\exists^D$ are defined by: $\forall^D\varphi$ is the abbreviation of $\forall x(D(x) \Rightarrow \varphi)$, and $\exists^D\varphi$ is the abbreviation of $\exists x(D(x) \wedge \varphi)$. They are sometimes written $(\forall x \in D)\varphi$, and $(\exists x \in D)\varphi$.

Example Write $(\forall x > 0)\varphi$ instead of $\forall x(x > 0 \Rightarrow \varphi)$. □

If φ is a proposition, its relativization φ^D is obtained by relativizing the quantifiers in φ under D.

Implementations

A theory $T \subseteq \mathrm{L}_\Sigma$ is to be implemented in another theory $T' \subseteq \mathrm{L}_{\Sigma'}$. A unary relational symbol D over Σ' is defined as the 'domain' of the implementation. Similarly, each symbol f or R (in Σ) is implemented by a symbol f' or R' (in Σ') of the same arity; these symbols can be defined in an extension of T'. This yields a translation γ of L_Σ in $\mathrm{L}_{\Sigma'}$ obtained by replacing f and R with f' and R', and by relativising the quantifiers under D. For each f', axioms 'D is non-empty' and 'f' operates over D' are added:

$$T' \vdash \exists x D,$$

$$T' \vdash D(x_1) \wedge \ldots \wedge D(x_n) \Rightarrow D(f'(x_1, \ldots, x_n)).$$

If T is implemented in T', $T' \vdash \gamma(\varphi)$ for each axiom of T.

In logic, this mechanism is mainly used in proofs of undecidability. In computer science, conveniently modified, it allows the definition of the implementation of a data structure using another data structure. Extensions and implementations are the two main construction methods for data structures.

Skolemization

Named after the logician T. Skolem, skolemization is a syntactic transformation of formulas, allowing the elimination of one of the two kinds of quantifiers. There therefore exist two kinds of transformation, one which constructs universal formulas, used for proofs by contradiction, the other which constructs existential formulas, used for direct proofs. Their use is widespread in mathematical reasoning, as is illustrated by the two examples of universal skolemization:

- '$\exists x\, P(x)$ was proven; let a such that $P(a)$, ...'
- '$\forall \varepsilon \exists n\, Q(\varepsilon, n)$ was proven; let $N(\varepsilon)$ such that $Q(\varepsilon, N(\varepsilon))$, ...'

In these examples, new symbols were added to the signature: a 'Skolem constant' a and a 'Skolem function' N.

To simplify, here is a description of a universal skolemization step for a *prenex* formula (this is an unnecessary restriction, and is in general a hindrance, which will be lifted below).

Let $\psi \in \mathrm{L}_\Sigma[x_1, \ldots, x_k, x_{k+1}]$ and $\varphi \equiv \forall x_1 \ldots \forall x_k \exists x_{k+1} \psi \in \mathrm{L}_\Sigma$. Introduce a new k-ary functional symbol F and define

$$\begin{aligned} \tilde{\Sigma} &= \Sigma \cup \{F\}, \\ \tilde{\varphi} &= \forall x_1 \ldots \forall x_k \psi[x_{k+1} := F x_1 \ldots x_k] \in \mathrm{L}_{\tilde{\Sigma}}. \end{aligned}$$

If φ is a prenex proposition containing p existential quantifiers, then p iterations of the transformation $\varphi \mapsto \tilde{\varphi}$ yield a prenex universal proposition φ^u over a signature Σ^u extended by p new symbols.

If φ is an arbitrary (non-prenex) proposition, there does not exist a unique prenex form equivalent to φ. Furthermore, certain prenex transformations can create parasitic dependencies in the skolemized form.

Example Let $\varphi = \forall x P(x) \wedge \exists y\, Q(y)$. Both of the prenex formulas $\forall x\, \exists y (P(x) \wedge Q(y))$ and $\exists y \forall x (P(x) \wedge Q(y))$ are equivalent to φ. Their skolemized forms are, respectively, $\forall x (P \wedge Q(Fx))$ and $\forall x (P \wedge Q(a))$, where F (respectively a) is a Skolem unary function (respectively constant). Both are consistent propositions, as is φ, but the second is easier to test than the first. □

It is therefore preferable to compute the skolemized form without using an intermediate prenex form. Instead, the skolemized propositions φ^u (universal) and φ^e (existential) of a proposition φ are defined by the rules in Table 4.10 where ^{s} is ^{u} or ^{e}, $s \mapsto s'$ exchanges u and e, L designates a literal, $\{u_1, \ldots, u_n\} = \mathrm{var}(\chi)$, and F is a new n-ary symbol.

Computing φ^u consists of eliminating each positive quantifier $\forall x$ or negative quantifier $\exists x$ and substituting in its scope the term $F u_1 \ldots u_n$ a x, where $u_1, \ldots, u_n$ are the universally quantified variables in the scope in which x is found.

$$\begin{array}{rcl} L^s & = & L \\ (\chi \wedge \psi)^s & = & \chi^s \wedge \psi^s \\ (\chi \vee \psi)^s & = & \chi^s \vee \psi^s \\ (\forall x \chi)^u & = & \forall x \chi^u \\ (\exists x \chi)^u & = & (\chi[x := F u_1 \ldots u_n])^u \\ (\forall x \chi)^e & = & (\chi[x := F u_1 \ldots u_n])^e \\ (\exists x \chi)^e & = & \exists x \chi^e \\ (\neg \chi)^s & = & \neg(\chi^{s'}) \\ (\chi \Rightarrow \psi)^s & = & \chi^{s'} \Rightarrow \psi^s \end{array}$$

Table 4.10 Skolemization

Example

$$\begin{array}{rcl} \text{Let } \varphi & = & \forall x(\forall y A(x,y) \Rightarrow \neg \exists z B(x,z)). \\ \text{Then } \varphi^u & = & \forall x(A(x, Fx) \Rightarrow \neg \exists z B(x,z)) \\ \text{and } \varphi^e & = & \forall y A(a,y) \Rightarrow \neg B(a,b), \end{array}$$

where F, a and b are symbols introduced by skolemization. □

In Chapter 4, it will be shown semantically that

1. $\vdash \varphi^u \Rightarrow \varphi$ and $\vdash \varphi \Rightarrow \varphi^e$;
2. φ is consistent if and only if φ^u is consistent; and
3. φ is a tautology if and only if φ^e is consistent.

Exercises[2]

1.★ Given a signature Σ, find Σ' such that $\mathrm{L}_\Sigma[X] = \mathrm{T}_{\Sigma'}$.

2. Show that if $\Gamma \vdash_M \varphi$, there exists a finite $\Gamma' \subseteq \Gamma$ such that $\Gamma' \vdash_M \varphi$.

3.★ Give the typed λ-terms corresponding to the minimal logic derivations of the formulas:

$$(\varphi \vee \psi) \Rightarrow (\varphi \Rightarrow \psi) \Rightarrow \psi,$$
$$(\varphi \Rightarrow (\psi \Rightarrow \chi)) \Rightarrow ((\varphi \wedge \psi) \Rightarrow \chi),$$
$$((\varphi \Rightarrow \psi) \vee (\varphi \Rightarrow \chi)) \Rightarrow \varphi \Rightarrow (\psi \vee \chi).$$

4.★ State and derive introduction and elimination rules for the connective $\Leftrightarrow$.

[2]Hints to the exercises labeled with a ★ can be found at the end of the book.

5. State and prove compatibility properties of proven equivalence for $\wedge$, $\vee$, $\Rightarrow$ and $\neg$.

6. Prove the Morgan axiom of minimal logic:

$$\vdash_M \neg(\varphi \vee \psi) \Longleftrightarrow (\neg\varphi \wedge \neg\psi).$$

7. Prove in minimal logic that $\varphi \Rightarrow \neg\neg\psi$, $\neg\psi \Rightarrow \neg\varphi$ and $\neg\neg\varphi \Rightarrow \neg\neg\psi$ are equivalent formulas. Prove they are all consequences of $\neg\neg(\varphi \Rightarrow \psi)$, which follows from $\varphi \Rightarrow \psi$, which in turn follows from $\neg\neg\varphi \Rightarrow \psi$. Prove that $\neg\neg(\varphi \Rightarrow \psi)$ is a consequence of $\neg\neg\varphi \Rightarrow \neg\neg\psi$ in intuitionistic logic. Prove that all these formulas are equivalent in classical logic.

8. Prove

$$\vdash_M \neg\neg\neg\varphi \Longleftrightarrow \neg\varphi,$$
$$\vdash_I \neg\neg(\neg\neg\varphi \Rightarrow \varphi).$$

9. Prove the minimal tautologies

$$\vdash_M \exists x \neg\varphi \Rightarrow \neg(\forall x \varphi),$$
$$\vdash_M \neg(\exists x \varphi) \Longleftrightarrow \forall x \neg\varphi,$$
$$\vdash_M \neg\neg\exists x(\neg\varphi) \Longleftrightarrow \neg\forall x(\neg\neg\varphi).$$

10. Prove the minimal tautologies

$$\vdash_M \exists x(\varphi \Rightarrow \psi(x)) \;\Rightarrow\; (\varphi \Rightarrow \exists x \psi(x)) \qquad \text{if } x \notin \operatorname{var} \varphi,$$
$$\vdash_M (\forall x \varphi(x) \vee \psi) \;\Rightarrow\; \forall x(\varphi(x) \vee \psi) \qquad \text{if } x \notin \operatorname{var} \psi.$$

11.★ Build derived rules, analogous to $\forall\forall$, for $\Rightarrow$, $\vee$ and $\wedge$ (they have no logical interest). Interpret them using the Curry–Howard correspondence. They lead to reduction rules over the typed λ-terms (the η rules, which are important in computation).

12. Derive an instantiation rule for the existential quantifier, analogous to $\forall\forall$, and explain the use of this rule at the level of the assumptions necessary to prove a formula.

13. Prove that the rules $(\neg\neg)$ and (C) are derivable from the axiom of the excluded middle and that (C) is derivable from $(\neg\neg)$.

14.★ Prove the classical tautologies

$$\vdash_C \neg(\forall x \varphi) \;\Rightarrow\; \exists x(\neg\varphi),$$
$$\vdash_C \neg(\varphi \wedge \psi) \;\Rightarrow\; \neg\varphi \vee \neg\psi.$$

15. Prove the classical tautologies

$$\vdash_C (\varphi \Rightarrow \exists x\psi(x)) \;\Rightarrow\; \exists x(\varphi \Rightarrow \psi(x)) \qquad \text{if } x \notin \operatorname{var}\varphi,$$
$$\vdash_C \forall x(\varphi(x) \vee \psi) \;\Rightarrow\; (\forall x\varphi(x) \vee \psi) \qquad \text{if } x \notin \operatorname{var}\psi.$$

16. Prove 'Peirce's axiom' $((\varphi \Rightarrow \psi) \Rightarrow \varphi) \Rightarrow \varphi$ in the NK system.

17.★ Derive $((\varphi \Rightarrow \chi) \Rightarrow \psi) \Rightarrow ((\psi \Rightarrow \varphi) \Rightarrow (\omega \Rightarrow \varphi))$ ('Łukasiewicz's axiom') in the sequent calculus.

18. Define a rewriting system to compute disjunctive forms.

19. Give an example of a formula having two distinct conjunctive normal forms.

20. Prove $\{Ax_E1, \forall xyz(Exy \wedge Ezy \Rightarrow Exz)\} \vdash Ax_E2 \wedge Ax_E3$.

21.★ State the formulas of L_Σ (Σ with equality) expressing the existence of at least (respectively at most, exactly) three elements.

22. Show that the equality axioms are derivable using the two rules

$$\frac{}{t = t} \qquad \text{and} \qquad \frac{t = t' \quad \alpha[x := t]}{\alpha[x := t']},$$

where t, t' are terms and α is an atomic formula.

23.★ Give the axioms of group theory using the signature $\Sigma' = \{e, ., =\}$, then using $\Sigma'' = \{e, m\}$, where m is a ternary relational symbol such that $m(x, y, z)$ states that $z = x.y$.

24. Give signatures and axioms for theories for commutative rings, for fields of characteristic p, $p \geqslant 0$ [the characteristic of a field is the least integer p such that $p.1 = 1 + \ldots + 1$ is null,] and for algebraically closed fields.

25. Give arithmetic formulas expressing that:

- x and y are relatively prime;
- x is the smallest prime strictly greater than y.

26.★ Write AE for elementary arithmetic. Prove that

- $AE \vdash s^m(0) = s^n(0)$, if $m = n$;
- $AE \vdash \neg(s^m(0) = s^n(0))$, if $m \neq n$;
- $AE \vdash s^m(0) + s^n(0) = s^p(0)$, if $m + n = p$;
- $AE \vdash s^m(0) * s^n(0) = s^p(0)$, if $m.n = p$.

27. Show that $\forall xy(x < y \vee x = y \vee y < x)$ is a theorem of Peano arithmetic.

28. Let p be an integer $\geqslant 0$. For which values of p is the theory of commutative fields of characteristic p inconsistent?

29. Prove that Γ is consistent if and only if every finite subset of Γ is consistent (see exercise 2).

30. Give the axioms of the theory of vector spaces as an extension of field theory.

31. Prove equations 4.1 and 4.2 of a definitional extension in the relational case (with R unary, to simplify).

Bibliographic notes

The Curry–Howard correspondence was made explicit by Howard [66] and forms the framework of Girard *et al.*'s book [54]. The correspondence is implemented in higher-order systems: Girard's system F [52, 54], and Martin-Löf's intuitionistic type theory [104].

Intuitionistic logic is studied by van Dalen in [154], and, more deeply, in the more recent book by Troelstra and van Dalen [150]. Natural deduction was analyzed by Prawitz in [125]. The sequent calculus and its links with computer science are treated in [47, 54].

Chapter 5

Models

5.1 The semantics of first-order logic

Interpretation schemes are given for each of the syntactic objects introduced in first-order logic: signatures, variables, terms, formulas and derivations.

5.1.1 Σ-algebras

Let Σ be a signature, i.e, a set with an arity function ar : $\Sigma \to \mathbf{N}$.

Functional signatures
A symbol of arity n is interpreted as a function of n arguments.

Definition 5.1 A Σ-*algebra* $\mathcal{A}$ consists of a set A, and for each $f \in \Sigma$ of arity n, of a mapping $f^{\mathcal{A}} : A^n \to A$. $\mathcal{A} = (A, (f^{\mathcal{A}})_{f \in \Sigma})$ is called an *interpretation* of Σ. The domain of interpretation $\mathcal{A}$ is A.

In particular, a constant symbol is interpreted by an element of A.

Example A singleton $\{a\}$ is the domain of a *trivial* Σ-algebra, each constant symbol being interpreted by a and the function symbols by constant functions of value a. A trivial interpretation equates everything. On the other hand, the free interpretations which follow equate nothing. □

If $\mathcal{A}$ and $\mathcal{B}$ are Σ-algebras, a Σ-*morphism* $m : \mathcal{A} \to \mathcal{B}$ is a mapping from A to B such that

$$m(f^{\mathcal{A}}(\vec{a})) = f^{\mathcal{B}}(m(\vec{a}))$$

for each $f \in \Sigma$ of arity n and each n-tuple $\vec{a}$ of elements of A. The category[1] of Σ-algebras and Σ-morphisms is written $\mathbf{Alg}_\Sigma$. Note that if $\Sigma \subseteq \Sigma'$, every Σ'-algebra

[1]Recall that a category is defined by a bunch of objects, and, for each pair (A, B) of objects, of a set of morphisms $f : A \to B$, with a morphism composition operator (if $f : A \to B$ and

$\mathcal{A}$ is automatically a Σ-algebra, written $\mathcal{A}{\restriction}\Sigma$, obtained by interpreting only the elements of Σ.

Term algebras

If X is an arbitrary set of variables, the set $\mathrm{T}_\Sigma[X]$ of terms has a canonical Σ-algebra structure $\mathcal{T}$ defined by

$$f^{\mathcal{T}}(t_1, \dots, t_n) = ft_1 \dots t_n$$

for $f \in \Sigma$, $\mathrm{ar}(f) = n$, and $t_1, \dots, t_n \in \mathrm{T}_\Sigma[X]$.

This Σ-algebra is also a free object in the category-theoretic sense, as free monoids and free groups. For every Σ-algebra $\mathcal{A}$ and every mapping $\xi : X \to A$, there exists a unique Σ-morphism $\hat{\xi}_\Sigma : \mathrm{T}_\Sigma[X] \to A$ such that $\hat{\xi}_\Sigma(x) = \xi(x)$ for every $x \in X$:

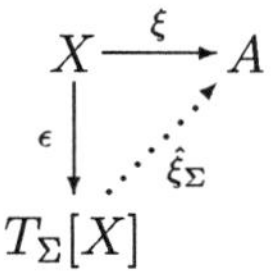

$$\begin{array}{ccc} X & \xrightarrow{\ \xi\ } & A \\ {\scriptstyle \epsilon}\downarrow & \nearrow_{\hat{\xi}_\Sigma} & \\ T_\Sigma[X] & & \end{array}$$

$\hat{\xi}_\Sigma$ is defined by induction over terms:

$$\begin{array}{rcl} \hat{\xi}_\Sigma(x) & = & \xi(x), \\ \hat{\xi}_\Sigma(ft_1 \dots t_n) & = & f^{\mathcal{A}}(\hat{\xi}_\Sigma(t_1), \dots, \hat{\xi}_\Sigma(t_n)). \end{array}$$

$\mathrm{T}_\Sigma[X]$ is called the *free Σ-algebra generated by X*.

In particular, this property applies to T_Σ, the algebra of closed terms, when $X = \varnothing$. The above property is more easily written as: for every Σ-algebra $\mathcal{A}$, there exists a unique Σ-morphism $\iota_{\mathcal{A}} : \mathrm{T}_\Sigma \to \mathcal{A}$, defined by:

$$\iota_{\mathcal{A}}(ft_1 \dots t_n) = f^{\mathcal{A}}(\iota_{\mathcal{A}}(t_1), \dots, \iota_{\mathcal{A}}(t_n)).$$

Because of this property, T_Σ is called the *initial* Σ-algebra.

Another important application of this property occurs when $\mathcal{A} = \mathcal{T}$. A unique endomorphism $\hat{\xi} : \mathrm{T}_\Sigma[X] \to \mathrm{T}_\Sigma[X]$ extending $\xi : X \to \mathrm{T}_\Sigma[X]$ is obtained. The *domain* of $\xi : X \to \mathrm{T}_\Sigma[X]$, or of $\hat{\xi}$, is the set

$$\mathrm{dom}(\xi) = \{x \in X;\ \xi(x) \neq x\}.$$

A previously defined concept reappears: if $\mathrm{dom}(\xi)$ is finite, ξ (or $\hat{\xi}$) is a *substitution*.

Algebras in normal form

Let $\mathcal{R}$ be a confluent and normalizing rewriting system, i.e, every term has a unique normal form. Consider the set N of normal terms for $\mathcal{R}$, and write $t{\downarrow} \in N$ for the normal form of t. Let $f \in \Sigma$, $\mathrm{ar}(f) = n \geqslant 0$. If $t_1, \dots, t_n \in N$ let

$$f^{\mathcal{N}}(t_1, \dots, t_n) = (ft_1 \dots t_n){\downarrow}\ .$$

A Σ-algebra structure has been defined over the set N.

$g : B \to C$, then $g \circ f : A \to C$ is defined) which is associative ($f \circ (g \circ h) = (f \circ g) \circ h$, for composable morphisms) and has a neutral element $1_A : A \to A$. See, for example, [5].

Logical connectives
Consider a signature of connectives, for example $\Sigma_c = \{\wedge, \vee, \bot, \Rightarrow\}$. The set Prop[$A$] of propositions over A is defined as the set $\mathrm{T}_{\Sigma_c \cup A}$. To give the Boolean meaning of the symbols in the signature, it is not an arbitrary Σ_c-algebra which must be used, but, rather, a *Boolean algebra*, and in particular the simplest among them, which is $\mathbf{2} = \{0, 1\}$. A semantics with two truth values will be given: *true* (for 1) and *false* (for 0). The operations min and max interpret $\wedge$ and $\vee$; $\bot$ is interpreted by 0; and $\Rightarrow^{\mathbf{2}} (a, b) = \max(1 - a, b)$.

Signatures of the predicate calculus
And now for first-order logic. Let $\Sigma = \Sigma_f \cup \Sigma_r$ be a predicate calculus signature. The relational symbols of the signature must be interpreted by relations. A Σ-algebra $\mathcal{A}$ consists of a *non-empty* set A along with the following:

- an element $c^{\mathcal{A}} \in A$ for each constant symbol $c \in \Sigma$;
- a mapping $f^{\mathcal{A}} : A^n \to A$ for each n-ary functional symbol $f \in \Sigma$;
- a relation $R^{\mathcal{A}} : A^n \to \{0, 1\}$ (i.e, $R^{\mathcal{A}} \subseteq A^n$) for each n-ary relational symbol $R \in \Sigma$,

The algebra $\mathcal{A} = (A, c^{\mathcal{A}}, \ldots , f^{\mathcal{A}}, \ldots , R^{\mathcal{A}}, \ldots)$ is called an interpretation of Σ. The set A is its *domain*.

In the definition of a Σ-morphism between two Σ-algebras $\mathcal{A}$ and $\mathcal{B}$, the condition

$$m(R^{\mathcal{A}}(\vec{a})) \subseteq R^{\mathcal{B}}(m(\vec{a}))$$

is added for every $R \in \Sigma$ of arity n and every n-tuple $\vec{a}$ of elements of A. The symbol $\mathbf{Alg}_\Sigma$ continues to be used for the category of Σ-algebras.

Examples For $\Sigma = \{0, s, +, ., \leqslant, =\}$, the set $\mathbf{N}$ with the usual 0, $s : n \mapsto n + 1$, +, ., $\leqslant$ and = is a Σ-algebra. But there are many others, since no property is imposed on $f^{\mathcal{A}}$ or $R^{\mathcal{A}}$.

If Σ is the signature consisting of the binary relational symbol R, a Σ-algebra is a simple graph: its domain is the set of vertices. □

Recall that if $\Sigma \subseteq \Sigma'$, i.e, $\Sigma_f \subseteq \Sigma'_f$ and $\Sigma_r \subseteq \Sigma'_r$, every Σ'-algebra $\mathcal{A}$ is automatically a Σ-algebra written $\mathcal{A}{\restriction}\Sigma$, obtained by interpreting only the elements of Σ: it is the *restriction* of $\mathcal{A}$ to Σ. A Σ-algebra $\mathcal{A}$ whose restriction to Σ_f is the initial Σ_f-algebra is called a *Herbrand Σ-algebra*. The set T_Σ underlying all Herbrand Σ-algebras is sometimes called the *Herbrand universe*. Note that the relational symbols can be interpreted in any manner. A Herbrand algebra $\mathcal{H}$ is defined by specifying a value in $\mathbf{2}$ for each atom $Rt_1 \ldots t_n$ or by the set of atoms $Rt_1 \ldots t_n$ such that $R^{\mathcal{H}}(t_1, \ldots , t_n) = 1$. Herbrand Σ-algebras can therefore be equated with subsets of the set At_Σ of closed terms, called the *Herbrand base*.

5.1.2 Terms

Variables are interpreted as 'varying' in a Σ-algebra. Write A^X for the set of mappings from X to A. The elements of A^X, written ξ, η, ... , are called *valuations*. In the semantics of programming languages, a valuation represents a memory state.

A term $t \in \mathrm{T}_\Sigma[X]$ is interpreted (or denoted) by a mapping $t^{\mathcal{A}} : A^X \to A$ defined by induction over terms by:

1. $x^{\mathcal{A}}(\xi) = \xi(x)$, if $x \in X$;
2. $c^{\mathcal{A}}(\xi) = c^{\mathcal{A}}$, if $c \in \Sigma$ is a constant symbol;
3. $(ft_1 \dots t_n)^{\mathcal{A}}(\xi) = f^{\mathcal{A}}(t_1^{\mathcal{A}}(\xi), \dots , t_n^{\mathcal{A}}(\xi))$, if $f \in \Sigma$ is an n-ary function symbol.

If $\mathrm{var}(t) \subseteq \{x_1, \dots , x_n\} \subseteq X$, $t^{\mathcal{A}}(\xi)$ depends only on $\xi(x_1), \dots , \xi(x_n)$. So the mapping can also be written $t^{\mathcal{A}} : A^n \to A$, whose value in $(a_1, \dots , a_n) \in A^n$ is obtained by 'substituting' $a_1, \dots , a_n$ for $x_1, \dots , x_n$ and $f^{\mathcal{A}}$ for f, $f \in \Sigma$. In particular, a closed term $t \in \mathrm{T}_\Sigma$ is interpreted by an element $t^{\mathcal{A}}$ of A.

Rational expressions Let Γ be an alphabet. Consider the signature Σ having:

- the elements of Γ along with the symbols $0, 1 \notin \Gamma$ as constants;
- * (unary, as superscript), $\cdot$ (binary, also written by simple juxtaposition) and $+$ (binary) as functional symbols.

Consider the closed terms over Σ, called *rational expressions* over the alphabet Γ. For example, for $\Gamma = \{a, b, c\}$, $(c + ab)^\star c$ is a closed term. Let $\mathcal{M}$ be the Σ-algebra $\mathcal{P}(\Gamma^\star)$ where $\Gamma^\star$ is the free monoid generated by Γ, with:

- $a^{\mathcal{M}} = \{a\}$, for $a \in \Gamma$;
- $0^{\mathcal{M}} = \varnothing$;
- $1^{\mathcal{M}} = \{\epsilon\}$, where ϵ is the empty string, the neutral element of $\Gamma^\star$;
- $+$ is implemented in $\mathcal{M}$ by union and $\cdot$ by concatenation, i.e,

$$LL' = \{uu'; u \in L, u' \in L'\};$$

- * is implemented by Kleene closure, i.e.,

$$L^\star = \{u_1 \dots u_n; n \geqslant 0, u_1, \dots , u_n \in L\}.$$

For $t = (c + ab)^\star c$, $t^{\mathcal{M}} = \{c, cc, abc, ccc, cabc, ababc, \dots\}$. The set $t^{\mathcal{M}}$ is called the rational language denoted by the rational expression t. The study of these languages forms one of the classical branches of formal language theory.

5.1.3 The propositional calculus

Since the language Prop$[A]$ of propositions formed over the set A of atoms is a set of terms, the *classical* semantics of the propositional calculus is a particular case of the above. The standard Boolean interpretation **2** is imposed on the signature Σ_c of connectives. An interpretation of Prop$[A]$ is a $\Sigma_c \cup A$-algebra whose restriction to Σ_c is the standard interpretation in **2**. It is therefore completely defined by an element I of $\mathbf{2}^A$, associating with each atom $a \in A$ its 'truth value' $I(a)$. This mapping I is extended to Prop$[A]$ with,

$$\begin{aligned} I(\varphi \vee \psi) &= \max(I(\varphi), I(\psi)), \\ I(\varphi \wedge \psi) &= \min(I(\varphi), I(\psi)), \\ I(\varphi \Rightarrow \psi) &= \max(1 - I(\varphi), I(\psi)). \end{aligned}$$

If $I(\varphi) = 1$, then $I \vDash \varphi$, read 'I is a model of φ'.

Example Let $I(p) = 1$ and $I(q) = 0$. Then $I \vDash (p \wedge q) \Rightarrow p$, since

$$\begin{aligned} I((p \wedge q) \Rightarrow p) &= \max(1 - \min(I(p), I(q)), I(p)) \\ &= \max(1 - \min(1, 0), 1) \\ &= 1. \end{aligned}$$

□

If A has n elements, n finite, there are only 2^n interpretations of A, so there exists an exhaustive algorithm to decide if a formula is a semantic tautology, i.e., valid for all interpretations. This algorithm is the well-known and inefficient *truth table*.

The chosen connectives have simple meanings, close to that of everyday language (although few people accept that 'false implies false' is 'true'!). A k-ary connective $\mathbf{c}$ is interpreted in the Boolean algebra **2** as a mapping $\mathbf{c}^\mathbf{2} : \mathbf{2}^k \to \mathbf{2}$. Nothing prevents a new k-ary connective from being defined simply by giving a function from $\mathbf{2}^k$ to **2**, without worrying about the meaning of the connective. It is not clear by looking at NK that there is a bound to the number of possible binary connectives. The semantics given here provides an exact bound: there are exactly $2^{(2^k)}$ k-ary connectives, in particular, 2 0-ary connectives ($\top$ and $\bot$), 4 unary connectives (only $\neg$ was used), and 16 binary connectives. An interesting question is, are there any other interesting connectives and do they have the same 'Boolean' characteristics that the usual connectors do? It was seen that some of the introduced connectives were sufficient to define all of them ($\bot$, $\top$, $\neg$, $\wedge$, $\vee$, $\Rightarrow$, $\Longleftrightarrow$) in classical logic. This is the case for $\vee$ and $\neg$, and $\bot$ and $\Rightarrow$, among others.

Let Σ_c be a fixed connective signature. A formula $\varphi \in$ Prop$[p_1, \ldots, p_k]$ has an interpretation $I(\varphi) \in \mathbf{2}$. Let $\tilde{\varphi}(I) = I(\varphi)$, thereby defining a mapping $\tilde{\varphi} : \mathbf{2}^k \to \mathbf{2}$, hence a k-ary connective $\mathbf{c}$ such that $\mathbf{c}^\mathbf{2} = \tilde{\varphi}$.

Σ_c is called a connective *basis* if for every n and every mapping $f : \mathbf{2}^n \to \mathbf{2}$, there exists a formula φ such that $f = \tilde{\varphi}$ (f is then said to be defined by φ).

Proposition 5.1 $\Sigma_c = \{\neg, \wedge, \vee\}$ is a connective basis.

Proof For $n = 1$, there are six functions from $\mathbf{2}$ to $\mathbf{2}$, defined by the formulas p, $\neg p$, $p \vee \neg p$, $p \wedge \neg p$.

Let $n \geqslant 1$ and $f : \mathbf{2}^{n+1} \to \mathbf{2}$. Define $f_i(x_1, \ldots, x_n) = f(x_1, \ldots, x_n, i)$ for $i \in \mathbf{2}$. By the inductive hypothesis, there exist formulas φ_0 and φ_1 such that $f_i = \tilde{\varphi}_i$. Define $\varphi = (p_{n+1} \wedge \varphi_1) \vee (\neg p_{n+1} \wedge \varphi_2)$. It is clear that $\tilde{\varphi} = f$, which proves that $\Sigma_c = \{\neg, \wedge, \vee\}$ is a connective basis. □

It is easily deduced that $\{\neg, \vee\}$ and $\{\neg, \wedge\}$ are also bases, by using Morgan's laws, whose algebraic expression is

$$\begin{aligned} \min(x, y) &= 1 - \max(1 - x, 1 - y), \\ \max(x, y) &= 1 - \min(1 - x, 1 - y), \end{aligned}$$

as well as $\{\bot, \Rightarrow\}$, since

$$\max(x, y) = \Rightarrow^{\mathbf{2}} (\Rightarrow^{\mathbf{2}} (x, \bot^{\mathbf{2}}), y).$$

There are also bases with one element. Let **nand** (also written |, Sheffer's bar) and **nor** (also written †, Peirce's dagger) be defined by the formulas $\neg(p_1 \wedge p_2)$ and $\neg(p_1 \vee p_2)$ respectively. The connectives are used in electronics to design logical gates: for example, **nor** tests if its two entries are simultaneously null, which is easy to implement.

5.1.4 Formulas

Closed formulas are interpreted by truth values (0 or 1) in an analogous manner to the propositional calculus. Because of variables, valuations must be used, as was done for terms, to interpret formulas as relations.

A formula $\varphi \in \mathrm{L}_\Sigma[X]$ is interpreted by a mapping $\varphi^{\mathcal{A}} : A^X \to \mathbf{2}$, i.e, by a subset of A^X, whose definition by induction over formulas is given by Table 5.1. In particular, for a proposition $\varphi \in \mathrm{L}_\Sigma$, $\varphi^{\mathcal{A}}$ is the constant 0 ('false') or 1 ('true') of $\mathbf{2}$. In a given interpretation, a formula is either true or false. Note that if $\varphi \in \mathrm{L}_\Sigma[x_1, \ldots, x_n]$, $\varphi^{\mathcal{A}}(\xi)$ depends only on $\xi(x_1)$, ..., $\xi(x_n)$ and hence defines a mapping, also written $\varphi^{\mathcal{A}} : A^n \to \mathbf{2}$.

Consider once again the extension $\hat{\xi}$ from $\mathrm{T}_\Sigma[X]$ to itself of a valuation $\xi : X \to \mathrm{T}_\Sigma[X]$. Write $\hat{\xi}$ also for the mapping $\mathrm{At}[X] \to \mathrm{At}[X]$ defined for an atomic formula $Rt_1 \ldots t_n$ by

$$\hat{\xi}(Rt_1 \ldots t_n) = R\hat{\xi}(t_1) \ldots \hat{\xi}(t_n).$$

The extension of $\hat{\xi}$ to formulas without quantifiers is immediate. But to extend $\hat{\xi}$ from atoms to quantified formulas, the domain of ξ must be finite, to allow the

$$\begin{array}{rcll}(Rt_1 \dots t_n)^{\mathcal{A}}(\xi) & = & R^{\mathcal{A}}(t_1^{\mathcal{A}}(\xi), \dots, t_n^{\mathcal{A}}(\xi)) & (R \in \Sigma_r) \\ (\varphi \vee \psi)^{\mathcal{A}}(\xi) & = & \max(\varphi^{\mathcal{A}}(\xi), \psi^{\mathcal{A}}(\xi)) & \\ (\varphi \wedge \psi)^{\mathcal{A}}(\xi) & = & \min(\varphi^{\mathcal{A}}(\xi), \psi^{\mathcal{A}}(\xi)) & \\ (\neg\varphi)^{\mathcal{A}}(\xi) & = & 1 - \varphi^{\mathcal{A}}(\xi) & \\ (\exists x\varphi)^{\mathcal{A}}(\xi) & = & \sup_\eta \varphi^{\mathcal{A}}(\eta) & \text{with } \eta(y) = \xi(y) \text{ for } y \neq x \\ (\forall x\varphi)^{\mathcal{A}}(\xi) & = & \inf_\eta \varphi^{\mathcal{A}}(\eta) & \text{with } \eta(y) = \xi(y) \text{ for } y \neq x\end{array}$$

Table 5.1 Interpretation of formulas

renaming of the bound variables outside of dom(ξ). So ξ is restricted to being a substitution.

5.1.5 Derivations

The syntactic objects remaining to be interpreted are the derivations NK.

Let $d \in \mathrm{D}_\Sigma$ be a derivation with conclusion $\varphi \in \mathrm{L}_\Sigma[X]$: it is a (binding) tree with nodes labeled by formulas, and with root labeled by φ. By substituting for these formulas their interpretations in a Σ-algebra $\mathcal{A}$, then for each $\xi \in A^X$, by taking their values at ξ, a tree $d^{\mathcal{A}}(\xi)$ labeled by truth values is obtained. The tree $d^{\mathcal{A}}(\xi)$ indicates the local behavior of d in A:

$$\frac{\varphi_1 \qquad \psi_2}{\chi} \qquad \mapsto \qquad \frac{\varphi^{\mathcal{A}}(\xi) \qquad \psi^{\mathcal{A}}(\xi)}{\chi^{\mathcal{A}}(\xi)}.$$

Lemma 5.2 If the unbound leaves of $d^{\mathcal{A}}(\xi)$ are labeled by 1, then the conclusion is $\varphi^{\mathcal{A}}(\xi) = 1$.

Proof By induction over derivations, thirteen rules must be examined. Consider the case where the conclusion φ follows from rule $(E_\vee)$:

$$(E_\vee) : \frac{\begin{matrix} d_0 \\ \chi \vee \psi \end{matrix} \qquad \begin{matrix} d_1[\chi] \\ \varphi \end{matrix} \qquad \begin{matrix} d_2[\psi] \\ \varphi \end{matrix}}{\varphi},$$

where d_1 and d_2 are derivations of φ containing, respectively, χ and ψ among their assumptions. The active assumptions of d are:

- those of d_0;
- those of d_1 without χ; and
- those of d_2 without ψ.

Suppose that the interpretations of the active assumptions of D all equal 1 at a given $\xi \in A^X$. The inductive hypothesis applied to d_0 shows that $(\chi \vee \psi)^{\mathcal{A}}(\xi) = 1$. By the definition of the interpretation of $\vee$, $\chi^{\mathcal{A}}(\xi) = 1$ or $\psi^{\mathcal{A}}(\xi) = 1$. Look at the first case, $\chi^{\mathcal{A}}(\xi) = 1$, and consider d_1. The active assumptions of d_1 also contain χ, and hence are all equal to 1 at ξ. The inductive hypothesis can then be applied to d_1 to deduce its conclusion, which is also the conclusion of d, that $\varphi^{\mathcal{A}}(\xi) = 1$.

□

5.2 Models of a theory

Now that interpretation schemes for all syntactic objects have been adopted, it is time to introduce the notion of model and the semantic properties of soundness and completeness.

5.2.1 The satisfiability relation $\models$

Let $\mathcal{A}$ be a Σ-algebra and $\varphi \in \mathrm{L}_\Sigma[X]$. The algebra $\mathcal{A}$ *satisfies* φ, written $\mathcal{A} \models \varphi$, if $\varphi^{\mathcal{A}}(\xi) = 1$ for all $\xi \in A^X$, i.e., if the formula φ is 'true in A by substituting any elements of A for the free variables'.

If $\mathcal{A} \models \varphi$ for every Σ-algebra $\mathcal{A}$, φ is *valid* or is a *semantic tautology*, written $\models \varphi$. Since numerous tautologies can be written using $\Longleftrightarrow$, note that $\mathcal{A} \models \varphi \Longleftrightarrow \psi$ if and only if $\varphi^{\mathcal{A}} = \psi^{\mathcal{A}}$.

Let Γ be a theory, i.e, a subset of $\mathrm{L}_\Sigma[X]$. The algebra $\mathcal{A}$ is called a *model* of Γ, written $\mathcal{A} \models \Gamma$, if $\mathcal{A} \models \varphi$ for every $\varphi \in \Gamma$. Write $\mathbf{Mod}(\Gamma)$ for the 'class' of models of Γ.

Example A model of the *axioms* of group theory is a group! See, however, the next section on equality. □

A model of a theory which is also a Herbrand algebra is called a *Herbrand model* of that theory. The importance of these models is shown by Herbrand's theorem and by its applications.

5.2.2 Equality

Let Σ be a signature with equality, the symbol E standing for equality. A Σ-algebra *with equality* is a Σ-algebra $\mathcal{A}$ such that $E^{\mathcal{A}}$ is the equality of A, i.e, the diagonal of $A \times A$.

The equality axioms Ax_E were given in Chapter 3. A Σ-algebra with equality obviously satisfies Ax_E, but not conversely. If $\mathcal{A} \models Ax_E$, then $E^{\mathcal{A}}$ is simply an equivalence relation over A (by $Ax_E 1, 2, 3$), compatible with $f^{\mathcal{A}}$ and $R^{\mathcal{A}}$ (by the following axioms). The quotient A' can be defined by $A' = A/E^{\mathcal{A}}$ and $\mathcal{A}'$ is then a Σ-algebra with equality, the projection $\pi : A \to A'$ being a Σ-morphism. By induction over formulas, one can prove that for $\varphi \in \mathrm{L}_\Sigma[X]$, and $\xi \in A^X$, the value $\varphi^{\mathcal{A}}(\xi)$ only depends on $\pi \circ \xi$, hence that $\varphi^{\mathcal{A}'}$ is the quotient of $\varphi^{\mathcal{A}}$ by $E^{\mathcal{A}}$.

$$\begin{array}{ccc} X & \xrightarrow{\xi} & A \\ & \searrow^{\xi'} & \downarrow \pi \\ & & A' \end{array}$$

The following proposition has therefore been demonstrated:

Proposition 5.3 Let $\Gamma \subseteq \mathrm{L}_\Sigma[X]$ with equality. Γ has a model with equality if and only if $\Gamma \cup Ax_E$ has a model.

Convention If Σ is with equality and $\Gamma \subseteq \mathrm{L}_\Sigma$, the phrase 'model of Γ' refers to a model with equality of Γ.

5.2.3 Soundness of deduction

Let $\Gamma \subseteq \mathrm{L}_\Sigma[X]$ be the theory generated by $\Delta \subseteq \mathrm{L}_\Sigma[X]$,

$$\Gamma = \{\varphi \in \mathrm{L}_\Sigma[X]\,;\, \Delta \vdash \varphi\}.$$

As $\Delta \subseteq \Gamma$, it is clear that $\mathbf{Mod}(\Gamma) \subseteq \mathbf{Mod}(\Delta)$, but is there equality?

Lemma 5.4 Let $\Delta \vdash \varphi$, $\mathcal{A}$ a Σ-algebra and $\xi \in A^X$. If $\psi^{\mathcal{A}}(\xi) = 1$ for every $\psi \in \Delta$, then $\varphi^{\mathcal{A}}(\xi) = 1$.

Proof It is a reformulation of lemma 5.2, since $\Delta \vdash \varphi$ means that Δ contains the set of active assumptions of a derivation of φ. □

Write $\Delta \models \varphi$ if for every Σ-algebra $\mathcal{A}$ and for every $\xi \in A^X$, $\Delta^{\mathcal{A}}(\xi) = 1$ implies $\varphi^{\mathcal{A}}(\xi) = 1$. It is a new relation over $\mathcal{P}(\mathrm{L}_\Sigma[X]) \times \mathrm{L}_\Sigma[X]$ (like $\vdash$), called *semantic consequence*. In the particular case of closed formulas, i.e., if $\Gamma \subseteq \mathrm{L}_\Sigma$ and $\varphi \in \mathrm{L}_\Sigma$, the following definition is obtained: $\Gamma \models \varphi$ if and only if for every Σ-algebra $\mathcal{A}$, $\mathcal{A} \models \Gamma$ implies $\mathcal{A} \models \varphi$. The lemma becomes the soundness theorem:

Theorem 5.5 (Soundness) If $\Gamma \vdash \varphi$ then $\Gamma \models \varphi$.

Corollary 5.6 The models of a theory are those of its axioms.

For example, the models of group theory (and not only of its axioms) are the groups.

5.2.4 Completeness of deduction

The completeness theorem is the converse of the soundness theorem. With these two theorems, the $\vdash$ and $\models$ relations are identical, which gives two perspectives for the same problems. For example, the following theorem equates consistency (the deductive point of view) with the existence of a model (the semantic point of view): it will be used to prove the completeness theorem.

Theorem 5.7 (Gödel completeness) A theory has a model iff it is consistent.

Proof

Necessary condition: immediate since φ and $\neg\varphi$ cannot have the same model.

Sufficient condition: this is the difficult part. An outline of L. Henkin's (1949 [60]) proof is given.

(1) An attempt is made to construct a 'syntactic model'. Suppose that Σ contains a constant, so that $T_\Sigma \neq \varnothing$. If $\Gamma \subseteq L_\Sigma[X]$, let $\mathcal{H}_\Gamma$ be the *canonical* Herbrand Σ-algebra of Γ, which interprets the relational symbols over domain T_Σ by:

$$R^{\mathcal{H}_\Gamma}(t_1, \ldots, t_n) = 1 \qquad \text{if and only if} \qquad \Gamma \vdash Rt_1 \ldots t_n.$$

Hence, for a closed atomic formula $\alpha \in \mathrm{At}_\Sigma$,

$$\mathcal{H}_\Gamma \models \alpha \qquad \text{if and only if} \qquad \Gamma \vdash \alpha.$$

But it does not hold for an arbitrary formula, since

(a) there are in general not enough closed terms to satisfy them: if $\Gamma \vdash \exists x\psi(x)$, there should exist $t \in T_\Sigma$ such that $\psi^{\mathcal{H}_\Gamma}(t) = 1$;

(b) there can exist $\varphi \in L_\Sigma$ such that $\Gamma \nvdash \varphi$ and $\Gamma \nvdash \neg\varphi$ while, necessarily, $\mathcal{H}_\Gamma \models \varphi$ or $\mathcal{H}_\Gamma \models \neg\varphi$.

(2) Assume that Γ satisfies the following properties:

(a) *Henkin property*: if $\Gamma \vdash \exists x\psi(x) \in L_\Sigma$, then there exists $t \in T_\Sigma$ such that $\Gamma \vdash \psi[x := t]$.

(b) *Consistency* and *completeness*: for every $\varphi \in L_\Sigma$, $\Gamma \vdash \varphi$ or else $\Gamma \vdash \neg\varphi$.

The canonical Σ-algebra is the sought model: for every $\varphi \in L_\Sigma$, $\mathcal{H}_\Gamma \models \varphi$ iff $\Gamma \vdash \varphi$. This is shown by induction over propositions. The case of atomic formulas is given by (1). There are three other cases: $\neg\psi$, $\varphi_1 \vee \varphi_2$ and $\exists x\psi$.

For negation, $\mathcal{H}_\Gamma \models \neg\psi$ if and only if $\mathcal{H}_\Gamma \not\models \psi$ if and only if, by the inductive hypothesis, $\Gamma \nvdash \psi$ if and only if, by assumption (b), $\Gamma \vdash \neg\psi$.

If $\mathcal{H}_\Gamma \models \varphi_1 \vee \varphi_2$, then $\mathcal{H}_\Gamma \models \varphi_1$ or $\mathcal{H}_\Gamma \models \varphi_2$, hence by the inductive hypothesis, $\Gamma \vdash \varphi_1$ or $\Gamma \vdash \varphi_2$, where by $\vee$ introduction, $\Gamma \vdash \varphi_1 \vee \varphi_2$. Conversely, if $\mathcal{H}_\Gamma \not\models \varphi_1 \vee \varphi_2$, then $\mathcal{H}_\Gamma \not\models \varphi_1$ and $\mathcal{H}_\Gamma \not\models \varphi_2$, hence by the inductive hypothesis, $\Gamma \nvdash \varphi_1$ and $\Gamma \nvdash \varphi_2$. By completeness, it follows that $\Gamma \vdash \neg\varphi_1$ and

$\Gamma \vdash \neg\varphi_2$, hence by $\wedge$-introduction, $\Gamma \vdash \neg\varphi_1 \wedge \neg\varphi_2$, and by a Morgan law, $\Gamma \vdash \neg(\varphi_1 \vee \varphi_2)$. By consistency, it follows that $\Gamma \nvdash \varphi_1 \vee \varphi_2$.

Consider, finally, a proposition $\varphi = \exists x\psi$. By the definition of $\models$ and $\mathcal{H}_\Gamma$, $\mathcal{H}_\Gamma \models \varphi$ is the same as $\psi^{\mathcal{H}_\Gamma}(t) = 1$ for some $t \in \mathrm{T}_\Sigma$, i.e, $\mathcal{H}_\Gamma \models \psi[x := t]$ for some $t \in \mathrm{T}_\Sigma$. By the inductive hypothesis, this is equivalent to $\Gamma \vdash \psi[x := t]$ for some $t \in \mathrm{T}_\Sigma$. Hence if $\mathcal{H}_\Gamma \models \varphi$, then by $\exists$-introduction, it follows that $\Gamma \vdash \exists x\psi$, i.e, $\Gamma \vdash \varphi$. Conversely, if $\Gamma \vdash \varphi$ by assumption (a) on Γ, there exists $t \in \mathrm{T}_\Sigma$ such that $\Gamma \vdash \psi[x := t]$, where by what has preceded, $\mathcal{H}_\Gamma \models \varphi$.

(3) Suppose only that Γ is consistent. Then a construction of an extension $\Gamma' \subseteq \mathrm{L}_{\Sigma'}[X]$ of Γ, $\Sigma \subseteq \Sigma'$, satisfying conditions (a) and (b) of (2) is needed.

First, a *conservative* extension is constructed: $\bar{\Gamma} \subseteq \mathrm{L}_{\Sigma'}[X]$ is an extension of Γ satisfying (a) by 'iteratively' adding new constants c_ψ to the signature and axioms using these constants. Start with $\Sigma_0 = \Sigma$ and $\Gamma_0 = \Gamma$, and construct Σ_{n+1} from Σ_n by adding a constant symbol $c_{\exists\psi}$, and Γ_{n+1} by adding to Γ_n an axiom $\exists\psi \Rightarrow \psi[x := c_{\exists\psi}]$ for each proposition $\exists\psi$ of L_{Σ_n}. Let $\Sigma' = \bigcup_{n\geqslant 0} \Sigma_n$ and $\bar{\Gamma} = \bigcup_{n\geqslant 0} \Gamma_n$.

All that is needed now is a consistent and complete extension Γ' of $\bar{\Gamma}$ over the same language $\mathrm{L}_{\Sigma'}[X]$. The existence (not the construction!) of this extension follows from the application of Zorn's lemma to the set of consistent extensions of $\bar{\Gamma}$. This set is non-empty, since $\bar{\Gamma}$ is consistent, and an increasing chain of consistent theories is consistent. There therefore exists a maximal element Γ' in the set. It is a consistent extension of $\bar{\Gamma}$, and if it were not complete, it could be extended in a consistent manner by adding a proposition φ such that $\Gamma' \nvdash \varphi$ and $\Gamma' \nvdash \neg\varphi$.

$\Gamma \subseteq \Gamma'$ and Γ' satisfy (a), since $\bar{\Gamma} \subseteq \Gamma' \subseteq \mathrm{L}_{\Sigma'}[X]$, and (b). So, $\mathcal{H}_{\Gamma'}$ is a model of Γ', hence also of Γ.

□

Theorem 5.8 If $\Gamma \models \varphi$ then $\Gamma \vdash \varphi$.

Proof It is known that $\Gamma \vdash \varphi$ if and only if $\Gamma \cup \{\neg\varphi\}$ is inconsistent. So if $\Gamma \nvdash \varphi$, $\Gamma \cup \{\neg\varphi\}$ is consistent and has a model $\mathcal{A}$ such that $\mathcal{A} \models \Gamma$ and $\mathcal{A} \models \neg\varphi$, i.e, for every $\xi \in A^X$, $\Gamma^{\mathcal{A}}(\xi) = 1$ and $\varphi^{\mathcal{A}}(\xi) = 0$, which proves $\Gamma \not\models \varphi$. □

Remark This proof is non-constructive because it does not construct the derivation of φ from Γ. Once the equivalence of $\Gamma \models \varphi$ and $\Gamma \vdash \varphi$ is assumed, two possible types of 'proof' are possible:

- deductive, by constructing *one* derivation of φ with assumptions in Γ;
- semantic, by showing that φ is true in *all* the models of Γ.

□

Deductive concepts such as the completeness of a theory can be studied semantically:

Example Field theory is consistent since **Q** is a field—one only need know one model to claim its consistency. It is not complete since $1+1=0$ is false in **Q** and $1+1 \neq 0$ is false in **Z**/2**Z**. Either one of these two propositions can be added to field theory to give a stronger theory (more theorems) which remains consistent. □

Exercises 6 and 7 are examples of very simple syntactic conditions (on the form of axioms) leading to semantic properties: universal propositions and subalgebras, universal–existential propositions and increasing chains. The converses of these properties are also true, but more difficult to prove: they are the Łoś–Tarski and Chang–Łoś–Suszko theorems.

True formulas and provable formulas According to the soundness theorem, every formula provable in Peano arithmetic (PA) is true in **N**. Conversely, every formula true in all models of PA is provable in PA. But it is not sufficient that it be true in **N**. Although 'standard', **N** is only one of many models of PA. It would seem reasonable that there be true formulas in **N** which are not provable in PA, in the same manner that there are formulas (for example, $\forall xy(x+y=y+x)$) true in the group $(\mathbf{Z},+)$ and not provable in group theory. The existence of such formulas is a consequence of Gödel's incompleteness theorem, proven in Chapter 7. There are, however, sufficiently simple formulas which are provable as soon as they are true: this is a property of closed atomic formulas, such as $s(0)+s(0)=s(s(0))$, propositions without quantifiers or with bounded quantifiers.

An arithmetic formula uses *bounded quantifiers* if all of its quantifiers are of the form $(\forall x < t)\varphi$ and $(\exists x < t)\varphi$, where $x \notin \mathrm{var}(t)$. Recall that relative quantifiers are, respectively, abbreviations of $\forall x(x < t \Rightarrow \varphi)$ and $\exists x(x < t \wedge \varphi)$. These formulas describing properties of bounded, hence finite, sets are sufficiently simple that they are provable as soon as they are true, and they are even provable in elementary arithmetic (EA):

Proposition 5.9 If φ is a proposition with bounded quantifiers, then

1. if $\mathbf{N} \models \varphi$, then $EA \vdash \varphi$;
2. if $\mathbf{N} \not\models \varphi$, then $EA \vdash \neg\varphi$.

Proof By induction over propositions with bounded quantifiers.

The case of atoms $t_1 < t_2$ and $t_1 = t_2$ is left as an exercise.

If $\mathbf{N} \models \psi\wedge\chi$, then $\mathbf{N} \models \psi$ and $\mathbf{N} \models \chi$, hence by the inductive hypothesis, $EA \vdash \psi$ and $EA \vdash \chi$, from which $EA \vdash \psi \wedge \chi$. If $\mathbf{N} \not\models \psi \wedge \chi$, then $\mathbf{N} \not\models \psi$ or $\mathbf{N} \not\models \chi$, hence by the inductive hypothesis, $EA \vdash \neg\psi$ or $EA \vdash \neg\chi$, from which $EA \vdash \neg\psi \vee \neg\chi$, and by a Morgan axiom, $EA \vdash \neg(\psi \wedge \chi)$.

Let t be a closed term $m = t^{\mathbf{N}}$. If $\mathbf{N} \models (\exists x < t)\psi$, then $\mathbf{N} \models \psi[x := s^i(0)]$ for an i, $0 < i < m$, hence by the inductive hypothesis, $EA \vdash \psi[x := s^i(0)]$. Now, $EA \vdash s^i(0) < s^m(0)$ and $EA \vdash t = s^m(0)$, so by the axioms of equality, $EA \vdash s^i(0) < t \wedge \psi[x := s^i(0)]$, and by $\exists$-introduction, $EA \vdash \exists x(x < t \wedge \psi(x))$.

If $\mathbf{N} \not\models (\exists x < t)\psi$, then $\mathbf{N} \not\models \psi[x := s^i(0)]$ for every i, $0 < i < m$, hence by the inductive hypothesis, $EA \vdash \neg\psi[x := s^i(0)]$ for each $i < m$. Now, $EA \vdash t = s^m(0)$ and $EA \vdash x < s^m(0) \iff x = 0 \vee \ldots \vee x = s^{m-1}(0)$. So, by $\vee$-elimination and by the equality axioms, $EA \vdash \forall x(x < t \Rightarrow \neg\psi(x))$, hence by the Morgan axioms, $EA \vdash \neg\exists x(x < t \wedge \psi(x))$.

The other cases are treated in an analogous manner. □

5.2.5 A few applications of model theory

Model theory has undergone substantial development. Some results are presented, particularly consequences of the compactness theorem, which give an overview of semantic *phenomena*. The essentially syntactic proof of the compactness theorem assumes the completeness theorem.

Theorem 5.10 (Compactness) A theory Γ has a model if and only if every finite subset of Γ has a model.

Proof Follows from the theorem of the existence of models and from the finitary character of derivations: if $\Gamma \vdash \bot$, there exists a finite subset Γ' of Γ such that $\Gamma' \vdash \bot$. □

Exercises 10 and 11 are classical applications of this result to field theory and to the existence of non-standard integers. In this manner one can show the existence of models, e.g., of arithmetic, of arbitrarily large cardinality. In the other direction, the following proposition shows that the consistent theories used here always have countable models. For example, set theory has a countable model, which seems paradoxical, since there are an uncountable number of sets.

Proposition 5.11 If $\Gamma \subseteq \mathrm{L}_\Sigma[X]$, Σ at most countable, then Γ has a countable model.

Proof Recall the proof of the model existence theorem. If Σ is countable then T_Σ is countable (easy). But, the constructed model $\mathcal{H}_{\Gamma'}$ has as domain $\mathrm{T}_{\Sigma'}$. It suffices to prove that Σ' is countable, which follows from the manner in which it was constructed: starting from Σ, a countable number of constants is added (since $\mathrm{L}_\Sigma[X]$ is countable) iteratively. □

This proposition is a strengthening of Gödel's theorem. There are other results on the cardinality of models, by Skolem–Löwenheim, Vaught, and others. Exercise 12 is one of those results.

A class $\mathcal{E}$ of Σ-algebras is called *axiomatizable*, (respectively *finitely axiomatizable*) or *elementary* if there exists a subset (respectively a finite subset) Γ of $\mathrm{L}_\Sigma[X]$ such that $\mathcal{E} = \mathbf{Mod}(\Gamma)$.

An important concept in model theory is that of elementary equivalence: two Σ-algebras $\mathcal{A}$, $\mathcal{B}$ are *elementarily equivalent*, written $\mathcal{A} \equiv \mathcal{B}$, if for every proposition

$\varphi \in \mathrm{L}_\Sigma$, $\mathcal{A} \models \varphi$ if and only if $\mathcal{B} \models \varphi$. This concept formalizes the intuitive idea of indistinguishability under observations: here, an observation is the value of an interpretation of a proposition. Exercise 15 allows this concept to be linked to the completeness of a theory.

A finer concept of elementary equivalence is that of elementary morphism or of elementary subalgebra. A morphism $m : \mathcal{A} \to \mathcal{B}$ of Σ-algebras is *elementary* if for every formula φ and every valuation $\xi \in A^X$, $\varphi^{\mathcal{A}}(\xi) = \varphi^{\mathcal{B}}(m \circ \xi)$. If m is inclusion, $\mathcal{A}$ is called an elementary subalgebra of $\mathcal{B}$, characterized by exercise 16: a formula with parameters in A, satisfiable in $\mathcal{B}$ by an element of B, is also satisfiable by an element of A.

5.2.6 Kripke interpretations

The semantics of classical logic is very simple: it reduces to an algebraic computation in a finite Boolean algebra. The semantics of intuitionistic logic is much more complex. Only the propositional case will be considered here. From a constructive point of view, knowing that a formula is true cannot be reduced to the result of the computation of a truth value: a result is never an explanation. But if the sequentiality of the computation can be controlled, then a constructive aspect can be given.

It is supposed that the computation of the truth value of a formula φ is done in several steps u, v, ... Let U be the (finite) set of these steps, called a *frame*. A notation is introduced to say that at step u, a formula φ has been verified. Let $I(p)$ be the finite set of steps where the atomic formula p is verified. This function $p \mapsto I(p)$ is called a *Kripke interpretation.* A relation $I, u \Vdash \varphi$ is defined stating that in the interpretation I, at step u, the formula φ is verified. If I is fixed, this relation is abbreviated to $u \Vdash \varphi$ (*u forces* φ).

To take account of the sequential character of the verification process, one must suppose that U is ordered, although not necessarily (totally) by time, since certain steps can be given to separate 'verifiers'. A natural condition is the *monotonicity* of the process: a verified fact is not put into question by a subsequent step, as is denoted at the atomic formula level by

$$\text{if } u \in I(p) \text{ and } u \leqslant v, \text{ then } v \in I(p).$$

In the inductive relation which follows from relation $\Vdash$, the cases of conjunction, disjunction and false are analogous to classical logic. However, extending the monotonicity condition to all formulas imposes a particular form on implication. The semantics requires that modus ponens be allowed at each step: if $u \Vdash \varphi \Rightarrow \psi$ and $v \geqslant u$, then by monotonicity $v \Vdash \varphi \Rightarrow \psi$, and if $v \Vdash \varphi$, by 'local modus ponens at v', $v \Vdash \psi$ is obtained. The following definition of $\Vdash$ satisfies this condition.

1. $u \Vdash p$, if p is an atom and $u \in I(p)$;

2. $u \Vdash \varphi \wedge \psi$ if $u \Vdash \varphi$ and $u \Vdash \psi$;
3. $u \Vdash \varphi \vee \psi$ if $u \Vdash \varphi$ or $u \Vdash \psi$;
4. $u \Vdash \varphi \Rightarrow \psi$ if for every $v \geqslant u$, if $v \Vdash \varphi$ then $v \Vdash \psi$;
5. $u \Vdash \bot$ is not true for any u.

Since $\neg\varphi$ is an abbreviation for $\varphi \Rightarrow \bot$, it follows that

(5′) $u \Vdash \neg\varphi$ if $v \Vdash \varphi$ for no $v \geqslant u$.

Interpretation I can be extended to all formulas by

$$I(\varphi) = \{u \in U; I, u \Vdash \varphi\}.$$

I satisfies φ, written $I \models \varphi$, if $I, u \Vdash \varphi$ for each step $u \in U$, i.e, if $I(\varphi) = U$. Formula φ is true in frame U, written $U \models \varphi$, if all the interpretations with values in U satisfy φ. Finally, φ is intuitionistically valid ($\models \varphi$) if $U \models \varphi$ for every frame U. Soundness and completeness theorems can be proven, as for the classical case.

It is easy to disprove a formula by finding a Kripke interpretation which does not satisfy it.

Example Let $\varphi \equiv p \vee \neg p$, and consider the interpretation I defined by $U = \{a, b\}$, $a \leqslant b$, $I(p) = \{b\}$. At step a, p is not verified since $a \notin I(p)$, nor is $\neg p$ since p is verified in $b \geqslant a$. □

In classical logic, the interpretation $I(\varphi)$ of a formula is an element of the Boolean algebra **2**. Here $I(\varphi)$ is a subset of U.

One might think that the interpretation I has values in the Boolean algebra $\mathcal{P}(U)$. However, $I(\varphi)$ is always an *open* subset of U, i.e., if $u \in I(\varphi)$ and $u \leqslant v$, then $v \in I(\varphi)$ (the open denomination effectively corresponds to open subsets of a topological space). Let $\mathcal{O}(U)$ be the set of open subsets. A Kripke interpretation is therefore a function with values in $\mathcal{O}(U)$, which while not being a Boolean algebra, does have an interesting algebraic structure, called *Heyting algebra*. In general, the set of open sets of a topological space X is an example of a Heyting algebra. The complement of an open set not being open in general, its interior must be taken: define $\sim G = (X \setminus G)^\circ$. Negation is interpreted by this operation and $\sim\sim G = G$ is not always true (only $G \subseteq \sim\sim G$ is guaranteed), which illustrates the behavior of intuitionistic negation. The role of a semantics is to offer intuitive views, as is done here.

Kripke interpretations are close to those defining the semantics of modal, temporal and epistemic logics, where the set U of steps is interpreted by a set of worlds, instants or knowledge states. These logics are used to study certain properties of the execution of sequential and concurrent programs such as termination, liveness and fairness. Artificial intelligence also uses modal logics.

5.3 Herbrand's theorem

Herbrand's theorem essentially consists of doing semantics using syntactic means. It is therefore the semantic technique most suited to computer science. However, since syntactic means are more restricted than general interpretations, semantics will be given through 'approximations': a formula of the predicate calculus will be approximated by its expansions, which are formulas of the propositional calculus.

5.3.1 Formula expansions

The problem is how to treat quantifiers. In semantics, they are interpreted by computing the least or greatest upper bound over an infinite set of elements, not a computable operation. For example, Fermat's theorem has the truth value, in **N**,

$$\inf_{n\geqslant 3}\ \inf_{x,y,z>0}\Big(\text{ if } x^n + y^n = z^n \text{ then } 0 \text{ else } 1\Big),$$

and no one knows how to compute the value. When the domain of an interpretation is finite, a universal quantifier (respectively existential) behaves as a conjunction (respectively disjunction). The quantifiers are 'interpreted' internally, i.e., by only working at the syntactic level of the formula, replacing the quantifiers with the connectives $\wedge$ and $\vee$.

Let D be a finite subset of the set T_Σ of closed terms. To the formula $\varphi \in \mathrm{L}_\Sigma[X]$ is associated its *expansion* under D, $\mathrm{Exp}(\varphi, D)$, which is the *formula* defined by induction in Table 5.2.

$$\begin{aligned}
\mathrm{Exp}(Rt_1\ldots t_n, D) &= Rt_1\ldots t_n\\
\mathrm{Exp}(\varphi_1 \wedge \varphi_2, D) &= \mathrm{Exp}(\varphi_1, D) \wedge \mathrm{Exp}(\varphi_2, D)\\
\mathrm{Exp}(\varphi_1 \vee \varphi_2, D) &= \mathrm{Exp}(\varphi_1, D) \vee \mathrm{Exp}(\varphi_2, D)\\
\mathrm{Exp}(\varphi_1 \Rightarrow \varphi_2, D) &= \mathrm{Exp}(\varphi_1, D) \Rightarrow \mathrm{Exp}(\varphi_2, D)\\
\mathrm{Exp}(\forall x\psi, D) &= \textstyle\bigwedge_{d\in D} \mathrm{Exp}(\psi[x := d], D)\\
\mathrm{Exp}(\exists x\psi, D) &= \textstyle\bigvee_{d\in D} \mathrm{Exp}(\psi[x := d], D)\\
\mathrm{Exp}(\neg\psi, D) &= \neg\, \mathrm{Exp}(\psi, D)
\end{aligned}$$

Table 5.2 The expansion under D, $\mathrm{Exp}(\varphi, D)$

It follows immediately that $\mathrm{Exp}(\varphi, D)$ is an open (quantifier-free) formula having the same free variables as φ.

Example Let $\varphi \equiv \forall x\,(x > 0 \Rightarrow \exists y\,(x = sy))$ and $D = \{0, ss0\}$. Then

$$\begin{array}{rcl}\mathrm{Exp}(\varphi, D) & = & (0 > 0 \Rightarrow (0 = s0) \vee (0 = sss0)) \\ & \wedge & (ss0 > 0 \Rightarrow (ss0 = s0) \vee (ss0 = sss0)).\end{array}$$

□

What is the relationship between a formula and its expansions? Consider the simple case of an atom with one quantified variable. By applying rules $E_\forall$ and $I_\wedge$ in the universal case and rules $I_\exists$ and $E_\vee$ in the existential case, the result is

$$\begin{array}{rcl}\forall x\,Rx & \vdash & Ra_1 \wedge \ldots \wedge Ra_n, \\ Ra_1 \vee \ldots \vee Ra_n & \vdash & \exists x\,Rx.\end{array}$$

Hence, an expansion is an 'approximation from above' in the first case, 'from below' in the second. It depends on the quantifier. In the case of an arbitrary formula, the place of quantifiers must be taken into account, and an analogous result is achieved only when the quantifiers all have the same force. A formula is called *universal* if all of its occurrences of $\forall$ are positive and all of its occurrences of $\exists$ are negative. Exchanging 'positive' and 'negative' gives the definitions of *existential* formulas.

In the preceding example, φ is neither universal nor existential, since the occurrences of $\forall$ and of $\exists$ are both positive. A prenex formula is universal (respectively existential) if all of its quantifiers are $\forall$ (respectively $\exists$). The transformation of a universal formula into an equivalent prenex formula is immediate: it suffices to place all the quantifiers at the front by replacing the $\exists$, necessarily negative, by $\forall$. One can then continue to 'normalize' the prenex form, for example by computing its conjunctive normal form.

Lemma 5.12 If φ is a universal (respectively existential) formula, then for all D,

$$\vdash \varphi \Rightarrow \mathrm{Exp}(\varphi, D) \qquad \text{(respectively} \quad \vdash \mathrm{Exp}(\varphi, D) \Rightarrow \varphi).$$

Rather than give a proof, consider the computation of the expansion and its behavior. Represent φ by a tree whose leaves are atoms and nodes are the connectives or quantifiers. Each node defines a subformula of φ. The expansion of φ is done in an ascending manner, i.e, by successively computing the expansions of its subformulas and by following the 'direction' of the implication between a formula and its expansion. The leaves hold atoms α and $\vdash \alpha \Leftrightarrow \mathrm{Exp}(\alpha, D) = \alpha$. When passing through a $\neg$ node, the direction of the implication is reversed. When passing through a $\wedge$ or $\vee$ binary node, the direction is preserved if it is the same for the two children. For an $\Rightarrow$ node the direction of the right-hand node is preserved if the directions of the children are different. When passing through $\forall$ (respectively $\exists$), the direction is preserved, if the direction is $\Rightarrow$ (respectively $\Leftarrow$). In the other cases there is no ascending transmission of relations between subformulas and their expansions. As for the sign of the subformula, it is transmitted from the root, where it is 0, towards the leaves (changing along the way on a $\neg$ or to the left of a $\Rightarrow$).

If φ is universal, the ascending propagation is possible and satisfies the invariant: the direction of implication is $\Rightarrow$ (respectively $\Leftarrow$) for a positive (respectively negative) subformula. It follows that at the root, which is a positive subformula, $\vdash \varphi \Rightarrow \mathrm{Exp}(\varphi, D)$.

In the terminology of attribute grammars, the sign of a formula is an inherited attribute while its expansion is a synthesized attribute.

To conclude, in the case of a universal or existential formula, its expansion is a syntactically simpler approximation for $\vdash$, from above or below.

5.3.2 Herbrand's theorem

In the following discussion, only closed formulas will be considered. If φ is closed, the expansion $\mathrm{Exp}(\varphi, D)$ is both open and closed, i.e., a formula of the propositional calculus over the set At_Σ of closed atoms. The expansion $\mathrm{Exp}(\varphi, D)$ has a value in **2** when the closed terms are interpreted in **2**, which means considering a Herbrand Σ-algebra. Recall that a Herbrand algebra is a subset B of At_Σ composed of the atoms of value 1 in this interpretation.

Let $\mathcal{A}$ be a Σ-algebra. Associate the Herbrand Σ-algebra $\mathcal{H}_\mathcal{A}$ defined by the set of closed atoms α such that $\mathcal{A} \models \alpha$. Write $Th_\forall(\mathcal{A})$ for the set of universal closed formulas φ such that $\mathcal{A} \models \varphi$.

Proposition 5.13

1. The Herbrand algebra $\mathcal{H}_\mathcal{A}$ is a model of theory $Th_\forall(\mathcal{A})$.
2. A universal theory has a model if and only if it has a Herbrand model.

Proof

(1) The Herbrand interpretation defined for the closed atoms can be extended to all the open formulas by

$$(\varphi^o)^{\mathcal{H}_\mathcal{A}}(\eta) = (\varphi^o)^{\mathcal{A}}(\imath_\mathcal{A} \circ \eta),$$

where $\imath_\mathcal{A}$ is the unique (initial) Σ_f-morphism from $\mathcal{H}_\mathcal{A}$ to $\mathcal{A}$.

Let $\varphi \in Th_\forall(\mathcal{A})$, assumed prenex to simplify: $\varphi = \forall \vec{x} \varphi^o$, where φ^o is open. By assumption, $\varphi^\mathcal{A} = 1$. Compute $\varphi^{\mathcal{H}_\mathcal{A}}$:

$$\begin{aligned} \varphi^\mathcal{H} &= \inf_{\eta \in \mathrm{T}_\Sigma^X} (\varphi^o)^\mathcal{H}(\eta) \\ &= \inf_{\eta \in \mathrm{T}_\Sigma^X} (\varphi^o)^\mathcal{A}(\imath_\mathcal{A} \circ \eta) \\ &\geqslant \inf_{\xi \in A^X} (\varphi^o)^\mathcal{A}(\xi) \\ &= 1, \end{aligned}$$

i.e, $\mathcal{H}_\mathcal{A} \models \varphi$.

(2) Let Γ be a theory formed of universal propositions. If Γ has a model $\mathcal{A}$, then $\mathcal{H}_\mathcal{A}$ is a model of Γ, since $\mathcal{H}_\mathcal{A} \models Th_\forall(\mathcal{A})$ and $\Gamma \subseteq Th_\forall(\mathcal{A})$. The converse is trivial. □

Theorem 5.14 (Herbrand) A universal proposition φ is unsatisfiable if and only if there exists an expansion $\text{Exp}(\varphi, D)$ which is unsatisfiable.

Proof To simplify the proof, suppose that φ is prenex, $\varphi = \forall \vec{x} \varphi^o$. In that case,

$$\text{Exp}(\varphi, D) = \bigwedge_{\theta:\text{var}(\varphi^o)\to D} \theta(\varphi^o)$$

is a conjunction of propositions taken from the set

$$\text{Exp}(\varphi) = \{\theta(\varphi^o); \theta : \text{var}(\varphi^o) \to \text{T}_\Sigma\}$$

of all the closed instances of φ^o.

Suppose that every expansion is satisfiable. The theory $\text{Exp}(\varphi)$ will be shown to be satisfiable. In fact, let $E = \{\theta_1(\varphi^o), \ldots, \theta_n(\varphi^o)\}$ be an arbitrary finite subset of $\text{Exp}(\varphi)$ and define $D = \bigcup_{i=1}^n Im(\theta_i)$, so that the θ_i take values in D. By assumption, $\text{Exp}(\varphi, D)$ has a model, which satisfies in particular each of the elements of E. Therefore, every finite subset of $\text{Exp}(\varphi)$ is satisfiable, which implies, by the compactness theorem, that $\text{Exp}(\varphi)$ is satisfiable. But it is a universal theory, which therefore has a Herbrand model, $\mathcal{H}$, from the preceding proposition. It remains to be shown that $\mathcal{H}$ is a model of φ, which follows from the following computation:

$$\begin{aligned}
\varphi^\mathcal{H} &= \inf_{\eta:\text{var}(\varphi^o)\to\text{T}_\Sigma} (\varphi^o)^\mathcal{H}(\eta) \\
&= \inf_{\eta:\text{var}(\varphi^o)\to\text{T}_\Sigma} (\eta(\varphi^o))^\mathcal{H} \\
&= \inf_{\psi\in\text{Exp}(\varphi)} \psi^\mathcal{H} \\
&= 1.
\end{aligned}$$

Conversely, φ being universal, $\vdash \varphi \Rightarrow \text{Exp}(\varphi, D)$, so every model of φ is a model of $\text{Exp}(\varphi, D)$. □

The 'dual' version of the Herbrand theorem, for an existential proposition φ, can be deduced from the universal case (exercise 19):

1. φ is valid if and only if φ is true in every Herbrand algebra;
2. φ is valid if and only if there exists a valid expansion of φ, i.e, for $\varphi = \exists \vec{x} \varphi^o$, there exist substitutions $\theta_i : \text{var}\, \varphi^o \to \text{T}_\Sigma$, $i = 1, \ldots, n$, such that $\bigvee_i \theta_i(\varphi^o)$ is a tautology of the propositional calculus.

The real Herbrand theorem is in fact a much stronger result: it states that the closed substitutions θ_i can be effectively computed from a derivation of φ and that, conversely, a derivation of φ can be effectively constructed from a derivation of $\bigvee_i \theta_i(\varphi^o)$, and that these constructions have a 'limited' complexity. The substitutions θ_i constitute a supplementary information on the truth of a formula: the Herbrand theorem allows the transformation of a validity proof into a 'computation' of the closed terms. A systematic implementation of this idea is used in logic programming.

5.3.3 Skolemization

Since the Herbrand theorem cannot be applied to an arbitrary formula, the formulas will be translated by skolemization before applying the theorem. Properties of the transformation, presented in Chapter 3, will be proven here.

To simplify the following proof, only prenex formulas will be used. Let $\psi \in \mathrm{L}_\Sigma[x_1, \ldots, x_k, x_{k+1}]$ and $\varphi \equiv \forall x_1 \ldots \forall x_k \, \exists x_{k+1} \, \psi \in \mathrm{L}_\Sigma$.

$$\begin{aligned} \tilde{\Sigma} &= \Sigma \cup \{F\}, \\ \tilde{\varphi} &= \forall x_1 \ldots \forall x_k \, \psi[x_{k+1} := F x_1 \ldots x_k] \in \mathrm{L}_{\tilde{\Sigma}} \end{aligned}$$

were defined, F a new k-ary function symbol.

The following lemma will be used to prove the Herbrand–Skolem theorem (next page).

Lemma 5.15

1. $\models \tilde{\varphi} \Rightarrow \varphi$.
2. Every Σ-algebra $\mathcal{A}$ can be extended to a $\tilde{\Sigma}$-algebra $\tilde{\mathcal{A}}$ such that $\varphi^{\tilde{\mathcal{A}}} = \varphi^{\mathcal{A}}$.
3. φ is satisfiable if and only if $\tilde{\varphi}$ is satisfiable.

Proof

(1) It suffices to prove that for every $\tilde{\Sigma}$-algebra $\mathcal{A}$, $\tilde{\varphi}^{\mathcal{A}} \leqslant \varphi^{\mathcal{A}}$ under order $0 \leqslant 1$ over $\{0, 1\}$:

$$\begin{aligned} \tilde{\varphi}^{\mathcal{A}} &= \inf_{a_1, \ldots, a_k} \psi^{\mathcal{A}}(a_1, \ldots, a_k, F^{\mathcal{A}}(a_1, \ldots, a_k)) \\ &\leqslant \inf_{a_1, \ldots, a_k} \sup_{a_{k+1}} \psi^{\mathcal{A}}(a_1, \ldots, a_k, a_{k+1}) \\ &= \varphi^{\mathcal{A}}. \end{aligned}$$

(2) Let $\mathcal{A}$ be a Σ-algebra. If $\varphi^{\mathcal{A}} = 0$, then for every extension $\mathcal{B}$ of $\mathcal{A}$, $\tilde{\varphi}^{\mathcal{B}} = 0$, from (1). If $\varphi^{\mathcal{A}} = 1$, it is because for every $a_1, \ldots, a_k \in A$, there exists $a_{k+1} \in A$ such that $\psi^{\mathcal{A}}(a_1, \ldots, a_k, a_{k+1}) = 1$. Define $F^{\mathcal{A}}(a_1, \ldots, a_k) = a_{k+1}$ and let $\tilde{\mathcal{A}}$ be the extension of $\mathcal{A}$ so defined. Then $\tilde{\varphi}^{\mathcal{A}} = 1$.

(3) If φ is satisfiable, so is $\tilde{\varphi}$, from (2). Conversely, every model of $\tilde{\varphi}$ is a model of φ, from (1).

□

Theorem 5.16 Let φ be a proposition, φ^u (respectively φ^e) the universal (respectively existential) skolemized form of φ. Then

1. $\vdash \varphi^u \Rightarrow \varphi$ and $\vdash \varphi \Rightarrow \varphi^e$;
2. φ is satisfiable if and only if φ^u is satisfiable;
3. φ is sound if and only if φ^e is sound;
4. φ is unsatisfiable if and only if there exists an unsatisfiable expansion of φ^u;
5. φ is sound if and only if there exists a sound expansion of φ^e.

5.3.4 Problems in automatic theorem-proving

Automatic theorem-proving is an applied area of computer science, in which algorithms from logic are used to solve problems involving deduction.

It has been shown that the statements '$\Gamma \vdash \varphi$', '$\Gamma \models \varphi$,' '$\Gamma \cup \{\neg\varphi\}$ inconsistent' and '$\Gamma \cup \{\neg\varphi\}$ unsatisfiable' are equivalent, as are those when Γ is finite: '$\vdash \varphi$', '$\models \varphi$', '$\neg\varphi$ inconsistent' and '$\neg\varphi$ unsatisfiable'. For these kinds of problems, of validity or unsatisfiability, typical methods are oriented towards proving inconsistency, i.e., work towards the negation $\neg\varphi$: $\perp$ is derived from $\neg\varphi$. The problem of satisfiability, 'does Γ have a model?', is the 'dual' of the preceding one.

The essential step is skolemization. Equivalence-preserving algebraic transformations are then made to facilitate deductive methods of refutation, in particular the so-called *resolution*. To a proposition φ is associated a set $S(\varphi)$ of clauses with the property that φ is satisfiable if and only if $S(\varphi)$ is. Herbrand's theorem is applied to $S(\varphi)$. A *clause* of a first-order language is a universally quantified disjunction of literals.

1. For every $\varphi \in \mathrm{L}_\Sigma$, a universal φ^u can be constructed such that φ is satisfiable if and only if φ^u is satisfiable.
2. φ^u is immediately transformed into an equivalent prenex formula $\forall\vec{x}\varphi^o$ by moving all the quantifiers to the head, the $\exists$ being replaced by $\forall$.
3. For every open formula φ^o, a conjunctive formula φ^c can be constructed such that $\vdash \varphi^o \iff \varphi^c$.
4. If $\varphi^c = \varphi_1 \wedge \ldots \wedge \varphi_m$, let

$$S(\varphi) = \{\forall\vec{x}\varphi_1, \ldots, \forall\vec{x}\varphi_m\}.$$

$S(\varphi)$ is a set of clauses, and

$$\vdash \varphi^u \iff \bigwedge_{C \in S(\varphi)} C.$$

Careful! The first transformation only preserves satisfiability, while the last three produce a formula equivalent to φ^u.

Proposition 5.17 Let φ be a proposition. Then:

1. $S(\varphi) \vdash \varphi$.
2. If φ is satisfiable, $S(\varphi)$ is as well.

Herbrand's theorem is the basis for unsatisfiability proofs: test the $\mathrm{Exp}(\varphi^u, D)$ for all the possible $D \subseteq \mathrm{T}_\Sigma$. It suffices to construct an increasing sequence of finite subsets D_n whose union is T_Σ to obtain a semi-decision procedure, called Herbrand's procedure, for the unsatisfiability problem of φ, since for every n there is a decision procedure for the unsatisfiability of $\mathrm{Exp}(\varphi^u, D_n)$, for example using truth tables. Let

$$\begin{aligned}
D_0 &= \Sigma_0, \\
D_{n+1} &= \{ft_1 \ldots t_p; f \in \Sigma_p, t_1, \ldots, t_p \in D_n\} \\
S(\varphi) &= \{C_1, \ldots, C_m\}, \\
u_p &= \operatorname{card} \Sigma_p, \qquad 0 \leqslant p \leqslant l, \\
v_i &= \operatorname{card} \operatorname{var}(C_i), \qquad 1 \leqslant i \leqslant m,
\end{aligned}$$

where Σ_p is the set of p-ary functional symbols of Σ and l is the maximum arity of symbols in Σ. For $n \geqslant 0$,

$$\begin{aligned}
d_n &= \operatorname{card} D_n, \\
h_n &= \operatorname{card} \mathrm{Exp}(\varphi^u, D_n),
\end{aligned}$$

must be evaluated, where this expansion is seen as a set of $m$clauses. Then (see exercise 22)

$$\begin{aligned}
d_0 &= u_0, && (5.1) \\
d_{n+1} &= u_0 + u_1 d_n + u_2 (d_n)^2 + \ldots + u_l (d_n)^l, && (5.2) \\
h_n &= (d_n)^{v_1} + \ldots + (d_n)^{v_m}. && (5.3)
\end{aligned}$$

In general the tremendous size of the sets of propositional clauses to test necessitates the use of finer methods.

5.4 The semantics of computation

The method taken with first-order logic was first to define a syntax (signature, terms, variables, formulas), then to define an interpretation scheme for the syntactic objects, associating values (algebras, valuations, interpretations) to these objects. The Boolean algebra **2** was used, and the connectives were interpreted as Boolean operations.

For computation, the same method must be used. The syntactic objects are programs written in programming languages. For the purposes of this book, they are the terms of the λ-calculus, of rewriting systems or of recursive definitions. The values associated with them will not be from a Boolean algebra, but, rather, from a complete partial order. Before introducing the interpretation schemes, it will be shown how an order relation can give meaning to a computation.

The ideas presented here are due to D. Scott, whose work came forty years after the formalization of computability by Church, Gödel and Turing, discussed in Chapter 7, and Tarski's semantic approach to logic. The originality of Scott's contribution cannot be denied. His work led to the first model of the λ-calculus. In computer science, he devised, along with C. Strachey, the denotational semantics of programming languages, which today constitutes a reference for other semantic approaches. Since the work of Scott, the natural framework for denotational semantics is that of complete partial orders.

Scott's main idea was to describe computation using a relation of information approximation. This idea is quite familiar in numerical analysis, but there it is more related to the nature of the objects themselves and the quality of the result, rather than the process itself in the computation.

5.4.1 Partial functions and complete orders

One of the essential operational problems of computation, such as it is practised in programming, or analyzed in the λ-calculus, is termination. The non-termination of computations will be interpreted using partial functions. A partial function from A to B is written $f : A \rightharpoonup B$. The same function is a (total) function from a subset $D(f) \subseteq A$, called the *domain* of f, which takes values in B. The notation $f(a)\downarrow$ means that the partial function f is defined at a, i.e, $a \in D(f)$. Otherwise, write $f(a)\uparrow$.

The following discussion uses partial functions from $\mathbf{N}^k$ to $\mathbf{N}$. Write $\mathcal{F}(\mathbf{N}^k, \mathbf{N})$, or simply $\mathcal{F}$, for their set.

The graph of $f : \mathbf{N}^k \rightharpoonup \mathbf{N}$ is the set

$$G(f) = \{(\vec{m}, n) \in \mathbf{N}^k \times \mathbf{N}; f(\vec{m})\downarrow \text{ and } f(\vec{m}) = n\}.$$

Its projection into $\mathbf{N}^k$ is the domain $D(f)$ of f. Write $f \subseteq g$ if $G(f) \subseteq G(g)$: this relation means that g is an extension of f, i.e, that if $f(\vec{m})\downarrow$ then $g(\vec{m})\downarrow$ and

$g(\vec{m}) = f(\vec{m})$. Information is gained in passing from f to g. Each of the lower bounds of g constitutes partial information about g.

Consider the ordered set $(\mathcal{F}, \subseteq)$. There is a smallest element $\varnothing$, the partial function with empty domain. The total functions are the maximal elements. Two arbitrary functions do not necessarily have a common upper bound, but should they do so, then they have a least upper bound.

More generally, a non-empty subset E of an ordered set X is *consistent* (respectively *directed*) if two arbitrary elements have a common upper bound in X (respectively in E).

Definition 5.2 A *complete partial order* (cpo) is an ordered set with a smallest element, generally written $\perp$, such that every directed subset E has a least upper bound $\sup E$.

Note that every increasing sequence in a cpo has a least upper bound.

For approximation to have meaning in computer science, it is normal to work in a complete space. It is very difficult in mathematics to do analysis seriously (in particular functional analysis) in non-complete metric spaces. For the example of $\mathcal{F}$ and others which will follow, the order relations model an intuitive notion of information approximation. So it seems natural that computational objects be elements of a complete partial order. The set $\mathcal{F}$ of partial functions is an important example (see exercise 23). Others are

- complete lattices, in particular the set $\mathcal{P}(X)$ of subsets of a set;
- the *flat* orders $X_\perp = X \cup \{\perp\}$, where $\perp \notin X$, under the order $x \leqslant y$ if $x = y$ or $x = \perp$;
- the sets of formal terms $\mathrm{T}^\infty_{\Sigma\cup\{\Omega\}}$.

The framework of complete partial orders is still too big. The set $\mathcal{F}$ satisfies other important properties, in particular, a concept of finiteness. To interpret the effective character of computation, the subject of Chapter 7, finite approximations are used (since only the finite is representable in extension; certain infinite sets are representable by intension, by formulas or by programs, and a semantics must be given to those non-elementary representations). The approximations of a partial function f are all the partial functions of which f is the extension. There are some which are finite: they are the partial functions of finite domain, which can be effectively described extensionally using a finite sequence of pairs $(\vec{m}, n)$. It is clear that a partial function is the least upper bound of its finite approximations. To compute a partial function is to reach it through its finite approximations.

The concept of finite object can be formalized without referring to the cardinality of a set. In a complete partial order, an element x is called *compact*, or isolated, if for every directed subset E, $x \leqslant \sup E$ implies the existence of $x_0 \in E$ such that $x \leqslant x_0$.

A complete partial order X is *algebraic* if for every $x \in X$, the set of its compact lower bounds is directed and has x as least upper bound. The set $\mathcal{F}$ is algebraic (see exercise 24). Suppose, furthermore, that the set of compact elements is countable, as always for reasons of computability. This is evidently the case for $\mathcal{F}$.

Definition 5.3 A *(Scott) domain* is a complete, algebraic, partial order in which the set of compact elements is countable.

A computation is described using domains. How is a function $f : X \to Y$ computed in this framework? For each $x \in X$, $f(x)$ is approximated from an approximation of x, so a definition of continuity over f is needed.

A function $f : X \to Y$ between complete ordered sets is *continuous* if for every directed subset E of X, $f(\sup E) = \sup f(E)$. In particular, a continuous function is monotone (take for E a pair $\{a, b\}$ with $a \leqslant b$). In the case where X is an algebraic order, a function is continuous if for every $x \in X$, $f(x)$ is the least upper bound of the set of images of the compact lower bounds of x: it suffices to compute over 'finite' objects.

The following result, along with the slogan

$$\textit{recursive definition} \quad = \quad \textit{fixpoint equation}$$

explains the utility of this modeling by itself.

Theorem 5.18 Let X be a complete ordered set and $\Phi : X \to X$ a continuous function. Then Φ has a least fixpoint, written $\mu\Phi$.

Proof Since Φ is monotone, and since $\perp$ is the smallest element of X, the sequence $\perp$, $\Phi(\perp)$, $\Phi^2(\perp)$, ... , $\Phi^n(\perp), \ldots$ is increasing, and hence directed. By the completeness of X, this sequence has a least upper bound $\mu\Phi$, and by the continuity of Φ,

$$\begin{aligned} \Phi(\mu\Phi) &= \Phi(\sup_{n \geqslant 0} \Phi^n(\perp)) \\ &= \sup_{n \geqslant 0} \Phi(\Phi^n(\perp)) \\ &= \sup_{n \geqslant 1} \Phi^n(\perp) \\ &= \mu\Phi. \end{aligned}$$

Hence $\mu\Phi$ is a fixpoint of Φ. If b is another fixpoint, $\perp \leqslant b$ implies $\Phi^n(\perp) \leqslant \Phi^n(b) = b$, hence $\mu\Phi \leqslant b$: $\mu\Phi$ is the least fixpoint. □

Consider the classical example of recursive definitions, to be studied in more detail in the next section. Consider the definition

$$f(x, y) = \begin{cases} y & \text{if } x = 0 \\ f(x-1, y+1) & \text{otherwise.} \end{cases}$$

Interpret this definition in $\mathcal{F}$. It is a fixpoint equation for the operator $\Phi : \mathcal{F} \to \mathcal{F}$ defined by

$$\Phi : f \mapsto (x, y) \mapsto \begin{cases} y & \text{if } x = 0 \\ f(x-1, y+1) & \text{otherwise.} \end{cases}$$

Φ is a continuous operator. The least fixpoint of Φ is the least upper bound of the sequence $\Phi^n(\varnothing)$.

5.4.2 Flat orders, strict and sequential functions

The preceding approach using partial functions is not satisfactory, compared to the semantics of first-order logic, where every formula receives a value (0 or 1) in a given interpretation, be it provable or not. If f is a partial function, and $n \in D(f)$, the value $f(n)$ exists. But if $n \notin D(f)$, $f(n)$ does not exist. A fairer treatment for both cases consists of introducing a value $\perp \notin \mathbf{N}$ to define $f(n) = \perp$ if $n \notin D(f)$.

For every set X, define the *flat* ordered set $X_\perp = X \cup \{\perp\}$, where $\perp \notin X$, under the order $x \leqslant y$ if $x = y$ or $x = \perp$. This ordered set is complete, having as directed subsets only the $\{\perp, x\}$ for $x \in X$ and the one-element subsets.

The symbol $\perp$, the smallest element of $X_\perp$ is the minimal information about a computation, expressing an undefined result, the result of a non-terminating computation, or, better, one *which has not yet terminated*. In the same way that the elements of $\mathcal{F}$ are *partial* functions, the elements of $\mathbf{N}_\perp$ can be considered as *partial* integers: an ordinary integer is a 'total' integer, completely computed, while $\perp$ is a (very) incompletely computed integer.

A function $u : X_\perp \to Y_\perp$ is monotone if for every $x \in X$, $\perp \leqslant x$ (the only possible inequality in $X_\perp$) implies $u(\perp) \leqslant u(x)$, hence $u(\perp) = u(x)$ or $u(\perp) = \perp$. In the first case u is constant, in the second, u is *strict*. In the case of flat orders, a monotone function is necessarily continuous.

The set of continuous functions from E to F is written $[E \to F]$. It is given the pointwise order: $f \leqslant g$ if for every $x \in E$, $f(x) \leqslant g(x)$. Now, it is $[\mathbf{N}_\perp \to \mathbf{N}_\perp]$ which plays the role of $\mathcal{F}(\mathbf{N}, \mathbf{N})$. Fixpoint equations can also be solved here, because of the following property:

Proposition 5.19 $[E \to F]$ is complete if E and F are complete partial orders.

Proof The constant function of value $\perp_F$ is the smallest element of $[E \to F]$. Furthermore, if D is a directed subset of $[E \to F]$, then the set of $f(d)$ for $d \in D$ is also directed, and its least upper bound is computed pointwisely: $(\sup D)(x) = \sup_{f \in D} f(x)$. □

A recursive definition

$$f(x) = \text{if } x = 0 \text{ then } 1 \text{ else } x * f(x-1)$$

is interpreted as a fixpoint equation for the operator $\Phi : [\mathbf{N}_\perp \to \mathbf{N}_\perp] \to [\mathbf{N}_\perp \to \mathbf{N}_\perp]$:

$$f \quad \mapsto \quad \lambda x \in \mathbf{N}_\perp . \, \text{if } x = 0 \text{ then } 1 \text{ else } x * f(x-1).$$

The successive approximations $f_n = \Phi^n(\perp_{\mathbf{N}_\perp \to \mathbf{N}_\perp})$ of $\mu\Phi$ are easy to compute:

$$\begin{aligned} f_0(x) &= \perp \quad \text{for every } x \in \mathbf{N}_\perp, \\ f_1(x) &= \begin{cases} 1 & \text{if } x = 0 \\ \perp & \text{otherwise,} \end{cases} \\ f_n(x) &= \begin{cases} x! & \text{if } x < n \\ \perp & \text{otherwise.} \end{cases} \end{aligned}$$

Scott's induction principle is generally used to prove properties about the least fixpoint:

Proposition 5.20 Let D be a complete order, P be a predicate over D stable under least upper bounds, i.e, if $E \subseteq D$ is directed and E satisfies P, then $\sup E$ satisfies P, and $\Phi : D \to D$ be a continuous function. If

(1) $P(\perp)$, and
(2) $P(x) \Rightarrow P(\Phi(x))$ for every $x \in D$,

then $P(\mu\Phi)$.

Proof From (1) and (2), and from the monotonicity of Φ, by induction over the integers, $P(\Phi^n(\perp))$ for every $n \geqslant 0$. Since P is stable under sup, it follows that $P(\sup_{n\geqslant 0} \Phi^n(\perp))$. If Φ is continuous, $\mu\Phi = \sup_{n\geqslant 0} \Phi^n(\perp)$ therefore satisfies P. □

There is a correspondence between partial functions and strict functions. If $f : X \rightharpoonup Y$ is a partial function between two sets X and Y, define the (total) function $f^s : X_\perp \to Y_\perp$, called the *strict extension* of f, by:

$$f^s(x) = \begin{cases} f(x) & \text{if } x \in D(f) \\ \perp & \text{otherwise.} \end{cases}$$

Not all functions from $X_\perp$ to $Y_\perp$ are strict extensions of partial functions. For example, there are two kinds of constant function: strict extensions c^s of constant functions from X to Y, and 'real' ones.

By replacing partial functions with their extensions over flat orders, new objects must now be considered, such as those non-strict constant functions. It is easy to give an operational meaning to such objects. A strict function can be evaluated using call by value: its argument is first evaluated, and if the value is defined, the result of the function is as well. However, the argument of a *non-strict* constant function must not be evaluated, since the result can be given immediately, and the evaluation of the argument is unnecessary, possibly dangerous if it does not terminate. Call by name must be used. Abstract interpretation, described below, can be used to determine when a function can be called by value without any danger.

If in the case of flat orders, non-strict monotone functions seem to be the exception (only the constant functions), there are many others in more 'rich' complete orders. Consider the case of the Cartesian product: if X and Y are complete orders, then the Cartesian product $X \times Y$, with the order $(x, y) \leqslant_{X \times Y} (x', y')$ if $x \leqslant_X x'$ and $y \leqslant_Y y'$, is also a complete order.

Definition 5.4 Let D_i be a complete order, and $f : D_1 \times \ldots \times D_n \rightarrow D_0$ a continuous function. Then f is strict in its i-th argument, written strict(i), if

$$f(x_1, \ldots, x_{i-1}, \bot_{D_i}, x_{i+1}, \ldots, x_n) = \bot_D.$$

It is important to be able to know that a function is strict in an argument: the evaluation of that argument can then be done *safely*, i.e., the non-termination of that evaluation will not prevent a value from being produced. The arithmetic operations of programming languages are generally strict in each of their arguments. The two arguments of an addition must be evaluated, and the result of an addition diverges as soon as one of the arguments diverges ($x + \bot = \bot + y = \bot$, for all $x, y \in \mathbf{N}_\bot$).

In most programming languages, where parameter passing is done by value, the functions defined by a program are strict. However, certain predefined functions, or special forms in the Lisp sense, have a different evaluation mode and are not strict.

For example, the conditional (`if _ then _ else _` in ML, (`_ ? _ : _`) in C) is strict in its first argument, which is always evaluated, and non-strict in the two last arguments. Only one will be evaluated, depending on the value of the first argument.

Each Boolean operator has a strict version and several non-strict versions. The `or` of Pascal always evaluates its two arguments, but the `or` of Lisp (as well as the 'short-cut' `or else` of Ada) always evaluates its first argument, and should it yield the value *true*, returns *true* without evaluating its second argument. The **and** does not evaluate its second argument if the first argument is *false*, which allows one to write expressions such as `if x != 0 and (1/x < 1) ...` . The Boolean operations therefore do not have the same semantics in all languages. Let $\mathbf{2}_\bot = \{\bot, 0, 1\}$ be the flat order over the Booleans 0 and 1. Every function from $\mathbf{2}^n$ to $\mathbf{2}$ has a unique strict extension and several other interesting extensions. Consider the case of conjunction. The strict extension is defined by $\bot \wedge^s x = x \wedge^s \bot = \bot$: it is the least defined extension. The extension corresponding to the **and** form of Lisp, called left sequential extension, satisfies

$$\begin{aligned} \bot \wedge^l x &= \bot, \\ 0 \wedge^l \bot &= 0, \\ 1 \wedge^l \bot &= \bot. \end{aligned}$$

It is strict only in its first argument. It is an example of a sequential function.

Definition 5.5 A function $f : D_1 \times \ldots \times D_n \to D_0$ is *sequential in* $x \in D_1 \times \ldots \times D_n$ if there exists an index i such that for every $y \geqslant x$ for which $y_i = x_i$, $f(x) = f(y)$. Then f is sequential if it is for all x.

We show that $\wedge^l$ is sequential. It is sequential in $(\perp, x)$ in the index 1, since for every y, $\perp \wedge^l x = \perp \wedge^l y (= \perp)$; in $(0, x)$, $(1, 0)$ and $(1, 1)$ in indices 1 and 2, since the result in **2** cannot be more defined for those arguments; and in $(1, \perp)$ in index 2.

One can design an extension of conjunction which is more defined than $\wedge^g$; it is the parallel extension of conjunction: $\perp \wedge^p 0 = 0 \wedge^p \perp = 0$ and $\perp \wedge^p 1 = 1 \wedge^p \perp = \perp \wedge^p \perp = \perp$. It is not sequential since in $(\perp, \perp)$ the result can be improved in two different manners. This extension is not implemented in sequential languages. A parallel implementation consists of evaluating the two arguments in parallel and stopping as soon as one of them has the value 0. A sequential simulation would evaluate the arguments in turn, which would ensure termination should one of the arguments obtain the value 0.

More generally, for every function $f : \mathbf{2}^n \to \mathbf{2}$, there is an extension f^s, strict in each of its arguments, which is the minimal continuous extension of f, and a maximal continuous extension f^p. To compute $f^p(x_1, \ldots, x_n)$ where $x_1, \ldots, x_n \in \mathbf{2}_\perp$, compute the set $F(x_1, \ldots, x_n)$ of all the $f(x'_1, \ldots, x'_n)$, where $x'_i = x_i$ if $x_i \in \mathbf{2}$, and $x'_i \in \mathbf{2}$ if $x_i = \perp$, and let

$$f^p(x_1, \ldots, x_n) = \begin{cases} a & \text{if } F(x_1, \ldots, x_n) = \{a\} \\ \perp & \text{if } \operatorname{card} F(x_1, \ldots, x_n) > 1. \end{cases}$$

Another useful construction is that of the *sum* of two domains. Let D_1 and D_2 be two domains whose smallest elements are $\perp_1$ and $\perp_2$. Define $D = D_1 + D_2 = D_1 \cup D_2 \cup \{\perp\}$, by extending the orders over D_1 and D_2 by $\perp < \perp_1$ and $\perp < \perp_2$. There are then several layers of partial information: if $d = \perp$, nothing is known about d; if $d = \perp_i$, it is known that $d \in D_i$. This operation is useful in programming, to define concrete data types and matching functions:

```
datatype D = one of D1 | two of D2 ;
```

Among the strict functions over D, the 'discriminating' functions only need to know the constructor of the argument:

```
fun f (one _) = 1 | (two _) = 2 ;
```

$f(\perp) = \perp$, but $f(\perp_1) = 1$ and $f(\perp_2) = 2$.

5.4.3 Infinite objects and approximations

Constructing a flat order introduced partial elements, which play an important role in Cartesian products. For other constructions, it is necessary to introduce infinite elements: the most important example is that of infinite lists.

Scott's methods are useful since domains can be constructed for all the usual data structures; a typical example is that of lists. (Finite) lists are finite terms over a many-sorted signature formed of a constant `[]` and a binary constructor `::` (`cons`). The concept of strict function is not precise enough. For example, the function `l` computing the length of a list is strict in the sense that $\mathtt{l}(\bot) = \bot_{\mathbf{N}}$, but it only requires a limited evaluation of its argument. Only its skeleton need be known:

$$\mathtt{l}(\bot :: x) = 1 + \mathtt{l}(x).$$

The `hd` function only requires the evaluation of the head of a list, and the `tl` function the rest of the list.

These remarks are of interest only if `cons` does not evaluate its arguments systematically. Programming languages have therefore been conceived where `cons`, unlike the predefined arithmetic operations, does not evaluate its arguments (or, at least, does not evaluate its second argument): a call `cons a l` can, for example, create a pair of pointers to closures containing the unevaluated arguments and the current environment, which are awoken subsequently by `hd` or `tl`. These are *lazy* pairs.

Suppose that the function `cons` is non-strict, and consider the function $x \mapsto 1 :: x$ from D to D. It is a continuous function as soon as the binary constructor $::$ is continuous. It has a fixpoint, the limit of the increasing sequence $\bot$, $1 :: \bot$, $1 :: 1 :: \bot$, $1 :: 1 :: 1 :: \bot$, $\ldots$. If the operation $::$ were strict in its second argument, then it would be the constant sequence $\bot$. Otherwise the limit *represents* the infinite list $1 :: 1 :: 1 :: \ldots$.

This problem will be examined in the more general framework of infinite terms. These were introduced in Chapter 1, by removing the finiteness restriction on the tree domain which had been imposed to ensure the description of finite terms. Let Σ be a signature. In this section, the elements of T_Σ are called *finite terms*, and those of T_Σ^∞ *formal terms*. Two possible definitions of approximation were given in Chapter 1: using a distance or using an order relation by introducing a constant Ω. The resulting structures are a complete metric space and a complete ordered set. Each of these structures allows the modeling of a computation and the proof of, for example, fixpoint theorems, which is important in studying the algebraic semantics of rewriting systems, and program schemes.

The principle of structural induction allows only proofs of properties for finite terms. To be able to work with infinite terms, the principle of *continuous induction* is used. It combines structural induction with finite partial terms with a passage to the limit, made possible by the algebraic character of $\mathrm{T}^\infty_{\Sigma\cup\{\Omega\}}$. Depending on the case, the true properties of all formal terms, finite or not, or simply of non-finite terms are of interest. In the second case, structural induction will be used over a subsignature formed of symbols of arity $\geqslant 1$ (since a constant is not a 'base case' for infinite terms!).

Proposition 5.21 Let P be a predicate over $\mathrm{T}^\infty_{\Sigma\cup\{\Omega\}}$, stable under least upper bound, and $\Sigma' \subseteq \Sigma$. If

1. $P(\Omega)$,

2. $P(M_1), \ldots, P(M_n)$ imply $P(fM_1 \ldots M_n)$ for every $f \in \Sigma'$of arity n,

then $P(M)$ for every $M \in T^{\infty}_{\Sigma' \cup \{\Omega\}}$.

Proof By structural induction, P is true over $T_{\Sigma' \cup \{\Omega\}}$. But every formal term $M \in T^{\infty}_{\Sigma' \cup \{\Omega\}}$ is the least upper bound of its finite approximations $\tau_k(M) \in T_{\Sigma' \cup \{\Omega\}}$. Since P is stable under least upper bound, it follows that $P(M)$ for every $M \in T^{\infty}_{\Sigma' \cup \{\Omega\}}$. □

Example Consider the following program for the concatenation of two lists:

```
fun append l l'
  = if (null l) then l' else (hd l)::(append (tl l) l') ;
```

It will be shown that if `::` is not strict, then `append l l' = l` if `l` is an infinite list. Structural induction is used over partial lists, i.e., lists of the form `[a;b;c;...;`Ω`]`: there are two cases, the base case for Ω, and the induction step for `l` to `a::l`. First, `if` is strict in its first argument, and `null` is strict; it follows that `append` Ω `l' =` Ω. Furthermore, if `append l l' = l`, then `append (a::l) l' = a::(append l l') = a::l`. On the other hand, it is not true that `append [] l' = []`, so this property does not hold for (real) non-partial finite lists. □

5.4.4 Models

As for first-order terms and formulas, the interpretation schemes are defined from the syntactic structure of expressions. We will examine two important cases, the λ-calculus and program schemes.

λ-calculus

The first attempt to give meaning to λ-calculus terms would be to interpret abstractions as functions and applications as applications of a function to an argument. Unfortunately, set theory does not allow the application of a function to itself, and so there can be no interpretation of a term such as $\lambda x \,.\, xx$.

The situation is simpler in the typed λ-calculus where self-application is not allowed by the syntax. It is easy to imagine an interpretation where each type τ is interpreted by a set D_τ, where $D_{\tau_1 \to \tau_2}$ is an appropriate set of functions from D_{τ_1} to D_{τ_2}. For example, $D_{\mathbf{num}} = \mathbf{N}$, $D_{\mathbf{num} \to \mathbf{num}} = \mathbf{N}^{\mathbf{N}}$, or else $D_{\mathbf{num}} = \mathbf{N}_\perp$, $D_{\mathbf{num} \to \mathbf{num}} = [\mathbf{N}_\perp \to \mathbf{N}_\perp]$.

In the case of the pure λ-calculus, the syntax must be followed more closely. Discussion will begin with combinatory logic, a variant of the λ-calculus with no bound variables.

When Σ is an applicative signature, formed of a unique binary symbol **App** and of a certain number of constant symbols, a Σ-algebra is called an *applicative algebra*. The applicative notation will still be used, omitting the **App** symbol, and associating parentheses to the left. In an applicative algebra $\mathcal{A}$, the 'product' ab of two elements a and b of A represents the application of the 'function' a to the 'argument' b. In the case of combinatory logic, the constants are $\boldsymbol{S}$, $\boldsymbol{K}$ and $\boldsymbol{I}$ and the first-order axioms are

$$\begin{aligned} \boldsymbol{S}xyz &= xz(yz), \\ \boldsymbol{K}xy &= x, \\ \boldsymbol{I}x &= x. \end{aligned}$$

A model of this theory, in the sense of first-order logic, in which the constants $\boldsymbol{S}$ and $\boldsymbol{K}$ are interpreted by two distinct elements, is called a *combinatory algebra*. The existence of these algebras is a consequence of the Church–Rosser theorem, which implies the consistency of the equational theory (two distinct normal forms are not interconvertible, by the Church–Rosser property). It is easy to construct a combinatory algebra in a standard manner by taking the quotient of the set of terms under an equivalence relation (see Chapter 5). This model does not tell us anything about the calculus: if in this model, the self-application $a(a)$ has a meaning, nothing says that the left a acts as a function.

The first model of the λ-calculus, D_∞, was constructed by Scott in 1969 (unpublished). A second, simpler model, $P\omega$, is due to Plotkin (1972 [121]).

Free variables are treated as usual using an environment. Write $[\![M]\!]\rho$ for the value of a term M in the environment ρ. This notation is standard in denotational semantics, and corresponds to the $M^{\mathcal{D}}(\rho)$ used in first-order logic. The value $[\![M]\!]\rho$ is defined using the syntax of the terms:

$$\begin{aligned} [\![x]\!]\rho &= \rho(x) \\ [\![MN]\!]\rho &= ([\![M]\!]\rho)([\![N]\!]\rho) \\ [\![\lambda x\, M]\!]\rho &= \lambda\!\!\lambda d \in D\,.\,[\![M]\!](\rho; (x = d)) \end{aligned}$$

where $\lambda\!\!\lambda d \in D\,.\,foo$ is the λ-notation for a function $d \mapsto foo$ (it is not a λ-term!), and $\rho; (x, d)$ designates the environment ρ extended by the binding $(x = d)$.

Here is the construction of the model $P\omega$ of Plotkin. Write $P\omega$ for the set of subsets of $\mathbf{N}$, ordered by set inclusion. It is a complete order (also a complete lattice) and even algebraic, a compact element being a finite subset of $\mathbf{N}$. A continuous function can therefore be computed using finite approximations: $f : P\omega \to P\omega$ is continuous if and only if for every $x \in P\omega$,

$$f(x) = \bigcup\{f(a); a \text{ finite } \subseteq x\}.$$

In other words, $n \in f(x)$ if there exists a finite subset a of x such that $n \in f(a)$ (in Chapter 1, a function $f : \mathcal{P}(X) \to \mathcal{P}(X)$ having this property was called *finitary*). One can then represent f by the set of pairs (a, n) such that $n \in f(a)$. But these pairs are *finite* objects which can be coded by integers in the standard manner (these encodings are defined and used in Chapter 7). A finite subset $\{i_1, \dots, i_n\}$

is coded by the integer $m = \sum_{k=1}^{p} 2^{i_k}$. Let $a = e_m$, where the elements of e_m are the positions occupied by 1's in the binary representation of m. There is also an encoding $\langle\rangle : \mathbf{N}^2 \to \mathbf{N}$ which codes the pair (e_m, n) by the pair of integers (m, n), then by the integer $\langle m, n\rangle$, and this encoding is bijective.

Hence, a continuous function from $P\omega$ to $P\omega$ is represented by a set of integers, i.e., by an element of $P\omega$: if $f \in [P\omega \to P\omega]$, define

$$G(f) = \{\langle m, n\rangle; n \in f(e_m)\}.$$

It is now easy to interpret application by a binary operation over $P\omega$: if $a, b \in P\omega$, define

$$ab = \{n \in \mathbf{N}; (\exists e_m \subseteq b)\langle m, n\rangle \in a\}.$$

In other words, $n \in ab$ if there exists an element $\langle m, n\rangle$ of a—seen as a function—allowing the computation of $a(b)$, i.e., in which m is a finite approximation of b. From this binary operation is derived a 'homothety' H: for $a \in P\omega$, $H(a)$ is the function $x \mapsto ax$. Conversely, the application is defined from H by $ab = H(a)(b)$. It is easily shown that $H(a)$ is continuous, that

$$\begin{aligned} G &: [P\omega \to P\omega] \to P\omega \\ H &: P\omega \to [P\omega \to P\omega] \end{aligned}$$

are continuous, and that

$$(H \circ G)(f) = f$$

for every $f \in P\omega$. It is now possible to interpret λ-terms in $P\omega$:

$$\begin{aligned} [\![x]\!]\rho &= \rho(x), \\ [\![MN]\!]\rho &= ([\![M]\!]\rho)([\![N]\!]\rho), \\ [\![\lambda x\, M]\!]\rho &= G(\lambda\!\!\lambda a \in P\omega \,.\, [\![M]\!](\rho; (x = a)) \\ &= \{\langle m, n\rangle; n \in [\![M]\!](\rho; (x = a))\}. \end{aligned}$$

Here are a few examples of interpretation:

$$\begin{aligned} [\![\lambda x\, x]\!] &= \{\langle m, n\rangle; n \in e_m\}, \\ [\![\lambda xy \,.\, x]\!] &= \{\langle m_1, \langle m_2, n\rangle\rangle; n \in e_{m_1}\}, \\ [\![\lambda xyz \,.\, xz(yz)]\!] &= \{\langle m_1, \langle m_2, \langle m_3, n\rangle\rangle\rangle; n \in e_{m_1}e_{m_3}(e_{m_2}e_{m_3})\}. \end{aligned}$$

Two examples will be used to show how this model $P\omega$ can be used to study the consistency of extensions of the λ-calculus.

Unsolvable terms What is the meaning of a non-normalizable term? It is tempting to identify 'non-normalizable' with 'undefined'. However, there do exist non-normalizable terms which have a useful and well-defined operational behavior, such as the fixpoint combinators. It is, furthermore, inconsistent to equate all non-normalizable terms. Suppose the two non-normalizable terms $M \equiv \lambda x \,.\, x\mathbf{K}\Omega$ and

$N \equiv \lambda x \,.\, x\mathbf{S}\Omega$ are equated, where $\Omega \equiv \omega\omega$, with $\omega \equiv \lambda x \,.\, xx$. If $M = N$, then $M\mathbf{K} = N\mathbf{K}$, i.e, $\mathbf{K} = \mathbf{S}$, which implies (exercise) that $x = y$ for arbitrary variables x and y.

The concept of a normalizable term is much too syntactic. A semantic condition characterizing 'undefined' terms is needed. In the model $P\omega$, an appropriate condition would be $[\![M]\!] = \varnothing$, the smallest element of $P\omega$. Exercise 31 shows that this is the case for Ω. It is easy to show (see exercise 30) that if $[\![M]\!] = \varnothing$, then $[\![MN]\!] = [\![\lambda x M]\!] = \varnothing$. This is clearly a different concept from non-normalizability, since MN can be normalizable without M being normalizable (see $Y(\mathbf{KI})$ in the following example). Barendregt and Wadsworth introduced a finer concept:

Definition 5.6 A closed term is *solvable* if there exist terms $N_1, \ldots, N_p$ such that $MN_1 \ldots N_p = \mathbf{I}$. An arbitrary term M is solvable if the closed term $(\lambda x_1 \ldots x_n \,.\, M)$ is solvable, with $\{x_1, \ldots x_n\} = \text{var}(M)$. A term is *unsolvable* if it is not solvable.

Example $\mathbf{K}$ is solvable since $\mathbf{KI}x = \mathbf{I}$. The term $Y \equiv \lambda f \,.\, (\lambda x \,.\, f(xx))(\lambda x \,.\, f(xx))$ is solvable since $Y(\mathbf{KI}) = \mathbf{KI}(Y(\mathbf{KI})) = \mathbf{I}$. The term Ω is unsolvable since ΩN can only yield by β-reduction a term of the form $\Omega N'$. □

From exercise 30, and since $[\![\mathbf{I}]\!] \neq \varnothing$, it is clear that M is unsolvable if $[\![M]\!] = \varnothing$: a proof of unsolvability is obtained from Ω by semantic means, without considering reductions. The converse, whose proof can be found in [8, theorem 19.1.10], is more difficult.

Theorem 5.22 M is unsolvable if and only if $[\![M]\!]^{P\omega} = \varnothing$.

There is no risk, as for the non-normalizables, of equating all terms, by equating the unsolvable terms: this equivalence relation is consistent with the λ-calculus. The unsolvable terms can be considered as undefined terms, all carrying the same amount of information, which is minimal. They also have a characterization with respect to β-reduction.

Definition 5.7 A term M is in *head normal form (hnf)* if it is of the form

$$\lambda x_1 \ldots x_p \,.\, yM_1 \ldots M_q,$$

with $p, q \geqslant 0$. Variable y is called the head variable. Terms which are not in hnf are necessarily of the form $\lambda x_1 \ldots x_p \,.\, (\lambda y M_0)M_1 \ldots M_q$. The redex $(\lambda y M_0)M_1$ is called the *head redex*. A term M has a hnf if there exists an M' in hnf such that $M = M'$.

A term is solvable if and only if it has a hnf [8, theorem 8.3.14]. Furthermore, the hnf of a term, if it exists, can be attained by a *head reduction*, i.e., by contracting the head redex at each step. For example, Y is not a hnf, having the head redex $(\lambda x \,.\, f(xx))(\lambda x \,.\, f(xx))$. Contracting yields $\lambda f \,.\, f((\lambda x \,.\, f(xx))(\lambda x \,.\, f(xx)))$, a hnf with f as head variable.

So, if unsolvable terms are considered to have no value, it suffices to attempt to reduce a term to hnf.

The λ-calculus with pairs As a second application of these semantic methods, *pairs* are added to the λ-calculus. Pairing combinators have already been defined:

$$[M,N] \equiv \lambda z\,.\,zMN,$$
$$\pi^1 \equiv \lambda p\,.\,p\mathbf{T},$$
$$\pi^2 \equiv \lambda p\,.\,p\mathbf{F}.$$

So

$$\pi^1[M,N] \xrightarrow{\star} M, \qquad \pi^2 \xrightarrow{\star} N,$$

but it is not true that $[\pi^1 P, \pi^2 P] \xrightarrow{\star} P$ for an arbitrary term M. One might ask what is the 'meaning' of the first projection of a term which is not a pair.

As a combinator satisfying the three preceding reduction properties does not exist (Barendregt 1984 [8]), the λ-calculus is *extended* by adding constant symbols π, π^1, and π^2 with the axioms

$$\pi^1(\pi xy) = x, \qquad \pi^2(\pi xy) = y, \qquad \pi(\pi^1 x)(\pi^2 x) = x.$$

Is the resulting theory, $\lambda\pi$, consistent, i.e, do these axioms imply the equality of all terms? The simplest proof of the consistency of the pure λ-calculus uses the Church–Rosser theorem: two new distinct terms cannot be interconverted. Unfortunately, the natural extension of β-reduction by the three rules

$$\pi^1(\pi xy) \longrightarrow x, \qquad \pi^2(\pi xy) \longrightarrow y, \qquad \pi(\pi^1 x)(\pi^2 x) \longrightarrow x,$$

is not confluent (see Klop [84]). A proof of consistency by semantic methods is needed. (A proof of syntactic consistency is, however, possible; see Klop and de Vrijer [85]).

In the same way that the axiom $1+1=0$ is proven to be consistent with field theory by showing that it is satisfied in the field $\mathbf{Z}/2\mathbf{Z}$, the consistency of $\lambda\pi$ is proven by showing that $P\omega$ is a model. Let

$$f(a,b) = \{2m; m \in a\} \cup \{2m+1; m \in b\},$$
$$f_1(c) = \{m; 2m \in c\},$$
$$f_2(c) = \{m; 2m+1 \in c\}.$$

These three functions satisfy

$$f_1(f(a,b)) = a, \qquad f_2(f(a,b)) = b, \qquad f(f_1(c), f_2(c)) = c.$$

However, to interpret the constants π, π^1 and π^2, elements of $P\omega$ are needed, while it was functions over $P\omega$ which were constructed. In fact, by a continuity argument, it can be shown that these functions are representable by elements of $P\omega$, which establishes the consistency of $\lambda\pi$.

The introduction of pairs in the λ-calculus is done more naturally using a type system. The development of categorical combinatory logic, due to P.-L. Curien [31, 70], and its interest for the compilation of CAML have shown that pairs constitute an essential data structure for the 'internal' manipulation of semantic objects linked to the evaluation of terms, such as environments and stacks.

Recursive definitions

Let Σ be a signature formed of primitive function symbols u, v, w, $\ldots$. Typically, for programs over the integers, Σ contains 0, the successor s, the $+$ and $*$ operations, and a ternary conditional *cond.* A *recursive definition of* Δ *over* Σ is a set of equations (one for each symbol of Δ),

$$f x_1 \ldots x_r = M_f,$$

where $f \in \Delta$, $r = \mathrm{ar}(f)$, $x_1, \ldots, x_r \in X$, $M_f \in \mathrm{T}_{\Sigma \cup \Delta}[x_1, \ldots, x_r]$.

In the case of the λ-calculus, the fact that $P\omega$ (as for D_∞) is a complete partial order does not appear to be fundamental, since λ-calculus programs are terms, not equations: the fixpoint operators allow the resolution of equations in an internal manner. However, the semantics of recursive definitions is based on the resolution of fixpoint equations: complete orders become natural.

Σ is interpreted in a complete partial order D, each symbol $u \in \Sigma$ of arity $n \geqslant 0$ being interpreted by a continuous function $u^{\mathcal{D}} \in [D^n \to D]$. By definition, $\mathcal{D}$ is thus a *complete* Σ*-algebra.* Each term $M \in \mathrm{T}_\Sigma[X]$ has an interpretation $M^{\mathcal{D}}(\rho) = [\![M]\!]\rho$ in the environment $\rho : X \to D$.

To simplify, consider $\Delta = \{f\}$, a symbol of arity r. If h is an arbitrary continuous function of D^r to D, the complete Σ-algebra $\mathcal{D}$ can be extended into a complete $\Sigma \cup \{f\}$-algebra $\mathcal{D}'$ by interpreting f by h. Each term $M \in \mathrm{T}_{\Sigma \cup \{f\}}[X]$, in an environment ρ, therefore has an interpretation $[\![M]\!]'\rho$ in this extended algebra. If $x_1, \ldots, x_n$ are variables of M, a continuous function $\tilde{M}(h) : D^n \to D$ is derived by

$$\tilde{M}(h)(d_1, \ldots, d_n) = [\![M]\!]'[x_1 := d_1; \ldots ; x_n := d_n].$$

Let this construction be applied to a recursive definition of f over Σ, where $f(x_1, \ldots, x_r) = M_f$. Let

$$\begin{aligned} \Phi : [D^r \to D] &\to [D^r \to D], \\ h &\mapsto \tilde{M}_f(h). \end{aligned}$$

The recursive definition is then interpreted as the fixpoint equation $\Phi(h) = h$. Since Φ is continuous, this equation has a least solution $f^{\mathcal{D}} = \mu\Phi \in [D^r \to D]$, which is, by definition, the interpretation of f *computed* in $\mathcal{D}$ by the recursive definition.

The complete Σ-algebra $\mathcal{D}$ is therefore extended into a complete $\Sigma \cup \{f\}$-algebra. So each term $M \in \mathrm{T}_{\Sigma \cup \{f\}}$ can be interpreted in $\mathcal{D}$, and so also can *programs* of the form

`let` $f(x_1, \ldots, x_r) \;=\; M_f$ `in` M `end ;`

Among all the interpretations, a particular role is played by the complete *initial* Σ-algebra. Since T_Σ is in a standard way a Σ-algebra, the set $H = \mathrm{T}^{\infty}_{\Sigma \cup \{\Omega\}}$ of formal partial terms, with the same interpretation of symbols, is a complete Σ-algebra, initial in the category of complete Σ-algebras. The initiality morphism

$i_{\mathcal{D}}$ from $\mathrm{T}^{\infty}_{\Sigma\cup\{\Omega\}}$ to a Σ-algebra $\mathcal{D}$ is constructed as in the case of T_{Σ}, by adding $i_{\mathcal{D}}(\Omega) = \perp_D$.

Computing the fixpoint of $\Phi : [H^r \to H] \to [H^r \to H]$ can be done formally in $\mathrm{T}^{\infty}_{\Sigma\cup\{\Omega\}}[X]$: $\mu\Phi$ is the least upper bound of the increasing sequence $(F_k)_{k\geqslant 0}$ of finite partial terms, called the *Kleene sequence*:

$$\begin{aligned} F_0 &= \perp_{[H^r\to H]}(x_1, \ldots, x_r) = \Omega, \\ F_{k+1} &= \Phi(F_k)(x_1, \ldots, x_r). \end{aligned}$$

The limit $F_\infty = \sup F_k$ is a formal term, which is by definition the *formal interpretation* of f. The initiality property allows the verification that for every complete Σ-algebra $\mathcal{D}$, $i_{\mathcal{D}}(F_\infty) = f^{\mathcal{D}}$.

Example Consider the recursive definition of f over $\Sigma = \{a, u, v\}$:

$$f(x) = u(x, a, f(v(x))).$$

The associated operator $\Phi : [H \to H] \to [H \to H]$ is defined by

$$\Phi(h)(M) = u(M, a, h(v(M)))$$

and the first elements of the Kleene sequence are

$$\begin{aligned} F_0 &= \perp_{[H\to H]}(x) = \Omega, \\ F_1 &= \Phi(F_0)(x) = u(x, a, \Omega), \\ F_2 &= \Phi(F_1)(x) = u(x, a, u(v(x), a, \Omega)), \\ F_3 &= \Phi(F_2)(x) = u(x, a, u(v(x), a, u(v(v(x)), a, \Omega))), \ldots . \end{aligned}$$

□

5.4.5 Abstract interpretations

A model of the λ-calculus must satisfy the theory of β-conversion, i.e., if two terms M and N are β-convertible ($M =_\beta N$), then $[\![M]\!]\rho = [\![N]\!]\rho$ for every environment ρ ($\mathcal{D} \models M = N$). The result is that the reduction mechanism is interpreted by equality. It means a considerable impoverishment of the operational content of the calculus—much as the deductive content of a logic disappears in its interpretations—hence a simplification from which one might hope to derive powerful methods. In fact, the interesting aspect of Scott's method is not to interpret everything in a standard way (in $\mathcal{F}$, $\mathbf{N}_\perp$, $\mathrm{T}^{\infty}_{\Sigma\cup\{\Omega\}}$ or $P\omega$). Rather, a semantic approach is taken, and, as for first-order logic, the power of this approach lies in the liberty of interpretation. Hence, other domains can be used, for example finite domains. The method of *abstract interpretation* fits in this framework.

It is often used for optimizing programs. An example from non-strict functional language compilation is examined here: strictness analysis of a function, due to A. Mycroft [113]. The case of recursive definitions is studied here.

Let $\mathcal{D}$ be the domain under which the standard interpretation of programs is applied, for example $\mathbf{N}_\perp$ and $[\mathbf{N}_\perp \to \mathbf{N}_\perp]$. A recursive definition of f defines an element $f^{\mathcal{D}} = \mu\Phi$, or f, in $[D^k \to D]$, as the least fixpoint of a continuous operator Φ. Since it is impossible to decide in a static manner if $f^{\mathcal{D}}$ is strict, abstract interpretations can be used to find sufficient conditions.

An abstract domain with two elements, $D^{\#} = \{0, 1\}$, is introduced. Ordered by $0 < 1$, it is 'abstract' since it is much simpler than $D = \mathbf{N}_\perp$. The abstraction function $abs : D \to D^{\#}$ is defined by

$$abs(d) = \begin{cases} 0 & \text{if } d = \perp \\ 1 & \text{otherwise.} \end{cases}$$

$D^{\#}$ must be made into a complete Σ-algebra. For $u \in \Sigma$, write $u^{\#}$ instead of $u^{\mathcal{D}^{\#}}$ for the interpretation of u, and $[\![M]\!]^{\#}$ for the interpretation of a term $M \in \mathrm{T}_\Sigma$. A recursive definition yields a continuous operator $\Phi^{\#} : [D^{\#} \to D^{\#}] \to [D^{\#} \to D^{\#}]$ whose least fixpoint $\mu\Phi^{\#}$ is written $f^{\#}$.

The abstract interpretation is *correct* if for every i, i is a strict index of $f^{\mathcal{D}}$ as soon as i is a strict index of $f^{\#}$:

$$f^{\#}(a_1, \ldots, 0, a_{i+1}, \ldots, a_r) = 0 \Rightarrow f^{\mathcal{D}}(d_1, \ldots, \perp, d_{i+1}, \ldots, d_r) = \perp.$$

Conditions over $\mathcal{D}^{\#}$ and *abs* ensuring this correctness property are needed.

A condition to impose is

$$abs(u(x_1, \ldots, x_n)) \leqslant_{D^{\#}} u^{\#}(abs(x_1), \ldots, abs(x_n))$$

for every $u \in \Sigma$, of arity n and $x_1, \ldots, x_n \in D$.

This condition is always verified if $u^{\mathcal{D}}$ is strict (it is not necessary to choose $u^{\#}$ strict). However, it imposes conditions on $u^{\#}$ if $u^{\mathcal{D}}$ is not strict. Consider the usual sequential conditional

$$\begin{aligned} cond(\perp, x, y) &= \perp, \\ cond(true, X, \perp) &= x, \\ cond(false, \perp, y) &= y. \end{aligned}$$

The second equality implies $x \leqslant cond^{\#}(1, x, 0)$ for $x \in \mathcal{D}^{\#}$, hence $cond^{\#}(1, 1, 0) = 1$ and the third equality implies $cond^{\#}(1, 0, 1) = 1$. The other values of $cond^{\#}$ can be arbitrary.

This property implies

$$abs([\![M]\!]\rho) \leqslant_{D^{\#}} [\![M]\!]^{\#}(abs \circ \rho)$$

for every term $M \in \mathrm{T}_\Sigma[X]$ and $\rho : X \to D$ (the proof is easy by induction). In particular,

$$abs([\![F_k]\!](x_i := d_i)) \leqslant_{D\#} [\![F_k]\!]^\#(x_i := abs(d_i))$$

for all the elements of the Kleene sequence of f. If it is also supposed that *abs* is continuous, more exactly

$$abs \sup d_n \leqslant \sup abs(d_n)$$

for every increasing sequence (d_n) of D, it follows by going to the limit that

$$abs(f^{\mathcal{D}}(d_1, \ldots, d_r)) \leqslant_{D\#} f^\#(abs(d_1), \ldots, abs(d_r)),$$

and more generally that

$$abs([\![M]\!]\rho) \leqslant_{D\#} [\![M]\!]^\#(abs \circ \rho),$$

for every term $M \in \mathrm{T}_{\Sigma \cup \{f\}}[X]$ and every environment $\rho : X \to D$.

So

$$abs(f^{\mathcal{D}}(d_1, \ldots, \bot, d_{i+1}, \ldots, d_r)) \leqslant f^\#(abs(d_1), \ldots, 0, abs(d_{i+1}), \ldots, abs(d_r)) = 0$$

if $f^\#$ is strict in i and if $abs(\bot) = 0$. If it is also supposed that $abs(d) = 0$ implies $d = \bot$, it follows that

$$f^{\mathcal{D}}(d_1, \ldots, \bot, d_{i+1}, \ldots, d_r) = \bot,$$

which proves that $f^{\mathcal{D}}$ is strict in i, and establishes the correctness of the abstract interpretation. So the following has been shown:

Proposition 5.23 Let $\mathcal{D}$ and $\mathcal{D}^\#$ be two interpretations of Σ and $abs : D \to D^\#$ a mapping satisfying:

1. $abs(d) = \bot_\#$ if and only if $d = \bot$;
2. $abs(u(x_1, \ldots, x_n)) \leqslant_{D\#} u^\#(abs(x_1), \ldots, abs(x_n))$, for every $u \in \Sigma$ and $x_1, \ldots, x_n \in D$;
3. $abs(\sup d_n) \leqslant \sup abs(d_n)$ for every increasing sequence in D.

Then the interpretation $\mathcal{D}^\#$ is correct, i.e, for every i, i is a strict index of $f^{\mathcal{D}}$ as soon as i is a strict index of $f^\#$.

The problem is to find the best abstract interpretation which is correct. Obviously, the best would be to choose $\mathcal{D}^\# = \mathcal{D}$: no information would be lost, but then no static calculus could be done! In the usual case where $D^\# = \{0, 1\}$, $u^\#$ is chosen as follows:

$$u^\#(a_1, \ldots, a_r) = \sup\{abs(u(d_1, \ldots, d_r)); \forall i\ abs(d_i) \leqslant a_i\}.$$

Consider abstract interpretations in $D^\# = \{0, 1\}$. By equating $D^\#$ with the set of Booleans, a continuous function is represented by a combination of the connectives $\wedge$ and $\vee$, and each constant by 1. The conditional is then interpreted by $(b, x, y) \mapsto b \wedge (x \vee y)$, and the arithmetic operations (which are strict) by $\wedge$. The conditions of the preceding proposition are verified.

Examples Consider the (non-recursive) definition

$$f(x, y) = \text{if } (x = 0) \text{ then } x + y \text{ else } x.$$

$f^{\#}(x, y) = (x \wedge 1) \wedge ((x \wedge y) \vee x) = x$, hence $f^{\#}(0, y) = 0$ and $f^{\#}(x, 0) = x$. So $f^{\#}$ is strict(1), and non-strict(2). So the only deduction is that f is strict(1).

One can interpret the recursive definition of the factorial

$$f(x) = \text{if } (x = 0) \text{ then } 1 \text{ else } x * f(x - 1)$$

in $D^{\#} = \{0, 1\}$. So $f^{\#}(x) = (x \wedge 1) \wedge (1 \vee (x \wedge f^{\#}(x \wedge 1)))$. This equation can be simplified to $f^{\#}(x) = x \wedge f^{\#}(x)$. The least fixpoint of this equation is computed by iteration and is obtained on its first iteration: it is $f^{\#}(x) = 0$, which proves that f is strict. □

Exercises[2]

1. Show that for every closed term t, $t^{\mathcal{A}} = \iota_{\mathcal{A}}(t)$, and that for every term t, $t^{\mathcal{A}}(\xi) = \hat{\xi}_{\Sigma}(t)$.

2.★ Show that $\{\textbf{nor}\}$ and $\{\textbf{nand}\}$ are bases for connectives, and that they are the only bases formed of a unique binary connective.

3.★ Show that $\{\top, \Rightarrow\}$ and $\{\wedge, \vee\}$ are not bases for connectives.

4. Show using $\models$ that the formula $\exists y \forall x (R(y, x) \Longleftrightarrow \neg R(x, x))$ does not have a model. An application is Russell's barber problem.

5. Show that $\models \exists x (\varphi(x) \wedge \psi) \Longleftrightarrow \exists x \varphi(x) \wedge \psi$ if $x \notin \text{var}(\psi)$.

6. If $\mathcal{A}$ is a sub-Σ-algebra of $\mathcal{B}$ (define this concept, using as an example the definition of subgroups), and if φ is a universally quantified prenex proposition (i.e., with only $\forall$), then $\mathcal{B} \models \varphi$ implies $\mathcal{A} \models \varphi$. Deduce that if Γ is formed of universal propositions, a subalgebra of a model of Γ is a model of Γ. Does this result apply to ring theory? to field theory?

7. If $(\mathcal{A}_n)_{n \geqslant 0}$ is an increasing sequence (for inclusion) of Σ-algebras, and if Γ is formed of prenex propositions of the form $\forall \vec{x} \exists \vec{y} \psi$, ψ without quantifier, then $A = \bigcup_{n \geqslant 0} A_n$ is the domain of a model $\mathcal{A}$ of Γ. Does this result apply to field theory? to the theory of discrete ordered sets?

8.★ Construct Kripke interpretations which do not satisfy the formulas $(p \Rightarrow q) \vee (q \Rightarrow p)$, $(p \Rightarrow (q \vee r)) \Rightarrow ((p \Rightarrow q) \vee (p \Rightarrow r))$, $(p \Rightarrow q) \Rightarrow (\neg p \vee q)$, 'Peirce's axiom'.

[2]Hints to the exercises labeled with a ★ can be found at the end of the book.

9. Show that a Kripke frame U satisfies $U \models p \vee \neg p$ for every interpretation in U if and only if the order over U is discrete, i.e, $u \leqslant v$ implies $u = v$.

10.★ Let φ be a proposition of the language of fields. Show that if $k \models \varphi$ for all the fields k of characteristic 0, then there exists an integer p such that $k \models \varphi$ for any field k of characteristic greater than p.

11. A constant c is added to the signature of arithmetic, as are the axioms $0 < c$, $s(0) < c$, $s(s(0)) < c$, ... to those of (elementary) arithmetic. Show, using the compactness theorem, that the resulting theory has a model different from the standard model $\mathbf{N}$.

12. Show that if Γ is a theory with arbitrarily large models, then Γ has an infinite model. Is the class of finite groups axiomatizable?

13. Show that a class $\mathcal{E}$ of algebras is finitely axiomatizable if and only if there exists φ such that $\mathcal{E} = \mathbf{Mod}(\{\varphi\})$.

14.★ Show that $\mathcal{E}$ is finitely axiomatizable if and only if $\mathcal{E}$ and $\mathbf{Alg}_\Sigma \setminus \mathcal{E}$ are axiomatizable.

15.★ Show that a theory Γ is complete if and only if two of its arbitrary models are elementarily equivalent.

16. Show that $\mathcal{A}$ is an elementary subalgebra of $\mathcal{B}$ if and only if for every formula $\varphi \in \mathrm{L}_\Sigma[x, y_1, \ldots, y_n]$, if $\varphi^{\mathcal{B}}(b, a_1, \ldots, a_n) = 1$ with $b \in B$ and $a_1, \ldots, a_n \in A$, there exists $a \in A$ such that $\varphi^{\mathcal{B}}(a, a_1, \ldots, a_n) = 1$.

17. If $(\mathcal{A}_n)_{n \geqslant 0}$ is an increasing elementary sequence of Σ-algebras, i.e, each $\mathcal{A}_n$ is an elementary subalgebra of $\mathcal{A}_{n+1}$, then each $\mathcal{A}_n$ is an elementary subalgebra of $\mathcal{A} = \bigcup_{n \geqslant 0} \mathcal{A}_n$.

18. Prove lemma 5.12 for the universal case, starting from a prenex formula in conjunctive form.

19. Deduce from the universal case the 'dual' version of Herbrand's theorem, for an existential proposition φ.

20.★ Compute the set of clauses $S(\varphi)$ for the formula φ

$$\forall \varepsilon(\varepsilon > 0 \Rightarrow \exists \delta(\delta > 0 \wedge \forall x(|x - a| < \delta \Rightarrow |f(x) - f(a)| < \varepsilon))).$$

21. The same for

$$\begin{aligned} &\forall x \exists z(\forall w(\exists y(w = z + y) \\ \Rightarrow\ &(\exists u(u * w = z) \Rightarrow w = 0 \vee w = 1)) \wedge \exists y(z = x + y)). \end{aligned}$$

22. Prove equations 5.1–5.3 of page 173.

23.★ Show that every consistent subset of the ordered set $\mathcal{F}$ of partial functions has a least upper bound. Show as well that each subset has a greatest lower bound.

24. Show that the compact elements of $\mathcal{F}$ are the partial functions of finite domain.

25.★ Find all the continuous solutions (including the least fixpoint) in $[[\mathbf{N}_\perp \to \mathbf{N}_\perp] \to [\mathbf{N}_\perp \to \mathbf{N}_\perp]]$ of the equations

$$\begin{aligned} f(x) &= f(x), \\ f(x) &= f(x+1), \\ f(x) &= \text{if } x = 0 \text{ then } 1 \text{ else } f(x+1). \end{aligned}$$

26. Find the least element and the directed subsets of the Cartesian product of two flat orders.

27. Show that a function which is strict in each of its arguments, a constant function and the conditional function are all sequential.

28. Show that for $M, N \in \mathrm{T}_{\Sigma\cup\{\Omega\}}$, $M \leqslant_\Omega N$ if and only if for every occurrence u of M, $M(u) = N(u)$ or $M(u) = \Omega$. Deduce that $M \leqslant_\Omega N$ is equivalent to the existence of $u_1, \ldots, u_k \in \mathcal{O}(M)$ and of terms $P_1, \ldots, P_k$ such that $M(u_i) = \Omega$ and $N = M[u_1 \leftarrow P_1, \ldots, u_k \leftarrow P_k]$.

29. Show that every consistent subset of $\mathrm{T}^\infty_{\Sigma\cup\{\Omega\}}$ has a least upper bound. Prove as well that every subset has a greatest lower bound, and that a partial term is compact if and only if it is finite.

30. Compute $\mathbf{N}x$ and $\varnothing x$, for $x \in P\omega$. In $P\omega$, if $[\![M]\!] = \varnothing$, show that $[\![MN]\!] = [\![\lambda x M]\!] = \varnothing$.

31. Compute the interpretation of Ω in $P\omega$:

- Show that if $n \in [\![\Omega]\!]$, there exists a sequence of integers p_i, $i \geqslant 0$, such that $\langle p_{i+1}, n\rangle \in e_{p_i}$ and $e_{p_{i+1}} \subseteq e_{p_i}$.
- Deduce that if $n \in [\![\Omega]\!]$, there exists p such that $\langle p, n\rangle \in e_p$.
- Show using the standard encodings that this implies $[\![\Omega]\!] = \varnothing$.

Bibliographic notes

The semantics of classical logic is considered in, among others, [154]. Kripke interpretations are studied in [154, 150]. The semantics of the λ-calculus is treated in [8] and in the recent book by Krivine [90]. Stoy's book [143] presents the denotational semantics of programming languages. [59] discusses recursive definitions of programs. Lazy pairs in functional programming are introduced in [46]. The abstract interpretation method is due to P. and R. Cousot [30]. Many applications of this method are presented in [2].

Chapter 6

Equational logic

6.1 Algebras and equations

Σ-algebras are the basic objects of semantics, as they occur in the predicate calculus, equational logic, logic programming, and even, with supplementary continuity conditions, in the semantics of computing.

The *algebraic structures* of mathematicians, such as (monoid, group, ring, ...), and the *data structures* (list, stack, record, ...) of computer scientists can all be formalized as Σ-algebras, also called structures by logicians.

These algebras are studied in universal algebra, which examines the general properties of such structures from the same vantage point. The group concept avoids repeating a proof of the same property, for example $(xy)^{-1} = y^{-1}x^{-1}$, for $\mathbf{Z}$, $\mathcal{S}_3$, or $\mathbf{SU}(2)$: not only is this property true for all groups, but the proof is the same regardless of the chosen group. Similarly, the algebra concept avoids the need to repeat proofs of certain structural properties, such as the kernel of a morphism (factorization of a morphism through a quotient, ...) for groups, vector spaces or rings. To a great extent, the study of these general properties has been unified even more by the use of categories and functors (universal objects, adjunction, ...), the vocabulary of which will sometimes be used here.

As new levels of abstraction are created, generalities with little depth can be ignored, so that one can concentrate on the particular cases, richer in content and difficulties. It is in fact in universal algebra that the first undecidable problems affecting the average mathematician were first discovered (the previous problems came more from metamathematics).

6.1.1 Modules in Standard ML

From a software engineering point of view, modularity is an essential attribute. It differentiates real languages such as Modula–2 and Common Lisp from school languages such as Pascal and Scheme. This is one reason for the ability of FORTRAN to survive, due to its quite rudimentary modularity (separate compilation), and the suspicion that was first held towards functional and logic languages, which had at first no concept of modularity. A modular structure meets several objectives: decomposition into simple elements, abstraction, separation into specification and implementation (ensuring independence from the choices of implementation and hardware), protection, separate compilation, reusability (use of libraries) and verification of compatibility, among others.

These objectives led to the design of two good modular languages: Modula–2, a descendant of Pascal, also designed by N. Wirth, and Ada, developed by a team led by J. Ichbiah. Another approach, motivated by modeling problems, can be found in object-oriented programming, whose concepts of class, instance and inheritance first appeared in K. Nygaard and O.J. Dahl's Simula, before being popularized by A. Kay's Smalltalk.

If modularity is an essential aspect of languages, the correctness of the resulting software—the most important objective from the user's point of view—can only be ensured through a rigorous approach, using verification or proof techniques, either integrated into the language, such as a compiler's type checker, or in production tools. There are two logical approaches to modularity: universal algebra and the λ-calculus's type system, and they do not cohabit easily.

Universal algebra favors specification by axioms, models and general functorial constructions. This approach is normally used in specification languages, some of which, such as the OBJ family, also allow computation.

Typed λ-calculi offer powerful type systems with properties such as genericity, type inclusion and dependent types, particularly adapted to functional and, sometimes, object-oriented programming.

As for logic programming languages, in which *ad hoc* modular structures were added to ensure industrial credibility (Prolog–II, M–Prolog), it is only recently that a logical status of modularity has been sought (see [107], where modules are presented as an extension of the Horn clause language).

It seems for the moment that Standard ML has, at the programming language level, the most evolved concepts of modularity. This success probably comes from the fact that its constructions can be viewed either as algebras or as dependent types in a two-level type system. Its three modular constructions are signatures, structures and functors.

Signatures and structures

What is written in logic as 'let Σ be a signature with two sorts s_1 and s_2, along with symbols $a : s_1$ and $f : s_1 \times s_2 \rightarrow s_1$', is written in ML as:

```
signature SIGMA =
  sig
    type s1 and s2
    val a : s1
```

```
    and f : s1 * s2 -> s1
  end ;
```

An example of a Σ-algebra $\mathcal{A}$, with two sets A_1 and A_2, of an element A_1 and a mapping from $A_1 \times A_2$ to A_1, would be written in ML as:

```
structure A =
  struct
    type s1 = int and s2 = bool
    val a = 3
    fun f(x,true) = x+1
     |  f(x,false) = 0
  end ;
```

and ML would answer as follows:

```
↷ structure A :
    sig
      eqtype s1
      eqtype s2
      val a : int
      val f : int * bool -> int
    end
```

In the terminology of language semantics, a structure is an environment, i.e, a set of bindings of names to values, which can be types, constants or functions, as in the preceding example, but also structures. ML's structures resemble Modula–2's `IMPLEMENTATION MODULE`*s* and the `body`*s* of Ada's `package`*s*. From the logical point of view, these are Σ-algebras. They are also the elements of a sum type, concept introduced by P. Martin-Löf in his Intuitionistic Type Theory, and one would write:

$$\sum_{s_1,s_2:\mathrm{Type}} [a : s_1, f : s_1 \times s_2 \to s_1].$$

The components `A.s1`, `A.s2`, `A.a`, `A.f` of `A` can be reached directly by opening `A`:

```
open A ;
```

A signature declares a set of names of types, typed values and structures, and, like Modula–2's `DEFINITION MODULE`*s* and the specifications of Ada's `package`*s*, acts as the interface for a structure. ML signatures are similar to those in logic: they can be seen as the 'type' of a structure.

A data structure, formed of one or more object sorts and of operations manipulating them, must be put together in a module. Here is an example of a stack:

```
structure Stack =
  struct
    datatype 'a stack = empty | push of 'a * 'a stack
    fun pop(push(_,p)) = p
    and top(push(x,_)) = x
  end ;
```

```
↝ std_in:19.5-20.26 Warning: match not exhaustive
          push (x,_) => ...
  std_in:19.5-20.26 Warning: match not exhaustive
          push (_,p) => ...
  structure Stack :
    sig
      datatype 'a  stack
        con empty : 'a stack
        con push : 'a * 'a stack -> 'a stack
      val pop : 'a stack -> 'a stack
      val top : 'a stack -> 'a
    end
```

An adequate signature for this structure is:

```
signature STACK =
  sig
    datatype 'a stack = empty | push of 'a * 'a stack
    val pop : 'a stack -> 'a stack
    and top : 'a stack -> 'a
  end ;
```

Note that the signature contains the *definition* of the `stack` type and not only the *declaration* of its name, in order to export the constructors `empty` and `push`. Other signatures are equally adequate for this structure, which allows different interfaces to be designed by hiding certain components of the structure, which is why one refers to *an adequate* signature and not to *the* signature of a structure. Another adequate signature is:

```
signature STACK1 =
  sig
    datatype 'a stack = empty | push of 'a * 'a stack
    val pop : 'a stack -> 'a stack
  end ;
```

```
↝ signature STACK1 =
    sig
      datatype 'a stack
        con empty : 'a stack
        con push : 'a * 'a stack -> 'a stack
      val pop : 'a stack -> 'a stack
    end
```

which restricts the exported operations (`top` disappeared):

```
structure Stack1 : STACK1 = Stack ;
```

In logic, this is the well-known operation of restriction: if $\Sigma \subseteq \Sigma'$, every Σ'-algebra $\mathcal{A}$ is automatically a Σ-algebra written $\mathcal{A}_{\restriction \Sigma}$, obtained by only interpreting the elements of Σ.

The complementary operation for restriction is extension. A test function is added to the `Stack` structure. A new structure is created by importing the preceding one and by defining the new operation:

```
structure Stack2 =
  struct
    structure S = Stack
    fun isempty(S.empty) = true
      | isempty(S.push(_,_)) = false
  end ;
```

```
↷ structure Stack2 :
    sig
      structure S : ...
      val isempty : 'a Stack.stack -> bool
    end
```

An adequate signature for `Stack2` is:

```
signature STACK2 =
  sig
    structure S : STACK
    val isempty : 'a S.stack -> bool
  end ;
```

```
↷ signature STACK2 =
    sig
      structure S : ...
      val isempty : 'a S.stack -> bool
    end
```

Functors

A functor is a 'function' associating a structure to a structure: it is used to define generic structures.

ML's type system allows the definition of polymorphic functions taking as argument other functions, simulating Ada's `generic packages` quite well. Functors can be used to handle (data) structures as such collectively.

Consider the following mathematical formulation: 'consider the lexical order $\leqslant^\star$ over the free monoid $A^\star$ constructed from an order $\leqslant$ over the alphabet A'. A structure $(A, \leqslant)$ is given, and a structure $(A, A^\star, \leqslant, \leqslant^\star)$ must be generated. The 'type' of what is given is the signature `RELATION` of ordered sets:

```
signature RELATION =
  sig
    eqtype t
    val order : t * t -> bool
  end ;
```

and the 'type' of the resulting structure is:

```
signature BIRELATION =
  sig
    structure A : RELATION
    eqtype m
    val lex : m * m -> bool
  end ;
```

The functor which does the transformation is:

```
functor MyLex(X : RELATION) : BIRELATION =
  struct
    structure A = X
    eqtype m = (A.t) list
    fun lex([], _) = true
     |  lex(x::u, y::v) =
          if A.order(x,y)
          then if A.order(y,x) then lex(u,v) else true
          else false
    end ;
```

When this functor is applied to the structure of ordered integers:

```
structure N =
  struct
    eqtype t = int
    fun order (x,y:t) = x>=y
  end ;
```

the result is the structure of lexically ordered integer words:

```
structure Nstar = MyLex(N) ;
```

This ML formulation is as natural, and as precise, as the initial mathematical formulation. A good programming language must have this 'faithful' capacity, be it to represent mathematical constructions, as here, or physical systems, or, more generally, 'knowledge'.

In algebra, a *functor* F between two categories $\mathcal{X}$ and $\mathcal{Y}$ associates with each object A of $\mathcal{X}$ an object $F(A)$ of $\mathcal{Y}$, and with each morphism $p : A \to B$ of the category $\mathcal{X}$ a morphism $F(p) : F(A) \to F(B)$ of $\mathcal{Y}$, satisfying the conditions $F(1_A) = 1_B$ and $F(p \circ q) = F(p) \circ F(q)$. Since there is no notion of morphism in ML, which is perhaps a defect, its `functor`*s* act only on objects, i.e., structures, and they can be seen as functors from a category $\mathbf{Alg}_\Sigma$ of Σ-algebras (Σ being the signature of its argument structure) to another, $\mathbf{Alg}_{\Sigma'}$.

The example of the `RELATION` signature of ordered set shows, however, the limits of the module system: the signature was given for sets with a binary relation, not sets with an order relation. Furthermore, the functor `MyLex` produces a binary relation over words, but the compiler is incapable of guaranteeing that it is an order, even if it is 'known' that the component `order` of its argument is an order. A type system would have to be devised in which the properties of objects can be given. A few experiments in this direction have been undertaken, but languages must be extended—possibly in the compiler—with proof techniques which are much more powerful than the ones used by existing type checkers. And of course, in the most general case, the static verification of these type–property pairs is undecidable. This problem calls into question the idea that a type is simply a set of values. So long as the examples of types are limited to the set of all integers, to Boolean strings or to arrays of pointers to Boolean records, this is not problematic, but it is when one wishes to refer to the types of even integers, or of character strings which are syntactically correct C programs or which form the definition of a non-ambiguous algebraic grammar. However, structures such as stacks, heaps and binary search trees, which are considered to be normal data structures, and which form the basis for the modular organization of software, are defined by properties of the same kind as even integers. Methods to handle them must be found.

6.1.2 Σ-algebras

After recalling the definitions of signatures, algebras and morphisms, as well as the principal algebraic constructions, a correspondence will be given between equational theories and classes of algebras.

Algebras, morphisms, products and subalgebras
A *signature* is a graded set $\Sigma = \bigcup_n \Sigma_n$, where Σ_n is the set of (functional) symbols of arity n. In the examples, the arithmetic signature $\Sigma = \{0, s, +, *\}$ will often be used, as will that of group theory (neutral element e, product written using juxtaposition, and inverse written $^{-1}$). A Σ-*algebra* $\mathcal{A}$ consists of:

- a set A, the *domain* of $\mathcal{A}$;
- for each $n \geqslant 0$ and each $f \in \Sigma_n$, of a mapping $f^{\mathcal{A}} : A^n \to A$, the *interpretation* of f.

A Σ-*morphism* $m : \mathcal{A} \to \mathcal{B}$ is a mapping of A to B such that for every n and every $f \in \Sigma_n$, $m(f^{\mathcal{A}}(a_1, \ldots, a_n)) = f^{\mathcal{B}}(m(a_1), \ldots, m(a_n))$. The category of Σ-algebras and of Σ-morphisms is written $\mathbf{Alg}_\Sigma$. Algebras and morphisms will be mentioned without referring to the signature, should no ambiguity arise.

Note that, unlike the Σ-algebras of the predicate calculus, the domain of an algebra can be empty. This possibility can only arise in signatures with no constant symbols ($\Sigma_0 = \varnothing$), in which case the set T_Σ of closed terms is empty and this is the unique empty algebra. The situation will be different when many-sorted signatures will be treated, in which some of the domains might be empty.

How are algebras constructed? Given a signature Σ and a set X, the algebra $\mathrm{T}_\Sigma[X]$ of Σ-terms with variables in X has already been presented.

Another simple construction is the product. Let $\mathcal{A}$ and $\mathcal{B}$ be two Σ-algebras. The *product* Σ-algebra $\mathcal{A} \times \mathcal{B}$ has as domain the product $A \times B$ of the domains of interpretation

$$f^{\mathcal{A}\times\mathcal{B}}((a_1,b_1),\dots,(a_n,b_n)) = (f^{\mathcal{A}}(a_1,\dots,a_n), f^{\mathcal{B}}(b_1,\dots,b_n)).$$

The projections of $\mathcal{A}\times\mathcal{B}$ to $\mathcal{A}$ and $\mathcal{B}$ are morphisms. Similarly for the product $\prod_{i\in I}\mathcal{A}_i$ of an arbitrary family $(\mathcal{A}_i)_{i\in I}$ of Σ-algebras. The particular case $I = \varnothing$ yields a trivial algebra whose domain has a single element; these algebras being isomorphic, an arbitrary one may be designated $\mathbf{1}_\Sigma$. It is *terminal* in the sense that for every Σ-algebra $\mathcal{A}$, there exists a unique Σ-morphism of $\mathcal{A}$ to $\mathbf{1}_\Sigma$.

The dual property for 'terminal' is 'initial'. The algebra T_Σ of closed terms is *initial*: for every Σ-algebra $\mathcal{A}$, there exists a unique Σ-morphism $\iota_{\mathcal{A}}$ from T_Σ to $\mathcal{A}$, the initial morphism.

$\mathcal{B}$ is a *subalgebra* of $\mathcal{A}$ if $B\subseteq A$ and $f^{\mathcal{B}}(b_1,\dots,b_n) = f^{\mathcal{A}}(b_1,\dots,b_n)$ for every $f\in\Sigma_n$ and $\vec{b}\in B^n$. In other words, the injection of B to A is a morphism. For example, the *image* $m(A)$ of a morphism $m:\mathcal{A}\to\mathcal{B}$ is a subalgebra of $\mathcal{B}$.

The intersection of an arbitrary family of subalgebras of $\mathcal{A}$ is a subalgebra: this property makes the set of subalgebras into a complete lattice with inclusion. In particular, a family of subalgebras has a least upper bound. If E is an arbitrary subset of A, the Σ-algebra *generated* by E is the intersection of the family of algebras containing E. This family is non-empty since it contains $\mathcal{A}$. The lattice of subalgebras of $\mathcal{A}$ has a greatest element ($\mathcal{A}$), and a least element (the minimal subalgebra of $\mathcal{A}$, the image of T_Σ by the initial morphism $\iota_{\mathcal{A}}$).

An algebra is *finitely generated*, or of *finite type*, if there exists a finite subset E of A such that $\mathcal{A}$ is the algebra generated by E: T_Σ is of finite type since it is generated by the empty set, and $\mathrm{T}_\Sigma[X]$ is of finite type if and only if X is finite.

Congruences

A *congruence* over a Σ-algebra $\mathcal{A}$ is an equivalence relation $\approx$ over A compatible with the operations: for every $f\in\Sigma_n$, if $a_i\approx b_i$ for $i=1,\dots,n$, then $f^{\mathcal{A}}(a_1,\dots,a_n)\approx f^{\mathcal{A}}(b_1,\dots,b_n)$. The set $A/{\approx}$ of equivalence classes is the domain of an algebra $\mathcal{A}/{\approx}$, the operations still working under the quotient because of the compatibility of $\approx$. The natural surjection $\varpi_\approx$ of A to $A/{\approx}$ is a morphism which will frequently be used in discussions.

Examples

- A^2 and $\Delta = \{(a,a); a\in A\}$ are congruences, with $A/A^2 = \mathbf{1}_\Sigma$ and $A/\Delta = A$.
- A congruence $\approx$ over a group G is determined by the class of the neutral element, which is a normal subgroup $H_\approx$ of G. $a\approx b$ if and only if $ab^{-1}\in H_\approx$.
- A congruence $\approx$ over a ring A is determined by the class of 0, which is an ideal $\mathbf{p}_\approx$ of A. $a\approx b$ if and only if $a-b\in\mathbf{p}_\approx$.

  In the two last examples, $G/H_\approx$ is a group, and $A/p_\approx$ is a ring, not only an algebra: this is not coincidental, as will be seen later.

The *kernel* of a morphism $m : \mathcal{A} \to \mathcal{B}$ is

$$\text{Ker } m = \{(x, y) \in A^2; m(x) = m(y)\}.$$

As a binary relation over A, Ker m is an equivalence. It is also a congruence, since if $(a_i, b_i) \in$ Ker m, it is because $m(a_i) = m(b_i)$. Hence

$$\begin{aligned} m(f^{\mathcal{A}}(a_1, \ldots, a_n)) &= f^{\mathcal{A}}(m(a_1), \ldots, m(a_n)) \\ &= f^{\mathcal{A}}(m(b_1), \ldots, m(b_n)) \\ &= m(f^{\mathcal{A}}(b_1, \ldots, b_n)), \end{aligned}$$

i.e, $(f^{\mathcal{A}}(a_1, \ldots, a_n), f^{\mathcal{A}}(b_1, \ldots, b_n)) \in$ Ker m.

If $\approx$ is a congruence, the kernel of the projection of $\mathcal{A}$ to $\mathcal{A}/\approx$ is precisely $\approx$. As in linear algebra, every morphism $m : \mathcal{A} \to \mathcal{B}$ is factorizable through the quotient by its kernel $\mathcal{A}/\text{Ker } m$, and its image $m(\mathcal{A})$ as the composition of natural surjection, an isomorphism, and inclusion:

$$\mathcal{A} \twoheadrightarrow \mathcal{A}/\text{Ker } m \xrightarrow{\sim} m(\mathcal{A}) \hookrightarrow \mathcal{B}.$$

The set of congruences over an algebra is also a complete lattice under inclusion, since the intersection of an arbitrary family of congruences is a congruence. It follows that there exists a congruence *generated* by a relation R (subset of A^2), namely the intersection of the congruences containing R. This lattice has the diagonal Δ as least element, and A^2 as largest element.

Unlike for term algebras and product algebras, for which every element is easily represented in a machine, quotient algebras are not immediately representable. It is easy to represent an equivalence class by an element of that class, but there remains the problem of determining if two elements belong to the same class. Since this problem is not always decidable, this construction must be seen as an abstract definition, which is made concrete when (for example, rewriting) 'algorithms' are available.

Free algebras, generators and relations

The free algebra generated by X was constructed over the set $\mathrm{T}_\Sigma[X]$ of closed terms with variables in X. Its fundamental property is that for every Σ-algebra $\mathcal{A}$ and for every mapping $\xi : X \to$A, there exists a unique Σ-morphism, $\hat{\xi}_\Sigma : \mathrm{T}_\Sigma[X] \to \mathcal{A}$, such that $\hat{\xi}_\Sigma(x) = \xi(x)$ for every $x \in X$, as seen in Figure 6.1.

The morphism $\hat{\xi}_\Sigma$ is defined by:

$$\begin{aligned} \hat{\xi}_\Sigma(x) &= \xi(x) \\ \hat{\xi}_\Sigma(ft_1 \ldots t_n) &= f^{\mathcal{A}}(\hat{\xi}_\Sigma(t_1), \ldots, \hat{\xi}_\Sigma(t_n)). \end{aligned}$$

The subscript $_\Sigma$ is normally left out, as also is the ^ of $\hat{\xi}_\Sigma$. The fact that $\mathrm{T}_\Sigma[X]$ is generated by X is exactly the definition by induction of the set of terms.

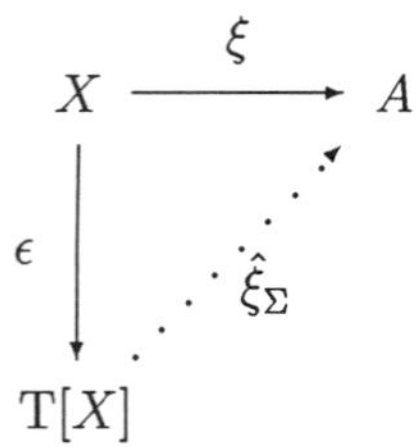

Figure 6.1 Definition of $\hat{\xi}_\Sigma$ from Σ

A universal problem This property of the free algebra has two characteristics found in other constructions: it is universal and relative.

The free algebra (or more precisely, the inclusion $X \hookrightarrow \mathrm{T}_\Sigma[X]$) is a solution of the following *universal problem*: given a fixed set X, find a Σ-algebra $\mathcal{F}$ and a mapping $f : X \to \mathcal{F}$ such that for every Σ-algebra $\mathcal{A}$ and for every mapping $\xi : X \to A$, there exists a unique morphism ('extension') $\hat{\xi} : \mathcal{F} \to \mathcal{A}$ satisfying $\hat{\xi} \circ f = \xi$.

A solution to this problem, if it exists, is necessarily unique up to isomorphism. So, let $f : X \to \mathcal{F}$ and $g : X \to \mathcal{G}$ be two solutions. Then there exist $\hat{g} : \mathcal{F} \to \mathcal{G}$ satisfying $\hat{g} \circ f = g$ and $\hat{f} : \mathcal{G} \to \mathcal{F}$ satisfying $\hat{f} \circ g = f$. The composition $\hat{f} \circ \hat{g}$ is a Σ-endomorphism of $\mathcal{F}$ satisfying $(\hat{f} \circ \hat{g}) \circ f = \hat{f} \circ g = f$. But the identity of $\mathcal{F}$ already satisfies this property, from which, by the required unicity of the extension of f, $\hat{f} \circ \hat{g}$ is the identity of $\mathcal{F}$. This proves that the two solutions $\mathcal{F}$ and $\mathcal{G}$ are isomorphic.

Many an abstract mathematical construction can be expressed as a solution of a universal problem: quotients, products, inductive and projective limits, complete metric spaces, and more. From a category-theoretic point of view, these solutions are always initial objects of appropriate categories, which allow powerful proof methods. However, the resulting objects are abstract, i.e., defined up to isomorphism, generally without an effective construction.

The other characteristic of the property is that it is 'relative', relating two kinds of structures, sets and algebras. A *mapping* ξ is given and a Σ-*morphism* $\hat{\xi}_\Sigma$ is obtained. Relative concepts have also had much success in algebra and geometry. Two variants of the construction of $\mathrm{T}_\Sigma[X]$ will be presented, where the relation algebra/set will be replaced by the relation Σ-algebra/Σ^0-algebra and by the relation $\mathcal{E}$-algebra/$\mathcal{E}^0$-algebra.

What happens if the construction already starts from a morphism between algebras? Let $\mathcal{A}$ be an algebra and form $\mathrm{T}_\Sigma[A]$, the free Σ-algebra generated by the domain of $\mathcal{A}$. For example, the elements of $\mathrm{T}_{\{0,s\}}[\mathbf{N}]$ are terms with 'parameters' in $\mathbf{N}$ such as 2 and $ss4$, in addition to the closed terms 0 and $sss0$. The relation stating the evaluation of these terms (such as $s3 = 4$) is the 'diagram' of $\mathbf{N}$.

In general, there is an evaluation morphism $\mathrm{ev}_\mathcal{A} : \mathrm{T}_\Sigma[A] \to \mathcal{A}$ which is the extension of the identity of A.

In other words,

$$\begin{aligned} \mathrm{ev}_{\mathcal{A}}(a) &= a, \\ \mathrm{ev}_{\mathcal{A}}(fM_1 \dots M_n) &= f^{\mathcal{A}}(\mathrm{ev}_{\mathcal{A}}(M_1), \dots, \mathrm{ev}_{\mathcal{A}}(M_n)). \end{aligned}$$

The kernel of $\mathrm{ev}_{\mathcal{A}}$ is a congruence, called the *(equational) diagram* of $\mathcal{A}$.

Example The diagram of T_Σ is the diagonal Δ of T_Σ, the evaluation morphism being the identity. □

Now, if $\xi : A \to B$ is also a morphism of $\mathcal{A}$ in $\mathcal{B}$, its extension $\hat{\xi} : \mathrm{T}_\Sigma[A] \to \mathcal{B}$ is nothing more than $\xi \circ \mathrm{ev}_{\mathcal{A}}$, since $(\xi \circ \mathrm{ev}_{\mathcal{A}})(a) = \xi(a)$, and $\hat{\xi}$ is the unique morphism satisfying this property.

Presentations The preceding construction shows that the set X appearing in $\mathrm{T}_\Sigma[X]$, whose elements are the variables in the syntax of terms, plays the role of generating an algebra. What is important is not a particular algebra $\mathrm{T}_\Sigma[X]$, such as T_Σ, but the functor $X \mapsto \mathrm{T}_\Sigma[X]$, in the same way that the *cosine* function is much more interesting than any particular value of $\cos x$.

This role is illustrated by the definition of an algebra by generators and relations. Start with an algebra $\mathcal{A}$ and a set $G \subseteq A$ generating it. Such a set always exists: one can always take $G = A$, but the smaller, the more interesting. The inclusion $\theta : G \hookrightarrow A$ extends to a morphism $\hat{\theta}_\Sigma : \mathrm{T}_\Sigma[G] \to \mathcal{A}$. The image of $\hat{\theta}_\Sigma$ is a subalgebra of $\mathcal{A}$ containing G, hence, by assumption on G, the image is $\mathcal{A}$. There is therefore a surjection $\mathrm{T}_\Sigma[G] \twoheadrightarrow \mathcal{A}$, and $\mathcal{A}$ is isomorphic to the quotient of $\mathrm{T}_\Sigma[G]$ by the kernel of $\hat{\theta}_\Sigma$: every algebra is therefore a quotient of a free algebra. In this construction, the elements of G appear more as constants than as variables.

The particular case where $\mathcal{A}$ is generated by $G = \varnothing$ is important: the initial morphism $\iota_{\mathcal{A}} : \mathrm{T}_\Sigma \to \mathcal{A}$ is then surjective and $\mathcal{A}$ is isomorphic to $\mathrm{T}_\Sigma/\mathrm{Ker}\ (\iota_{\mathcal{A}})$. This surjectivity ensures that the elements of $\mathcal{A}$ are all reachable, i.e., they can be designated by closed terms.

There is therefore an interesting method to define an algebra. A *presentation by generators and relations* of an algebra $\mathcal{A}$ is a set G and a subset R of $\mathrm{T}_\Sigma[G] \times \mathrm{T}_\Sigma[G]$ such that $\mathcal{A}$ is isomorphic to $\mathrm{T}_\Sigma[G]/\approx$, where $\approx$ is the congruence generated by R. From above, every algebra has such a presentation (it suffices to take $G = A$ and for R the kernel of an evaluation morphism $\mathrm{ev}_{\mathcal{A}}$). This concept is particularly interesting in the case where G and R are finite sets: a method to define an algebra with finite means becomes possible, as for the definition of formal language, typically infinite, using a finite grammar. An algebra with a finite presentation is of finite type.

However, it is not always convenient to write finite presentations. Consider the case of $B^\star$, the set of binary words, i.e, over the alphabet $B = \{0, 1\}$. It is a Σ-algebra for Σ formed of a constant symbol ϵ (the empty word) and a binary symbol (concatenation). There are obviously two generators 0 and 1. The relations state that ϵ is neutral ($\epsilon 0 = 0 \epsilon = 0$ and $\epsilon 1 = 1 \epsilon = 1$) and that concatenation is associative

$((00)0 = 0(00), (00)1 = 0(01), (01)0 = 0(10), \dots)$. It would be preferable to write one general law for associativity rather than eight particular laws. This will be done by introducing equational propositions.

Varieties and equalities
To express general laws, a countably infinite set, distinct from the signature Σ, of variables is introduced, along with a binary relational symbol $=$ for equality. Choosing X to be infinite allows laws using an arbitrary number of variables. The formulas to be manipulated are *equational propositions*, or *equations*, of the form $\forall x_1 \dots x_k(M = N)$, where $\{x_1, \dots, x_k\} = \mathrm{var}(M) \cup \mathrm{var}(N)$. Since all formulas are of this form, the quantifiers will always be omitted (only $M = N$ will be written). They were made explicit here only for the purposes of justifying the semantics of equations in the general framework of the predicate calculus. Hence associativity, a property which was awkardly expressed using several relations between generators $((00)1 = 0(01), \dots)$ to specify the free monoid $B^\star$, can now be expressed using a single equation between terms with variables: $(xy)z = x(yz)$. Here the variables find their intuitive role of 'metaterm' standing for an arbitrary *term*. But semantics gives them a stronger meaning, by making them stand for an arbitrary *value*.

An algebra $\mathcal{A}$ is a model of, or satisfies, equation $M = N$, written $\mathcal{A} \models M = N$, if for every $\xi \in A^X$, $\hat{\xi}(M) = \hat{\xi}(N)$. The elements of A^X are called valuations.

The examples will show that equations have two roles:

- to state properties, typically in an axiomatization of a class of structures, such as that of groups:

$$(groups)\quad \begin{cases} ex &= x \\ x^{-1}x &= e \\ (xy)z &= x(yz); \end{cases}$$

- to define operations in a manner close to that of a recursive program; here is a definition of addition by induction, using the signature $\{0, s, +\}$:

$$(+)\quad \begin{cases} x + 0 &= x \\ x + s(y) &= s(x + y). \end{cases}$$

The set $\mathbf{N}$, with the usual interpretations of 0, s, and $+$ is a model of $(+)$, but there are others ...

The correspondence **Mod** $\leftrightarrows$ **Eq** If $\mathcal{E}$ is a set of equations, $\mathbf{Mod}(\mathcal{E})$ is the class of algebras which are models of each of the equations of $\mathcal{E}$. The models of $\mathcal{E}$ are also called *$\mathcal{E}$-algebras*, or *$(\Sigma, \mathcal{E})$-algebras.* Note that every set of equations has at least one model: the trivial algebra $\mathbf{1}_\Sigma$. This was not the case for the predicate calculus.

Conversely, with every class $\mathcal{C}$ of algebras and every set X is associated the *equational theory* of $\mathcal{C}$ over X, which is the set $\mathbf{Eq}_X(\mathcal{C})$ of equations over X which are satisfied by all the algebras of $\mathcal{C}$:

$$\mathbf{Eq}_X(\mathcal{C}) = \{(M,N) \in \mathrm{T}_\Sigma[X] \times \mathrm{T}_\Sigma[X]; \mathcal{A} \models M = N \text{ for every } \mathcal{A} \in \mathcal{C}\}.$$

X will often be the countably infinite set of variables, but not necessarily. In particular $\mathbf{Eq}_\varnothing(\mathcal{C})$, formed of equations between closed terms satisfied in $\mathcal{C}$, is of interest. When it is not necessary to state X, simply write $\mathbf{Eq}(\mathcal{C})$. In the case of a class formed of a single algebra $\mathcal{A}$, write $\mathbf{Eq}(\mathcal{A})$ instead of $\mathbf{Eq}(\{\mathcal{A}\})$.

These two transformations **Mod** and **Eq** satisfy the following obvious properties:

$$\mathcal{E}_1 \subseteq \mathcal{E}_2 \;\Rightarrow\; \mathbf{Mod}(\mathcal{E}_2) \subseteq \mathbf{Mod}(\mathcal{E}_1), \tag{6.1}$$

$$\mathcal{C}_1 \subseteq \mathcal{C}_2 \;\Rightarrow\; \mathbf{Eq}(\mathcal{C}_2) \subseteq \mathbf{Eq}(\mathcal{C}_1), \tag{6.2}$$

$$\mathcal{E} \subseteq \mathbf{Eq}(\mathbf{Mod}(\mathcal{E})), \tag{6.3}$$

$$\mathcal{C} \subseteq \mathbf{Mod}(\mathbf{Eq}(\mathcal{C})). \tag{6.4}$$

These four properties characterize the pair $\mathbf{Mod} \leftrightarrows \mathbf{Eq}$ as a *Galois correspondence* between the sets ordered by inclusion of the sets of equations and the classes of algebras. They will often be used below, as will the following equations, which follow directly (exercise 5):

$$\mathbf{Mod}(\mathbf{Eq}(\mathbf{Mod}(\mathcal{E}))) = \mathbf{Mod}(\mathcal{E}), \tag{6.5}$$

$$\mathbf{Eq}(\mathbf{Mod}(\mathbf{Eq}(\mathcal{C}))) = \mathbf{Eq}(\mathcal{C}), \tag{6.6}$$

$$\mathbf{Mod}(\bigcup_i \mathcal{E}_i) = \bigcap_i \mathbf{Mod}(\mathcal{E}_i), \tag{6.7}$$

$$\mathbf{Mod}(\bigcap_i \mathbf{Eq}(\mathbf{Mod}(\mathcal{E}_i))) = \mathbf{Mod}(\mathbf{Eq}(\bigcup_i \mathbf{Mod}(\mathcal{E}_i))). \tag{6.8}$$

The name 'Galois correspondence' evidently came from the famous Galois theory, which establishes a correspondence between subgroups of the Galois group of a field extension and the intermediate fields of this extension.

Definition 6.1 A set of equations $\mathcal{E}$ is an *equality* if there exists a class of algebras $\mathcal{C}$ such that $\mathcal{E} = \mathbf{Eq}(\mathcal{C})$. A class of algebras is a *variety* if there exists a set of equations $\mathcal{E}$ such that $\mathcal{C} = \mathbf{Mod}(\mathcal{E})$.

Examples

- The class of algebras is a variety, associated with $\mathcal{E} = \varnothing$.
- The class formed of trivial algebras (i.e., of which the domain has only one element), and possibly of the empty algebra, is the variety associated with an equation $x = y$.
- Algebraic structures give numerous familiar and non-trivial examples of varieties: monoids, groups, abelian groups, rings, Boolean algebras, etc.

- However, the class of fields is not a variety. This does not come as a surprise, since the usual axioms of field theory include a formula $\forall x(x \neq 0 \Rightarrow \exists y(xy = 1))$ which is not equational, but this does not imply the non-existence of equational axioms for fields. For example, the (non-equational) axiom $\forall x \exists y(xy = 1)$ is sometimes given for groups, but it does not prevent the class of groups from being a variety. Other examples will be given for the specification of data structures in computer science.

The properties $\mathbf{Mod}(\mathbf{Eq}(\mathbf{Mod}(\mathcal{E}))) = \mathbf{Mod}(\mathcal{E})$ and $\mathbf{Eq}(\mathbf{Mod}(\mathbf{Eq}(\mathcal{C}))) = \mathbf{Eq}(\mathcal{C})$ show that for every equality $\mathcal{E}$, and for every variety $\mathcal{C}$,

$$\begin{aligned} \mathbf{Mod}(\mathbf{Eq}(\mathcal{C})) &= \mathcal{C}, \\ \mathbf{Eq}(\mathbf{Mod}(\mathcal{E})) &= \mathcal{E}. \end{aligned}$$

So, the transformations **Mod** and **Eq** establish a bijection between equalities and varieties. Similar definitions can be given in the general framework of the predicate calculus, where one would refer to complete theories and axiomatized classes, and they also lead to a bijection. In the equational case, there are also simple characterizations of equalities and varieties, using stability properties.

Equalities and stable congruencies Let $\mathcal{E} = \mathbf{Eq}(\mathcal{C})$ be an equality. First, it is a congruence, since its definition can be reformulated as

$$\mathbf{Eq}(\mathcal{C}) = \bigcap_{\mathcal{A} \in \mathcal{C}} \bigcap_{\xi \in A^X} \text{Ker } \hat{\xi},$$

and the kernel of a morphism is a congruence, as is the intersection of congruences. One can then write $\mathbf{Eq}(\mathcal{C}) = \text{Ker } \theta_{\mathcal{C}}$ where $\theta_{\mathcal{C}}$ is the morphism

$$\begin{aligned} \theta_{\mathcal{C}} : \mathrm{T}_\Sigma[X] &\rightarrow \prod_{\mathcal{A} \in \mathcal{C}} \prod_{\xi \in A^X} \mathcal{A} \\ M &\mapsto ((\hat{\xi}(M))_{\xi \in A^X})_{\mathcal{A} \in \mathcal{C}}. \end{aligned}$$

Equalities are not only congruences, they also satisfy the following stability property:

Definition 6.2 A binary relation R over $\mathrm{T}_\Sigma[X]$ is *stable under substitutions* if $(M, N) \in R$ implies $(\hat{\sigma}M, \hat{\sigma}N) \in R$, for every substitution $\sigma : X \rightarrow \mathrm{T}_\Sigma[X]$.

That $\mathbf{Eq}(\mathcal{C})$ be stable under substitutions follows intuitively from the universal quantification of equational propositions: if $(M, N) \in \mathbf{Eq}(\mathcal{C})$ and σ is a substitution, $(\hat{\sigma}M, \hat{\sigma}N) \in \mathbf{Eq}(\mathcal{C})$ must be proven, i.e, it must be proven that $\hat{\xi}(\hat{\sigma}M) = \hat{\xi}(\hat{\sigma}N)$ for every $\xi \in A^X$, which follows from $\hat{\xi} \circ \hat{\sigma} = \widehat{\hat{\xi} \circ \sigma}$:

Proposition 6.1 Let $\mathcal{E}$ be a set of equations and $\mathcal{E}'$ be the congruence stable under substitutions generated by $\mathcal{E}$. Then $\mathcal{E}$ and $\mathcal{E}'$ have the same models and $\mathcal{E}'$ is the equational theory of the models.

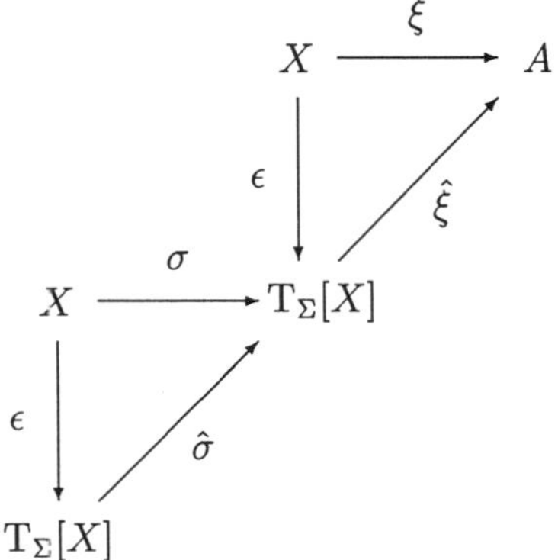

Proof If $\mathcal{E}$ is an arbitrary set of equations, then the equality $\mathcal{E}' = \mathbf{Eq}(\mathbf{Mod}(\mathcal{E}))$ is a congruence stable under substitutions. According to the general properties of the correspondence $\mathbf{Mod} \leftrightarrows \mathbf{Eq}$,

$$\mathbf{Mod}(\mathcal{E}') = \mathbf{Mod}(\mathbf{Eq}(\mathbf{Mod}(\mathcal{E}))) = \mathbf{Mod}(\mathcal{E}) :$$

$\mathcal{E}$ and $\mathcal{E}'$ therefore have the same models.

Conversely, it will be shown that a congruence $\mathcal{E}$ over $\mathrm{T}_\Sigma[X]$, stable under substitutions, is an equality, i.e, an equational theory. A class $\mathcal{C}$ of algebras must be found such that $\mathcal{E} = \mathbf{Eq}(\mathcal{C})$. Obviously, $\mathcal{C} = \mathbf{Mod}(\mathcal{E})$. Form the quotient $\mathrm{T}_\Sigma[X]/\mathcal{E}$, written $\mathrm{T}_{\Sigma,\mathcal{E}}[X]$, or $\mathrm{T}_\mathcal{E}[X]$, and let $\varpi : \mathrm{T}_\Sigma[X] \twoheadrightarrow \mathrm{T}_\mathcal{E}[X]$ be the natural surjection.

First, it must be shown that $\mathrm{T}_\mathcal{E}[X]$ is a model of $\mathcal{E}$. Let $(M, N) \in \mathcal{E}$ and $\xi : X \to \mathrm{T}_\mathcal{E}[X]$. The morphism $\hat{\xi}$ factors through $\mathrm{T}_\Sigma[X]$, by using an arbitrary right inverse (or section) s of the surjection ϖ. (There is no reason to believe that the inverse is a morphism. It is simply supposed that the composed mapping $\varpi \circ s$ is the identity of $\mathrm{T}_\mathcal{E}[X]$.) Define, then, $\eta = s \circ \xi : X \to \mathrm{T}_\Sigma[X]$. Its extension $\hat{\eta}$ satisfies $\varpi \circ \hat{\eta} = \hat{\xi}$, by the unicity of $\hat{\xi}$ since $\varpi \circ \eta = \varpi \circ s \circ \xi = \xi$.

Hence, $\hat{\xi}(M) = \varpi(\hat{\eta}(M)) = \varpi(\hat{\eta}(N)) = \hat{\xi}(N)$, since $\mathcal{E}$ is stable under substitutions and $(\hat{\eta}(M), \hat{\eta}(N)) \in \mathcal{E}$ if $(M, N) \in \mathcal{E}$. This proves $\mathrm{T}_\mathcal{E}[X] \models \mathcal{E}$.

To prove that $\mathcal{E}$ is an equality, it suffices to show the inclusion $\mathbf{Eq}(\mathbf{Mod}(\mathcal{E})) \subseteq \mathcal{E}$, since the converse inclusion always holds. Let $(M, N) \in \mathbf{Eq}(\mathbf{Mod}(\mathcal{E}))$. Since it has been shown that $\mathrm{T}_\mathcal{E}[X]$ is a model of $\mathcal{E}$ if $\mathcal{E}$ is stable under substitutions, then in particular $\mathrm{T}_\mathcal{E}[X] \models M = N$. Note that if $\xi : X \to \mathrm{T}_\mathcal{E}[X]$ associates its class modulo $\mathcal{E}$ with each variable, then $\hat{\xi}$ is the natural surjection ϖ. So, $\mathrm{T}_\mathcal{E}[X] \models M = N$ implies $\varpi(M) = \varpi(N)$, i.e, $(M, N) \in \mathcal{E}$, which proves $\mathbf{Eq}(\mathbf{Mod}(\mathcal{E})) \subseteq \mathcal{E}$.

Let $\mathcal{E}$ be an arbitrary set of equations, and $\mathcal{E}_1$ be an equality containing $\mathcal{E}$. From $\mathcal{E} \subseteq \mathcal{E}_1$, it follows that $\mathbf{Eq}(\mathbf{Mod}(\mathcal{E})) \subseteq \mathbf{Eq}(\mathbf{Mod}(\mathcal{E}_1))$. But, $\mathcal{E}_1$ being an equality, $\mathbf{Eq}(\mathbf{Mod}(\mathcal{E}_1)) = \mathcal{E}_1$. On the other hand, it was seen that $\mathcal{E}' = \mathbf{Eq}(\mathbf{Mod}(\mathcal{E}))$ is an equality. It follows that $\mathbf{Eq}(\mathbf{Mod}(\mathcal{E}))$ is the least equality containing $\mathcal{E}$: it is the *equality generated* by $\mathcal{E}$. It is also, according to the given characterization of equalities, the least congruence stable under substitutions containing $\mathcal{E}$. □

Varieties and stable classes In the same manner that equalities were characterized as congruences stable under substitutions, varieties are the classes of algebras satisfying certain stability properties.

Definition 6.3 A class $\mathcal{C}$ of algebras is called

- *stable under isomorphisms* if $\mathcal{C}$ contains every algebra which is isomorphic to an algebra in $\mathcal{C}$;
- *stable under subalgebras* if $\mathcal{C}$ contains all the subalgebras of its algebras;
- *stable under quotients* if $\mathcal{C}$ contains every algebra which is a quotient (by a congruence) of an algebra in $\mathcal{C}$;
- *stable under products* if for every family $\mathcal{A}_i$ of algebras in $\mathcal{C}$, the product $\prod_{i\in I}\mathcal{A}_i$ is in $\mathcal{C}$.

It is easy to prove that every variety is stable under isomorphisms, subalgebras, quotients and products (exercise 6).

In the case of the variety of groups, these properties are well-known: a subalgebra of a group, a quotient of a group (by a particular subgroup) and a product of groups are all groups, not just algebras. Stability under subalgebras is a property of the models of universal theories. Stability under products is a property of models of Horn theories.

G. Birkhoff showed, conversely, that a class of algebras $\mathcal{C}$, stable under isomorphisms, subalgebras, quotients and products, is a variety. If $\mathcal{C}$ satisfies these stability properties, a set of equations $\mathcal{E}$ must be constructed such that $\mathcal{C} = \mathbf{Mod}(\mathcal{E})$.

An infinite set X having been chosen, the equational theory $\mathcal{E} = \mathbf{Eq}_X(\mathcal{C})$ is formed, followed by the quotient $\mathrm{T}_{\mathcal{E}}[X]$. By definition, $\mathbf{Eq}_X(\mathcal{C}) = \mathrm{Ker}\ \theta_{\mathcal{C}}$ where $\theta_{\mathcal{C}}$ is the morphism from $\mathrm{T}_\Sigma[X]$ to the product algebra $\Pi_{\mathcal{C}} = \prod_{\mathcal{A}\in\mathcal{C}}\prod_{\xi\in A^X}\mathcal{A}$, associating the family $((\hat{\xi}(M))_{\xi\in A^X})_{\mathcal{A}\in\mathcal{C}}$ with a term M. By passing through the quotient by $\mathcal{E}$, an isomorphism of $\mathrm{T}_{\mathcal{E}}[X]$ to a subalgebra $\mathcal{U}$ of $\Pi_{\mathcal{C}}$ is obtained. By assumption, $\mathcal{C}$, being stable under products, contains $\Pi_{\mathcal{C}}$. Being stable under subalgebras, it contains $\mathcal{U}$, a subalgebra of $\Pi_{\mathcal{C}}$. Being stable under isomorphisms, it contains $\mathrm{T}_{\mathcal{E}}[X]$, which is isomorphic to $\mathcal{U}$.

So $\mathcal{C}$ has been shown to contain $\mathrm{T}_{\mathcal{E}}[X]$, where $\mathcal{E}$ is the equational theory of $\mathcal{C}$. It remains to be shown that $\mathbf{Mod}(\mathcal{E}) \subseteq \mathcal{C}$. Let $\mathcal{A}$ be a model of $\mathcal{E}$. Choose a set Y, containing X, large enough that there exists a surjection η from Y to A, hence a surjective morphism from $\mathrm{T}_\Sigma[Y]$ to $\mathcal{A}$. But it is easy to see that $\mathbf{Mod}(\mathbf{Eq}_X(\mathcal{C})) \subseteq \mathbf{Mod}(\mathbf{Eq}_Y(\mathcal{C}))$ if $X \subseteq Y$.

Hence, $\mathcal{A}$, which is a model of $\mathcal{E} = \mathbf{Eq}_X(\mathcal{C})$, is also a model of $\mathbf{Eq}_Y(\mathcal{C})$. But, $\mathrm{T}_\Sigma[Y]/\mathbf{Eq}_Y(\mathcal{C})$ is the free model generated by the Y of the theory $\mathbf{Eq}_Y(\mathcal{C})$. There is therefore a (unique) extension $\bar{\eta} : \mathrm{T}_\Sigma[Y]/\mathbf{Eq}_Y(\mathcal{C}) \to \mathcal{A}$ of η, which is surjective, like η. What was proved for $\mathrm{T}_{\mathcal{E}}[X]$ also holds for the algebra $\mathrm{T}_\Sigma[Y]/\mathbf{Eq}_Y(\mathcal{C})$: it is contained in $\mathcal{C}$. So, there is a surjective morphism $\bar{\eta}$ from an algebra of $\mathcal{C}$ to $\mathcal{A}$, hence an isomorphism between $\mathcal{A}$ and a quotient (by the kernel of $\bar{\eta}$) of an algebra of $\mathcal{C}$. By using the assumptions of stability under isomorphisms and quotients once again, it follows that $\mathcal{A}$ is contained in $\mathcal{C}$. This proves that $\mathbf{Mod}(\mathcal{E}) \subseteq \mathcal{C}$, and, with the inverse inclusion, that $\mathcal{C}$ is the variety $\mathbf{Mod}(\mathcal{E})$.

To conclude, here is Birkhoff's theorem:

Theorem 6.2 (Birkhoff stability) 1. A binary relation over $\mathrm{T}_\Sigma[X]$ is an equality, i.e, the equational theory of a class of algebras, if and only if it is a congruence that is stable under substitutions.

2. A class of algebras is a variety, i.e., the class of models of a set of equations, if and only if it is stable under isomorphisms, subalgebras, quotients and products.

This theorem, difficult to prove, is easily applied, especially to prove that a given class of algebras is not a variety. For example, the class of fields cannot be a variety, since a subalgebra (in particular, a subring) of a field is not necessarily a field. It is therefore useless to look for an equational definition of fields.

Many-sorted signatures

Recall a definition given in Chapter 1. A *many-sorted signature* is a pair (S, Σ), where

- S is a set of *sort* symbols;
- Σ is a set of typed constant and function symbols:

$$\begin{aligned} c &: s \\ f &: s_1 \dots s_n \rightarrow s, \end{aligned}$$

  where $s_1, \dots, s_n, s \in S$ and $n \geqslant 1$.

An *S-set* is a family $A = (A_s)_{s \in S}$ of sets indexed by S, and an *S-mapping* $m : A \rightarrow B$ between two S-sets A and B is a family of mappings $m_s : A_s \rightarrow B_s$, $s \in S$. Introduce an S-set $X = (X_s)_{s \in S}$ of variables. The rules for term formation are:

1. a variable of sort s is a term of sort s;
2. a constant of sort s is a term of sort s;
3. if f is a symbol of type $s_1 \dots s_n \rightarrow s$ and if M_i is a term of sort s_i, $i = 1, \dots, n$, then $fM_1 \dots M_n$ is a term of sort s.

The set of terms of sort s is $\mathrm{T}_\Sigma[X]_s$ and the set of all terms is $\mathrm{T}_\Sigma[X] = \bigcup_{s \in S} \mathrm{T}_\Sigma[X]_s$.

From a semantics point of view, each sort $s \in S$ is interpreted by a set (its domain) A_s. An (S, Σ)-algebra (or Σ-algebra) $\mathcal{A}$ is an S-set A with, for each symbol $f : s_1 \dots s_n \rightarrow s$ of Σ, a mapping $f^{\mathcal{A}} : A_{s_1} \times \dots \times A_{s_n} \rightarrow A_s$. A Σ-morphism $m : A \rightarrow B$ is an S-mapping commuting with the interpretations of the elements of Σ. All the definitions and properties given in this section are also valid for many-sorted algebras. The same notations are used for $\mathrm{T}_\Sigma[X]$, the Σ-algebra freely generated by an S-set X, and T_Σ, the Σ-algebra of closed terms, which is an initial Σ-algebra.

Many-sorted signatures give a natural framework to describe algebraic structures such as vector spaces, which have two sorts of objects: scalars and vectors. The specification of data structures is another application domain, which has been studied in detail since the middle of the 1970s.

Example The specification of a set structure will be given by a signature with two sorts *element* and *set*, containing a constant and an insertion operation of an element in a set:

$$\begin{aligned} \varnothing &: \ set \\ \# &: \ element \times set \rightarrow set \end{aligned}$$

and the axioms

$$(\#) \quad \left\{ \begin{array}{rcl} a\#(a\#x) & = & a\#x \\ a\#(b\#x) & = & b\#(a\#x). \end{array} \right.$$

□

6.1.3 The inference system of equational logic

From equality axioms to rules

From a deductive point of view, consider a set $\mathcal{E}$ of equational axioms and the equational propositions derived from $\mathcal{E}$.

Consider a general inference system, such as the NK system of natural deduction. Axioms, already given, must be added to state that $=$ is the equality:

$(Ax_{=}1)$	$\forall x \; x = x$
$(Ax_{=}2)$	$\forall xy \; x = y \Rightarrow y = x$
$(Ax_{=}3)$	$\forall xyz \; x = y \wedge y = z \Rightarrow x = z$
$(Ax_{=}f)$	$\forall x_1 \ldots x_n y_1 \ldots y_n \; x_1 = y_1 \wedge \ldots \wedge x_n = y_n \Rightarrow f x_1 \ldots x_n = f y_1 \ldots y_n$

Apart from the first, these axioms are not all equational. However, they all have a restricted form: they are equational Horn clauses, also called conditional clauses, which share with equations a certain number of properties which are found in logic programming.

To manipulate them, the rules of NK dealing with the connectives $\Rightarrow$ and $\wedge$ and the quantifier $\forall$ are used. It follows that formulas which are not equations must be used to prove an equation from other equations. This detour is anything but natural. To avoid it, a solution consists of combining the use of equality axioms and the rules of NK in the derivation of new specialized rules. Here, for example, is how the symmetry rule

$$(\mathbf{S}) : \frac{\forall \vec{x}(M = N)}{\forall \vec{x}(N = M)}$$

can be derived in NK:

$$(I_\forall):\dfrac{(E_\Rightarrow):\dfrac{(E_\forall):\dfrac{\forall\vec{x}(M=N)}{M=N}\qquad (E_\forall):\dfrac{(Ax_=2):\dfrac{}{\forall xy(x=y\Rightarrow y=x)}}{M=N\Rightarrow N=M}}{N=M}}{\forall\vec{x}(N=M)}.$$

From semantics to rules

The second approach is semantic and exploits the results obtained. Instead of deriving the rules in a general inference system such as NK, without any intuition, an inference system will be constructed. The relation $\vdash$ of deductive consequence will be equivalent to the relation $\models$ of semantic consequence: $\mathcal{E}\models M=N$ if and only if $\mathcal{E}\vdash M=N$. This equivalence will be stated as the completeness and the soundness of the inference system.

Fix an infinite set X of variables for what follows, and let $\mathcal{E}$ be a set of equations over these variables. By using the correspondence $\mathbf{Mod}\leftrightarrows\mathbf{Eq}$, $\mathcal{E}\models M=N$ means exactly $M=N\in\mathbf{Eq}(\mathbf{Mod}(\mathcal{E}))$. Then let $\mathcal{E}'=\mathbf{Eq}(\mathbf{Mod}(\mathcal{E}))$: it was seen that $\mathcal{E}$ and $\mathcal{E}'$ have the same models, that $\mathcal{E}'$ is the equality generated by $\mathcal{E}$, i.e, from the characterization of equalities, $\mathcal{E}'$ is the least congruence stable under substitutions containing $\mathcal{E}$.

The problem comes back to finding the equality generated by a set of equations, which is of purely syntactic nature. The rules of Table 6.1 define the generated stable congruence exactly. Write $\vdash_{eq}$ for the deduction relation defined by this inference system.

$$(\mathbf{R}):\dfrac{}{M=M}\qquad(\mathbf{S}):\dfrac{M=N}{N=M}$$

$$(\mathbf{T}):\dfrac{M=M'\qquad M'=M''}{M=M''}$$

$$(\mathbf{cong}):\dfrac{M_1=N_1\qquad\ldots\qquad M_r=N_r}{fM_1\ldots M_r=fN_1\ldots N_r}$$

$$(\mathbf{subst}):\dfrac{M=N}{\sigma(M)=\sigma(N)}$$

Table 6.1 The inference system of equational logic

The following lemma is easily checked, and it leads directly, along with the preceding discussion, to Birkhoff's completeness theorem.

Lemma 6.3 If $\mathcal{E}'$ is the equality generated by $\mathcal{E}$, then $(M,N)\in\mathcal{E}'$ if and only if $\mathcal{E}\vdash_{eq} M=N$, i.e, the judgment $M=N$ is derivable from the axioms $M=N$, for $(M,N)\in\mathcal{E}$.

Theorem 6.4 (Birkhoff completeness) $\mathcal{E} \vdash_{eq} M = N$ if and only if $\mathcal{E} \models M = N$.

It will be convenient to use a more general rule for context passing which can be derived from the preceding ones. It operates over an arbitrary context $\mathcal{C}[\ ]$, i.e, a term with a 'hole', in contrast to the rules $(f, i)_{1 \leqslant i \leqslant n}$, which deal only with 'superficial' contexts:

$$(\textbf{cont}) : \frac{M = N}{\mathcal{C}[M] = \mathcal{C}[N]}.$$

The many-sorted case

The many-sorted case is not identical to the single-sorted one from the deductive point of view. If the preceding inference system is applied directly, and if the signature allows certain interpretation domains to be empty, invalid equations might be derivable, making the inference system unsound, and this would be unacceptable.

Example Consider a signature with two sorts s and s' with a symbol $f : s \to s'$ and the equation $f(x) = y$. From $f(x) = y$, deduce $f(x) = z$ using the substitution $[y := z]$; then by symmetry and transitivity, $y = z$. Consider the algebra $\mathcal{A}$ with $\mathcal{A}_s = \varnothing$ and $\mathcal{A}_{s'} = \{a, b\}$, the interpretation of f being the empty function from $\mathcal{A}_s$ to $\mathcal{A}_{s'}$ (there is no other choice for $f^{\mathcal{A}}$). So $\mathcal{A} \models f(x) = y$: a universal proposition is always true in an empty domain! But $\mathcal{A} \not\models y = z$ since $a \neq b$. The inference $f(x) = y \vdash y = z$ is therefore unsound. □

This situation cannot occur in the single-sorted case, since all the equations are valid in the empty algebra. In the many-sorted case, either a restriction must be made to signatures so that empty domains are not allowed, or the inference system must be modified so that it no longer deduces invalid equations.

It is very convenient to use algebras, some of whose domains can be empty. For example, the initial algebra over the signature of lists has the empty set for domain of sort *element*. This is perfectly normal, since a list needs generators to exist and the initial algebra has no generator.

The inference system will therefore be modified; the first part to be modified is the form of judgments. For each sort s of the signature introduce a predicate H_s stating that the sort s is inhabited, i.e., H_s means that $\exists x : s$. This existential in an equational framework is not problematic, since the $\exists$ will have a negative sign in the judgment, hence a universal force.

Judgments are *guarded* equations

$$H_{s_1}, \ldots, H_{s_p} \Rightarrow M = N.$$

Such a judgment is written $H \Rightarrow M = N$, where H is its guard, $M = N$ the equation and $s_1, \ldots, s_p$ are the guarded sorts. A variable, whether or not it appears in the equation, is guarded if its sort is guarded.

So a new form of satisfaction of judgments is defined:

$$\mathcal{A} \models H_{s_1}, \ldots, H_{s_p} \quad \text{if} \quad A_{s_i} \neq \varnothing \text{ for } i = 1, \ldots, p.$$

Then

$$\mathcal{A} \models H \Rightarrow M = N \quad \text{if} \quad \mathcal{A} \models H \text{ implies } \mathcal{A} \models M = N.$$

The axioms of an equational theory must be written under a guarded form, by guarding the sorts of all the variables. Since this guard is systematic, it can be considered as implicit—for the axioms only.

Here are the rules:

$$(\mathbf{R}) : \frac{}{H \Rightarrow M = M},$$

where H is formed of guards over the sorts of variables of M;

$$(\mathbf{S}) : \frac{H \Rightarrow M = N}{H \Rightarrow N = M}, \qquad (\mathbf{T}) : \frac{H \Rightarrow M = N \qquad H \Rightarrow N = P}{H \Rightarrow M = P},$$

$$(\mathbf{subst}) : \frac{H \Rightarrow M = N}{H' \Rightarrow \sigma(M) = \sigma(N)},$$

where H' is formed of guards over the sorts of the variables of the image by σ of the guarded variables of the premise;

$$(\mathbf{cont}) : \frac{H \Rightarrow M = N}{H' \Rightarrow \mathcal{C}[M] = \mathcal{C}[N]},$$

where H' is H extended by the guards over the sorts of the variables figuring in the context $\mathcal{C}[\;]$.

Example (cont.) Start from the axiom $H_s, H_{s'} \Rightarrow f(x) = y$. Deduce by substitution $\left[\begin{smallmatrix} y \\ z \end{smallmatrix}\right]$: $H_s, H_{s'} \Rightarrow f(x) = z$ (the guard H_s has no reason to disappear, even if no other variable is of sort s in the inferred equation, since the variables of sort s are guarded in the premise and invariant by this substitution), then by symmetry and transitivity, $H_s, H_{s'} \Rightarrow y = z$, always with the same guard. From the semantic point of view, $\mathcal{A} \models H_s, H_{s'} \Rightarrow f(x) = y$ for the simple reason that $\mathcal{A} \not\models H_s$, i.e, $A_s = \varnothing$, and similarly for the formulas that are derived from it: $\mathcal{A} \models H_s, H_{s'} \Rightarrow y = z$. On the other hand, $\mathcal{A} \not\models H_{s'} \Rightarrow y = z$, but this judgment was not inferred. □

The guard is conserved or extended by (**R**), (**S**), (**T**), and (**cont**). One would want to infer judgments with the smallest possible guard. Only the rule (**subst**) can reduce the guard (this rule can also extend it).

A weakening rule can be derived:

$$(\mathbf{weak}) : \frac{H \Rightarrow M = N}{H, H_s \Rightarrow M = N}, \tag{6.9}$$

which is used essentially so that the two premises of the rule (**T**) have the same guard. It is more interesting to have a reinforcement rule:

$$(\mathbf{reinf}) : \frac{H_s, H \Rightarrow M = N}{H \Rightarrow M = N},$$

which eliminates an unnecessary guard, i.e., a guard over a sort s for which no variable of $M = N$ is of that sort. Suppose that there exists a closed term P of sort s, and apply the rule **(subst)** with the substitution σ associating P with all the variables of sort s. Then σ leaves M and N invariant, so that the conclusion is $H \Rightarrow M = N$. The same manipulation can be done by using for P a term of sort s whose variables are of sorts already guarded by H.

Example Consider a signature with two sorts s and s', with two unary symbols $f : s \to s'$ and $g : s' \to s$ (and no constant, which means that the initial algebra has its two empty domains). From $H_s, H_{s'} \Rightarrow f(x) = y$, it follows as for the previous example that $H_s, H_{s'} \Rightarrow y = z$. Now, use the substitution σ associating with every variable of sort s the term $g(u)$, where u is a variable of sort s'. By applying **(subst)**, infer $H_{s'} \Rightarrow y = z$, since the image by σ of the guarded variables of $H_s, H_{s'} \Rightarrow y = z$ (i.e. the variables of sort s and s', including y and z) is reduced to y, z and u, all of sort s'. The unnecessary guard H_s has been eliminated. □

Goguen proposed a condition over a signature allowing the elimination of unnecessary guards, and thereby allowing work with an unguarded inference system, which is more pleasant: for every pair of sorts s, s', $(\mathrm{T}_\Sigma[x_{s'}])_s$ is non-empty, i.e, there exists a term of sort s with a variable $x_{s'}$ of sort s'. This condition is natural and so commonly satisfied in practice that one might say that signatures which do not satisfy it are bad.

6.1.4 Free models

The algebra $\mathrm{T}_\mathcal{E}[X]$ is an *$\mathcal{E}$-free algebra generated by X*. It is an $\mathcal{E}$-algebra freely generated by X in the category $\mathbf{Alg}_\mathcal{E}$ of $(\Sigma, \mathcal{E})$-algebras and of Σ-morphisms, just as $\mathrm{T}_\Sigma[X]$ is an object freely generated by X in the category $\mathbf{Alg}_\Sigma$. It is the solution, unique up to isomorphisms, of the following universal problem (ϖ is the projection of $\mathrm{T}_\Sigma[X]$ in $\mathrm{T}_\mathcal{E}[X]$).

Lemma 6.5 For every $\mathcal{A} \in \mathbf{Mod}(\mathcal{E})$ and every mapping $\xi : X \to A$, there exists a unique morphism $\bar{\xi}_\mathcal{E} : \mathrm{T}_\mathcal{E}[X] \to \mathcal{A}$ such that $\bar{\xi}_\mathcal{E}(\varpi(x)) = \xi(x)$ for every $x \in X$.

Proof It is important that $\mathrm{T}_\Sigma[X]$ be free in $\mathbf{Alg}_\Sigma$. The morphism $\bar{\xi}_\mathcal{E}(\dot{M})$ must be defined if $\dot{M} \in \mathrm{T}_\mathcal{E}[X]$. But, if $\varpi(M) = \varpi(N)$, i.e, $(M, N) \in \mathcal{E}$, then $\hat{\xi}(M) = \hat{\xi}(N)$, since $\mathcal{A}$ is a model of $\mathcal{E}$. One can then define $\bar{\xi}_\mathcal{E}(\dot{M}) = \hat{\xi}(s(\dot{M}))$, where $\dot{M}$ is the equivalence class of M, and s is a right-inverse of ϖ. By the unicity of $\hat{\xi}$, $\bar{\xi}_\mathcal{E} \circ \varpi = \hat{\xi}$. The unicity of $\bar{\xi}_\mathcal{E}$ follows from this relation and from the surjectivity of ϖ. See Figure 6.2. □

In particular, there is, for $X = \varnothing$, an *initial model*. By writing $\mathrm{T}_\mathcal{E}$ instead of $\mathrm{T}_\mathcal{E}[\varnothing]$, the initiality property in $\mathbf{Mod}(\mathcal{E})$ can be stated as follows: for every model $\mathcal{A}$ of $\mathcal{E}$, there exists a unique morphism $\iota_\mathcal{A} : \mathrm{T}_\mathcal{E} \to \mathcal{A}$. (The initiality property in

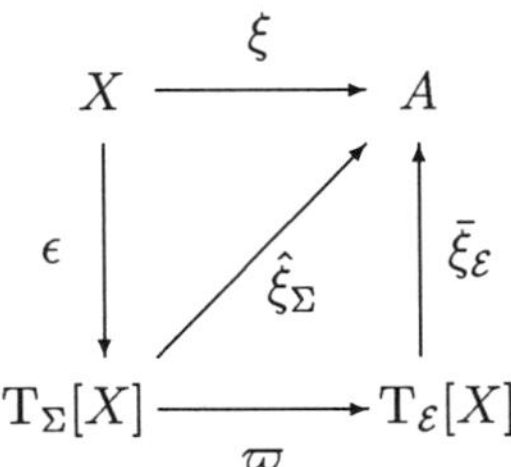

Figure 6.2 The free $\mathcal{E}$-algebra

$\mathbf{Alg}_\Sigma$ was: for every Σ-algebra, there exists a unique morphism $\iota_\mathcal{A} : T_\Sigma \to \mathcal{A}$.) The initial model will play an important role in the case of data structures.

The existence of free models is a very strong property of varieties. By reconsidering the examples of algebraic structures, the above discussion ensures the existence, for every set X, of the free monoid $X^\star$, of the free commutative monoid $X^{(\mathbf{N})}$, of the free group $\mathbf{L}(X)$, of the free abelian group $\mathbf{Z}^{(X)}$, of the free ring, etc., generated by X. In the case of data structures, the free models allow the construction of parametrized structures.

Definitions by generators and relations

The existence of free models allows the definition of a Σ-algebra by generators and relations in $\mathbf{Alg}_\mathcal{E}$, and not only in $\mathbf{Alg}_\Sigma$. The axioms of the variety state the general laws such as associativity, while the relations state particular properties. Finite presentations are particularly interesting. The varieties most studied using presentations are monoids and groups.

Here is the case of groups, which has received much attention due to its practical applications to other branches of mathematics, such as algebraic topology. Let G be a finite set of 'generators': the group freely generated by G is $\mathbf{L}(G) = T_\mathcal{E}[G]$, where $\mathcal{E}$ is the congruence stable under substitutions generated by the three usual axioms of left-neutral element, left-inverse and associativity. One can represent $\mathbf{L}(G)$ by the set of words over the alphabet of g, $\bar{g}$, with $g \in G$, and $\bar{g}$ designating the inverse of g, which do not contain any subwords of the form $g\bar{g}$ or $\bar{g}g$, since these subwords would simplify. Introduce R, a finite set of elements of $T_\Sigma[G]$, the 'relators' (in the case of groups, a relation under the form of a pair of terms $M = N$, is equivalent to $MN^{-1} = e$, i.e. to the relator MN^{-1}). Let N_R be the normal subgroup of $\mathbf{L}(G)$ generated by the image $\varpi(R)$ of the relators by the natural surjection. The group defined by the presentation $(G; R)$ and by generators and relators, is then the quotient group $\mathbf{L}(G; R) = \mathbf{L}(G)/N_R$.

Example The free group with one generator g is isomorphic to $\mathbf{Z}$, by sending g to 1. By introducing the relator g^6, i.e, the relation $g^6 = e$, one obtains the presentation group $(g; g^6 = e)$, which is the additive group $\mathbf{Z}/6\mathbf{Z}$. This group allows other presentations. Here is one with two generators and three relations

$(h, k; h^3 = k^2 = h^{-1}k^{-1}hk = e)$. One can go from one to the other by defining $h = g^4$, $k = g^3$ and $g = hk$. □

In the general case, let $\mathcal{E}$ be an equational theory, G a set of generators and $R \subseteq \mathrm{T}_\Sigma[G] \times \mathrm{T}_\Sigma[G]$ a set of relations. The $\mathcal{E}$-algebra defined by $(G; R)$ is the quotient $\mathrm{T}_\mathcal{E}[G; R]$ de $\mathrm{T}_\mathcal{E}[G]$ by the congruence generated by the image of R.

In the case of monoids, a *Thue congruence* is a congruence generated by a Thue system, i.e., by a set R of pairs of words over an alphabet A. The monoid defined by $(A; R)$ is the quotient of the free monoid $A^\star$ by the congruence generated by R.

Example Correct parenthesization is formalized using the alphabet

$$P_n = \{p_1, \bar{p}_1, \ldots, p_n, \bar{p}_n\}$$

formed of n pairs of parentheses (for $n = 3$, one could use $P_3 = \{(,), \{,\}, [,]\}$), and the relations $R = \{p_i\bar{p}_i = \epsilon; i = 1, \ldots, n\}$, which state simplification rules. `[(()}{[]{}})]{[]()}` is a correct parenthesization, while `[])(` is not. The class of the empty word is the set of correct parenthesizations. This set is called the *Dyck language* with n pairs of parentheses. □

At first, the *word problem* was defined in the case of algebras defined by generators and relations in a variety, first in the case of groups by Dehn [37]. Let $\mathcal{E}$ be an equational theory, and $(G; R)$ be a finite presentation. The word problem for $(G; R)$ is, given a pair $(M, N) \in \mathrm{T}_\Sigma[G] \times \mathrm{T}_\Sigma[G]$, determine if $\mathrm{T}_\mathcal{E}[G; R] \models M = N$.

For example, for the presentation of group $(h, k; h^3 = k^2 = h^{-1}k^{-1}hk = e)$ (of $\mathbf{Z}/6\mathbf{Z}$, but this is not needed), the problem $h^5k^9 = k^3h^{-1}$ has a positive solution: since the generators h and k commute, this equation is equivalent to $h^6k^6 = e$, which is satisfied, since $h^6 = (h^3)^2 = e$ and $k^6 = (k^2)^3 = e$. This equational reasoning is not always simple: in 1955 Novikov and Boone [114] found a finite presentation of a group for which the word problem is undecidable. The word problem for a free group (this is the case where $R = \varnothing$) is decidable, as will be seen, but not in the whole variety of groups.

Genericity of free models

From a semantic point of view, the free objects of a variety play a generic role, in the sense that the equations between generators are satisfied by all the algebras of the variety.

Proposition 6.6 $\mathbf{Eq}_X(\mathrm{T}_\mathcal{E}[X]) = \mathcal{E}$ if $\mathcal{E}$ is an equality over $\mathrm{T}_\Sigma[X]$.

Proof Let X be a set of variables and $M = N$ an equation with variables in X. Suppose that $\mathrm{T}_\mathcal{E}[X] \models M = N$. It must be shown that, for every model $\mathcal{A}$ of $\mathcal{E}$, $\mathcal{A} \models M = N$ must also hold. To prove this, $\hat{\xi}(M) = \hat{\xi}(N)$ must be proven for every $\xi \in A^X$. Since $\mathrm{T}_\mathcal{E}[X]$ is the $(\Sigma, \mathcal{E})$-algebra freely generated by X, there is an extension $\bar{\xi} : \mathrm{T}_\mathcal{E}[X] \to \mathcal{A}$ of ξ satisfying $\bar{\xi} \circ \varpi = \hat{\xi}$. But, $\mathrm{T}_\mathcal{E}[X] \models M = N$ implies

$\varpi(M) = \varpi(N)$ since ϖ is the extension of the inclusion of X in $\mathrm{T}_{\mathcal{E}}[X]$. It follows that $\hat{\xi}(M) = \bar{\xi}(\varpi(M)) = \bar{\xi}(\varpi(N)) = \hat{\xi}(N)$, which is the desired result. □

Hence, to establish validity in **Mod**($\mathcal{E}$) of an arbitrary equation (which here means with an arbitrary number of variables), it suffices to work in a unique algebra, which is $\mathrm{T}_{\mathcal{E}}[X]$, where X is an *infinite* set fixed once and for all. Therefore, the study of a variety of algebras can be done through a single algebra, of a syntactic nature, which will allow equational reasonings.

In particular, this result is applicable if $X = \varnothing$: if $M = N$ is an equation between closed terms satisfied by the initial model $\mathrm{T}_{\mathcal{E}}$, then it is also satisfied by all of the models of $\mathcal{E}$. However, the assumption 'equations between generators' is crucial. For example, if $\mathrm{T}_{\mathcal{E}} \models M = N$, where $M = N$ is an equation between terms containing two variables x and y, one cannot deduce that $M = N$ is satisfied by all of the models of $\mathcal{E}$. To obtain this conclusion, $\mathrm{T}_{\mathcal{E}}[x, y] \models M = N$ would have had to be proven, not just $\mathrm{T}_{\mathcal{E}} \models M = N$.

Consider the case of groups. The initial model of the variety of groups, which is the free group with 0 generators $\mathbf{L}(\varnothing)$, is the trivial group $\{e\}$: it satisfies any equation! The situation is hardly improved by adding a generator: the free group with one generator $\mathbf{L}(\{g\})$ is isomorphic to $\mathbf{Z}$. It satisfies at least the commutativity relation (and many others), which is not satisfied by all groups. It turns out that the equational theory of the free group with two generators coincides with the equational theory of groups, but that is a specific property of the variety of groups.

The situation is generally less caricatural in the case of the initial model of a data structure whose equational theory is of real interest.

Addition is specified by the following equations:

$$(+) \quad \left\{ \begin{array}{rcl} x + 0 & = & 0 \\ x + s(y) & = & s(x + y). \end{array} \right.$$

Their 'standard' model is $\mathbf{N}$, which is the initial model. A non-standard model will be constructed, the algebra $\mathbf{N}'$ of colored integers, to show that there are equations satisfied by $\mathbf{N}$ which are not consequences of the addition axioms. The domain of $\mathbf{N}'$ is formed of two copies of $\mathbf{N}$, the black integers $(\spadesuit, n)$ and the red integers $(\heartsuit, n)$. The constant 0 is interpreted by the black integer $(\spadesuit, 0)$. The operations are normally interpreted over the integers, with the added stipulation that the successor conserves the color of its argument and addition conserves the color of its first argument:

$$\begin{array}{rcl} s(\gamma, n) & = & (\gamma, n + 1) \\ (\gamma, n) + (\gamma', n') & = & (\gamma, n + n'), \end{array}$$

where γ is either the color $\spadesuit$ or the color $\heartsuit$. This algebra is a model of additive arithmetic.

For $(\gamma, n), (\gamma', n') \in \mathbf{N}'$,

$$\begin{aligned}(\gamma,n)+(\spadesuit,0) &= (\gamma,n)\\ (\gamma,n)+s(\gamma',n') &= (\gamma,n)+(\gamma',s(n'))\\ &= (\gamma,n+s(n'))\\ &= (\gamma,s(n+n'))\\ &= s(\gamma,n+n').\end{aligned}$$

However, the equations $0+x=x$ and $x+y=y+x$ are not satisfied in $\mathbf{N}'$:

$$\begin{aligned}(\spadesuit,0)+(\heartsuit,3) &= (\spadesuit,3)\neq(\heartsuit,3)\\ (\spadesuit,1)+(\heartsuit,2) &= (\spadesuit,3)\\ &\neq (\heartsuit,3)\\ &= (\heartsuit,2)+(\spadesuit,1).\end{aligned}$$

As is standard, this example suffices to show that commutativity is not a consequence of the equations defining addition.

Free extensions without axioms

Given the signature of lists and an arbitrary set A of 'atoms', the algebra freely generated by A is the set of lists of elements of A. If A is not only a set, but an algebra with its own operations, one should be able to pass these operations as parameters. This is what the `functor` construction of Standard ML implements.

Starting with a signature Σ^0 and an *extension* $\Sigma = \Sigma^0 \cup \Sigma^1$ of Σ^0, a 'functor' is constructed: it associates with every Σ^0-algebra $\mathcal{X}$ a Σ-algebra $F(\mathcal{X})$. The idea is that the operations of Σ^0 which are already interpreted in $\mathcal{X}$ keep this interpretation, while the operations of Σ^1 are 'free'. So the evaluation takes place in $\mathrm{T}_\Sigma[X]$, and everything which can be evaluated in $\mathcal{X}$ is associated with its value. The elements of $\mathrm{T}_\Sigma[X]$ which can be first evaluated are those of $\mathrm{T}_{\Sigma^0}[X]$, using the evaluation morphism $\mathrm{ev}_{\mathcal{X}} : \mathrm{T}_{\Sigma^0}[X] \to \mathcal{X}$, whose kernel is the equational diagram of $\mathcal{X}$. To evaluate a term of $\mathrm{T}_\Sigma[X]$, evaluate its subterms, which are in $\mathrm{T}_{\Sigma^0}[X]$, which is expressed by forming the Σ-congruence of $\mathrm{T}_\Sigma[X]$ generated by the diagram of $\mathcal{X}$. Let $\mathrm{T}_{\Sigma/\Sigma^0}[\mathcal{X}]$ be the quotient of $\mathrm{T}_\Sigma[X]$ by this congruence. This Σ-algebra is the one sought, and can be written in ML:

```
functor T (X : SIGMA0) = TΣ/Σ0[𝒳] : SIGMA
```

By taking the most evaluated term in each class, the domain of $\mathrm{T}_{\Sigma/\Sigma^0}[\mathcal{X}]$ can be equated with the subset of $\mathrm{T}_\Sigma[X]$ composed of terms for which every subterm belonging to $\mathrm{T}_{\Sigma^0}[X]$ is in fact an element of X, which gives an injection from $\mathrm{T}_{\Sigma/\Sigma^0}[\mathcal{X}]$ to $\mathrm{T}_\Sigma[X]$, which is a right-inverse of the natural surjection. This injection is not a Σ-morphism, since the operations of Σ^0 are interpreted in $\mathrm{T}_{\Sigma/\Sigma^0}[\mathcal{X}]$, and not in $\mathrm{T}_\Sigma[X]$, but it is a Σ^1-morphism. Nevertheless, this allows the elements of $\mathrm{T}_{\Sigma/\Sigma^0}[\mathcal{X}]$ to be interpreted as terms—assuming that those of $\mathcal{X}$ can be implemented.

The property which justifies this construction is similar to that of the free algebra. It has simply been relativized to Σ^0: for every Σ^0-morphism $\xi : \mathcal{X} \to \mathcal{A}$, there exists a unique Σ-morphism $\hat{\xi}_{\Sigma/\Sigma^0} : \mathrm{T}_{\Sigma/\Sigma^0}[\mathcal{X}] \to \mathcal{A}$ such that for every $x \in X$, $\hat{\xi}_{\Sigma/\Sigma^0}(x) = \xi(x)$.

As for the free algebra, it is not the algebra $\mathrm{T}_{\Sigma/\Sigma^0}[\mathcal{X}]$ which is important, but the functor $\mathrm{T}_{\Sigma/\Sigma^0} : \mathcal{X} \mapsto \mathrm{T}_{\Sigma/\Sigma^0}[\mathcal{X}]$, for which there is an implementation in Standard ML's system of modules.

Free extensions and axioms

The functor $\mathrm{T}_{\Sigma/\Sigma^0}$ gives a satisfying answer to the parametrization problem in the absence of axioms. If what is desired is not a list of `foo` but, rather, a set of `foo`, there are axioms which state that insertion into a set is 'idempotent', 'associative' and 'commutative', and hence axioms over the produced algebra. If one also wishes to transform a binary relation over the elements into an operation over sets of elements, similarly to the iterators of functional programming, this operation itself must be commutative and associative, otherwise, the transformation would have no meaning. Therefore axioms must be imposed on a parametrized algebra.

Frequently, the base type is the initial algebra of a class $\mathbf{Alg}_\Sigma$, for which the signature is the set of constructors or generators: this is the case for the natural integers (constructors 0 and s) and the Booleans (constructors *true* and *false*). Rational integers are generated by 0, s and p, with $psx = x$ and $spx = x$.

Starting from a signature Σ^0, a set of axioms $\mathcal{E}^0$ over Σ^0, an extension $\Sigma = \Sigma^0 \cup \Sigma^1$ and $\mathcal{E} = \mathcal{E}^0 \cup \mathcal{E}^1$, where $\mathcal{E}^1$ is a set of equations over Σ (not only over Σ^1), one constructs a functor $\mathrm{T}_{\mathcal{E}/\mathcal{E}^0} : \mathcal{X} \mapsto \mathrm{T}_{\mathcal{E}/\mathcal{E}^0}[\mathcal{X}]$ from the category of $\mathcal{E}^0$-algebras to that of $\mathcal{E}$-algebras.

Let $\mathcal{X}$ be a $\mathcal{E}^0$-algebra. The set $\mathrm{T}_{\mathcal{E}/\mathcal{E}^0}[\mathcal{X}]$ is the quotient of $\mathrm{T}_{\mathcal{E}}[X]$ by the congruence generated by the kernel of the Σ^0-morphism of $\mathrm{T}_{\mathcal{E}}[X]$ in $\mathcal{X}$ extending the identity of X. This algebra verifies the universal property: for every $\mathcal{E}$-algebra $\mathcal{A}$ and every Σ^0-morphism $\xi : \mathcal{X} \to \mathcal{A}$, there exists a unique Σ-morphism $\hat{\xi}_{\mathcal{E}/\mathcal{E}^0} : \mathrm{T}_{\mathcal{E}/\mathcal{E}^0}[\mathcal{X}] \to \mathcal{A}$ such that for every $x \in X$, $\hat{\xi}_{\mathcal{E}/\mathcal{E}^0}(x) = \xi(x)$.

Examples Here are a few extensions starting from $\Sigma^0 = \{0, s, p\}$, $\mathcal{E}^0 = \{psx = x, spx = x\}$ and $\mathcal{X} = \mathbf{Z}$.

- $\Sigma^1 = \{+\}$ and $\mathcal{E}^1 = \{x + 0 = x, x + s(y) = s(x + y), x + p(y) = p(x + y)\}$: $\mathrm{T}_{\mathcal{E}/\mathcal{E}^0}[\mathcal{X}] \approx \mathbf{Z}$: this is a good situation, and one can say that $\mathcal{E}$ is a good specification of addition over the integers;
- $\Sigma^1 = \varnothing$ and $\mathcal{E}^1 = \{ssx = x\}$: $\mathrm{T}_{\mathcal{E}/\mathcal{E}^0}[\mathcal{X}] \approx \{0, 1\}$: the axiom implies the identification of elements of the domain to integers modulo 2; the extension is not 'hierarchically consistent';
- $\Sigma^1 = \{+\}$ and $\mathcal{E}^1 = \{x + 0 = x, x + s(y) = s(x + y)\}$: $\mathrm{T}_{\mathcal{E}/\mathcal{E}^0}[\mathcal{X}] \not\approx \mathbf{Z}$: there are new elements such as $0 + p(0)$ which cannot be evaluated to an integer; the specification of $+$ is not 'sufficiently complete'.

□

The deduction problem

In general, the problem is to know, given a class of algebras $\mathcal{C}$ and an equation $M = N$, whether $\mathcal{C} \models M = N$. Since the simplest way of producing a class of algebras is to fix a set $\mathcal{E}$ of axioms, and to take for $\mathcal{C}$ the variety defined by $\mathcal{E}$, the particular case $\mathbf{Mod}(\mathcal{E}) \models M = N$ should suffice. But this problem can be brought back to the study of a single algebra $\mathrm{T}_{\mathcal{E}}[X]$, X infinite, thanks to the genericity property of free models. This problem will be handled using a rewriting system obtained by orienting the equations of $\mathcal{E}$, when this is possible.

However, the apparently simpler situation, where an algebra $\mathcal{A}$ is taken instead of a class, is in general much more complex, except if $\mathcal{A} = \mathrm{T}_{\mathcal{E}}[X]$, where X is infinite. However, this situation presents itself naturally with $\mathcal{A} = \mathrm{T}_{\mathcal{E}}$, which satisfies more equations than the free model, as the colored integer example showed. There is no finite and complete inference system in that case, and one must then use induction proofs.

6.2 Unification

Unification plays a central role, to prove as well as to compute. It also allows the resolution and propagation of constraints, hence the transfer of information.

The history of unification goes back to 1930 with Herbrand's A property [62]. Used by Davis in 1963 [33] in the implementation of a proof procedure extending that of Prawitz (1960, [124]), unification was rediscovered in 1965 by Robinson [128] when he introduced the resolution principle. It became the subject of major research with the contributions of Plotkin (1972, [121]) and Huet (1976, [67]).

The practical importance of unification is due to the fact that it is an elementary operation: the power of 5-th generation machines is measured in kLIPS, for thousands of logical inferences per second, each logical inference corresponding to exactly one unification.

6.2.1 Equation resolution

Let Σ be a signature, $M_1, M_2 \in \mathrm{T}[X]$, and $\mathcal{A}$ be a Σ-algebra. A *solution* of the equation $M_1 = M_2$ *in* $\mathcal{A}$ is a valuation $\xi : \mathrm{var}(M_1) \cup \mathrm{var}(M_2) \to A$ such that $\hat{\xi}(M_1) = \hat{\xi}(M_2)$. To solve an equation is to search for its solution, which means to find in A values for the variables of M_1 and M_2 which satisfy the equation.

Example The equation $x^3 - 2 = 0$ does not have a solution in $\mathbf{Q}$, has a solution in $\mathbf{R}$ ($\xi(x) = \sqrt[3]{2}$) and has three solutions in $\mathbf{C}$ ($\xi_k(x) = i^k \sqrt[3]{2}$, $k = 0, 1, 2$). $\square$

The very definition of the problem uses the algebra $\mathcal{A}$ upon which depend the existence and the number of eventual solutions. Many constructions, such as fields of fractions, algebraic extensions, localization, completion of a metric space and distributions, only exist to guarantee the existence, and in certain conditions the unicity, of solutions of a certain class of equations.

Solving equations remains an abstract problem, except when equality over A is decidable, and in particular when the conditions required to answer this question all hold (see Chapter 7). To unify is to solve equations over a syntactic domain. Depending on the domain, solving equations is called:

- finite unification, in the algebra of terms $\mathrm{T}[X]$;
- rational unification, in the algebra $\mathrm{T}^r[X]$ of rational terms;
- unification modulo $\mathcal{E}$, in the free model $\mathrm{T}_{\mathcal{E}}[X]$ of a theory $\mathcal{E}$;
- higher-order unification, in the set of typed λ-terms.

In the equational case, since the necessary condition of decidability is not always verified, only some particularly interesting cases are studied: the associative–commutative (AC), associative (which corresponds to solving equations in the free monoid) and Boolean ring cases.

Most of this section will deal with finite unification, which is a fundamental operation for all 'symbolic' (functional, logic and equational) programming. A solution of $M_1 = M_2$ is in this case a substitution σ, of domain included in $\mathrm{var}(M_1) \cup \mathrm{var}(M_2)$, called a unifier, such that $\sigma M_1 = \sigma M_2$. It then has universal validity: if $\mathcal{A}$ is a Σ-algebra, the $\hat{\xi} \circ \sigma$ are solutions in $\mathcal{A}$ for every valuation ξ with values in A. In particular, for $\mathcal{A} = \mathrm{T}[X]$, all the instances $\mu \circ \sigma$ of σ by an arbitrary substitution μ are also solutions in $\mathrm{T}[X]$, i.e, unifiers. But this infinity of solutions can be described as the set of instances of one among them, the least one.

Although it is easy to prove, this is a fundamental result, related to the computation of the least upper bound in the term semi-lattice, which is constantly used in logic programming and rewriting systems. Proven for the finite and rational cases, this property is rare for unification modulo $\mathcal{E}$ (it has been known only since 1986 for the theory of Boolean rings [102]).

This property does not even have any meaning in non-syntactic domains: for example, the equation $e^x = x + 2$ has two solutions in $\mathbf{R}$, neither of which can reasonably be considered to be an instance of the other. In the case of vector spaces, a linear equation $Lu = 0$ has in general a set of infinite solutions. But (in finite dimension, for example), if a finite *basis* $B = \{e_1, \ldots, e_n\}$ is known, all of the solutions are obtained by linear combinations of the e_i. This situation appears for some equational theories, such as AC. In other cases, only infinite bases exist, or no basis whatsoever. In the general context of unification modulo $\mathcal{E}$, finite unification is only a particular case, that of the empty theory.

6.2.2 Systems of equations and substitutions

We begin by examining the relations between substitutions and systems of equations for which the substitutions are solutions. Write $\mathbb{S}$ for the set of substitutions and $\vec{M} = \vec{N}$ for a system of equations $M_i = N_i$, $i = 1, \ldots, p$.

Definition 6.4

- Let E be a system of equations $\vec{M} = \vec{N}$ and $\sigma \in \mathbb{S}$. The substitution σ is a *unifier* of E or σ *unifies* E if σ is a solution of E in $\mathrm{T}[X]$, i.e, for every i, $\sigma M_i = \sigma N_i$. Write $\mathrm{U}(E)$ for the set of unifiers of E.
- E is *unifiable* if $\mathrm{U}(E) \neq \varnothing$.
- Two systems of equations E and E' are *equivalent* if $\mathrm{U}(E) = \mathrm{U}(E')$.

The systems which are easiest to solve are obviously those which have already been solved! For example, if a solution to the equation $x = a$ is sought, the obvious solution is $x = a$, or, to conform to the above definition, the substitution $\sigma = \left[\begin{smallmatrix} x \\ a \end{smallmatrix}\right]$. It is therefore natural to associate a substitution with a system of variables having only distinct variables on the left-hand sides.

Definition 6.5

- A system of equations E is in *reduced form* if it is of the form $\vec{x} = \vec{M}$, the x_i variables being pairwise distinct;
- if E is in reduced form, its *associated* substitution, written σ_E, is $\left[\begin{smallmatrix} x_1 & \cdots & x_n \\ M_1 & \cdots & M_n \end{smallmatrix}\right]$,

$$\left\{ \begin{array}{rcll} \sigma_E x_i & = & M_i, & 1 \leqslant i \leqslant n \\ \sigma_E y & = & y & \text{if } y \notin \{x_1, \ldots, x_n\} \end{array} \right.$$

The error would be to believe that, as soon as E is in reduced form, σ_E is a solution of E. Consider three examples:

$$(E_1) \left\{ \begin{array}{rcl} x & = & f(z) \\ y & = & g(b, z), \end{array} \right. \qquad (E_2) \left\{ \begin{array}{rcl} x & = & y \\ y & = & a, \end{array} \right. \qquad (E_3) \left\{ \begin{array}{rcl} x & = & f(y) \\ y & = & g(x). \end{array} \right.$$

For E_1 everything goes well and $\sigma_{E_1} = \left[\begin{smallmatrix} x & y \\ f(z) & g(b,z) \end{smallmatrix}\right]$ is a solution of E_1.

For E_2, things are not as simple: σ_{E_2} is not a solution of the first equation. The value of x is not obtainable immediately, since y appears on the left-hand side of the second equation. The substitution σ_{E_2} must be iterated a second time to obtain a unifier $\theta = \sigma_{E_2}^2 = \left[\begin{smallmatrix} x & y \\ a & a \end{smallmatrix}\right]$.

Finally, for E_3, σ_{E_3} does not satisfy either of the two equations. Furthermore, it is obvious that the iterating method used for E_2 leads to an infinite loop. This occurs because of the cyclic dependencies between x and y, while in E_2 there were no cycles. In fact, E_3 is not unifiable in $\mathrm{T}[X]$, but, rather, in $\mathrm{T}^r[X]$ by $\theta = \left[\begin{smallmatrix} x & y \\ (fg)^\omega & (gf)^\omega \end{smallmatrix}\right]$, which is the 'limit' of the sequence of iterations of σ_{E_3}: σ_{E_3}, $\sigma_{E_3}^2$, $\sigma_{E_3}^3, \ldots$.

So a stronger condition must be imposed on systems of equations so that they need not be iterated to obtain their solutions.

Definition 6.6 A system of equations E is in *(finite) solved form* if it is in reduced form, with $\{x_1, \ldots, x_n\} \cap \bigcup_{i=1}^n \mathrm{var}(M_i) = \varnothing$.

It will be shown that every unifiable system is equivalent to a system in solved form. If E is in solved form, it will be shown that $\sigma_E \in \mathrm{U}(E)$ and that it is *the* most general solution of E, under a preorder which will be introduced.

The preorder of substitutions

Definition 6.7 Let σ and θ be two substitutions. The substitution σ is called an *instance* of θ, or θ is *more general* than σ, written $\sigma \geqslant \theta$, if there exists a substitution μ such that $\sigma = \mu\theta$.

This relation $\geqslant$ is clearly a preorder over $\mathbb{S}$. Another way of defining a preorder over $\mathbb{S}$ consists of extending to substitutions that of $\mathrm{T}[X]$, by extensionality. Write $\sigma \geqslant_e \theta$ if $\sigma M \geqslant \theta M$ for every $M \in \mathrm{T}[X]$.

Proposition 6.7

- If $\sigma \geqslant \theta$, then $\sigma \geqslant_e \theta$.
- If $\sigma \geqslant_e \theta$, and if Σ is non-linear (i.e., there exists at least one $f \in \Sigma$ of arity > 1), then $\sigma \geqslant \theta$.

Proof The first point is trivial. To prove the converse, note that if Σ is non-linear, there exists for every $W \subseteq X$, W finite, a term $M \in \mathrm{T}_\Sigma[X]$ such that $\mathrm{var}(M) \supseteq W$. Suppose that $\sigma \geqslant_e \theta$, and take M such that $\mathrm{var}(M) \supseteq \mathrm{dom}(\sigma) \cup \mathrm{dom}(\theta) \cup \mathrm{im}(\theta)$. Since $\sigma(M) \geqslant \theta(M)$, there exists μ such that $\sigma M = \mu(\theta M) = (\mu\theta)M$. It will be shown that $\sigma = \mu \circ \theta$. It has already been shown that $\sigma\lceil_{\mathrm{var}(M)} = (\mu\theta)\lceil_{\mathrm{var}(M)}$. Furthermore, suppose that $\mathrm{dom}(\mu) \subseteq \mathrm{var}(\theta M)$. But this last set of variables is itself bounded by: $\mathrm{var}(\theta M) \subseteq \mathrm{var}(M) \cup \mathrm{im}(\theta)$. Hence, if $x \notin \mathrm{var}(M)$, x does not belong to any of $\mathrm{dom}(\theta)$, $\mathrm{dom}(\sigma)$ or $\mathrm{dom}(\mu)$, so $\sigma x = \mu\theta x = x$. Hence, $\sigma = \mu\theta$. □

The preorder can be used to formalize the following intuitive fact: one can always instantiate a solution of a system of equations.

Proposition 6.8 Every instance of a unifier is a unifier.

This means that $\mathrm{U}(E)$ is a *filter* of $(\mathbb{S}, \geqslant)$. The most general unifier property will show that the filter is *general*, i.e., generated by a single element σ, in the sense that $\theta \in \mathrm{U}(E)$ if and only if $\theta = \mu\sigma$ for some μ.

To analyze a substitution σ, associate with it the following three sets of variables:

- $\mathrm{dom}(\sigma) = \{x \in X; \sigma x \neq x\}$ is the *domain* of σ;
- $\mathrm{range}(\sigma, V) = \bigcup_{x \in V} \mathrm{var}(\sigma x)$ is the *range* of σ over V;
- $\mathrm{im}(\sigma) = \mathrm{range}(\sigma, \mathrm{dom}(\sigma))$ is the *image* of σ.

Recall that, by definition, $\mathrm{dom}(\sigma)$ is always a finite set. Note that $\mathrm{range}(\sigma, V)$ is infinite if and only if V is infinite and that $\mathrm{im}(\sigma)$ is therefore finite. A substitution is essentially represented by an application of $\mathrm{dom}(\sigma)$ in $\mathrm{T}[\mathrm{im}(\sigma)]$, but also by taking an arbitrary V larger than $\mathrm{dom}(\sigma)$, by a mapping from V to $\mathrm{T}[\mathrm{range}(\sigma, V)]$. The following property holds: $\mathrm{dom}(\sigma) \cup \mathrm{range}(\sigma, V) = V$.

Idempotent substitutions A substitution θ is *idempotent* if $\theta^2 = \theta$. The three following statements are equivalent:

(1) θ is idempotent;
(2) $\mathrm{dom}(\theta) \cap \mathrm{im}(\theta) = \varnothing$;
(3) $\mathrm{dom}(\theta) \cap \mathrm{range}(\theta, X) = \varnothing$.

Another way to state (2) is to say that $\begin{bmatrix} x_1 & \cdots & x_n \\ M_1 & \cdots & M_n \end{bmatrix}$ is idempotent if $\{x_1, \ldots, x_n\} \cap \bigcup_{i=1}^{n} \mathrm{var}(M_i) = \varnothing$. It follows that, for a system of equations E in reduced form, θ_E is idempotent if and only if E is in solved form.

In the preorder over $\mathbb{S}$, the idempotent substitutions play a key role.

Lemma 6.9 θ is idempotent if and only if for every $\sigma \in \mathbb{S}$, $\sigma \geqslant \theta$ is equivalent to $\sigma = \sigma\theta$.

Proof Suppose $\theta^2 = \theta$. If $\sigma \geqslant \theta$, there exists μ such that $\sigma = \mu \circ \theta$, so $\sigma \circ \theta = (\mu \circ \theta) \circ \theta = \mu \circ \theta^2 = \mu \circ \theta = \sigma$. Conversely, if $\sigma = \sigma\theta$, then, by the definition of $\geqslant$, $\sigma \geqslant \theta$.

If equivalence holds, then $\theta \geqslant \theta$ implies $\theta = \theta\theta = \theta^2$. □

The link between the preorder over $\mathbb{S}$ and the resolution of equations in reduced form is given by the following lemma, where (2) follows directly from the preceding lemma:

Lemma 6.10 Let x be a variable, M a term and σ a substitution.

1. $\sigma \in \mathrm{U}(x = M)$ if and only if $\sigma\begin{bmatrix} x \\ M \end{bmatrix} = \sigma$;
2. $\begin{bmatrix} x \\ M \end{bmatrix}$ is idempotent if and only if $\mathrm{U}(x = M) = \{\sigma; \sigma \geqslant \begin{bmatrix} x \\ M \end{bmatrix}\}$.

The first properties of substitutions lead directly to the desired result.

Proposition 6.11 If the system $(E) : \vec{x} = \vec{M}$ is in solved form, then θ_E is the least element of $\mathrm{U}(E)$.

Proof It has already been seen that $\theta_E \in \mathrm{U}(E)$. Let $\sigma \in \mathrm{U}(E)$. Then $\sigma x_i = \sigma M_i$ for every $i = 1, \ldots, n$, so $\sigma \circ \begin{bmatrix} x_i \\ M_i \end{bmatrix} = \sigma$, $i = 1, \ldots, n$ according to the lemma, from which $\sigma \circ \begin{bmatrix} x_1 \\ M_1 \end{bmatrix} \circ \ldots \circ \begin{bmatrix} x_n \\ M_n \end{bmatrix} = \sigma$. But, since the x_i are pairwise distinct and do not appear in the M_j, it is clear by induction that

$$\begin{bmatrix} x_1 \\ M_1 \end{bmatrix} \circ \ldots \circ \begin{bmatrix} x_n \\ M_n \end{bmatrix} = \begin{bmatrix} x_1 & \cdots & x_n \\ M_1 & \cdots & M_n \end{bmatrix} = \theta_E$$

and so $\sigma \circ \theta_E = \sigma$, i.e, $\sigma \geqslant \theta_E$. □

θ_E is called a *most general unifier* (*mgu*) of E. Of course, since $\geqslant$ is only a preorder of $\mathbb{S}$, and not an order, this unifier is only defined up to equivalence. However, when it does not create problems, we will refer to *the* mgu.

6.2.3 Transformations of systems of equations

The search for solutions in a system of equations means searching for an equivalent solved form. The Herbrand unification algorithm to be presented consists of a sequence of steps which either transform the system into an equivalent system closer to the solved form, or detect that the system does not have a solution.

Finite unification
Transformations $E \leadsto_t E'$ are defined between systems of equations. Here, a system of equations will refer either to a set of equations, as before, or a symbol $\perp$ representing a system without solution. Write $E \leadsto E'$ if $E \leadsto_t E'$ for one of the transformations t (see Figure 6.3). The transition $\overset{\star}{\leadsto}$ is the reflexive transitive closure of $\leadsto$. Solving E_0 consists of applying a sequence of transformations $E_0 \overset{\star}{\leadsto} E_n$ so that E_n is in solved form, or else $E_n = \perp$.

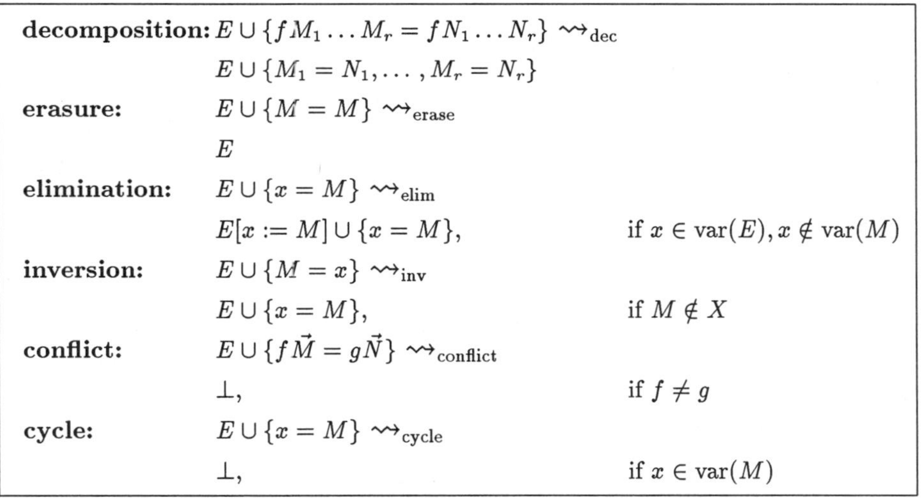

decomposition:	$E \cup \{f M_1 \ldots M_r = f N_1 \ldots N_r\} \leadsto_{\text{dec}}$	
	$E \cup \{M_1 = N_1, \ldots, M_r = N_r\}$	
erasure:	$E \cup \{M = M\} \leadsto_{\text{erase}}$	
	E	
elimination:	$E \cup \{x = M\} \leadsto_{\text{elim}}$	
	$E[x := M] \cup \{x = M\}$,	if $x \in \text{var}(E), x \notin \text{var}(M)$
inversion:	$E \cup \{M = x\} \leadsto_{\text{inv}}$	
	$E \cup \{x = M\}$,	if $M \notin X$
conflict:	$E \cup \{f\vec{M} = g\vec{N}\} \leadsto_{\text{conflict}}$	
	$\perp$,	if $f \neq g$
cycle:	$E \cup \{x = M\} \leadsto_{\text{cycle}}$	
	$\perp$,	if $x \in \text{var}(M)$

Figure 6.3 Finite unification

The last two transformations lead immediately to the system $\perp$ to which no more transformations are applicable. They are failures. Elimination is the least elementary transformation, since substitution is not an elementary operation. The *broadcasting* of a binding

$$E \cup E' \cup \{x = M\} \underset{\text{diff}}{\leadsto} E[x := M] \cup E' \cup \{x = M\}$$

is a general transformation, which is more primitive when E is formed of a single equation. When $E' = \varnothing$, the transformation is elimination.

Example

$$
\begin{array}{ll}
 & \{f(x,g(x),k(y)) = f(h(z),t,k(z)), x = h(y)\} \\
\rightsquigarrow_{\text{dec}} & \{x = h(z), g(x) = t, k(y) = k(z), x = h(y)\} \\
\rightsquigarrow_{\text{elim}} & \{x = h(z), g(h(z)) = t, k(y) = k(z), h(z) = h(y)\} \\
\rightsquigarrow_{\text{inv}} & \{x = h(z), t = g(h(z)), k(y) = k(z), h(z) = h(y)\} \\
\rightsquigarrow_{\text{dec}} & \{x = h(z), t = g(h(z)), y = z, h(z) = h(y)\} \\
\rightsquigarrow_{\text{dec}} & \{x = h(z), t = g(h(z)), y = z, z = y\} \\
\rightsquigarrow_{\text{elim}} & \{x = h(z), t = g(h(z)), y = z, z = z\} \\
\rightsquigarrow_{\text{erase}} & \{x = h(z), t = g(h(z)), y = z\}.
\end{array}
$$

The last system is in solved form, defining the idempotent substitution:

$$\theta = \begin{bmatrix} x & y & t \\ h(z) & z & g(h(z)) \end{bmatrix}.$$

□

Lemma 6.12

- If $E \rightsquigarrow_{\text{dec}} E'$, then E and E' have the same solutions in $\mathrm{T}[X]$ (and in $\mathrm{T}^r[X]$).
- If $E \rightsquigarrow_{\text{erase}} E'$, $E \rightsquigarrow_{\text{elim}} E'$ or $E \rightsquigarrow_{\text{inv}} E'$, then E and E' have the same solutions in every algebra.
- If $f \neq g$, equation $f\vec{M} = g\vec{N}$ has no solution in $\mathrm{T}[X]$ (nor in $\mathrm{T}^r[X]$).
- If $x \in \mathrm{var}(M)$, equation $x = M$ has no solution in $\mathrm{T}[X]$.

Proof Consider the case of $\rightsquigarrow_{\text{elim}}$. Let $\mathcal{A}$ be an arbitrary algebra: it must be shown that $\{P = Q, x = M\}$ and $\{P[x := M] = Q[x := M], x = M\}$ have the same solutions $\xi : X \to A$. A solution of one system or the other satisfies $\xi(x) = \hat{\xi}(M)$. It follows, by induction on P, that $\hat{\xi}(P) = \hat{\xi}(P[x := M])$, which establishes the equivalence. □

Note the role of the algebra in each case. Decomposition and failure by conflict preserve solutions since the work is being done in $\mathrm{T}[X]$ or $\mathrm{T}^r[X]$. In the general case, every solution of E' is a solution of E, but not the other way around (in **N**, $x + 2 = 1 + y$ does not imply $x = 1$ and $y = 2$!). It is at this level that unification in algebras is simpler than unification in an equational theory. The equivalence between equation $M = N$ and system $M_1 = N_1, \ldots, M_r = N_r$ also shows that if one starts with a single equation to be solved, one must then solve a system of equations.

The elimination of a variable is nothing but the ubiquitous 'by replacing x in equation (3) by the its value found in (6), the result is ... ', valid reasoning in any algebra.

The only transformation which differentiates $\mathrm{T}[X]$ and $\mathrm{T}^r[X]$ is the failure by cycle, when an *occur check* $x \in \mathrm{var}(M)$ is positive.

Proposition 6.13

- Relation $E \rightsquigarrow E'$ is well-founded.
- If $E \overset{\star}{\rightsquigarrow} \bot$, then E is not unifiable.
- If $E \overset{\star}{\rightsquigarrow} E' \neq \bot$, and if no transformation is applicable to E', then E' is a system equivalent to E, in solved form, and such that $\mathrm{var}(E') = \mathrm{var}(E)$, and $\theta_{E'}$ is a most general idempotent unifier of E.

Proof According to lemma 6.12, all the transformations considered preserve the set of solutions. Note also that the set of variables of the system is also preserved by the first four transformations.

The termination is proven by supposing that the algorithm does not terminate in failure. A 'solved' variable is a variable x which has only one occurrence in E, and that in the left-hand side. Use ν for the number of unsolved variables at a given step of the algorithm. It is clear that, at each elimination, the substituted variable x goes from the status of unsolved variable to that of solved variable, since the occur check has been done. Hence each $\rightsquigarrow_{\text{elim}}$ decreases ν. It is also clear that the status of solved variable is acquired permanently, and that the other transformations cannot increase ν. Furthermore, each decomposition or erasure decreases the total size of E, while each inversion leaves it untouched. Write τ for its size. Finally, write δ for the number of 'inverted' equations of the form $M = x$ with M non-variable. Each inversion decreases δ. Hence, each transformation reduces the triplet (ν, τ, δ) using the lexical order over $\mathbf{N}^3$. Since this order is well-founded, the relation $\rightsquigarrow$ is also well-founded, and so only a finite number of transformations can be done.

If the algorithm terminates in failure, then E is not unifiable, according to the lemma.

If no transformation is applicable to the result E', then E' is in solved form. □

This proposition allows the design of a 'non-deterministic' unification algorithm (Herbrand's algorithm, rediscovered by Martelli and Montanari). As in the case of the λ-calculus or of rewriting systems, there are rules, applicable in an undetermined order; here, systems of equations play the role of terms. This process can therefore be seen as a 'computation by reduction', which satisfies the strong normalization property (there is no confluence, but that is not important), which allows a don't care non-determinism. There are, however, two other hidden properties, of soundness ($\theta_{E'}$ is a unifier) and of completeness (every unifier is an instance of $\theta_{E'}$), which are much more difficult to show for other unification problems. Furthermore, there are failure cases, which yield an immediate response here (non-unifiable system). In other situations, backtracking would have to be done to try other transformations: this is don't know non-determinism. It is in fact the first example of 'computation by resolution', another computation paradigm which will be studied in the next chapter. Proposition 6.13 can be stated under the form:

Theorem 6.14 If a system of equations E is unifiable, there exists $\theta \in \mathrm{U}(E)$ such that

- $\sigma \in \mathrm{U}(E)$ if and only if $\sigma \geqslant \theta$;
- $\mathrm{dom}(\theta) \cup \mathrm{im}(\theta) = \mathrm{var}(E)$;
- θ is idempotent.

Rational unification

The search for solutions in $\mathrm{T}^r[X]$ is distinguished by the absence of the occur check. The concept of solved form is therefore different.

A system of equations E is in *rational solved form* if it is in reduced form and it does not include any cycle $x_{i_1} = x_{i_2}, \ldots, x_{i_p} = x_{i_1}$.

If the occur check is omitted from $\rightsquigarrow_{\mathrm{elim}}$ and the failure transformation $\rightsquigarrow_{\mathrm{cycle}}$ is omitted, then relation $\rightsquigarrow$ is no longer well-founded:

$$\{x = f(x), y = x\} \underset{\mathrm{elim}}{\rightsquigarrow} \{x = f(x), y = f(x)\} \underset{\mathrm{elim}}{\rightsquigarrow} \{x = f(x), y = f(f(x))\} \underset{\mathrm{elim}}{\rightsquigarrow} \cdots .$$

Uncontrolled, elimination is dangerous, though useful. For example, if a system $\{x = f(a, y), x = f(z, b)\}$ is obtained, x must be eliminated to allow the continuation with $\{x = f(a, y), f(a, y) = f(z, b)\}$ by decomposition, ... But one can always start by applying decompositions, inversions and erasures. Colmerauer then proposed another transformation, *confrontation*, or *merge* (see Figure 6.4), and he limited elimination of x to the case of an equation $x = y$, where x and y are distinct variables, a condition which is a degenerate form of the occur check! The choice of equation conserved by a merge is made using an order over the equations.

merge:	$E \cup \{x = M, x = N\}$	if $M \notin X$ and $x = N \succ M = N$
	$\rightsquigarrow_{\mathrm{conf}} E \cup \{x = M, M = N\}$	
coalescence:	$E \cup \{x = y\}$	if $y \neq x$
	$\rightsquigarrow_{\mathrm{elim}} E[x := y] \cup \{x = y\}$.	

Figure 6.4 Rational unification

To handle rational unification, the six transformations (erasure, inversion, decomposition, conflict, merge and coalescence) constitute a complete system which also satisfies the termination property.

6.2.4 Unification algorithms

The unification algorithms used by most Prolog interpreters are really inefficient. Most use Robinson's algorithm, which represents a system of equations as a stack, and which operates on the equation at the top of the stack unless no rule applies to it: it has an exponential maximal complexity, as is shown by the following example:

$$\begin{aligned} M_p &= f(x_1, x_2, \ldots, x_p), \\ N_p &= f(g(x_0, x_0), g(x_1, x_1), \ldots, g(x_{p-1}, x_{p-1})). \end{aligned}$$

The image of x_p, which is a term of size $2^{p+1} - 1$, with 2^p occurrences of variable x_0, is constructed explicitly, then searched to ensure that it does not contain x_p. The explicit construction and the search lead to exponential complexities: spatial and temporal repsectively.

It was first supposed that omitting the occur check would avoid the exponential temporal complexity. But the unification of

$$\begin{aligned} M'_p &= f(g(x_0, x_0), g(x_1, x_1), \ldots, g(x_{p-1}, x_{p-1}), y_1, \ldots, y_p, y_p), \\ N'_p &= f(x_1, \ldots, x_p, g(y_0, y_0), g(y_1, y_1), \ldots, g(y_{p-1}, y_{p-1}), x_p), \end{aligned}$$

begins by constructing the images of x_p and y_p, which are terms similar to those of the preceding example, then it must traverse them to unify y_p and x_p, which requires exponential time to show 2^p times that $x_0 = y_0$.

This problem occurs because of an unshared implementation of terms. Corbin and Bidoit [26] showed that an implementation of terms by dags reduced the complexity to $O(p^2)$, where p is the number of symbols to unify in the terms.

Unification is in fact a linear problem, as was shown by Paterson and Wegman [117], with an asymptotically linear algorithm which is not efficient for typical cases. Pseudo-linear algorithms have been given by Huet [67] and Martelli and Montanari [100], Escalada-Imaz and Ghallab [41] proposed an algorithm in the same spirit as that of Huet, but using smaller data structures.

As in logic programming, it is often important to fail as soon as possible. Unification algorithms privilege one of the two possible causes of failure: conflict or cycle. The Peterson–Wegman algorithm fails quickly in the case of cycle—a quite rare case in practice. Those of Huet and Escalada-Imaz–Ghallab privilege failure by conflict by delaying the occur check. The latter two adapt readily to rational terms. Those of Colmerauer and Martelli–Rossi [101] are explicitly designed for rational terms.

Systems of multiequations Martelli and Montanari structure a system of equations using multiequations by grouping terms by classes for the equivalence generated by the equations of the system.

Example The system $\{x = z, y = f(x, x), f(x, a) = f(y, z), z = y\}$ defines two classes of terms: $\{x, y, z, f(x, x)\}$ and $\{f(x, a), f(y, z)\}$. It is convenient to distinguish the variables by writing the two multiequations $\{x, y, z\} = \{\!\{f(x, x)\}\!\}$ and $\{w\} = \{\!\{f(x, a), f(y, z)\}\!\}$, where w is a new variable. □

A *multiequation* is an expression $V = S$, where V is a non-empty set of variables and S is a multiset of non-variable terms. A *merge* of two multiequations $V = S$

and $V' = S'$ such that $V \cap V' \neq \varnothing$ consists of replacing these two multiequations by $V \cup V' = S \uplus S'$. The right-hand side of a multiequation is to be decomposed as soon as it contains several terms. Martelli and Montanari suggested that several decomposition steps be combined by computing the *common part* $\mathcal{C}(S)$ and the *border* $\mathcal{F}(S)$ of a multiset S of terms.

Example The common part of $S = \{\!\{f(x, g(x), k(y)), f(h(z), t, k(z))\}\!\}$ is the term $\mathcal{C}(S) \equiv f(x, t, k(y))$. Its border is the system $(\mathcal{F}(S))$ formed of three multiequations: $\{x\} = \{\!\{h(z)\}\!\}$, $\{t\} = \{\!\{g(x)\}\!\}$, $\{y, z\} = \varnothing$. However, $\{\!\{f(x, b), f(a, a)\}\!\}$ does not have a common part. □

If the common part does not exist, there is failure by conflict. A decomposition step consists of replacing a multiequation $V = S$ by $V = \{\!\{\mathcal{C}(S)\}\!\}$ and by all the multiequations of $\mathcal{F}(S)$.

It is useful to work with two systems T and U. Initially, to unify terms M_1 and M_2, containing variables $x_1, \ldots, x_n$, one defines

$$\begin{aligned} U &= \{\{u\} = \{\!\{M_1, M_2\}\!\}\}, \\ T &= \{\{x_1\} = \varnothing, \ldots, \{x_n\} = \varnothing\}, \end{aligned}$$

where u is a new variable.

The objective is to obtain $U = \varnothing$ by transferring the multiequations of U to T. During the decomposition of $V = S$, the principle is to place $V = \{\!\{\mathcal{C}(S)\}\!\}$ in T, possibly after a merge, and to add $\mathcal{F}(S)$ to U. A multiequation of U is selected and it is then merged as much as possible with the multiequations of T. If the result contains at least two terms on the right-hand side, it is decomposed as indicated; otherwise, it is placed in T.

Example

$$\begin{aligned}
& U = \{\{u\} = \{\!\{f(x, g(x), k(y)), f(h(z), t, k(z))\}\!\}, \{v\} = \{\!\{x, h(y)\}\!\}\} \\
& T = \{\{x\} = \varnothing, \{y\} = \varnothing, \{z\} = \varnothing, \{t\} = \varnothing\} \\
\rightsquigarrow\ & U = \{\{x\} = \{\!\{h(z)\}\!\}, \{t\} = \{\!\{g(x)\}\!\}, \{y, z\} = \varnothing, \{v\} = \{\!\{x, h(y)\}\!\}\} \\
& T = \{\{u\} = \{\!\{f(x, t, k(y))\}\!\}, \{x\} = \varnothing, \{y\} = \varnothing, \{z\} = \varnothing, \{t\} = \varnothing\} \\
\rightsquigarrow\ & U = \{\{v\} = \{\!\{x, h(y)\}\!\}\} \\
& T = \{\{u\} = \{\!\{f(x, t, k(y))\}\!\}, \{x\} = \{\!\{h(z)\}\!\}, \{t\} = \{\!\{g(x)\}\!\}, \{y, z\} = \varnothing\} \\
\rightsquigarrow\ & U = \{\{x\} = \{\!\{h(y)\}\!\}\} \\
& T = \{\{u\} = \{\!\{f(x, t, k(y))\}\!\}, \{v, x\} = \{\!\{h(z)\}\!\}, \{t\} = \{\!\{g(x)\}\!\}, \{y, z\} = \varnothing\} \\
\rightsquigarrow\ & U = \{\{y, z\} = \varnothing\} \\
& T = \{\{u\} = \{\!\{f(x, t, k(y))\}\!\}, \{v, x\} = \{\!\{h(z)\}\!\}, \{t\} = \{\!\{g(x)\}\!\}, \{y, z\} = \varnothing\} \\
\rightsquigarrow\ & U = \varnothing \\
& T = \{\{u\} = \{\!\{f(x, t, k(y))\}\!\}, \{v, x\} = \{\!\{h(z)\}\!\}, \{t\} = \{\!\{g(x)\}\!\}, \{y, z\} = \varnothing\}.
\end{aligned}$$

□

The final system T, in which the equations containing u and v can be ignored, is not in solved form: x must still be eliminated. Nevertheless this 'triangular' form can be considered satisfactory as it is more compact than a solved form. Differentiating eliminations allows one to handle unification in $T^r[X]$ without worrying about cycles. However, if the search is restricted to solutions in $T[X]$, tests must be made to ensure that there are no cycles. The simple occur check is not only inefficient, but insufficient, so long as the eliminations have not been done (for example: $x = f(y)$, $y = g(x)$). The existence of cycles is tested by associating a counter to each multiequation, of T only, if the test is final, enumerating the number of occurrences of variables of its left-hand side appearing in the other equations of T and U. Those counters are modified each time the system is transformed. The value of the counter is written in square brackets.

Example For the preceding system, the initial counters are:

$$U = \left\{ \begin{array}{l} \{u\} = \{\!\{f(x, g(x), k(y)), f(h(z), t, k(z))\}\!\}, \\ \{v\} = \{\!\{x, h(y)\}\!\} \end{array} \right\},$$

$$T = \left\{ \begin{array}{l} [3]\{x\} = \varnothing, \\ [2]\{y\} = \varnothing, \\ [2]\{z\} = \varnothing, \\ [1]\{t\} = \varnothing \end{array} \right\}.$$

□

A system is in solved form if and only if all of its counters are zero. Conversely, if there is no cyclic dependency, there is at least one multiequation $[c]\, V = S$ whose counter c is zero, and the variables of V can be eliminated—without making any substitution, since they have no occurrences elsewhere. One can then ignore this multiequation, thereby decrementing the counters of the variables appearing in its right-hand side S.

Example (cont.) The final system, without its auxiliary variables u and v, is:

$$\begin{array}{rl} T = \{ & [1] \quad \{x\} = \{\!\{h(z)\}\!\}, \\ & [0] \quad \{t\} = \{\!\{g(x)\}\!\}, \\ & [1] \; \{y, z\} = \varnothing\}. \end{array}$$

Start by eliminating t, then x, whose counter has become zero, and y, z last, giving systems T:

$$\begin{array}{rl} [0]: & \{x\} = \{\!\{h(y)\}\!\} \\ [0]^1: & \{t\} = \{\!\{g(x)\}\!\} \\ [1]: & \{y, z\} = \varnothing \end{array} \quad \dots \quad \begin{array}{rl} [0]^2: & \{x\} = \{\!\{h(y)\}\!\} \\ [0]^1: & \{t\} = \{\!\{g(x)\}\!\} \\ [0]^3: & \{y, z\} = \varnothing. \end{array}$$

Written as exponent of the counter is the order in which the multiequations are eliminated. This order allows one to construct the solved form explicitly: first $y = z$, then $x = h(y) = h(z)$, then $t = g(x) = g(h(z))$. □

If it is important to fail as early as possible by cycle detection, this simple computation can be done without having reached the final form of the system: a multiequation having been selected in U, then merged with multiequations in T, the decomposition is done only if the associated counter is zero.

The representative of a variable Escalada-Imaz and Ghallab's algorithm constructs a graph whose vertices are tagged by terms. To each variable x corresponds a vertex u_x. Every connected component corresponds to a multiequation of the form $\{\vec{x}\} = \{\!\{M\}\!\}$, or else $\{\vec{x}\} = \varnothing$: it contains at most one non-variable vertex, and, if it contains one, has no cycle. The initial graph G_0 is formed of one vertex per variable, with no edges.

In a given configuration of the graph, to each variable x is associated its representative $\text{rep}(x)$: if u_x does not have a successor, then $\text{rep}(x) = x$. Otherwise, by following the path starting at u_x, $\text{rep}(x)$ is the last variable if the path does not contain any cycles, or the last variable visited, should there be a cycle. To find this representative practically, a marking of visited vertices is necessary. This marking is also useful when the graph is traversed from one connected component to another, to simulate the elimination of a variable.

There are four possibilities, which are written in the following manner:

x:	u_x has no successor;
$x \twoheadrightarrow \text{rep}(x)$:	$u_{\text{rep}(x)}$ has no successor;
$x \twoheadrightarrow \text{rep}(x) \rightarrow X$:	$u_{\text{rep}(x)}$ has a successor u_X, with X non-variable;
$x \twoheadrightarrow \text{rep}(x) \circlearrowleft$:	the graph has a cycle, the successor of $u_{\text{rep}(x)}$ is the vertex of a variable already traversed, starting from x.

Write $x \ldots$ for the first of these cases and $x \twoheadrightarrow \text{rep}(x) \ldots$ for one of the last three cases.

Transformations operate over pairs (E, G) of a system of equations and a graph. Decomposition is only used for the system of equations:

$$\begin{array}{llll} \textbf{decomposition:} & (E \cup \{f\vec{M} = f\vec{N}\}, G) & \rightsquigarrow_{\text{dec}} & (E \cup \{\vec{M} = \vec{N}\}, G). \\ \textbf{conflict:} & (E \cup \{f\vec{M} = g\vec{N}\}, G) & \rightsquigarrow_{\text{conflict}} & \bot, \text{if } f \neq g. \end{array}$$

When an equation $x = M$ is selected in E, there is, depending on the case, either graph construction or merging.

If M is non-variable, there are three cases, depending on x:

$$\begin{array}{ll} x & x \rightarrow M \\ \left.\begin{array}{l} x \twoheadrightarrow \text{rep}(x) \\ x \twoheadrightarrow \text{rep}(x) \circlearrowleft \end{array}\right\} & \boxed{x \twoheadrightarrow \text{rep}(x)} \rightarrow M \\ x \twoheadrightarrow \text{rep}(x) \rightarrow X & \boxed{x \twoheadrightarrow \text{rep}(x)} \rightarrow X \\ & \text{and } E \cup \{x = M\} \rightsquigarrow E \cup \{X = M\}. \end{array}$$

If M is a variable y, there are three more cases, depending on y (plus the symmetric cases in x/y):

$$\begin{array}{lll}
x & y\ldots & x \to y\ldots \\
\left.\begin{array}{l} x \twoheadrightarrow \mathrm{rep}(x) \\ x \twoheadrightarrow \mathrm{rep}(x) \circlearrowleft \end{array}\right\} & y\ldots & \boxed{x \twoheadrightarrow \mathrm{rep}(x)} \to y \\
x \twoheadrightarrow \mathrm{rep}(x) \to X & y \twoheadrightarrow \mathrm{rep}(y) \to Y & \boxed{x \twoheadrightarrow \mathrm{rep}(x) \to y \twoheadrightarrow \mathrm{rep}(y)} \to Y \\
 & & \text{and } E \cup \{x = M\} \leadsto E \cup \{X = Y\}.
\end{array}$$

In each case where a representative must be computed, a box $\boxed{x\ldots z}$ indicates that the path $x\ldots z$ must be compressed, i.e., replaced by a tree of depth 1 and root z, the other vertices on the path being linked to z. In the cases where a cycle $\circlearrowleft$ is detected, it must be unwound.

By applying these transformations a finite number of times, the pair (E, G_0) is transformed into $(\varnothing, G)$.

Example Suppose two terms

$$\begin{array}{rcl}
M & \equiv & f(h(x_1, x_2, x_3), h(x_6, x_7, x_8), x_3, x_6), \\
N & \equiv & f(h(g(x_4, x_5), x_1, x_2), h(x_7, x_8, x_6), g(x_5, a), x_5),
\end{array}$$

must be unified. The processing of equation $M/1 = N/1$ gives a first component of the graph:

$$x_3 \quad\to\quad x_2 \quad\to\quad x_1 \quad\to\quad g(x_4, x_5).$$

Equation $M/2 = N/2$ gives the cycle

$$x_6 \quad\to\quad x_7 \quad\to\quad x_8 \circlearrowleft [x_6].$$

Equation $M/3 = N/3$ requires finding the representative of x_3, which is x_1, compressing the path $x_3\ldots x_1$, which after merging the equation gives $g(x_4, x_5) = g(x_5, a)$, which creates a third component:

$$x_4 \quad\to\quad x_5 \quad\to\quad a.$$

The last equation $x_6 = x_5$ requires finding the representative of x_6, which is x_8, compressing the path $x_6\ldots x_8$, and merging the second and third components. The resulting graph has two connected components:

$$\begin{array}{ccc}
x_3 & & \\
\downarrow & & \\
x_1 & \to & g(x_4, x_5) \\
\uparrow & & \\
x_2 & &
\end{array}
\qquad\qquad
\begin{array}{ccccc}
x_6 & & & & \\
\downarrow & & & & \\
x_8 & \to & x_5 & \to & a \\
\uparrow & & \uparrow & & \\
x_7 & & x_4 & &
\end{array}$$

□

The graph G having been constructed, the absence of cycle must still be tested, and the unifier θ must be computed, i.e., for each variable x, $\theta(x)$. To each vertex

u of the graph is associated a value $\mathrm{val}(u)$, which can be a (pointer to a) term, the symbols *nil* or *cycle*: this value is used to memoize the computed results and to mark the traversed vertices for the detection of cycles. Initially, $\mathrm{val}(u_x) = nil$ for all variables:

- if $\mathrm{val}(u_x) = M$: return M;
- if $\mathrm{val}(u_x) = cycle$: halt, failure by cycle;
- otherwise, find $\mathrm{rep}(x)$; let t be a local variable:
  - if $\mathrm{val}(u_{\mathrm{rep}(x)}) = M$: let $t := M$;
  - if $\mathrm{val}(u_{\mathrm{rep}(x)}) = cycle$: halt, failure by cycle;
  - otherwise, in cases $x \twoheadrightarrow \mathrm{rep}(x)$ and $x \twoheadrightarrow \mathrm{rep}(x) \circlearrowleft$, let $t := \mathrm{rep}(x)$;
  - if $x \twoheadrightarrow \mathrm{rep}(x) \to M$, with M closed, let $t := M$;
  - if $x \twoheadrightarrow \mathrm{rep}(x) \to M$, with M not closed, then:
    - mark each vertex u of the path $u_x \dots u_{\mathrm{rep}(x)}$ by $\mathrm{val}(u) := cycle$,
    - compute $\theta(y)$ for each $y \in \mathrm{var}(M)$,
    - let $t := \theta(M)$;

  for each vertex u of the path $u_x \dots u_{\mathrm{rep}(x)}$, let $\mathrm{val}(u) := t$ and return t.

Example (cont.) The computation of $\theta(x_2)$ determines the representative of x_2 (x_1), marks the vertices x_2 and x_1 by *cycle*, and computes $\theta(x_4)$ and $\theta(x_5)$, both equaling a. After this computation, the vertices x_2 and x_1 receive the value $g(a, a)$, and x_4, x_5 the value a. The graph need not be traversed to compute $\theta(x_1)$, $\theta(x_4)$ and $\theta(x_5)$. □

6.2.5 Critical pairs

The Knuth–Bendix confluence criterion, an application of unification, is important to the study of rewriting systems.

Let P and P' be two terms. Term P' is *superposable* upon P at occurrence $u \in \mathcal{O}(P)$ if $P(u) \notin X$ and if P/u and P' are unifiable.

Let $P \to Q$ and $P' \to Q'$ be two separate rewriting rules, i.e, their sets of variables are disjoint. The rewriting rule $P' \to Q'$ is *superposable* upon $P \to Q$ at occurrence $u \in \mathcal{O}(P)$ if P' is superposable upon P at u. For every unifier σ of P/u and P', there are the two reductions $\sigma P \xrightarrow{\epsilon} \sigma Q$ and $\sigma P \xrightarrow{u} \sigma P[u \leftarrow \sigma Q']$.

Definition 6.8 The pair $(\theta Q, \theta P[u \leftarrow \theta Q'])$ is a *critical pair* of $\mathcal{R}$, when θ is a most general unifier of P/u and P'.

Note that there are finitely many critical pairs in a finite rewriting system.

Example The associativity rule $(xy)z \to x(yz)$ superposes upon itself at occurrence 1, xy and $(x'y')z'$ being unifiable, giving the two reductions:

$$((x'y')z')z \xrightarrow{\epsilon} (x'y')(z'z),$$
$$((x'y')z')z \xrightarrow{1} (x'(y'z'))z,$$

hence the critical pair $((x'y')(z'z), (x'(y'z'))z)$. □

The existence of critical pairs is a sign of the 'ambiguity' of a rewriting system.

Theorem 6.15 (Knuth–Bendix, Huet) If $M_1 \leftarrow M \to M_2$, then

- either there exist reductions $M_1 \xrightarrow{\star} M' \xleftarrow{\star} M_2$
- or there exists a critical pair (C_1, C_2) of $\mathcal{R}$, a context $\mathcal{K}[\,]$ and a substitution μ such that $M_i \equiv \mathcal{K}[\mu C_i]$ for $i = 1, 2$.

Proof Let $M \xrightarrow{u_1,\rho_1} M_1$ and $M \xrightarrow{u_2,\rho_2} M_2$. Distinguish the relative positions of occurrences u_1 and u_2.

- If u_1 and u_2 are independent, then M_1 and M_2 rewrite to a term M' by the reductions $M_1 \xrightarrow{u_2,\rho_2} M'$ and $M_2 \xrightarrow{u_1,\rho_1} M'$.
- If $u_1 \leqslant u_2$, it can be supposed that $u_1 = \epsilon$. Then $M \equiv \sigma_1 P_1$, hence $M \xrightarrow{\epsilon} \sigma_1 Q_1$, and $M/u_2 \equiv \sigma_2 P_2$, $M \xrightarrow{u_2} M[u_2 \leftarrow \sigma_2 Q_2]$. There are two cases according to the nature of u_2.
  - If u_2 is an internal occurrence of P_1, i.e, $u_2 \in \mathcal{O}(P_1)$ and $P_1(u_2) \notin X$. Suppose that the rules ρ_1 and ρ_2 are separate. Since $(\sigma_1 P_1)/u_2 \equiv \sigma_2 P_2$, the substitution σ, sum of σ_1 and σ_2, is a unifier of P_1/u_2 and P_2. By assumption on u_2, ρ_2 superposes itself on ρ_1, which yields the critical pair (C_1, C_2) where $C_1 \equiv \theta Q_1$ and $C_2 \equiv \theta(P_1[u_2 \leftarrow Q_2])$, θ being a most general unifier of P_1/u_2 and P_2. Let μ be a substitution such that $\sigma = \mu\theta$. Then $M_1 \equiv \sigma_1 Q_1 \equiv \mu C_1$ and $M_2 \equiv M[u_2 \leftarrow \sigma_2 Q_2] \equiv \mu C_2$.
  - If $u_2 \in \mathcal{O}(\sigma P_1)$ is not an internal occurrence of P_1, then $u_2 = w'u'$, with $w' \in \mathcal{O}(P_1)$, $P_1/w' \equiv x \in X$, and $u' \in \mathcal{O}(\sigma(x))$. Let W_1 be the set of occurrences of x in P_1, and $U_2 = \{wu'; w \in W_1\} \subset \mathcal{O}(M)$. Application of rule $\sigma P_2 \to \sigma Q_2$ to occurrence u_2, then to other occurrences, pairwise independent, of U_2, yields successively:

    $$M \equiv \sigma P_1 \xrightarrow{u_2} M_2 \equiv (\sigma P_1)[u_2 \leftarrow \sigma Q_2]. \xrightarrow{\star} M_2' \equiv (\sigma P_1)[U_2 \leftarrow \sigma Q_2].$$

    Let σ' be the substitution obtained by replacing $\sigma(x)$ by

    $$\sigma'(x) \equiv \sigma(x)[u' \leftarrow \sigma Q_2].$$

    Then $M_2' \equiv \sigma' P_1$. Rule $P_1 \to Q_1$ can then be applied:

    $$M_2' \equiv \sigma' P_1 \xrightarrow{\epsilon} M' \equiv \sigma' Q_1.$$

M' can also be obtained from M_1,

$$M_1 \equiv \sigma Q_1 \xrightarrow{\star} (\sigma Q_1)[U_1 \leftarrow \sigma Q_2] \equiv \sigma' Q_1,$$

where U_1 is the set of occurrences of σP_2 in σQ_1. Hence $M_1 \xrightarrow{\star} M' \xleftarrow{\star} M_2$. This case is illustrated by Figure 6.5.

□

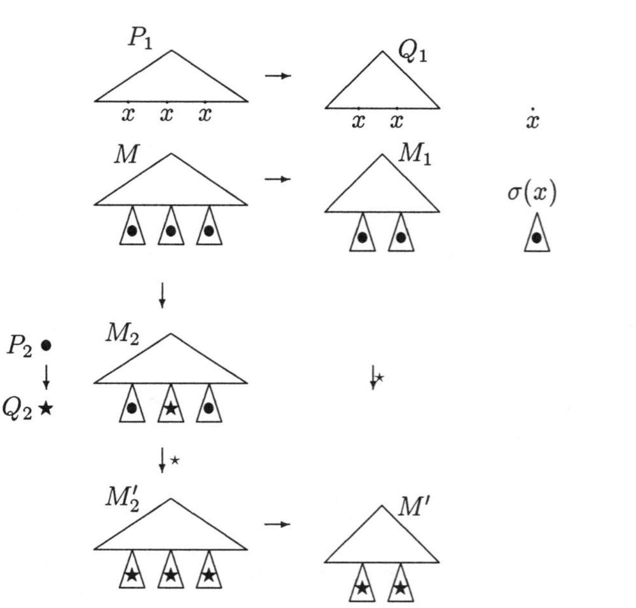

Figure 6.5 Proof of the Knuth–Bendix theorem

Corollary 6.16 A rewriting system is locally confluent if and only if every critical pair is confluent.

Combining Newman's lemma with the preceding corollary yields D. E. Knuth and P. B. Bendix's original statement:

Corollary 6.17 A noetherian rewriting system is confluent if and only if every critical pair is confluent.

This theorem will be used in the third part of the chapter to normalize equational proofs by 'completing' a rewriting system. Another application of the concept of critical pair is the Huet–Lévy confluence criterion, a simplified form of which is given here: a rewriting system whose left members are linear terms (at most one occurrence of each variable) and which has no critical pairs is confluent.

6.3 Equational proofs

6.3.1 Equations and rewriting

Since a set of equations $\mathcal{E}$ defines an equivalence relation over terms, the problem $\mathcal{E} \vdash M = N$ can be viewed as determining if M and N belong to the same equivalence class. Should there be a canonical representative for each class, and if for each element one can associate the canonical representative of its class, then the equivalence problem simply consists of determining the identity of canonical representatives. For example, to decide whether two fractions are equivalent, it suffices to simplify them, then to test the equality of numerators and denominators.

This aim is fulfilled by a rewriting system, assuming it satisfies the weak normalization property (every term has a normal form), that it be confluent (hence that the normal form be unique) and that there exists an effective strategy to attain the normal form of a term. This last condition is satisfied if the system is noetherian. A confluent and noetherian rewriting system is *convergent* or *canonical.*

To use a canonical rewriting system $\mathcal{R}$ in equational proofs, the following properties must be satisfied:

- $\mathcal{R}$ is *sound* relative to $\mathcal{E}$ if $M \overset{\star}{\longleftrightarrow}_{\mathcal{R}} N$ implies $M =_{\mathcal{E}} N$;
- $\mathcal{R}$ is *complete* relative to $\mathcal{E}$ if $M =_{\mathcal{E}} N$ implies the syntactic equality of the normal forms $\bar{M}$ and $\bar{N}$ for $\mathcal{R}$;
- $\mathcal{R}$ is *canonical* relative to $\mathcal{E}$ if it is convergent, sound and complete relative to $\mathcal{E}$.

Given $\mathcal{E}$, a canonical $\mathcal{R}$ relative to $\mathcal{E}$ is sought.

Example The axioms of group theory are:

$$(groups) \quad \left\{ \begin{array}{rcl} ex & = & x \\ x^{-1}x & = & e \\ (xy)z & = & x(yz). \end{array} \right.$$

A canonical rewriting system for this theory is formed by the following ten rules:

$$\begin{array}{rcl@{\qquad}rcl} ex & \rightarrow & x & xe & \rightarrow & x \\ x^{-1}x & \rightarrow & e & xx^{-1} & \rightarrow & e \\ (xy)z & \rightarrow & x(yz) & e^{-1} & \rightarrow & e \\ x^{-1}(xy) & \rightarrow & y & x(x^{-1}y) & \rightarrow & y \\ (x^{-1})^{-1} & \rightarrow & x & (xy)^{-1} & \rightarrow & y^{-1}x^{-1}. \end{array}$$

□

It is easy to construct a sound system, simply by orienting the equations of $\mathcal{E}$ into rules. Such a system will rarely be complete. For example, the natural orientation of the axioms for groups, $ex \rightarrow x$, $x^{-1}x \rightarrow e$, $(xy)z \rightarrow x(yz)$ is not complete, since

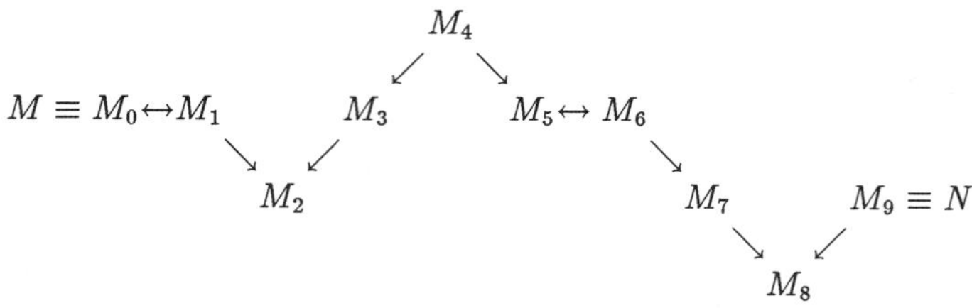

Figure 6.6 A proof of $M = N$

the term xx^{-1} is irreducible, although $xx^{-1} = e$ is a semantic consequence of the axioms. It is noetherian, but not confluent, since critical pairs exist. A sound and complete system would be obtained by taking the two possible orientations of each relation, the two relations $\overset{\star}{\longleftrightarrow}_{\mathcal{R}}$ and $=_{\mathcal{E}}$ then being identical. But such a system is never noetherian, since there are cycles $P \longrightarrow Q \longrightarrow P$. An intermediate solution is therefore required.

Mixed proofs The Knuth–Bendix completion procedure attempts to transform a system of equations $\mathcal{E}_0$ into a rewriting system $\mathcal{R}$ which is convergent, sound and complete relative to $\mathcal{E}$. Since the procedure constructs $\mathcal{R}$ step by step, it operates over pairs $(\mathcal{E}, \mathcal{R})$ starting from $(\mathcal{E}_0, \varnothing)$, and it succeeds if it reaches $(\varnothing, \mathcal{R}_\infty)$. At each step, $\mathcal{R}$ is sound relative to $\mathcal{E}_0$ and is noetherian. Better, all the $\mathcal{R}$ are included in the same well-order.

The resulting proofs will be *mixed*, using a pair $\Gamma = (\mathcal{E}, \mathcal{R})$ of sets of equations and of rules. Write $\pi :_\Gamma M = N$ if π is a derivation of $M = N$ in Γ. Rather than use the equational inference system freely, only 'linearized' derivations will be used (this is an easy result of proof normalization): a derivation of M is represented by a sequence of terms $\pi = (M_0 \equiv M,\ M_1,\ \ldots\ ,\ M_p \equiv N)$, with, for each pair (M_i, M_{i+1}), $0 \leqslant i < p$, an 'elementary' derivation of $M_i \longrightarrow_{\mathcal{R}} M_{i+1}$, of $M_{i+1} \longrightarrow_{\mathcal{R}} M_i$, or of $M_i \longleftrightarrow_{\mathcal{E}} M_{i+1}$. Each of these elementary derivations must be specified by a rule or an equation, the occurrence where it is applied and the matching substitution.

6.3.2 Knuth–Bendix completion

Complexity of proofs Two proofs of $M = N$ are represented in Figures 6.6 and 6.7. A proof *by confluence* is simpler than an arbitrary proof.

If $\mathcal{E}$ is empty, and if $\mathcal{R}$ is confluent, the Church–Rosser property states that any proof of $M = N$ can be transformed into a proof by confluence. This lemma is a result of proof normalization. Similar results have already been evoked: cut-elimination in sequent calculus and normalization in natural deduction. In the last

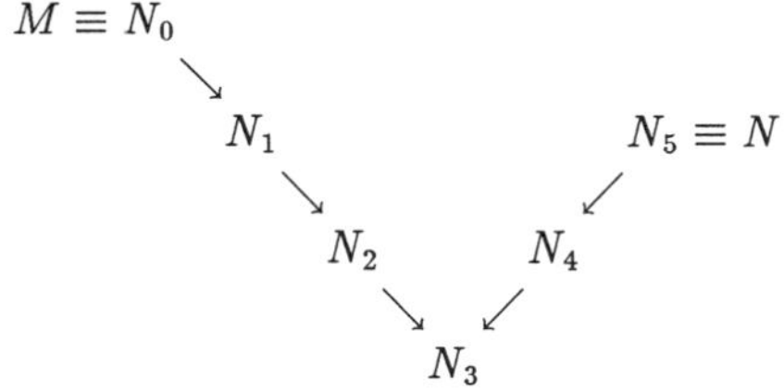

Figure 6.7 A proof by confluence of $M = N$

case, the proofs were typed λ-terms, by the Curry–Howard correspondence, and they were normalized by contracting the redexes (i.e., an introduction followed by an elimination).

Here, the normalization of the proof of an equation consists essentially of replacing an effluence $M_1 \leftarrow M_0 \rightarrow M_2$ by a confluence $M_1 \xrightarrow{\star} M' \xleftarrow{\star} M_2$. This is not possible if $\mathcal{E}$ is non-empty and if $\mathcal{R}$ is not confluent. Bachmair, Dershowitz and Hsiang's idea was to see the completion procedure as a transformation of the axiom system in order to enable the writing of more and more normal proofs. It is well known that the same theory can have, depending on the set of axioms chosen, proofs of various complexities: for example, the equational axiom $x^{-1}x = e$ allows simpler proofs than $\forall x \exists y(yx = e)$.

The complexity $\|\pi\|$ of an equational proof π is defined uniquely through its 'shape'. The complexity of an elementary proof is a multiset of terms:

$$\begin{aligned} \|M \rightarrow_{\mathcal{R}} N\| &= \{\!\{M\}\!\}, \\ \|M \longleftrightarrow_{\mathcal{E}} N\| &= \{\!\{M, N\}\!\}. \end{aligned}$$

The *complexity* of a proof $(M_0, \ldots, M_p)$ of $M_0 = M_p$ is the multiset of the complexities of the elementary proofs making it up:

$$\|M_0, \ldots, M_p\| = \{\!\{\|M_0, M_1\| \ldots \|M_{p-1}, M_p\|\}\!\}.$$

Example The complexity of the proof in Figure 6.6 is

$$\{\!\{\{\!\{M_0, M_1\}\!\}, \{\!\{M_1\}\!\}, \{\!\{M_3\}\!\}, \{\!\{M_4\}\!\}, \{\!\{M_4\}\!\}, \{\!\{M_5, M_6\}\!\}, \{\!\{M_6\}\!\}, \{\!\{M_7\}\!\}, \{\!\{M_9\}\!\}\}\!\}.$$

That of the proof by confluence in Figure 6.7 is:

$$\{\!\{\{\!\{N_0\}\!\}, \{\!\{N_1\}\!\}, \{\!\{N_2\}\!\}, \{\!\{N_4\}\!\}, \{\!\{N_5\}\!\}\}\!\}.$$

□

Note that an equational proof has a complexity which lies between that of a proof by effluence and a proof by confluence.

Since a reduction order (i.e., a well-founded order compatible and stable under substitutions) $\succ$ has been defined over terms, it is extended to multisets of terms, then to the multisets of those multisets. Define $\pi \succ^p \pi'$ if $\|\pi\| \ggg \|\pi'\|$: it is a well-founded order, since it is constructed using extensions to multisets of the well-founded order $\succ$.

Normalization of proofs Normalizing a proof requires contraction schemes which reduce its complexity. For example, a proof $\ldots M \longleftrightarrow M \ldots$, using axiom $P = P$, can be simplified to $\ldots M \ldots$, which is $\prec^p \ldots M \longleftrightarrow M \ldots$. This axiom can even be removed without affecting the equality $=_{\mathcal{E}}$. In other cases, either $\mathcal{E}$ or $\mathcal{R}$ will have to be extended before proofs can be simplified.

The idea is to define a family of transformations $(\mathcal{E}, \mathcal{R}) \rightsquigarrow (\mathcal{E}', \mathcal{R}')$ such that:

- if $\pi :_\Gamma M = N$ is a non-normal proof, there exists a transformation $\Gamma \rightsquigarrow \Gamma'$ and a proof $\pi' :_{\Gamma'} M = N$ such that $\pi \succ^p \pi'$;
- for every transformation $\Gamma \rightsquigarrow \Gamma'$ and every $\pi :_\Gamma M = N$, there exists $\pi' :_{\Gamma'} M = N$ such that $\pi \succcurlyeq^p \pi'$.

Once these transformations have been chosen, a completion procedure for $\mathcal{E}_0$ will construct a sequence $\Gamma_0, \Gamma_1, \ldots$ of pairs $(\mathcal{E}_i, \mathcal{R}_i)$, with $\Gamma_0 = (\mathcal{E}_0, \varnothing)$, such that for every non-normal proof $\pi :_{\Gamma_i} M = N$, there exists $\pi' :_{\Gamma_j} M = N$ for $j \geqslant i$ such that $\pi \succ^p \pi'$:

First group of transformations The first group operates over the structure of proofs:

orientation	$\mathcal{E} \cup \{P = Q\}, \mathcal{R} \rightsquigarrow \mathcal{E}, \mathcal{R} \cup \{P \to Q\}$	if $P \succ Q$
	$\mathcal{E} \cup \{P = Q\}, \mathcal{R} \rightsquigarrow \mathcal{E}, \mathcal{R} \cup \{Q \to P\}$	if $Q \succ P$
erasure	$\mathcal{E} \cup \{M = M\}, \mathcal{R} \rightsquigarrow \mathcal{E}, \mathcal{R}$	
effluence	$\mathcal{E}, \mathcal{R} \rightsquigarrow \mathcal{E} \cup \{M_1 = M_2\}, \mathcal{R}$	if $M_1 \swarrow M_0 \searrow M_2$
normalization	$\mathcal{E} \cup \{M = N\}, \mathcal{R} \rightsquigarrow \mathcal{E} \cup \{M' = N'\}, \mathcal{R}$	if $\begin{cases} M \xrightarrow{+}_{\mathcal{R}} M' \\ N \xrightarrow{+}_{\mathcal{R}} N' \end{cases}$

Note first that

- each of the transformations preserves equality $\xleftrightarrow{\star}_{\mathcal{E},\mathcal{R}}$;
- the rewriting systems are all noetherian for the same order $\succ$; and
- the transformations can only make $\mathcal{R}$ grow.

Each transformation $\Gamma \rightsquigarrow \Gamma'$ defines a transformation over proofs, associating with each proof π of $M = N$ in Γ a proof π' of the same equation in Γ', obtained by 'contracting' a subproof (analogous to the contraction of a redex in the λ-calculus). If π does not use an axiom affected by the transformation, then $\pi' = \pi$. Otherwise, an equational proof

$$\ldots \mathcal{C}[\theta P] \longleftrightarrow \mathcal{C}[\theta Q] \ldots$$

is replaced, for each transformation, by:

$$\begin{array}{ll} \text{erasure:} & \ldots \mathcal{C}[\theta P] \ldots \\ \text{normalization:} & \ldots \mathcal{C}[\theta P] \rightarrow \mathcal{C}[\theta P'] \longleftrightarrow \mathcal{C}[\theta Q] \ldots \\ \text{orientation:} & \ldots \mathcal{C}[\theta P] \rightarrow \mathcal{C}[\theta Q] \ldots \end{array}$$

and an effluence proof $\ldots M_1 \leftarrow M_0 \rightarrow M_2 \ldots$ by the equational proof $\ldots M_1 \longleftrightarrow M_2$. The proof π' obtained is strictly simpler, since:

$$\begin{array}{rcll} \{\!\{\ldots \{\!\{P, P\}\!\} \ldots\}\!\} & \succ\!\!\succ & \{\!\{\ldots\}\!\}, & \\ \{\!\{\ldots \{\!\{M, N\}\!\} \ldots\}\!\} & \succ\!\!\succ & \{\!\{\ldots \{\!\{M\}\!\}, \{\!\{M', N\}\!\} \ldots\}\!\}, & \text{since } M \succ M', \\ \{\!\{\ldots \{\!\{P, Q\}\!\} \ldots\}\!\} & \succ\!\!\succ & \{\!\{\ldots \{\!\{P\}\!\} \ldots\}\!\}, & \\ \{\!\{\ldots \{\!\{M_0\}\!\}, \{\!\{M_0\}\!\} \ldots\}\!\} & \succ\!\!\succ & \{\!\{\ldots \{\!\{M_1, M_2\}\!\} \ldots\}\!\}. & \end{array}$$

In all cases, $\pi \succeq^p \pi'$.

What is required is to strictly reduce the complexity of π if π is not a proof by confluence. This is possible by supposing that at least one equation used by π, both sides of which are distinct and irreducible by $\mathcal{R}$, is orientable. Sure enough, at least one equational subproof of π can be contracted, by erasure, normalization or orientation of an equation and every subproof by effluence can be transformed into an equation.

This last transformation is, however, unnecessary. Consider a subproof $M_1 \leftarrow M_0 \rightarrow M_2$ of π. If there exists a proof by confluence $M_1 \xrightarrow{\star} M_0' \xleftarrow{\star} M_2$ in $\mathcal{R}$, it is strictly simpler, since M_0 is strictly greater than each of the terms which appear in the reductions $M_1 \xrightarrow{\star} M_0' \xleftarrow{\star} M_2$, by assumption on the reduction order $\succ$. In this case, no transformation of the axioms is necessary. Otherwise, according to the Knuth–Bendix theorem, then $M_1 \equiv \mathcal{C}[\theta P]$, $M_2 \equiv \mathcal{C}[\theta Q]$, where (P, Q) is a critical pair of $\mathcal{R}$. After adding the equation $P = Q$ to $\mathcal{E}$, $\mathcal{C}[\theta P] \leftarrow M_0 \rightarrow \mathcal{C}[\theta Q]$ can be replaced by $\mathcal{C}[\theta P] \longleftrightarrow \mathcal{C}[\theta Q]$, which is strictly simpler, since $M_0 \succ M_1, M_2$.

Fairness An equation (or rule, or critical pair) is called *persistent* if it appears in all of the Γ_i above a certain rank. A sequence $(\Gamma_n)_{n \geqslant 0}$ is *fair* for the first group of transformations if no equation is persistent, and if every critical pair created is later transformed into an equation. In other words, every equation of $\mathcal{E}_i$ will later be erased, normalized or oriented, and every critical pair created by the sequence will ultimately be transferred to the equational part, and then erased, normalized or oriented. Fairness is an important property of processes, as are confluence and termination, in the case of non-determinism, to avoid the possibility of a choice being offered an infinite number of times without it ever being taken.

Proposition 6.18 If the sequence $(\Gamma_n)_{n\geqslant 0}$ is fair and if $M =_{\Gamma_i} N$, there exists a step $k \geqslant i$ such that $M = N$ admits a proof by confluence in $\mathcal{R}_k$.

The proof of a stronger theorem will be given later.

Corollary 6.19 If the sequence $(\Gamma_n)_{n\geqslant 0}$ is fair, then the rewriting system $\mathcal{R}_\infty = \bigcup_{n\geqslant 0} \mathcal{R}_n$ is sound, canonical and complete for $\mathcal{E}_0$.

Proof The system $\bigcup_{n\geqslant 0} \mathcal{R}_n$ is sound relative to $\mathcal{E}_0$, since each $\mathcal{R}_n$ is. It is noetherian since the $\mathcal{R}_n$ are uniformly noetherian, i.e, for every n, $M \rightarrow_{\mathcal{R}_n} N$ implies $M \succ N$, the reduction order $\succ$ being independent of n.

If (P, Q) is a critical pair of $\bigcup_{n\geqslant 0} \mathcal{R}_n$, there exists an i such that (P, Q) is a critical pair of $\mathcal{R}_i$, and the proposition shows that (P, Q) join for an $\mathcal{R}_k$, hence $\bigcup_{n\geqslant 0} \mathcal{R}_n$. Confluence is a result, according to the Knuth–Bendix theorem, and so $\mathcal{R}_\infty$ is convergent.

If $M =_{\mathcal{E}_0} N$, the proposition shows that there exists a k such that $M =_{\mathcal{R}_k} N$, hence $M =_{\mathcal{R}_\infty} N$. It follows that the normal forms of M and N in $\mathcal{R}_\infty$ are identical. This proves the completeness of $\mathcal{R}_\infty$ relative to $\mathcal{E}_0$. □

A completion procedure for $\mathcal{E}$ which constructs a fair sequence is therefore a semi-decision procedure. Let M and N be two terms. Since each system $\mathcal{R}_i$ is noetherian and finite, let $\bar{M}_i$ (respectively $\bar{N}_i$) be the finite set of irreducible terms obtained by reduction of M (respectively N) in $\mathcal{R}_i$, and $P_i = \bar{M}_i \cap \bar{N}_i$. If $M =_{\mathcal{E}} N$, there exists an i and a term P such that $M \xrightarrow{*} P \xleftarrow{*} N$ in $\mathcal{R}_i$, hence P_i is non-empty. The semi-decision procedure consistin of computing P_i until $P_i \neq \varnothing$: it halts if and only if $M =_{\mathcal{E}} N$.

Inter-simplifications The first group of transformations allows an inefficient, non-deterministic completion procedure. It is preferable to simplify the created rules on the fly. Otherwise, the probable result will be a redundant system.

A rule $P \rightarrow Q$ is reduced in $\mathcal{R}$ if P is irreducible for $\mathcal{R} \setminus \{P \rightarrow Q\}$, and Q is irreducible for $\mathcal{R}$. A rewriting system $\mathcal{R}$ is *reduced* if each of its rules is reduced. What can one do with an unreduced system?

If $Q \xrightarrow{+} Q'$ (in particular $Q \succ Q'$), the rule $P \rightarrow Q$ should be replaced by $P \rightarrow Q'$, which is simpler.

If P is reducible by a rule $R \rightarrow S$, then a subterm of P is an instance of R: $P/u \equiv \theta R$. The critical pair $(Q, P[u \leftarrow \theta S])$ is therefore obtained, but it is a particular case of computing critical pairs, where the substitution θ links only variables of R. It follows that the rule $R \rightarrow S$, which is applicable each time that $P \rightarrow Q$ is applicable, is simpler than $P \rightarrow Q$.

The search for reduced systems leads to the use of the strict order $\succ^R$ over rules, defined by: $P \rightarrow Q \succ^R R \rightarrow S$ if $P \equiv R$ and $Q \succ S$, or if a subterm of P is an instance of R and $P \not\equiv R$. It is a well-founded order, since matching and the relation 'is a subterm of' are well-founded. A system is reduced as soon as every rule is minimal under $\succ^R$.

A second group of transformations is therefore used. It operates by inter-simplification of the rules to obtain a reduced system:

inter-r	$\mathcal{E}, \mathcal{R} \cup \{P \to Q\}$	$\leadsto$	$\mathcal{E}, \mathcal{R} \cup \{P \to Q'\}$	if $Q \xrightarrow{+}_{\mathcal{R}} Q'$
inter-l	$\mathcal{E}, \mathcal{R} \cup \{P \to Q\}$	$\leadsto$	$\mathcal{E} \cup \{P' = Q\}, \mathcal{R}$	if $P \xrightarrow{\rho} P'$, $P \to Q \succ^R \rho$

Note that one can always reduce the right-hand side of a rule, while the reduction of a left-hand side transforms the rule into an equation, which will often be deleted and sometimes oriented in the opposite direction to the original rule. As for the first group of transformations, these allow the contraction of subproofs.

A proof by rewriting π,

$$\ldots \mathcal{C}[\theta P] \to \mathcal{C}[\theta Q] \ldots,$$

is replaced, for each transformation, by π':

$$\begin{array}{ll} \text{inter-r:} & \ldots \mathcal{C}[\theta P] \to \mathcal{C}[\theta Q'] \xleftarrow{+} \mathcal{C}[\theta Q] \ldots \\ \text{inter-l:} & \ldots \mathcal{C}[\theta P] \xrightarrow{+} \mathcal{C}[\theta P'] \longleftrightarrow \mathcal{C}[\theta Q'] \ldots . \end{array}$$

However, $\pi \succ^p \pi'$ does not necessarily hold in the order on proofs. This order will be refined to consider not only the shape of the proofs, but also their contents, by using the order $\succ^R$ over the rules.

The *fine complexity* of an elementary proof is a pair formed of a multiset and a rule or _ (irrelevant value):

$$\begin{array}{rcl} |||M \xrightarrow{\rho} N||| & = & (\{\!\{M\}\!\}, \rho), \\ |||M \longleftrightarrow_{\mathcal{E}} N||| & = & (\{\!\{M, N\}\!\}, \_). \end{array}$$

These pairs are compared using the lexical ordering, using the extension to multisets of the order $\succ$ over terms in the first component, and the order $\succ^R$ over the rules in the second component. The fine complexity of a proof $(M_0, \ldots, M_p)$ of $M_0 = M_p$ is still the multiset of the complexities of the elementary proofs making it up:

$$|||M_0, \ldots, M_p||| = \{\!\{|||M_0, M_1|||, \ldots, |||M_{p-1}, M_p|||\}\!\}.$$

The fine complexities of proofs are compared by extension to multisets of the order $\succ$ over pairs. Define $\pi \succ^P \pi'$ if $|||\pi||| \succ\!\!\succ |||\pi'|||$. It is also a well-founded order, since it is constructed by lexical product and extensions to multisets of the well-orders $\succ$ and $\succ^R$. It is then easy to show:

Proposition 6.20

- $\pi \succ^p \pi'$ implies $\pi \succ^P \pi'$;
- if π is transformed to π', then $\pi \succeq^P \pi'$;

- if π uses an unreduced rule, then π can be transformed into π' such that $\pi \succ^P \pi'$.

When the two groups of transformations are used jointly, transfers between equations and rules are made in two directions. Since certain rules disappear by inter-simplification, the definition of a fair sequence of transformations must be modified. A sequence $(\Gamma_n)_{n\geqslant 0}$ is *fair* for the two groups of transformations if no equation is persistent, if every rule whose right member is reducible is simplified at a later step, and if every *persistent* critical pair is transformed into an equation at a later step. It is unnecessary to handle a critical pair between equations which will disappear.

Theorem 6.21 If the sequence $(\Gamma_n)_{n\geqslant 0}$ is fair and if $M =_{\Gamma_i} N$, there exists a step $k \geqslant i$, such that $M = N$ admits a proof by confluence in $\mathcal{R}_k$ using only reduced rules.

Proof By well-founded induction over $\succ^P$. Let $\pi :_{\Gamma_i} M = N$, containing an equational subproof or a subproof by effluence, or using an unreduced rule. An equational subproof uses an equation which, by fairness, will have disappeared in Γ_{j_1}, $j_1 > i$. A subproof by effluence can be transformed into a subproof by confluence in Γ_i, or else comes out of a critical pair of $\mathcal{R}_i$. If a critical pair of $\mathcal{R}_i$ is not persistent, it will disappear in Γ_{j_2}, $j_2 > i$ by inter-simplification of one of the rules which created it by superposition. Otherwise, it is transferred into the equational part of Γ_{j_3}, $j_3 > i$. If π uses a rule whose left member is reducible, the corresponding critical pair will be handled as previously. If it is the right member which is reducible, it will be simplified in Γ_{j_4}, $j_4 > i$.

In all cases, π is transformed into π', in a Γ_j, with $j \geqslant i$ and $\pi \succ^P \pi'$. By the inductive hypothesis, there exists a step k with $k \geqslant j$ such that $M = N$ admits a proof by confluence in $\mathcal{R}_k$. □

6.3.3 An example: the equational theory of groups

A completion procedure takes as input a set $\mathcal{E}_0$ of equations and a reduction order $\succ$, and generates a sequence $(\Gamma_n)_{n\geqslant 0}$ of mixed systems, starting from $\Gamma_0 = (\mathcal{E}_0, \mathcal{R}_0)$. The elementary operations used are:

- *superposition*(R_i, R_j, u): superpose the left member of R_i on that of R_j at occurrence u and produce the corresponding critical pair;
- *normalization*$(R_i, \ldots)$: normalize the two members of a given equation by applying the rules $R_i, \ldots$ of the current rewriting system;
- *erasure*(R_i): erase equation R_i;
- *orientation*: transform a given equation into a rule by comparing its two rules by $\succ$;
- *left-simplification*(R_i, R_j): transform the rule R_i into an equation after having reduced its left member by R_j.

The example of the equational theory of groups is the best known. The order of reduction used is an *rpo* order, the product being greater than the constant e, and having a lexical status. Here is the trace of a completion procedure, starting from the three usual equations, which are oriented in the natural manner. The complete system is formed of rules 1, 2, 3, 4, 8, 10, 11, 12, 13 and 16.

$1 : ex \rightarrow x$		
$2 : x^{-1}x \rightarrow x$		
$3 : (xy)z \rightarrow x(yz)$		
	superposition$(2, 3, 1)$:	$x^{-1}(xy) = ey$
	normalization(1):	$x^{-1}(xy) = y$
	orientation:	
$4 : x^{-1}(xy) \rightarrow y$		
	superposition$(1, 4, 2)$:	$y = e^{-1}y$
	orientation:	
$5 : e^{-1}y \rightarrow y$		
	superposition$(2, 4, 2)$:	$e = (x^{-1})^{-1}e$
	orientation:	
$6 : (x^{-1})^{-1}e \rightarrow e$		
	superposition$(5, 4, 2)$:	$y = (e^{-1})^{-1}y$
	orientation:	
$7 : (e^{-1})^{-1}y \rightarrow y$		
	superposition$(2, 7, \epsilon)$:	$e^{-1} = e$
	orientation:	
$8 : e^{-1} \rightarrow e$		
	left-simplification$(5, 8)$:	$ey = y$
	normalization(1):	$y = y$
	erasure(5)	
	left-simplification$(7, 8)$:	$e^{-1}y = y$
	normalization$(8, 1)$:	$y = y$
	erasure(7)	
	superposition$(4, 4, 2)$:	$xy = (x^{-1})^{-1}y$
	orientation:	
$9 : (x^{-1})^{-1}y \rightarrow xy$		
	left-simplification$(6, 9)$:	$xe = x$
	erasure(6):	
	orientation:	
$10 : xe \rightarrow x$		
	superposition$(9, 10, \epsilon)$:	$(x^{-1})^{-1} = xe$
	normalization(10):	$(x^{-1})^{-1} = e$
	orientation:	
$11 : (x^{-1})^{-1} \rightarrow x$		
	left-simplification$(9, 11)$:	$xy = xy$
	erasure(9)	

	superposition(11, 2, 1):	$e = xx^{-1}$
	orientation:	
$12 : xx^{-1} \to e$		
	superposition(11, 4, 1):	$y = x(x^{-1}y)$
	orientation:	
$13 : x(x^{-1}y) \to y$		
	superposition(12, 3, ϵ):	$x(y(xy)^{-1}) = e$
	orientation:	
$14 : x(y(xy)^{-1}) \to e$		
	superposition(14, 4, 2):	$y(xy)^{-1} = x^{-1}e$
	normalization(10):	$y(xy)^{-1} = x^{-1}$
	orientation:	
$15 : y(xy)^{-1} \to x^{-1}$		
	left-simplification(14, 15):	$xx^{-1} = e$
	normalization(12):	$e = e$
	erasure(14)	
	superposition(15, 4, 2):	$(yx)^{-1} = x^{-1}y^{-1}$
	orientation:	
$16 : (yx)^{-1} \to x^{-1}y^{-1}$		
	left-simplification(15, 16):	$y(y^{-1}x^{-1}) = x^{-1}$
	normalization(13):	$x^{-1} = x^{-1}$
	erasure(15).	

Exercises[1]

1. Let $p, q : \mathcal{A} \to \mathcal{B}$ be two morphisms. Show that the set of $x \in A$ such that $p(x) = q(x)$ is a subalgebra $\mathcal{E}$ of $\mathcal{A}$, and that every morphism $r : \mathcal{X} \to \mathcal{A}$ such that $p \circ r = q \circ r$ factors through the inclusion $\mathcal{E} \hookrightarrow \mathcal{A}$.

2. Compare the subalgebras of $\mathcal{A}$ and the congruences of $\mathcal{A} \times \mathcal{A}$.

3. Let $p, q : \mathcal{A} \to \mathcal{B}$ be two morphisms. Define a quotient $l : \mathcal{B} \to \mathcal{C}$, such that every morphism $r : \mathcal{B} \to \mathcal{X}$ satisfying $r \circ p = r \circ q$ factors through l.

4. Determine the class of models for each of the following axioms:

$$x = x, \qquad x = y, \qquad f(x) = f(y), \qquad f(x) = y.$$

5. Prove the properties of the Galois correspondences, equations (6.1)–(6.8).

6. Show that every variety is stable under isomorphism, subalgebras, quotients and products.

---

[1]Hints to the exercises labeled with a ★ can be found at the end of the book.

7. Determine whether the following classes are varieties: finite groups, infinite groups, abelian groups, rings of characteristic p and rings of characteristic dividing p.

8. Derive the rules (**R**) and (**T**) from the equality axioms in NK:

$$(\mathbf{R}): \frac{}{\forall \vec{x}(M = M)} \qquad (\mathbf{T}): \frac{\forall \vec{x}(M = M') \qquad \forall \vec{x}(M' = M'')}{\forall \vec{x}(M = M'')}.$$

9. Show that the rule (**cong**) is derivable from (**T**), (**subst**) and the unary rules of context passing:

$$((f, i)_{1 \leqslant i \leqslant r}) : \frac{M_i = N_i}{f x_1 \ldots x_{i-1} M_i x_{i+1} \ldots x_r = f x_1 \ldots x_{i-1} N_i x_{i+1} \ldots x_r}, f \in \Sigma_r.$$

10. Derive the weakening rule, equation (6.9), of the guarded equations from the (**subst**) rule.

11.★ Show that $(a, b; a^3 = b^2 = e, ab = ba^2)$ and $(u, v; u^2, v^2, (uv)^3)$ are two presentations of the same group. Which one?

12.★ Construct a model of $(+)$ where addition is not associative.

13.★ Show that $\mathrm{T}_{\Sigma'/\Sigma} \circ \mathrm{T}_{\Sigma/\Sigma^0} = \mathrm{T}_{\Sigma'/\Sigma^0}$. What are the values of $\mathrm{T}_{\Sigma/\Sigma}$, $\mathrm{T}_{\Sigma/\varnothing}$ and $\mathrm{T}_{\Sigma/\Sigma^0}[\mathrm{T}_{\Sigma^0}]$?

14. Show by a counterexample that the assumption 'Σ non-linear' is necessary to obtain the second point of proposition 6.7.

15.★ Show, by completion, that the equation $x(x^{-1}y) = y$ is a consequence of $x^{-1}(xy) = y$ (without any other axioms).

16.★ Construct by completion a canonical system for idempotent semigroups, with axioms $(xy)z = x(yz)$ and $xx = x$.

17.★ Let E be a system of equations. Show that $E \cup \{M_1 \neq N_1, \ldots, M_p \neq N_p\}$ is unsolvable if and only if there exists i such that $E \cup \{M_i \neq N_i\}$ is unsolvable. A substitution σ is the solution of the inequality $M \neq N$ if $\sigma(M) \not\equiv \sigma(N)$.

18.★ Show by an example how to associate a pure λ-term M with a simple equation between simple types with variables, so that M is typable if and only if this system of equations has a solution.

Bibliographic notes

The modules of Standard ML are described in [111]. For universal algebra, a standard reference is Cohn's book [22]. The use of categories in computer science is described in [5].

Unification is the theme of volume 8 (1989) of the *Journal of Symbolic Computation.* In it can be found a general presentation by Siekmann [137]. Another survey is [86]. Robinson's algorithm is presented in [128], those of Huet in [67], of Paterson and Wegman in [117], of Martelli and Montanari in [100], of Corbin and Bidoit in [26], of Martelli and Rossi in [101], and of Escalada-Imaz and Ghallab in [41].

The proof of the Knuth–Bendix theorem comes from Huet [68]. The Knuth–Bendix completion procedure is given by its authors in [87]. A complete proof for it is given in [69]. The presentation given here comes from Bachmair, Dershowitz and Hsiang's article [6].

Chapter 7

Resolution

7.1 Logic programs

Prolog was first implemented at the University of Marseille in 1973, under the direction of A. Colmerauer [23]; the first logical interpretation of the Prolog mechanism was given by R. Kowalski in 1974 [88]. Logic programming developed through the conjunction of the two events. Although logic programming is not simply Prolog, and Prolog programs are not always logical, logic programming will first be presented through 'pure' Prolog programs. A few typical examples are given here to show the role of unification.

7.1.1 A Prolog interpreter

Notation The *Edinburgh syntax* will be used. Terms are written using parenthesized notation. They are formed of variables, identifiers beginning with a capital letter, and of symbols, identifiers starting with a small letter. Numbers are considered symbols of zero arity. So `r(a,X)` is used instead of the usual $R(a, x)$ of logic. The queries and rules are terminated by a '`.`'. Most of the examples are written in C–Prolog[1] or NU–Prolog.[2]

The Marseille syntax, used by Prolog–II+,[3] and occasionally used here, is very different. Examples presented will use one-character identifiers for variables and identifiers with at least two characters for signature symbols: hence `rr(aa,x)` corresponds to the `r(a,X)` of Edinburgh. Queries and rules are terminated by a '`;`'.

[1]EdCAAD, University of Edinburgh, Department of Architecture.

[2]Machine Intelligence Project, Department of Computer Science, University of Melbourne, Australia.

[3]Groupe d'Intelligence Artificielle de Luminy, Université d'Aix-Marseille II, distributed by PrologIA

Unification A Prolog interpreter consists mainly of a unification machine. A program does not have to be written before a *query* such as

```
?- f(X,g(X),Y) = f(h(Z),T,Z).
```

is made. Prolog variables are *unknown* in the mathematical sense, and the interpreter attempts to give them a value. An interpreter would give the following solution to the above query:

```
↷ T = g(h(Z)),
  Z = Z,
  Y = Z,
  X = h(Z)
```

However, Prolog does not evaluate terms, unlike functional language interpreters:

```
?- 2 + 2 = 3 + 1.

↷ fail
```

This non-evaluation and the answer `fail` are typical of Prolog, where failure does not necessarily mean an error.

Data structures The data manipulated by Prolog are first-order terms.

Types can be defined, along with their constructors, but the concept of type is not explicit in Prolog: it is simply a way of reading certain terms. For example, the unary integers can be defined using the constants `nat`, `0` and `s`, and the *definite program*

```
nat(0).
nat(s(X)) :- nat(X).
```

which states that `0` is an integer and that `s(X)` is an integer if `X` is an integer. Note that these constants were not declared, and that the interpreter infers their arity, just as the ML interpreter infers the type of expressions. The set of the two rules constitutes the *definition of predicate* `nat/1`. The `/1` indicates that `nat` is of arity 1. Several predicates can use the same symbol, so long as they are distinguished by their arity. There is therefore no arity checking akin to ML's type checking.

A set of predicate definitions constitutes a *logic program*. Once supplied with this program, the interpreter can respond to queries:

```
?- nat(s(0)).

↷ true
```

Should there be several solutions, they are given successively in response to a '`; RETURN`' typed by the user. The interpreter stops looking in response to a simple '`RETURN`'. It enumerates the integers in response to the query:

```
?- nat(X).

↷ X = 0   ;RETURN
```

```
X = s(0)    ;RETURN

X = s(s(0))   RETURN
```

Other recursive data structures, more useful than the integers, such as lists and trees, can be defined similarly.

Relations Here is a program, using the same integers, defining addition:

```
p(X,0,X).
p(X,s(Y),s(Z)) :- p(X,Y,Z).
```

Even if the values of a *function* are to be computed, a *relation* must be programmed, with an argument for the result. So ordinary computations such as

```
?- p(s(0),s(0),Z).

↷ Z = s(s(0))
```

can be done, but also queries asking for the arguments of a relation such as:

```
?- p(X,Y,s(s(0))).

↷ X = s(s(0)),
  Y = 0  ;

  X = s(0),
  Y = s(0)   ;

  X = 0,
  Y = s(s(0))   ;

  fail
```

The final `fail` means that there are no more solutions.

A request can be composed of several literals. The search for X and Y, such that $X+Y=4$ and $X+2=Y$, can be written by the query:

```
?- p(X, Y, s(s(s(s(0))))), p(X, s(s(0)), Y).

↷ Y = s(s(s(0))),
  X = s(0)
```

Even though it is not visible, a standard Prolog interpreter starts by computing a solution for the first literal, then checks if it is a solution to the second literal. Here, it will first find $X=4, Y=0$, which does not satisfy `p(X, s(s(0)), Y)`. It will then return to the first literal to examine the other solutions. Thus $X=3, Y=1$, then $X=2, Y=2$ are obtained, but they are still not solutions to the second literal. Finally, $X=1, Y=3$ is a solution to both literals, the result is obtained, and the interpreter continues to search. The declarative reading of the query, where a system of equations is to be solved, and the operational reading just outlined are very different. This search mechanism is the strength of Prolog in situations where an exhaustive search of graphs, state spaces and similar structures is necessary, since it need not be programmed explicitly. This strength is also a weakness of Prolog, since it is too easy to generate inefficient programs.

Lists As in Lisp and ML, lists form a data structure with a predefined syntax. Table 7.1 indicates the correspondence between the notation in the three languages.

Prolog	*Standard ML*	*Lisp*
`[a,b,c]`	`[a,b,c]`	`(a b c)`
`[a|L]`	`a::L`	`(a . L)`
`[a,b,c|L]`	`a::b::c::L`	`(a b c . L)`
`[]`	`[]`	`()`

Table 7.1 Lists in different languages

Numerous programs manipulating lists come from the Lisp tradition, such as the concatenation of two lists, the reversal of a list and quicksort:

```
append([],Y,Y).
append([H|T],Y,[H|TY]) :- append(T,Y,TY).

reverse([],[]).
reverse([H|T],Y] :- reverse(T,RT), append(RT,[H],Y).
```

The predicate `quicksort/2` sorts a list of numbers by decomposing it into two lists of numbers smaller than or equal and larger than the first element (the 'pivot'), respectively, using the predicate `partition/4`, and calls itself recursively for each of the sublists.

```
quicksort([],[]).
quicksort([H|T],Y) :-
   partition(T,H,Littles,Bigs),
   quicksort(Littles,S_Littles),
   quicksort(Bigs,S_Bigs),
   append(S_Littles,[H|S_Bigs],Y).

partition([],_,[],[]).
partition([H|T],P,[H|Littles],Bigs) :-
   H =< P,
   partition(T,P,Littles,Bigs).
partition([H|T],P,Littles,[H|Bigs]) :-
   H > P,
   partition(T,P,Littles,Bigs).
```

The set, more exactly the sorted list, of answers to a query is obtained using the predefined predicate `solutions/3` of NU–Prolog:

```
solutions([X|Y], p(X,Y,s(s(0))),L).
```

```
↝ L = [[0| s(s(0))], [s(0)| s(0)], [s(s(0))| 0]]
```

The logic variable and unification The data structures of functional programming contain 'values'. Here, they can contain constants and 'variables'. A closed term is completely known. If it contains variables, it is a partially specified object. The interpreter will make it 'rise' in the lattice of terms by assigning values to the variables so that it satisfies the query in which it occurs, and will make it rise the least amount possible so that it remains the most general possible.

Terms can be used, as for the concrete types of ML, to define structured values. A term `rectangle(width(2), length(3))`, or simply `rectangle(2,3)`, represents a particular rectangle, and `rectangle(X,Y)` an arbitrary rectangle. But no ordinary language can be used to specify squares as constrained rectangles. In Prolog, `rectangle(X,X)` is an arbitrary square.

The essential mechanism is unification, which, by itself, does a number of basic functions which are handled by distinct mechanisms in traditional languages:

- parameter passing, both input and output; unification is more powerful than call by pattern matching, since it eliminates the difference between input and output parameters;
- data destructuring (access to the fields in records), but without *mutable* objects, the modification of fields is not efficient;
- data construction, in a manner similar to C's `malloc`;
- variable assignment, similar to the 'single-assignment' variables of functional programming languages;
- testing the value of variables, as for call by pattern matching.

D-lists D-lists (from difference lists) are an original data structure, typical of logic programming, first used by Colmerauer, even before Prolog came to be, in his research on syntactic analysis, and can be used to represent term contexts. A context is a term with a 'hole', written '?', ready to receive a graft of a term. It is tempting to represent such a hole by a variable. Doing that would have no meaning in functional programming, but it is possible in logic programming, even though it is inefficient, since the graft needs to go through the entire term to reach the variable. In sequential programming, the problem would be handled by holding a pointer to the hole, as in the standard implementation of files. The logic variable allows such an access: all the occurrences of the same variable denote the same object. Hence, to reach the end of the list `[a,b,c,d]` quickly, it suffices to have a variable `X` having an occurrence at the end of the list, and another at the 'surface', which is immediately accessible. This can be done by using, for example, the infix constant `-` to form the term: `[a,b,c,d|X]-X` represents the context $[a, b, c, d|?]$.

A pair of lists L, L' is a D-list if $L = [L_1, \ldots, L_p|M]$ and L' is unifiable with M. In that case, if θ is the most general unifier, then the pair L, L' represents the list context $[\theta L_1, \ldots \theta L_p|?]$.

Example The D-lists `[a,b,c,d]-[c,d]`, `[a,b,c]-[c]`, `[a,b,c|T]-[X|T]`, and `[X,b,a|T]-[X|T]` all represent the context $[a, b|?]$. The simplest representations are `[a,b]-[]`, of no interest, and `[a,b|X]-X`, the 'standard' representation. But `[a,b]-[c,d]` is not a D-list, since `[c,d]` does not unify with any suffix of `[a,b]`. □

With this data structure, concatenation is implemented using a single unification, by the program:

```
append_dl(L-M,M-N,L-N).
```

To concatenate the lists `[a,b]` and `[c,d]`, query

```
?- append_dl([a,b|X]-X,[c,d|Y]-Y,Z).
```

is solved by unifying `M` with `[c,d|Y]`, `N` with `Y`, `L` with `[a,b,c,d|Y]` and `Z` with `[a,b,c,d|Y]-Y`. In fact, concatenation is often done implicitly, without even calling `append_dl`.

Occur check Most Prolog implementations do not implement unification in $T[X]$ completely, for unconvincing efficiency reasons. Only Prolog–II+ implements real unification, but in $T^r[X]$, and not in $T[X]$, and so one can program with rational terms, i.e, cyclical objects. In the simplest case

```
?- X = f(X).
```

```
↷ X = f(f(f(f(f(f(f(f(f(f(f(f(f(f(f(f(f(f((f(f(f(f(f(f(^C
```

C–Prolog's answer is incorrect in the term algebra, but correct if it is interpreted as a rational term. Prolog–II responds better:

```
eq(x, ff(x));
```

```
↷ x=ff(*1)
```

In other cases, such as

```
?- f(X,g(X),X) = f(g(Y),Y,Y).
```

a standard interpreter would give no response, since it loops! But in Prolog–II:

```
>eq(ff(x,gg(x),x), ff(gg(y),y,y));
```

```
↷ x=gg(gg(*2)), y=gg(gg(*2))
```

If the values of variables are not asked for, there is no way of knowing if a bad unification took place:

```
test :- p(X,X).
p(X,f(X)).
?- test.
```

```
↷ true
```

Prolog gives an answer which is logically incorrect. This lack of occur check takes away substantially from the logical credibility of Prolog, even if in practice it occurs rarely.

D-lists are, however, a major exception, with seemingly useful examples giving incorrect answers. The following query, which asks whether the list [a] is the result of concatenating two empty lists, succeeds!

```
?- append_dl(L-L,M-M,[a|N]-N).
```

The program

```
rep(L-E,X) :- append(X,E,L).
```

computes the list represented by a D-list. However, the following query loops with an erroneous answer, even though the solution X = [a,b] is expected:

```
?- rep([a,b|E]-E,X).
```

7.1.2 Views of logic programming

Logic programming is attractive because of the many ways in which it can be viewed: logical, procedural, rewriting, equation resolution, theorem-proving, networks of communicating processes, etc.

Logical formulas
Kowalski's view [88] initiated logic programming. A *literal* is an atom (positive literal) or the negation of an atom (negative literal). A formula is *clausal* if it is of the form $\forall\vec{x}\varphi^o$, where φ^o is a disjunction of literals. The universal closure of a disjunction of literals, of which at most one is an atom, is called a *Horn clause*. It is standard to distinguish the following three forms,

$$\forall\vec{x}A,$$
$$\forall\vec{x}(A \vee \neg A_1 \vee \ldots \vee \neg A_p),$$
$$\forall\vec{x}(\neg L_1 \vee \ldots \vee \neg L_q),$$

where the A, A_i, L_i are atoms, classically equivalent to:

$\forall\vec{x}A$	unit clause, or fact,
$\forall\vec{x}(A_1 \wedge \ldots \wedge A_p \Rightarrow A)$	definite clause, or rule,
$\neg\exists\vec{x}(L_1 \wedge \ldots \wedge L_q)$	negative clause, or goal.

Note that a unit clause is a particular case of a definite clause, when $p = 0$. The Prolog notations for fact and rule are, respectively:

$$A.$$
$$A \texttt{ :- } A_1, \ldots, A_p.$$

If C is the definite clause $A \texttt{ :- } A_1, \ldots, A_p$, write C^+ for the atom A, and C^- for the set of literals $A_1, \ldots, A_p$: C^+ is the *head* of C, and C^- is its *body*. A *definite program* is a sequence of definite clauses. Its logical meaning is the conjunction of the clauses which make it up. Note that this logical meaning is not affected by the order or repetition of literals and clauses, so that bodies and programs can be considered as sets rather than sequences. The implicit universal quantification makes every variable local to the clause in which it appears: there are no 'global variables' in Prolog.

The Prolog interpreter responds to *queries* of the form

$$\texttt{?-}\ L_1, \ldots, L_q,$$

where the logical meaning is that of formula

$$\exists \vec{x}(L_1 \wedge \ldots \wedge L_q),$$

which is the negation of a goal.

It is useful to extend the notation used for terms, occurrences and subterms to queries. If R is the query ?- $L_1, \ldots, L_q$, write R/i for literal L_i, $R(i)$ for the head symbol of L_i, and $R \setminus i$ for the query ?- $L_1, \ldots, L_{i-1}, L_{i+1}, \ldots, L_q$, the result of erasing L_i. For $q = 0$, an empty query is obtained, written $\top$, which is equated with the connective $\top$ (true). To avoid confusion, when R is the query ?- $L_1, \ldots, L_q$, R is also used for the formula $L_1 \wedge \ldots \wedge L_q$, and $\forall(\neg R)$, or $R^{\vee}$, for the goal $\forall \vec{x}(\neg L_1 \vee \ldots \vee \neg L_q)$.

A Horn formula is a conjunction of Horn clauses. The conjunction of program rules and of a goal is therefore a Horn formula. The importance of the class of Horn clauses will be explained later.

Submitting a query R to a program P, consists, from a logical point of view, of proving the existential formula $\exists \vec{x} R$ from P, i.e, of refuting $P \wedge R^{\vee}$. A *solution* to this query is a substitution σ such that $P \models \sigma(R)$. The interpreter proves $\exists \vec{x} R$ in a constructive manner by generating solutions.

This interpretation of programs and queries is only declarative, and gives no interpretation of the execution of programs as proof procedures. Natural deduction could be used instead of methods particular to clauses, such as the resolution principle and its refinements. Kowalski introduced SLD-derivation as a particular case of the resolution principle, thereby giving an operational semantics to logic programming. This subject will be treated in depth.

Intuitionistic sequents

The definite clause A :- $A_1, \ldots, A_p$ can be seen not only as a formula $\forall \vec{x}(A_1 \wedge \ldots \wedge A_p \Rightarrow A)$, but also as a *sequent* $A_1, \ldots, A_p : A$ composed of atomic formulas, which has the same declarative meaning. On the other hand, a query ?- $L_1, \ldots, L_q$ corresponds to a sequent $L_1, \ldots, L_q$:. In both cases they are sequents of the intuitionistic calculus (at most one formula in the right member of the sequent) restricted to atomic formulas. We therefore have a good idea of what logic to use. Recall that the rules of the sequent calculus are divided into the structural rules (rearranging of formulas), the logical rules (for connectives and quantifiers), the cut rule and the identity axiom. The logical rules were proposed by Gentzen, as were the rules of NK, to make the meaning of the logical symbols explicit. But it is possible to do logic without using any logical symbols! Formulas without logical symbols are atoms. Natural deduction can do practically nothing as it only manipulates atoms. However, the sequent calculus can work with sequents formed from atoms, and there are (structural, identity and cut) rules using no logic symbols, allowing a very combinatoric reasoning with axioms.

With only the identity axiom, the Hauptsatz shows that the cut can be eliminated, but then only structural properties can be proven (for example, to derive $A, B : A$). If other axioms $\Gamma : \Delta$ formed of axioms are allowed, the cuts can no longer be eliminated; they are essential. This is what happens in logic programming, where axioms are definite clauses. A proof in LJ, from the axioms and all of their instances, may use no other rule than the cut rule, one of the premises being the instance of an axiom.

$$(\text{cut}) : \frac{A_1, \ldots, A_p : A \qquad A, B_2, \ldots, B_q :}{A_1, \ldots, A_p, B_2, \ldots, B_q :}.$$

Therein lies the principle of SLD-resolution.

Rewrite rules

This is Colmerauer's view, which inspired the Prolog*s* of Marseille. The rules are written A -> $A_1, \ldots, A_p$;, and read from left to right, as in 'A rewrites to $A_1, \ldots, A_p$', instead of '$A_1, \ldots, A_p$ imply A'. A query is a sequence of terms—the difference between function and relation symbols is eliminated—to be erased using the rules of the program. There is a striking analogy with context-free grammars where non-terminals must be eliminated by rewriting. A rule can be applied if a system of equations, which grows as the literals are erased, can be solved. There is no reason to limit oneself to equations between terms. Prolog–II+ treats equations and inequalities ($\neq$) over rational terms. Other languages work with other domains (Booleans, finite sets, rational, algebraic or real numbers): logic programming then becomes constraint programming. In these languages, such as Prolog–III, CHIP, or the family of CLP (*Constrained Logic Programming*), the interpreter is capable of solving systems of equations and inequalities ($\neq$) between rational terms or Booleans, and inequalities ($\leqslant$) between integers or reals.

$$\begin{array}{rcll}
f(x, g(a)) & = & f(g(x), y) & \\
x & \neq & y, & x, y \quad \text{rational terms} \\
2n_1 + 3n_2 & > & 14 & \\
n_1 - n_2 & = & 7, & n_1, n_1 \quad \text{integers} \\
(b_1 \vee b_2) \wedge b_3 & = & 1 & \\
b_1 \vee b_2 & \neq & b_3, & b_1, b_2, b_3 \quad \text{Booleans.}
\end{array}$$

7.2 Semantics of definite programs

There is now a well-established logical basis, from the works of Kowalski, Apt, van Emden, Clark, Lassez and Maher, which is synthesized in Lloyd's book [95]. The logical approach can be split into two parts. The declarative semantics determines what can be computed. The operational semantics describes what can be executed.

7.2.1 Declarative semantics of definite programs

It is doubtful whether one can consider the predicate calculus as a programming language: a formula computes nothing. However, it is possible to isolate fragments, containing a subset of the predicate calculus, a semantics and appropriate proof methods, in which one can compute effectively. Even though several fragments have been considered, only the 'Horn' fragment has been developed into a real programming language. Two essential ideas are implemented: given the problem $\varphi \vdash \psi$, program φ specifies a search for a proof of ψ, and from this proof a result is extracted. The presentation will begin by justifying Horn clauses.

Horn clauses and product algebras
Just as in equational theories, Horn clauses satisfy an important stability property, which explains their use in programming: stability under product.

First, the definition of the product $\mathcal{M} = \prod_{i \in I} \mathcal{M}_i$ of a family of Σ-algebras $\mathcal{M}_i$, already given for the functional case, is extended to predicate calculus signatures by defining $R^{\mathcal{M}} = \prod_i R^{\mathcal{M}_i}$, for every relational symbol R: for an atom A and a valuation $\xi : X \to \prod_i M_i$ of components $\xi_i : X \to M_i$, then $\hat{\xi}(A) = 1$ if $\hat{\xi}_i(A) = 1$ for every i.

Proposition 7.1 The class of models of a set of Horn clauses is stable under product.

Proof Let $(\mathcal{M}_i)_i$ be a family of models of a Horn clause $C \equiv A \vee \neg A_1 \vee \ldots \vee \neg A_p$ and $\xi = (\xi_i)_i : X \to \prod_i M_i$ a valuation. It must be shown that $C^{\mathcal{M}}(\xi) = 1$, i.e, $A^{\mathcal{M}}(\xi) = 1$, $A_1^{\mathcal{M}}(\xi) = 0, \ldots$, or $A_p^{\mathcal{M}}(\xi) = 0$. Suppose that $A^{\mathcal{M}}(\xi) \neq 1$. This implies that there exists i such that $A^{\mathcal{M}_i}(\xi_i) = 0$. Since $\mathcal{M}_i \models C$, $A^{\mathcal{M}_i}(\xi_i) = 1$, $A_1^{\mathcal{M}_i}(\xi_i) = 0, \ldots$, or $A_p^{\mathcal{M}_i}(\xi_i) = 0$. There therefore exists k_i such that $A_{k_i}^{\mathcal{M}_i}(\xi_i) = 0$, which implies $A_{k_i}^{\mathcal{M}}(\xi) = 0$, hence $C^{\mathcal{M}}(\xi) = 1$.

If Γ is a set of Horn clauses and $\mathcal{M}_i \models \Gamma$ for every i, then $\mathcal{M}_i \models C$ for every i and every $C \in \Gamma$. Hence, by the above, $\prod_i \mathcal{M}_i \models C$ for every C, i.e, $\prod_i \mathcal{M}_i \models \Gamma$. □

An arbitrary clause does not satisfy this property, as the following property proves. Let $C \equiv p(x) \vee q(x)$, and the algebras $\mathcal{M}_1$, $\mathcal{M}_2$ be of domain $\{a, b\}$, $p^{\mathcal{M}_1}(a) = q^{\mathcal{M}_1}(b) = 1$, $p^{\mathcal{M}_1}(b) = q^{\mathcal{M}_1}(a) = 0$, and the contrary for $\mathcal{M}_2$. Then $\mathcal{M}_1, \mathcal{M}_2 \models C$, but $\mathcal{M}_1 \times \mathcal{M}_2 \not\models C$. The converse of proposition 7.1 is true: it is the McKinsey theorem.

Extraction of a result
One of the underlying ideas of logic programming is to consider a computation as the extraction of a result from a proof. For example, from a derivation $\exists x \psi$,

a term t would be extracted such that $\psi[x := t]$ were derivable. The Herbrand theorem leans in this direction.

Consider two propositions, one universal $\varphi = \forall\vec{x}\varphi^o$, and the other existential $\psi = \exists\vec{x}\psi^o$, written in prenex form for simplicity. The formula $\varphi \Rightarrow \psi$ being existential, the existential version of the Herbrand theorem states the equivalence between $\vdash \varphi \Rightarrow \psi$ and the existence of a domain D such that $\vdash \mathrm{Exp}(\varphi \Rightarrow \psi, D)$. Recall that the formula $\mathrm{Exp}(\varphi, D)$ is the expansion of φ over D, a finite set of closed terms. By the definition of the expansion,

$$\mathrm{Exp}(\varphi \Rightarrow \psi, D) = \mathrm{Exp}(\varphi, D) \Rightarrow \mathrm{Exp}(\psi, D).$$

Since φ is universal (lemma 5.12), $\vdash \varphi \Rightarrow \mathrm{Exp}(\varphi, D)$, hence $\vdash \varphi \Rightarrow \mathrm{Exp}(\psi, D)$. By making the expansion explicit, closed substitutions θ_i are obtained such that

$$\varphi \vdash \theta_1\psi^o \vee \ldots \vee \theta_n\psi^o.$$

According to the strong version of the Herbrand theorem—not proven here—these substitutions can be extracted effectively from a derivation of ψ—if one can be constructed, which the theorem does not indicate. Furthermore, the result obtained is a disjunction, and one cannot necessarily conclude that $\varphi \vdash \theta_i\psi^o$ for some i, which would really allow one to refer to the result of a computation. Otherwise, there is an indeterminate result.

Example Let $\varphi = p(a) \vee p(b)$, and $\psi = \exists x p(x)$. Then $\vdash \varphi \Rightarrow \psi$, but there is no term t such that $\vdash \varphi \Rightarrow p(t)$. □

This situation is not satisfactory, although it does lead the calculus to non-determinism in an interesting manner. If determinate results are wanted, then either the logic must be changed to a more constructive logic or the class of formulas must be restricted. Standard logic programming plays both games: its proofs use intuitionistic logic and its formulas are restricted to Horn clauses. More precisely, with respect to the above situation, logic programming handles problems $\varphi \vdash \psi$ where φ^o is a conjunction of definite clauses, and ψ is the existential closure of a conjunction of atoms, i.e., the negation of a negative clause.

Theorem 7.2 ([48]) Let φ be a universal Horn proposition, and let there be n existential propositions $\psi_i = \exists\vec{x}\psi_i^o$, with $\psi_i^o = (B_i^1 \wedge \ldots \wedge B_i^{p_i})$, $i = 1, \ldots, n$, where the B_i^k are atoms:

1. $\varphi \models \psi_1 \vee \ldots \vee \psi_n$ if and only if there exist i such that $\varphi \models \psi_i$;
2. $\varphi \models \psi_i$ if and only if there exists a closed term t such that $\varphi \models \psi_i^o[\vec{x} := \vec{t}]$.

Proof (1) By contradiction: Suppose $\models \varphi \Rightarrow (\psi_1 \vee \ldots \vee \psi_n)$ and $\varphi \not\models \psi_i$ for every i. There therefore exist algebras $\mathcal{M}_i$ such that $\mathcal{M}_i \models \varphi$ and $\mathcal{M}_i \models \neg\psi_i$. Since $\neg\psi_i$ is equivalent to $\forall\vec{x}(\neg B_i^1 \vee \ldots \vee \neg B_i^{p_i})$, $\mathcal{M}_i \models \neg\psi_i$ means that for every $\xi_i : X \rightarrow M_i$ there is an index k_i such that $(\neg B_i^{k_i})^{\mathcal{M}_i}(\xi_i) = 1$, i.e, $(B_i^{k_i})^{\mathcal{M}_i}(\xi_i) = 0$.

Let $\mathcal{M} = \prod_{1 \leqslant i \leqslant n} \mathcal{M}_i$. Then $(B_i^{k_i})^{\mathcal{M}}(\xi) = 0$, by the definition of the product algebra, hence $(\neg B_i^{k_i})^{\mathcal{M}}(\xi) = 1$, for every $\xi : X \to M$. This proves that $\mathcal{M} \models \neg\psi_i$ for every i. Furthermore, by using the property of stability under product of Horn clauses, $\mathcal{M} \models \varphi$. It follows that $\mathcal{M} \models \varphi \wedge \neg\psi_1 \wedge \ldots \wedge \neg\psi_n$, which contradicts $\models \varphi \Rightarrow (\psi_1 \vee \ldots \vee \psi_n)$.

(2) According to the Herbrand theorem, if $\varphi \models \psi_i$, there exist closed substitutions θ_j such that $\varphi \models \theta_1\psi_i^o \vee \ldots \vee \theta_n\psi_i^o$. This implies, by (1), that $\varphi \models \theta_j\psi^o$ for a j. □

This theorem allows one to give a declarative semantics to existential queries. What are computable are the substitutions. A closed solution of a query $\exists\vec{x}\psi$ is a closed substitution σ such that $P \vdash \sigma(\psi)$.

As in similar situations, such as unification and typing in λ-calculus, rather than enumerating the set of solutions, the most general of the possible solutions is sought. In natural deduction, all the derivations considered have their assumptions among the propositions forming the program P. If there is a derivation of $p(t)$ for a closed term t, one can infer $\exists x p(x)$ by losing the information contained in t, which is certainly not in the spirit of logic programming. In the case of a non-closed term t, for example $f(a, z)$, one can always infer $\exists x p(x)$ from $p(f(a, z))$, but, better still, one can infer $\forall z p(f(a, z))$ by $\forall$-introduction, variable z not being free in any assumption since all the assumptions are propositions. The semantics of a query must therefore be changed by searching for σ with values in $\mathrm{T}_\Sigma[X]$ such that $P \vdash \forall(\sigma\psi)$. This appears surprising, since the result is a universal formula in the form of a query, while to apply the Herbrand theorem, what is needed is 'universal $\vdash$ existential'. However, the $\forall$ only plays a superficial role, stating that solutions are unique up to renaming of variables.

Definition 7.1 The *solutions* to query R for program P are the substitutions σ : $\mathrm{var}(R) \to \mathrm{T}_\Sigma[X]$ such that $P \vdash \forall(\sigma R)$. Their set is written $[\![R]\!]_P$.

Examples Let $P \equiv p(a) \wedge \forall x(p(x) \Rightarrow (f(x)))$, the program

```
p(a).
p(f(X)) :- p(X).
```

Then $[\![\exists x p(x)]\!]_P = \{\left[{x \atop f^n(a)}\right]; n \geqslant 0\}$.

For a closed term A, either $[\![A]\!]_P = \varnothing$, if $P \nvdash A$, or $[\![A]\!]_P = \{\mathit{1}\}$, if $P \vdash A$, where *1* is the unit substitution. □

Herbrand models

It would be tempting to associate with a set of Horn clauses the product of 'all' its models, but it is a gigantic algebra! However, for universal formulas, the satisfiability problem can be restricted to Herbrand algebras. They are algebras whose restriction to the functional part of the signature is the algebra of closed terms. A Herbrand algebra $\mathcal{H}$ is defined by its *fact base*, a subset $B(\mathcal{H})$ of the set At_Σ of closed terms (called Herbrand base): $R^{\mathcal{H}}(t_1, \ldots, t_n) = 1$ if $Rt_1 \ldots t_n \in B(\mathcal{H})$.

Furthermore, if $\mathcal{H}$ and $\mathcal{H}'$ are two Herbrand algebras, there exists an algebra morphism from $\mathcal{H}$ to $\mathcal{H}'$ if and only if $B(\mathcal{H}) \subseteq B(\mathcal{H}')$. This morphism is then unique, i.e., the identity, which makes $\mathcal{H}$ a subalgebra of $\mathcal{H}'$ (see exercise 4).

A product of a family of Herbrand algebras $(\mathcal{H}_i)_{i\in I}$ is not a Herbrand algebra since its domain is $(\mathrm{T}_\Sigma)^I$ and not T_Σ. But the diagonal $\mathcal{D}_I$ of this product has an algebraic structure, isomorphic to the Herbrand algebra defined by the intersection of the fact bases $B(\mathcal{H}_i)$:

$$B(\mathcal{D}_I) = \cap_{i\in I} B(\mathcal{H}_i).$$

Since only Herbrand algebras are being considered, the correct notion of product is not the ordinary product of algebras, but, rather, the intersection of the bases of facts, which satisfies the same algebraic properties as a product. Because of all these good properties, Herbrand algebras are equated with subsets of At_Σ, as are products of Herbrand algebras with intersection in $\mathcal{P}(\mathrm{At}_\Sigma)$.

Every definite program has a Herbrand model: At_Σ itself satisfies all the definite clauses in an obvious way. Starting from a family of Herbrand *models* of a set Γ of Horn clauses, it is known (proposition 7.1) that a product of models is a model. The diagonal of the product is also a model, since a subalgebra of a model of a universal theory is a model of it. From an algebraic point of view, the category of Herbrand models of a set of Horn clauses *has products*. In particular, it has a terminal object, which is At_Σ. It also has an initial object, which is defined by the intersection of the Herbrand models of Γ.

Let P be a definite program, i.e, a finite set of definite clauses. Write $\mathcal{M}_P$ for the initial Herbrand model, the intersection of all the Herbrand models of P.

Proposition 7.3 $\mathcal{M}_P = \{A \in \mathrm{At}_\Sigma; P \models A\}$.

Proof P and A being universal, $P \models A$ if and only if $P \Rightarrow A$ is satisfied by all the Herbrand algebras, i.e., A is true in all the Herbrand models of P, which means that A belongs to the intersection of these models, which is P. □

The T_P operator

R. Kowalski and M. van Emden associate with a definite program P a monotone operator T_P over $\mathcal{P}(\mathrm{At}_\Sigma)$. Each clause A :- $A_1, \ldots, A_p$ of P generates the inference rules over At_Σ,

$$\frac{\theta A_1 \quad \ldots \quad \theta A_p}{\theta A},$$

for all closed substitutions θ. To this inference system corresponds a monotone operator $T_P : \mathcal{P}(\mathrm{At}_\Sigma) \to \mathcal{P}(\mathrm{At}_\Sigma)$,

$$T_P(\mathcal{H}) = \{\theta C^+; \theta : X \to \mathrm{T}_\Sigma \text{ and } \mathcal{H} \models \theta C^-\}.$$

Since the rules only have a finite number of premises, the operator T_P is finitary, hence continuous, and its least fixpoint is therefore

$$\mu T_P = T_P^{\uparrow\omega} = \bigcup_{n\geqslant 0} T_P^n(\varnothing).$$

(see proposition 2.6 and lemma 2.5).

Lemma 7.4 A Herbrand algebra $\mathcal{H}$ is a model of P if and only if $T_P(\mathcal{H}) \subseteq \mathcal{H}$.

Proof Let C be the clause A :- $A_1, \ldots, A_p$. $\mathcal{H} \models C$ if for every valuation ξ taking values in T_Σ, $A_1^{\mathcal{H}}(\xi) = 1, \ldots, A_p^{\mathcal{H}}(\xi) = 1$ imply $A^{\mathcal{H}}(\xi) = 1$. But, $A_i^{\mathcal{H}}(\xi) = 1$ means $\hat{\xi}(A_i) \in \mathcal{H}$. Hence $\mathcal{H} \models C$ if and only if $T_P(\mathcal{H}) \subseteq \mathcal{H}$. □

Proposition 7.5 $\mathcal{M}_P = \mathrm{Ind}(T_P) = T_P^{\uparrow\omega}$.

Proof Since T_P is continuous, $\mathrm{Ind}(T_P) = T_P^{\uparrow\omega}$. By definition,

$$\mathrm{Ind}(T_P) = \bigcap\{\mathcal{H}; T_P(\mathcal{H}) \subseteq \mathcal{H})\}.$$

But, $T_P(\mathcal{H}) \subseteq \mathcal{H}$ means that $\mathcal{H}$ is a model of P by lemma 7.4, and $\mathcal{M}_P$ is the intersection of the models of P. □

The operator T_P therefore allows the enumeration of the consequences of P by 'forward chaining'. It is an effective method of computation, but not efficient, since the essential aspect is to allow proofs by induction over P. Being monotone, T_P also has a greatest fixpoint $\bar{\mu}T_P$, obtained by transfinite induction of T_P starting from above (i.e, At_Σ): $\bar{\mu}T_P = T_P^{\downarrow\alpha}$ for a countable ordinal α, which can be strictly greater than ω.

Searching for a proof

The standard interpretation of logic programming is in conflict with the Curry–Howard correspondence, according to which a program corresponds to a proof and its specification corresponds to a formula. Here, a 'program' is a formula! But this nomenclature is in conflict with ordinary programming, since these programs are not executable. A query is needed to produce a result, although a program *could* be run by forward chaining, without a query, to obtain the set of its logical consequences. Rather, a logic program should be seen as a specification of the search for a proof, and it is normal that this specification be done with a formula. It is queries which play the role of programs, viewed as terms to be reduced. In fact, for a Curry–Howard correspondence of logic programs to exist would probably require unifying logic programming and functional programming, and, for this to be possible, would require a radical change of viewpoint.

Miller, Nadathur and Scedrov [109] have proposed a characterization of the deduction relations, adapted to logic programming.

A program specifies the search for a proof (not how the search must be done, but what the properties characterizing the search for a proof are) and must allow the extraction of a result from the proof:

($\wedge$) $P \vdash \varphi \wedge \psi$ if and only if $P \vdash \varphi$ and $P \vdash \psi$;
($\vee$) $P \vdash \varphi \vee \psi$ if and only if $P \vdash \varphi$ or $P \vdash \psi$;

($\Rightarrow$) $P \vdash \varphi \Rightarrow \psi$ if and only if $P, \varphi \vdash \psi$;
($\exists$) $P \vdash \exists x \varphi$ if and only if there exists a term t such that $P \vdash \varphi[x := t]$;
($\forall$) $P \vdash \forall x \varphi$ if and only if $P \vdash \varphi[x := y]$ for every bound variable y in P or $\forall x \varphi$.

These conditions are satisfied by Prolog, but only vacuously in the case of ($\vee$), ($\Rightarrow$) and ($\forall$) since queries do not use these connectives. Any fragment of first-order logic satisfying these requirements might be a candidate for extending logic programming.

7.2.2 Operational semantics of definite programs

The operational semantics of definite programs is given by SLD-resolution: it is the logical meaning given to them by Kowalski, as a refinement of the resolution principle, introduced by Robinson in 1965 [128], and since studied extensively as a method for automatic theorem-proving. SLD means *Selection, Linear, Definite*. Discovered and implemented independently by Colmerauer in the first Prolog, SLD-resolution has other non-logical readings, being similar to a form of rewriting controlled by equation resolution.

The resolution rule
In clausal logic, and in the restricted framework of logic programming, the resolution rule is (the formulas are implicitly universally quantified):

$$\frac{\neg L_1 \vee \ldots \vee \neg L_q \qquad A \Leftarrow A_1 \wedge \ldots \wedge A_p}{\neg\theta L_1 \vee \ldots \vee \neg\theta L_{s-1} \vee \neg\theta A_1 \vee \ldots \vee \neg\theta A_p \vee \neg\theta L_{s+1} \vee \ldots \neg\theta L_q},$$

where the L_i, A and A_j are atoms and θ is a most general unifier of L_s and A. Using the language of definite programs, if R is the query `?-` $L_1, \ldots, L_q$ and C is the clause A `:-` $A_1, \ldots, A_p$, the application of this rule consists of globally instantiating the query R with θ, of deleting the selected literal L_s and replacing it with the body of the clause, also instantiated with θ:

$$\frac{\texttt{?-}\ L_1, \ldots, L_q \qquad A\ \texttt{:-}\ A_1, \ldots, A_p}{\texttt{?-}\ \theta L_1, \ldots, \theta L_{s-1}, \theta A_1, \ldots, \theta A_p, \theta L_{s+1}, \ldots, \theta L_q}.$$

From a logical point of view, it is easy to derive this rule, for example in natural deduction, or more directly in the sequent calculus. Hence,

$$\begin{aligned}&\neg L_1 \vee \ldots \vee \neg L_q,\\ &A \Leftarrow A_1 \wedge \ldots \wedge A_p,\\ &\quad\vdash \neg\theta L_1 \vee \ldots \vee \neg\theta L_{s-1} \vee \neg\theta A_1 \vee \ldots \vee \neg\theta A_p \vee \theta L_{s+1} \vee \ldots \neg\theta L_q,\end{aligned}$$

and by contraposition

$$C \vdash \forall\vec{x}\theta(L_1 \wedge \ldots \wedge L_{s-1} \wedge A_1 \wedge \ldots \wedge A_p \wedge L_{s+1} \wedge \ldots \wedge L_q) \Rightarrow \exists\vec{x}(L_1 \wedge \ldots \wedge L_q).$$

There is even a strong soundness result, true in minimal logic, which has been expected since theorem 7.2.

Lemma 7.6 If $\sigma L_s = \sigma C^-$, then

$$C \vdash_M \forall(\sigma(L_1 \wedge \ldots \wedge L_{s-1} \wedge C^- \wedge L_{s+1} \wedge \ldots \wedge L_q)) \Rightarrow \forall(\sigma(L_1 \wedge \ldots \wedge L_q)).$$

Proof The notation is simplified to show in natural deduction that

$$C \vdash_M \forall(\sigma C^- \wedge \sigma L') \Rightarrow \forall(\sigma L \wedge \sigma L')$$

as soon as $\sigma L = \sigma C^+$:

$$\dfrac{\dfrac{\dfrac{\dfrac{\forall(C^- \Rightarrow C^+)}{\sigma C^- \Rightarrow \sigma C^+} \quad \dfrac{\dfrac{\forall(\sigma C^- \wedge \sigma L')}{\sigma C^- \wedge \sigma L'}}{\sigma C^-}}{\sigma C^+} \quad \dfrac{\dfrac{\forall(\sigma C^- \wedge \sigma L')}{\sigma C^- \wedge \sigma L'}}{\sigma L'}}{\sigma L \wedge \sigma L'}}{\forall(\sigma L \wedge \sigma L')}.$$

□

Constraint resolution

This inference will be described as a procedure to solve equations, *à la* Colmerauer, by making the environment explicit, as was done for functional programming. The rules of a program are used to construct incrementally a system of equations whose solution will be that of an initial query. A solution state is a pair $E \cdot R$ formed of a system E of equations to solve and a query R to delete. These states must be transformed, starting from $\varnothing \cdot R_0$, to obtain a state $E_n \cdot \top$, the solutions of E_n being the solutions of R_0. There are, *a priori*, two ways to advance towards a solution: transform a subset of the query into equations, and transform the system of equations into a more solved form.

Example Consider the addition program

```
p(U,0,U).
p(U,s(V),s(W)) :- p(U,V,W).
```

and the query

```
?- p(X,s(Y),s(s(0))).
```

By using the second rule, and then the first,

$$\begin{array}{l} \quad \varnothing \cdot p(X, s(Y), s(s(0))) \\ \rightarrow_P \{p(X, s(Y), s(s(0))) = p(U_1, s(V_1), s(W_1))\} \cdot p(U_1, V_1, W_1) \\ \rightarrow_P \{p(X, s(Y), s(s(0))) = p(U_1, s(V_1), s(W_1)), p(U_1, V_1, W_1) = p(U_2, 0, U_2)\} \cdot \top \end{array}$$

A solution for the last system is $X = U_1 = U_2 = W_1 = s(0)$, $Y = V_1 = 0$, which answers the query ?- $p(X, s(Y), s(s(0)))$ with the substitution $\left[\begin{smallmatrix} X & Y \\ s(0) & 0 \end{smallmatrix}\right]$. □

It is, however, awkward to form these systems blindly, waiting for $\top$ to be derived before starting to solve them. It is better to add the equation $L = A$ to E only if the system $E \cup \{L = A\}$ has a solution. This general scheme forms the basis of 'constraint programming'. In the particular case where the equations are solved over a syntactic domain, each system can be transformed into solved form before adding a new equation.

Example (cont.) The first system obtained,

$$\{p(X, s(Y), s(s(0))) = p(U_1, s(V_1), s(W_1))\},$$

is transformed into the equivalent system, in solved form:

$$\{X = U_1, Y = V_1, W_1 = s(0)\}.$$

Does the second equation, $p(U_1, V_1, W_1) = p(U_2, 0, U_2)$, have a solution in this 'environment'? This question is equivalent to solving $p(U_1, V_1, s(0)) = p(U_2, 0, U_2)$, which means solving $U_1 = U_2 = s(0)$, $V_1 = 0$. □

The only pairs $E \cdot R$ to be considered are therefore those where E is in solved form, i.e, is an idempotent substitution ρ. Furthermore, only those bindings in ρ which are of interest need be preserved, i.e., those pertaining to the variables of the original query. The others can be eliminated by applying them to the current query. One can then suppose the dom(ρ) and var(R) are disjoint.

Example (cont.)

$$\begin{aligned} & \{X = U_1, Y = V_1, W_1 = s(0)\} \cdot p(U_1, V_1, W_1) \\ \equiv\ & \{X = U_1, Y = V_1\} \cdot p(U_1, V_1, s(0)). \end{aligned}$$

□

If the query has several literals, as in

```
?- p(X,Y,s(s(s(s(0))))), p(X,s(s(0)),Y).
```

there is another degree of freedom: will the equations be added one by one by selecting one literal each time, or will the equations defined by all the literals be formed before they are solved? In the following example, the second rule is applied to each of the two literals, and the system is then solved globally:

$$\begin{aligned} & \varnothing \cdot p(X, Y, s(s(s(s(0))))), p(X, s(s(0)), Y). \\ \longrightarrow_P\ & \{p(X, Y, s(s(s(s(0))))) = p(U_1, s(V_1), s(W_1)), \\ & p(X, s(s(0)), Y) = p(U_2, s(V_2), s(W_2))\} \\ & \cdot p(U_1, V_1, W_1), p(U_2, V_2, W_2) \\ \longrightarrow_P\ & \{Y = s(V_1)\} \cdot p(X, V_1, s(s(s(0)))), p(X, s(0), V_1). \end{aligned}$$

This situation is comparable to that of choosing a redex to be contracted in the λ-calculus. The *confluence* property showed that this choice, so long as it remains free at each step, does not affect the result. However, the reduction strategy was not neutral with respect to termination. Here there is a *switching* property, to be proven later on, which follows from the fact that a system of equations can be solved in any order. It is easier to work by selecting a single literal each time, and one can even impose a selection rule for all queries—SLD-resolution—which will be defined using the following concepts.

Definition 7.2

- A *(logical) environment* ρ is an idempotent substitution. Write $\mathit{1}$ for the identity substitution of empty domain.
- A *query* is a list of atoms, written ?- $L_1, \ldots, L_q$. The empty query is written $\top$.
- A *solution state* is a pair $\rho \cdot R$, where ρ is an environment, R is a query and $\mathrm{dom}(\rho) \cap \mathrm{var}(R) = \varnothing$. A state is *initial* if it is of the form $\mathit{1} \cdot R$ and is a *success state* if it is of the form $\rho \cdot \top$.

The component ρ represents a partial computation of the solution, and the derived query R indicates in which direction ρ must grow to reach a solution.

Normalizing the resolution procedure in this way is a simplification which will have to be abandoned in less sequential implementations, where a unification can be 'suspended' so long as a certain condition is not satisfied. The most general unifier cannot then be computed immediately, and the bindings are only 'broadcast' to certain variables.

SLD-derivations
Let P be a definite program, i.e, a list of definite clauses. The program P defines a transition relation between states tagged by pairs (s, C), where s is a literal occurrence and C a clause. This relation will be written $\longrightarrow_P$, or simply $\longrightarrow$.

A *separate variant* of a clause is a clause renamed using variables hereto unused: sometimes phrases such as 'separate from $\mathrm{var}(R)$', or simply 'separate from R', will be used. In what follows, let V be a fixed set of variables. It will be used as the set of variables of an initial query.

Definition 7.3 Let $\rho \cdot R$ be a state with $R \neq \top$; $R/s = L_s$ be the literal of R at occurrence s; and C be a variant of a clause of P, separate from R and V, with body $C^- = A_1, \ldots, A_p$ and head C^+. Form the system $\rho \cup \{R/s = C^+\}$, equating the substitution ρ to the system in reduced form which defines it and for which it is a most general unifier.

If the system has no solution, there does not exist a tagged transition (s, C) from $\rho \cdot R$.

Otherwise, every solution of $\rho \cup \{R/s = C^+\}$ is a solution of ρ, hence an instance $\mu\rho$ of ρ. By the separation assumption on C, and by the definition of states, ρ does not operate over $R/s = C^+$. Hence μ is a solution of $R/s = C^+$. So a most general unifier of $\rho \cup \{R/s = C^+\}$ is of the form $\theta\rho$, where θ is a most general unifier of $R/s = C^+$. Then define $\rho \cdot R \xrightarrow{s,C}_P \rho' \cdot R'$, where

$$\begin{aligned} \rho' &= \theta\rho, \\ R' &= \theta L_1, \ldots, \theta L_{s-1}, \theta A_1, \ldots, \theta A_p, \theta L_{s+1}, \ldots, \theta L_q. \end{aligned}$$

If $q = 1$ and $p = 0$ let $R' = \top$; it is a success state. The literals $\theta A_1, \ldots, \theta A_p$ are *introduced* by this transition, and θL_i, $i \neq s$ is the *residue* of L_i.

Note that this concept of residue is much simpler than that of β-reduction, where there can also be reduction and multiplication of redexes.

This rule is written in an abbreviated manner:

$$\frac{\{R/s = C^+\} \overset{\star}{\rightsquigarrow} \theta}{\rho \cdot R \longrightarrow_P \theta\rho \cdot \theta(R[s \leftarrow C^-])}.$$

Some of the terminology of reduction relations (from Chapter 2) will be used.

An *(SLD-)derivation* is a finite or infinite sequence

$$\rho_0 \cdot R_0 \xrightarrow{s_0, C_0}_P \ldots \xrightarrow{s_{n-1}, C_{n-1}}_P \rho_n \cdot R_n \xrightarrow{s_n, C_n}_P \ldots .$$

Note that each $\rho_i \cdot R_i$ of this derivation is of the form:

$$\theta_i \ldots \theta_1 \rho_0 \cdot \theta_i \ldots \theta_1 \rho_0 R_0 [s_0 \leftarrow C_0^-] \ldots [s_{i-1} \leftarrow C_{i-1}^-].$$

A derivation can be infinite. If it is not, it succeeds if it halts in a success state $\rho_n \cdot \top$, where n is the length of the derivation.

Definition 7.4 The *result substitution* of a successful derivation $\mathit{1} \cdot R_0 \xrightarrow{\star} \rho_n \cdot \top$, or *answer* to R_0, is the restriction to $\mathrm{var}(R_0)$ of ρ_n. The *result query* is $\rho_n R_0$.

In the definition of a resolution step, it would be possible to restrict the environment component of each state to V, but the following presentation would be unnecessarily complicated.

Fundamental properties

The $\longrightarrow_P$ relation satisfies several important properties. The logical ones are strong (constructive) soundness and completeness, and the others are particular to this form of computation: generalization and switching.

Soundness

Theorem 7.7 (Strong soundness) If $\rho \cdot R \xrightarrow{\star}_P \rho'\rho \cdot R'$, then $P \vdash_M \forall(R') \Rightarrow \forall(\rho' R)$. In particular, the result substitution ρ of a successful derivation is a solution of R.

Proof By induction on the length of the derivation. The null length derivation case is trivial. Let

$$\rho \cdot R \xrightarrow{s,C}_P \theta\rho \cdot \theta\rho R[s \leftarrow C^-] \xrightarrow{\star}_P \rho'\theta\rho \cdot R'$$

be a derivation. By the inductive hypothesis,

$$P \vdash_M \forall(R') \Rightarrow \forall(\rho'\theta\rho R[s \leftarrow C^-]).$$

Furthermore, θ is a solution of $R/s = C^+$, as is $\rho'\theta$. Lemma 7.6 applies to the first transition, giving

$$P \vdash_M \forall(\rho'\theta\rho R[s \leftarrow C^-]) \Rightarrow \forall(\rho'\theta\rho R).$$

By transitivity, $P \vdash_M \forall(R') \Rightarrow \forall(\rho'\theta\rho R)$. □

Note that the soundness of SLD-resolution does not depend on the choice of most general unifier: every solution to the equations $R/s = C^+$ is satisfactory.

This result justifies the concept of a qualified answer to a query, which can be used for the analysis of programs (*debugging*), or, in the case of a failed derivation, to find the last constructed state. If $\mathit{1} \cdot R \xrightarrow{\star} \rho \cdot R'$, then the pair (ρ, R') is a *qualified* (or conditional) *answer*: if $\forall(R')$ is true, ρ is a solution.

By ignoring the result ρ, the theorem states that a successful derivation of R constitutes a proof of $P \models \exists\vec{x}R$, i.e. it is a refutation of $P \wedge \neg(\exists\vec{x}R)$, or a refutation of the clausal formula $P \wedge R^\vee$. By omitting the program, a successful derivation from $\mathit{1} \cdot R$ is called a *refutation* of the goal $R^\vee$.

A successful derivation of R computes a solution to R by successive approximations, using the increasing sequence $R \leqslant \rho_1 R \leqslant \rho_2 R \leqslant \ldots \leqslant \rho_n R$ until ρ_n is a solution.

Expansions of a query A successful derivation provides a solution to a query. An arbitrary finite derivation can only provide a conditional solution. An infinite derivation allows a finer and finer approximation, assuming it is 'fair'.

Definition 7.5 A derivation is *fair* if it is finite, or if for every n and every literal L appearing in it, L or one of its residues is ultimately selected.

The Herbrand expansion $\mathrm{Exp}(\exists(R), \mathrm{T}_\Sigma)$ would be the conjunction of all the closed instances of the literals of R, which is infinite if T_Σ is infinite. Instead, the set of instances is used, written $\mathrm{Exp}(R)$: it is the *expansion* of query R. Since the formula is existential, $\mathrm{Exp}(R) \vdash_M \exists(R)$.

Proposition 7.8

1. If the successful derivation $\mathit{1} \cdot R \xrightarrow{\star} \rho_m \cdot \top$ is of length m, $\mathrm{Exp}(\rho_m R) \subseteq T_P^{\uparrow m}$.
2. If the infinite derivation $\mathit{1} \cdot R \longrightarrow \ldots \longrightarrow \rho_n \cdot R_n \longrightarrow \ldots$ is fair, then for every m there exists n such that $\mathrm{Exp}(\rho_n R) \subseteq T_P^{\downarrow m}$.

Proof

(1) By induction on m.

If $m = 0$, then $R \equiv \top$ and $\rho_0 = \mathit{1}$. Hence $\mathrm{Exp}(\top) = \varnothing = T_P^{\uparrow 0}$.

If $m > 0$, consider the derivation of length m,

$$\mathit{1} \cdot R \xrightarrow{s,C} \theta \cdot \theta R[s \leftarrow C^-] \xrightarrow{\star} \rho\theta \cdot \top.$$

Then $\mathrm{Exp}(\rho\theta R) = \mathrm{Exp}(\rho\theta(R \setminus s)) \cup \mathrm{Exp}(\rho\theta C^+)$. By the inductive hypothesis applied to the successful derivation from $\theta R[s \leftarrow C^-]$, and by the monotonicity of T_P,

$$\mathrm{Exp}(\rho\theta(R \setminus s)) \subseteq \mathrm{Exp}(\rho\theta R[s \leftarrow C^-]) \subseteq T_P^{\uparrow m-1} \subseteq T_P^{\uparrow m}$$
$$\mathrm{Exp}(\rho\theta C^-) \subseteq \mathrm{Exp}(\rho\theta R[s \leftarrow C^-]) \subseteq T_P^{\uparrow m-1},$$

T_P can be applied to obtain C^+ from C^-,

$$\mathrm{Exp}(\rho\theta C^+) \subseteq T_P(T_P^{\uparrow m-1}) = T_P^{\uparrow m}.$$

It follows that $\mathrm{Exp}(\rho\theta R) \subseteq T_P^{\uparrow m}$.

(2) By induction on m.

The case $m = 0$ is trivial: $\mathrm{Exp}(R) \subseteq T_P^{\downarrow 0} = \mathrm{At}_\Sigma$.

Let $m > 0$. Let i be one of the literal occurrences of R. By the fairness assumption, there exists a p such that R/i is selected at the p-th step of the derivation,

$$\mathit{1} \cdot R \xrightarrow{\star} \rho_p \cdot \rho_p R_p \xrightarrow{s,C} \theta\rho_p \cdot \theta\rho_p R_p[s \leftarrow C^-,$$

with $\theta\rho_p R_p/i = \theta C^+$. By the inductive hypothesis applied to the (also fair) derivation from the $(p+1)$-th step and to $m-1$, there exists q such that

$$\begin{aligned}\theta\rho_p \cdot \theta\rho_p R_p[s \leftarrow C^-] &\xrightarrow{\star} \rho'_q\theta\rho_p \cdot R'_{p+1+q},\\ \mathrm{Exp}(\rho'_q\theta\rho_p R_p[s \leftarrow C^-]) &\subseteq T_P^{\downarrow m-1}.\end{aligned}$$

It follows that:

$$\mathrm{Exp}(\rho'_q\theta\rho_p C^-) \subseteq \mathrm{Exp}(\rho'_q\theta\rho_p R_p[s \leftarrow C^-]) \subseteq T_P^{\downarrow m-1},$$
$$\mathrm{Exp}(\rho'_q\theta\rho_p(R/i)) \subseteq T_P(\mathrm{Exp}(\rho'_q\theta\rho_p C^-)) \subseteq T_P(T_P^{\downarrow m-1}) = T_P^{\downarrow m}.$$

Let $n_i = p+1+q$, so that $\mathrm{Exp}(\rho_{n_i}(R/i)) \subseteq T_P^{\downarrow m}$, with $\rho_{n_i} = \rho'_q\theta\rho_p$. Let $n = \max_i n_i$: since $\rho_{n_i} \leqslant \rho_n$, $\mathrm{Exp}(\rho_n(R/i)) \subseteq \mathrm{Exp}(\rho_{n_i}(R/i))$, hence:

$$\mathrm{Exp}(\rho_n R) = \bigcup_i \mathrm{Exp}(\rho_n(R/i)) \subseteq T_P^{\downarrow m}.$$

□

Since $\mathrm{Exp}(A) = \{A\}$ if A is a closed atom, the following corollary is obtained.

Corollary 7.9 Let $A \in \mathrm{At}_\Sigma$.

- If A has a successful derivation, then $A \in T_P^{\uparrow\omega}$.
- If A has a fair infinite derivation, then $A \in T_P^{\downarrow\omega}$.

Switching So the derivations $\mathbf{1} \cdot R \xrightarrow{\star}_P \rho \cdot \top$ must be constructed. This solution search is *non-deterministic*: at each step, a literal and a clause must be selected. The following theorem ensures that the non-determinism in the choice of literal does not matter: this is called *don't care* non-determinism.

Theorem 7.10 (Switching) If in a query R, the literals L_{s_1} and the residue of L_{s_2} are successively deleted, they can also be deleted in the reverse order, and the derived states are the same up to renaming of variables.

Proof Suppose that the literals L_{s_1} and L_{s_2} of R are selected in the following order:

$$\begin{array}{rl} & \rho \cdot \ldots, L_{s_1}, \ldots, L_{s_2}, \ldots, \\ \xrightarrow{1}_P & \theta_1\rho \cdot \theta_1(\ldots, C_1^-, \ldots, L_{s_2}, \ldots), \\ \xrightarrow{2}_P & \theta_2\theta_1\rho \cdot \theta_2\theta_1(\ldots, C_1^-, \ldots, C_2^-, \ldots). \end{array}$$

One can start by selecting L_{s_2} in R, to show that $\theta_2\theta_1 L_{s_2} = \theta_2 C_2^+ = \theta_2\theta_1 C_2^+$ holds (since, by assumption, R, C_1 and C_2 are separate, hence θ_1 leaves C_2^+ invariant): $\theta_2\theta_1$ is a solution to the equation $L_{s_2} = C_2^+$, which therefore has a most general unifier θ_2', and there exists a substitution μ_2 such that

$$\theta_2\theta_1 = \mu_2\theta_2'. \tag{7.1}$$

The selection of L_{s_2} in R succeeds:

$$\begin{array}{rl} & \rho \cdot \ldots, L_{s_1}, \ldots, L_{s_2}, \ldots, \\ \xrightarrow{2}_P & \theta_2'\rho \cdot \theta_2'(\ldots, L_{s_1}, \ldots, C_2^-, \ldots). \end{array}$$

One can then select $\theta_2' L_{s_1}$. Because $\theta_1 L_{s_1} = \theta_1 C_1^+$, from (7.1), and by the separation assumption, it follows that:

$$\mu_2\theta_2' L_{s_1} = \theta_2\theta_1 L_{s_1} = \theta_2\theta_1 C_1^+ = \mu_2\theta_2' C_1^+ = \mu_2 C_1^+.$$

Hence, μ_2 is a solution to $\theta_2' L_{s_1} = C_1^+$, which therefore has a most general unifier θ_1', and there exists a substitution ν_1 such that

$$\mu_2 = \nu_1\theta_1'. \tag{7.2}$$

Hence the selection of $\theta_2' L_{s_1}$ succeeds in turn:

$$\begin{array}{rl} & \rho \cdot \ldots, L_{s_1}, \ldots, L_{s_2}, \ldots, \\ \xrightarrow{2}_P & \theta_2'\rho \cdot \theta_2'(\ldots, L_{s_1}, \ldots, C_2^-, \ldots), \\ \xrightarrow{1}_P & \theta_1'\theta_2'\rho \cdot \theta_1'\theta_2'(\ldots, C_1^-, \ldots, C_2^-, \ldots). \end{array}$$

The two derived queries,

$$\theta_2\theta_1\rho \cdot \theta_2\theta_1(\ldots, C_1^-, \ldots, C_2^-, \ldots),$$

and

$$\theta_1'\theta_2'\rho \cdot \theta_1'\theta_2'(\ldots, C_1^-, \ldots, C_2^-, \ldots),$$

must still be shown to be variants.

So, by equations (7.1) and (7.2): $\theta_2\theta_1 = \mu_2\theta_2' = \nu_1\theta_1'\theta_2'$, i.e, $\theta_2\theta_1 \geqslant \theta_1'\theta_2'$.

Conversely, the reasoning leading to μ_2 and ν_1 can be done symmetrically. First, $\theta_1'\theta_2'L_{s_1} = \theta_1'C_1^+ = \theta_1'\theta_2'C_1^+$: $\theta_1'\theta_2'$ is a solution of $L_{s_1} = C_1^+$, hence

$$\theta_1'\theta_2' = \mu_1\theta_1 \tag{7.3}$$

for a substitution μ_1. Second, from $\theta_2'L_{s_2} = \theta_2'C_2^+$, from (7.3), and by the separation assumption, it follows that:

$$\mu_1\theta_1L_{s_2} = \theta_1'\theta_2'L_{s_2} = \theta_1'\theta_2'C_2^+ = \mu_1\theta_1C_2^+ = \mu_1C_2^+,$$

i.e, μ_1 is a solution of $\theta_1L_{s_2} = C_2^+$, hence

$$\mu_1 = \nu_2\theta_2 \tag{7.4}$$

for a substitution ν_2. Therefore, from (7.3) and (7.4), $\theta_1'\theta_2' = \mu_1\theta_1 = \nu_2\theta_2\theta_1$, i.e, $\theta_1'\theta_2' \geqslant \theta_2\theta_1$.

To conclude, $\theta_2\theta_1$ and $\theta_1'\theta_2'$ are variants. □

Remark The switching property is weaker than confluence. Let P be the program $p(a, a)$, $p(b, b)$. From the same query, the two derivations:

$$\boldsymbol{1} \cdot p(a, X), p(b, X) \xrightarrow{1,1}_P \begin{bmatrix} X \\ a \end{bmatrix} \cdot p(b, a),$$

$$\boldsymbol{1} \cdot p(a, X), p(b, X) \xrightarrow{2,2}_P \begin{bmatrix} X \\ b \end{bmatrix} \cdot p(a, b),$$

are not confluent. □

However, the choice of clause is done by a *don't know* non-determinism. Since all the possible successful derivations must be accounted for, they are organized in the form of a tree.

A *search state* is a pair $(\rho \cdot R, s)$, where $\rho \cdot R$ is a resolution state and s is a literal occurrence in R. When $R = \top$, there is no occurrence to indicate. A *search* (or resolution) *tree* for R_0 is a tree whose nodes are tagged by the search states. The initial state $\boldsymbol{1} \cdot R_0$ appears at the root and a non-terminal $(\rho \cdot R, s)$ has as child all of the $\rho' \cdot R'$ such that $\rho \cdot R \xrightarrow{s,C}_P \rho' \cdot R'$ for each clause C of P. Since P is a *list* of definite clauses, the tree is ordered. A terminal node is either a success state $\rho \cdot \top$, or a failure state $(\rho \cdot R, s)$ in the case where there is no transition $\rho \cdot R \xrightarrow{s,C}_P \ldots$, i.e, for each clause C, equation $R/s = C^+$ has no solution. The same query can have search trees of many forms, finite or infinite. Figure 7.3 on page 277 shows a 'standard' search tree for a program computing paths in a graph.

Completeness It will be shown that all the solutions of a query R can be obtained by constructing an arbitrary search tree for R.

Lemma 7.11 (Generalization) If $\mathbf{1} \cdot \lambda R \xrightarrow{\star} \rho \cdot \rho\lambda R'$, then there exists a substitution $\sigma \leqslant \rho\lambda$ and a derivation $\mathbf{1} \cdot R \xrightarrow{\star} \sigma \cdot \sigma R'$.

Proof By induction on the length of the derivation from $\mathbf{1} \cdot \lambda R$.

The property is trivial for a derivation of length zero, with $\sigma = \mathbf{1} \leqslant \lambda$.

Consider a derivation

$$\mathbf{1} \cdot \lambda R \xrightarrow{\star} \rho \cdot \rho\lambda R' \xrightarrow{s,C}_P \theta\rho \cdot \theta\rho\lambda R'[s \leftarrow C^-].$$

By the inductive hypothesis, there exists a derivation

$$\mathbf{1} \cdot R \xrightarrow{\star} \sigma \cdot \sigma R', \tag{7.5}$$

with $\sigma \leqslant \rho\lambda$, i.e, $\rho\lambda = \mu\sigma$, for a particular substitution μ. Rename the variables of C outside the domain of μ, so that $\mu C^+ = C^+$. From, $\theta\rho\lambda(R'/s) = \theta C^+ = \theta\mu C^+$, it follows that the equation $\sigma(R'/s) = C^+$ has a solution, $\theta\mu$. If θ' is a most general solution for it, $\theta\mu = \nu\theta'$ for a particular substitution ν. By equation (7.5), the derivation can therefore be extended by

$$\sigma \cdot \sigma R' \xrightarrow{s,C}_P \theta'\sigma \cdot \theta'\sigma R'[s \leftarrow C^-],$$

with $\theta\rho\lambda = \theta\mu\sigma = \nu\theta'\sigma \geqslant \theta'\sigma$. □

For the $R' = \top$ case, there is

Corollary 7.12 If $\mathbf{1} \cdot \lambda R \xrightarrow{\star} \rho \cdot \top$, then $\mathbf{1} \cdot R \xrightarrow{\star} \sigma \cdot \top$ for a $\sigma \leqslant \rho\lambda$.

The following lemma states the completeness of SLD-resolution, restricted to closed atoms.

Lemma 7.13 If $L \in \mathcal{M}_P$, there exists a successful derivation $\mathbf{1} \cdot L \xrightarrow{\star}_P \mathbf{1} \cdot \top$.

Proof From proposition 7.5, $\mathcal{M}_P$ is the subset of At_Σ inductively defined by P, which allows a proof by induction over P.

If $L \equiv \theta A$ is an instance of a unit clause of P, then $\mathbf{1} \cdot L \xrightarrow{1}_P \mathbf{1} \cdot \top$.

If $L \equiv \theta A$, for a clause A :- $A_1, \ldots, A_p$, and if $P \models \theta A_1, \ldots, \theta A_p$, there exists, by the inductive hypothesis, successful derivations $\mathbf{1} \cdot \theta A_i \xrightarrow{\star} \mathbf{1} \cdot \top$ for each i. By combining these successful derivations in an arbitrary order, one obtains:

$$\begin{aligned} & \mathbf{1} \cdot \theta A_1, \theta A_2, \ldots, \theta A_p \\ \xrightarrow{\star} \; & \mathbf{1} \cdot \theta A_2, \ldots, \theta A_p \\ \xrightarrow{\star} \; & \mathbf{1} \cdot \theta A_p \\ \xrightarrow{\star} \; & \mathbf{1} \cdot \top. \end{aligned}$$

Let $\theta' = \theta_{\restriction \mathrm{var}(A)}$. It is the most general unifier of θA and A. So there is a transition

$$\mathit{1} \cdot \theta A \longrightarrow \theta' \cdot \theta' A_1, \ldots, \theta' A_p.$$

Since $\theta' \leqslant \theta$, by the generalization lemma, there exists a derivation

$$\mathit{1} \cdot \theta' A_1, \ldots, \theta' A_p \xrightarrow{\star} \mathit{1} \cdot \top,$$

and by combining these derivations,

$$\mathit{1} \cdot \theta A \longrightarrow \theta' \cdot \theta' A_1, \ldots, \theta' A_p \xrightarrow{\star} \theta' \cdot \top,$$

where the substitution θ' has as domain a set of variables not appearing in θA, and so can replaced by *1* in the final result. □

This lemma can be extended to $L \in \mathrm{At}_\Sigma[X]$, by considering the variables of L as constants, which means working in $\Sigma \cup \mathrm{var}(L)$ instead of Σ:

Lemma 7.14 Let $L \in \mathrm{At}_\Sigma[X]$. If $P \models L$, there exists a successful derivation $\mathit{1} \cdot L \xrightarrow{\star} \mathit{1} \cdot \top$.

Theorem 7.15 (Strong completeness) If $\theta \in [\![R]\!]_P$, there exists a successful derivation $\mathit{1} \cdot R \xrightarrow{\star} \rho \cdot \top$ such that $\rho \leqslant \theta$.

Proof By definition, $\theta \in [\![R]\!]_P$ means that $P \models \theta R$. If $R \equiv L_1, \ldots, L_q$, it follows that $P \models \theta L_i$ for every i. From lemma 7.14, there exists for each i a successful derivation $\mathit{1} \cdot \theta L_i \xrightarrow{\star} \mathit{1} \cdot \top$. Combining these successful derivations yields: $\mathit{1} \cdot \theta R \xrightarrow{\star} \mathit{1} \cdot \top$. From the generalization lemma, there exists a successful derivation $\mathit{1} \cdot R \xrightarrow{\star} \rho \cdot \top$, with $\rho \leqslant \theta$. □

To obtain the existence of a result $\rho \leqslant \theta$ in an arbitrary search tree, it suffices to apply the switching theorem.

Figure 7.1 illustrates the completeness theorem by representing, for a query R, three answers θR, $\theta' R$ and $\theta'' R$, partial solutions $\rho_1 R$ and $\rho_2 R$, and a solution σR which is an instance of one of the answers.

7.3 Control structures

Going from the general framework of logic programming to Prolog consists of choosing a selection rule and a search strategy. Once those choices are known, a programmer can think procedurally, and even ignore the logical basis.

7.3.1 Selection and searching

A *selection rule* is a mapping associating with each derivation $\mathit{1} \cdot R \xrightarrow{\star} \rho \cdot R'$ a literal occurrence in R'. Each selection rule S defines a unique search tree, whose

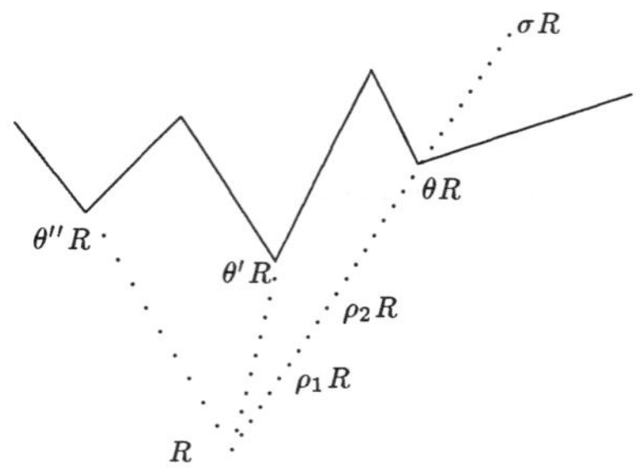

Figure 7.1 Solution space of a query

internal nodes are tagged by $(\rho \cdot R, S(\varpi))$, where ϖ is the derivation corresponding to the path from the root of the tree to the node.

By the switching theorem, adopting a selection rule is not a restriction: if there is a successful SLD-derivation for R, there is also a successful derivation using S.

The standard Prolog rule selects the left literal of each query: $R/1$. It is a *local* rule: its result depends only on the final state of the derivation. Note that if a query reappears in a derivation, it is always the same literal which is selected. Although an important simplification, it tends to—when combined with depth-first traversal of the tree—create infinite loops, as does call by value in functional programming.

Although an important simplification, this rule tends to create infinite branches in the search tree. Any traversal of such a tree implies the non-termination of the query-solving; moreover, an 'eager' traversal, which steadily follows an infinite branch, prevents any answer from being issued, as does call by value in functional programming. Completeness can only be guaranteed by traversing the search tree in an exhaustive manner, to enumerate all of the answers: this is performed, for instance, by breadth-first traversal.

A *search strategy* is necessary to define how the search tree is to be traversed. In Prolog, the standard strategy is 'depth-first'. This choice allows one not to construct the entire tree, and to use a stack whose elements are the successively pushed states of a branch of the tree. They are popped while *backtracking*.

This choice of search strategy makes the completeness property of SLD-resolution irrelevant for the user: by traversing the tree depth-first, no solution to the right of an infinite branch will be found. To avoid this situation, the order of clauses and of literals within a clause can sometimes be changed.

7.3.2 The activation-box model

L. Byrd introduced an execution model which is used by most Prolog systems in *trace* and *debug* mode.

To each node $(\rho \cdot R, s)$ of the search tree corresponds a call R/s of the procedure associated with the predicate $R(s)$ (in the standard case, $s = 1$). Each call R/s is represented as an *activation box* with four gates: `call`, `exit`, `redo` and `fail`, through which the control flow passes, as shown in Figure 7.2. For a given box, the trace of the flow belongs to the language defined by the rational expression `call(exit|redo)*fail`.

`call` the (initial) activation of the procedure, before unification with a clause head; a number is assigned to the box;

`exit` successful exit from the procedure, all the calls created after the `call` entry were successful; corresponds to a partial derivation of the form $(\rho \cdot R, s) \xrightarrow{\star} (\theta\rho \cdot \theta(R \setminus s), \_)$;

`redo` call reentering the (already active) procedure triggering a unification attempt of R/s with a new clause of the procedure;

`fail` (final) exit from the procedure.

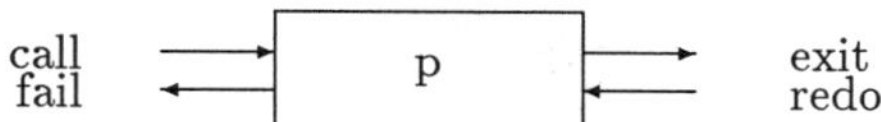

Figure 7.2 An activation box

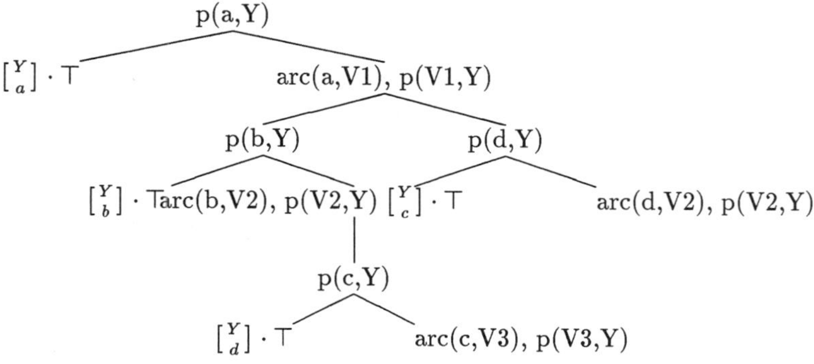

Figure 7.3 A standard search tree

Graphs The search tree in Figure 7.3 is a standard example from the relational calculus. A predicate `arc/2`, specifying a graph, is given extensionally, for example:

```
arc(a,b).
arc(b,c).
arc(a,d).
```

The predicate path/2 states the existence of a (directed) path between two vertices of the graph:

```
path(U,U).
path(U,W) :- arc(U,V), path(V,W).
```

In this definition, the order of the rules, and for the second rule, the order of the literals in its right-hand side, affect the results:

```
?- path(a,Y).
```

The search tree is finite. The execution is visible in *trace* mode:

```
        (0) 0 call: path(a, _VGML)?
        (0) 0 exit: path(a, a)?
    Y = a ;
        (0) 0 redo: path(a, _VGML)?
        (1) 1 call: arc(a, _VGTW)?
        (1) 1 exit: arc(a, b)?
        (2) 1 call: path(b, _VGML)?
        (2) 1 exit: path(b, b)?
        (0) 0 exit: path(a, b)?
    Y = b ;
        (0) 0 redo: path(a, _VGML)?
        (2) 1 redo: path(b, _VGML)?
        (3) 2 call: arc(b, _VHBX)?
        (3) 2 exit: arc(b, c)?
        (4) 2 call: path(c, _VGML)?
        (4) 2 exit: path(c, c)?
        (2) 1 exit: path(b, c)?
        (0) 0 exit: path(a, c)?
    Y = c ;
        (0) 0 redo: path(a, _VGML)?
        (2) 1 redo: path(b, _VGML)?
        (4) 2 redo: path(c, _VGML)?
        (5) 3 call: arc(c, _VHJY)?
        (5) 3 fail: arc(c, _VHJY)?
        (4) 2 fail: path(c, _VGML)?
        (3) 2 redo: arc(b, _VGML)?
        (3) 2 fail: arc(b, _VHBX)?
        (2) 1 fail: path(b, _VGML)?
        (1) 1 redo: arc(a, _VGTW)?
        (1) 1 exit: arc(a, d)?
        (2) 1 call: path(d, _VGML)?
        (2) 1 exit: path(d, d)?
        (0) 0 exit: path(a, d)?
    Y = d ;
```

```
    (0) 0 redo: path(a, _VGML)?
    (2) 1 redo: path(d, _VGML)?
    (3) 2 call: arc(d, _VHBX)?
    (3) 2 fail: arc(d, _VHBX)?
    (2) 1 fail: path(d, _VGML)?
    (1) 1 redo: arc(a, _VGTW)?
    (1) 1 fail: arc(a, _VGTW)?
    (0) 0 fail: path(a, _VGML)?
fail.
```

The integer appearing in parentheses uniquely designates a predicate call. The following integer is the number of ancestors in its derivation. Variables starting with a _ are newly generated variables. Each line corresponds to the input or output of a process through one of the four gates.

Should the second rule be changed to be left-recursive,

```
path(U,W) :- path((U,V), arc(V,W).
```

the search tree has an infinite branch formed of queries

$$\texttt{path}(a, V_n), \texttt{arc}(V_n, V_{n-1}), \ldots, \texttt{arc}(V_2, V_1), \texttt{arc}(V_1, Y)$$

such that the other branches (always to its left) fail as soon as $n > 2$, since there is no path of length $\geqslant 2$ in this example. After having enumerated the four solutions, the interpreter chooses the infinite branch and stays there. If the two clauses defining **path** are permuted, the tree is transformed by reflection, and the infinite branch is now leftmost, so the interpreter will give no answer.

7.3.3 Cuts

It is the cut which ensured Prolog's success, and it is as controversial as the GOTO and COMMON of FORTRAN. Attempts have been made to replace it with more elaborate control structures (coroutines in Prolog–II+ and NU–Prolog), more symmetric ones (commitment in concurrent programming), or more logical ones (Lloyd's declarative pruning procedure).

The syntax for a cut is !, and it can appear anywhere in the body of a rule. It has no declarative meaning. The choice of word, cut, is perhaps not the best, since the basic mechanism of Prolog is the cut of the sequent calculus.

The cut is a global control structure, which only affects the search tree. At the local transition level, a cut succeeds as soon as it is selected.

In the following discussion, the environment component will be omitted, as it plays no role. Let $R_0 \to R_1$ be a transition such that the left literal of R_1 triggers a rule A :- $\ldots, !, \ldots$. Then $R_0 \to R_1 \to (\ldots, !, \ldots), R_2$. If there exists a derivation

$$R_0 \to R_1 \to (\ldots, !, \ldots), R_2 \xrightarrow{\star} !, R_n$$

(all the queries to the left of ! have been deleted), then the search continues normally below $!, R_n$, whose unique descendent is R_n. If the standard traversal backtracks to R_n, then the control immediately passes to R'_1m which is the query derived from R_0 which follows R_1, according to the standard selection rules. The preceding clauses between R_1 and R_n are not active: the ! therefore prunes certain branches of the tree, and short-circuits the normal traversal by passing from R_n to R'_1 (Figure 7.4). Should one of the literals to the left of the ! not succeed, the

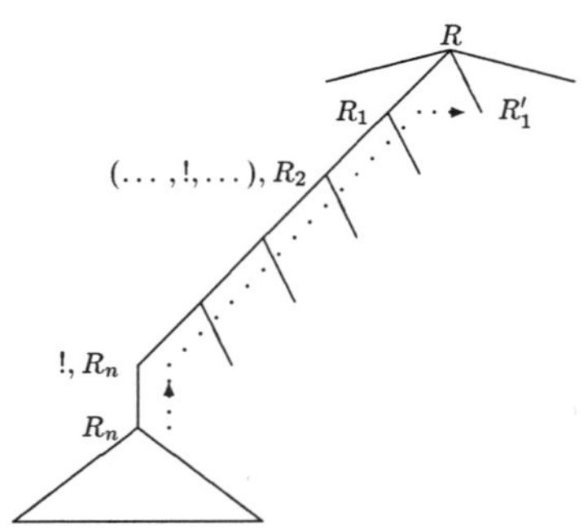

Figure 7.4 The cut

search continues normally.

Example In the first position of a rule A :- $!, \dots$, the rules which follow which could be activated are not:

```
member(X, [X|_]) :- !.
member(X, [_|Y]) :- member(X,Y).
```

□

Although all the solutions to a query can be found, it sometimes occurs that only the existence of a solution is necessary: the search can then be stopped as soon as the first solution is found.

A cut is called *green* it it does not prune any branches leading to solutions; otherwise, it is *red*. A green cut is logically correct: it is a way to limit the don't know non-determinism of Prolog and to make programs more efficient. A red cut modifies the declarative meaning of programs. Unfortunately, the 'color' of a cut cannot be determined statically. Modern Prologs offer constructions which respect the declarative meaning of programs and fulfill some of the functions of the cut.

Example Two vertices `a` and `b` of a graph can be shown to be in the same connected component with predicate `path(X,L,Y)`:

```
path(A,[],A).
path(A,[B|L],C) :- arc(A,B), path((B,L,C).
```

which constructs all of the paths L leading to X from Y. It is silly to simply call path(a,L,b), since the values of L are of no interest. Rather than use a cut, it is preferable to call the explicitly quantified query:

```
?- some L path(a,L,b).
```

which is possible in NU–Prolog. □

7.3.4 Non-standard control

Several Prologs do not use the standard control and give ways of specifying the conditions under which a literal will be selected. IC–Prolog offers variable annotations, MU–Prolog the wait, NU–Prolog the when and Prolog–II+ the freeze. In addition, concurrent logic programming offers new control structures (guards, protected variables).

Coroutines
In NU–Prolog, it is in general the left literal which is selected. However, if certain conditions are not satisfied, this selection is suspended, and another literal is selected. The suspended literal will be awoken as soon as the conditions are satisfied, through the resolution of the other literals which will have 'broadcast' their value to those variables. This mechanism is employed when a predicate is the object of a suspension declaration (when), for certain predefined predicates, for evaluating predicates and for the treatment of negation and quantifiers. The basic idea is to suspend a call if it is not 'safe'.

Often, for the same program, certain queries have a satisfactory behavior, while others do not.

Example The list concatenation program

```
append([],Y,Y).
append([H|T],Y,[H|TY]) :- append(T,Y,TY).
```

works correctly for queries concatenating two lists or computing the decompositions of the third argument, as in

```
?- append([a],[b],Z)
```

or

```
?- append(X,Y,[1,2,3])
```

but give an infinite number of results in response to the query

```
?- append(X,[a],Z)
```

```
↷ X = [] ,
  Z = [a] ;
```

```
↷ X = [_U₁, ..., _Uₙ] ,
  Z = [_U₁, ..., _Uₙ, a] ; ...
```

But it is difficult to master the form of a call to append when the literal appears in the middle of a computation. Consider, for example, the program to concatenate three lists:

```
append3(X, Y, Z, XYZ) :-
   append(X, Y, XY),
   append(XY, Z, XYZ ).
```

A query to decompose a list immediately introduces an append with three variables:

append3(X,Y,Z,[a,b,c]) $\longrightarrow$ append(X,Y,XY), append(XY,Z,[a,b,c])

which will get lost in an infinite branch once the ten decompositions of [a,b,c] have been made. In this case, the second literal is safe and should be selected before the first. The standard rule is avoided by using a when declaration:

```
?- append(A, B, C) when A or C.
```

which suspends a selection of append so long as its first and third arguments are variables, and so allows it if the first or third is not a variable (the predicate append is predefined with this declaration). □

Fairness

A selection rule is fair if all of the derivations which it defines are fair. The standard rule is not fair: with the left-recursive clause p :- p, q, there is an unfair infinite derivation $p \longrightarrow p, q \longrightarrow p, q, q \longrightarrow \ldots$, since the literal q is never selected. The search tree is infinite. It will be shown that fairness is a crucial property in the theoretical treatment of negation.

Here is a fair rule: select at each step the leftmost residual redex, and if there are none, select one of the introduced radicals. For this rule, the derivation $p \longrightarrow p, q$ fails immediately if there is no head clause q and the search tree is finite.

Nondeterminism

The don't know non-determinism of Prolog appears because of the choice of clause, and from the existence of multiple solutions to a query.

For a given program, the predicates can be distinguished according to their degree of activity. An active predicate tends to instantiate its arguments, hence to produce solutions. A passive predicate acts as a test which checks that a relation is satisfied or fails without instantiating its arguments, hence without producing any value. The same predicate can be active and passive.

```
member(X, [X|L]).
member(X, [Y|L]) :- member(X,L).
intersection(L,M) :- member(X,L), member(X,M).
```

In the definition of intersection, the first call member(X,L) is active and binds X successively to each element of L, while member(X,M) is passive, simply testing if the value bound to X is an element of M. This is the 'generate and test' principle, whose abuse ensured the success of Prolog as a declarative language, as well as its undeserved reputation for inefficiency.

Unlike the definitions using call by pattern matching in functional programming, a logic program can contain several clauses whose heads unify with a literal. In certain situations, this non-determinism is inevitable. A typical example is that of the comparison of two binary trees (with leaves nil and nodes tagged by atoms) up to reflections:

```
iso(nil,nil).
iso(t(L1,X,R1),t(L2,X,R2)) :- iso(L1,R1), iso(L2,R2).
iso(t(L1,X,R1),t(L2,X,R2)) :- iso(L1,R2), iso(L2,R1).
```

It is impossible to determine which clause leads to a solution, so long as the solution is not known! The search tree must therefore be constructed, even though only one solution is sought.

There are other circumstances where the construction of a tree can be avoided.

Definition 7.6 A predicate p is called

- *locally deterministic* if for every literal L, there is at most one rule head defining p which unifies with L;
- *deterministic* if every search tree with root p is linear.

A query about a deterministic predicate can be executed more efficiently. The preprocessor nac of NU–Prolog is capable of adding when declarations to force a predicate to be locally deterministic, by suspending the selection of an insufficiently instantiated literal which would unify with several clause heads.

Example The following program computes the sum of (numeric) tags of the preceding binary trees, after having transformed them into lists by flatten:

```
sum_tree(T,N) :- flatten(T,L), sum_list(L,N).
?- flatten(t(A, B, C), D) when A.
flatten(t(t(XX,XU,XY),U,Y),L) :- flatten(t(XX,XU,t(XY,U,Y)),L).
flatten(t(nil,U,Y),[U|L]) :- flatten(Y,L).
flatten(nil,[]).
sum_list(L,S) :- sum_list1(L,0,S).
?- sum_list1(A, B, C) when A.
sum_list1([],S,S).
sum_list1([N|L],A,S) :- plus(A,N,AN), sum_list1(L,AN,S).
```

Note that the ?- is part of the program. The when declaration over sum_list1 requires the instantiation of the first argument, which makes it locally deterministic. Since plus is deterministic, sum_list1 is as well, hence sum_list too. flatten is, due to its `when` declaration. So `sum_tree` is deterministic since it is locally deterministic (only one clause) and it calls two deterministic clauses. □

Concurrent logic programming
Concurrent Prologs introduce two new structures: guards and variable protection. Unification is subject to restriction in order to fill two new necessary functions,

- synchronization, and
- communication

between the different processes. The flow of data is no longer bidirectional. The differences between the existing dialects CP (*Concurrent Prolog*), *Flat* CP, GHC (*Guarded Horn Clauses*), *Flat* GHC, PARLOG, etc., are mostly about the different restrictions imposed on unification.

7.3.5 Evaluating predicates

It was stated right from the beginning that terms are not evaluated by the interpreter, unlike in functional programming. A real programming language must implement numeric computations efficiently and so it is out of the question to implement numbers as terms. Every Prolog contains an applicative subset, which evaluates arithmetic expressions, without resolution, using the underlying machine arithmetic. This applicative subset is interfaced with resolution by the 'evaluating predicates', such as the order (`=<`, `<`, ...) and assignment (`is`) predicates.

Example The literal $M \leqslant N$ succeeds if M and N are closed arithmetic expressions, and if their evaluation produces numbers m and n such that $m \leqslant n$. If M or N is not an arithmetic expression, $M \leqslant N$ fails. Finally, if they are arithmetic expressions, but one of the terms contains variables, the evaluation does not take place. Standard Prolog interpreters then terminate by triggering an error. Hence

```
?- X = 1, X + 1 =< 3.

↷ X = 1
```

succeeds, while the inverse order of the literals provokes an error:

```
?- X + 1 =< 3, X = 1.

↷ Error in arithmetic expression: not a number
      (1) 1 Fail: _0+1=<3
  no
```

□

This behavior calls into question the switching property of pure logic programming: resolution is subordinate to the applicative subset, and unification appears to be another form of assignment, with the problems of uninitialized variables in sequential programming!

The only solution is to suspend the evaluation of non-closed arithmetic expressions until the variables are bound to numeric values. So the left literal selection rule is not used, even though the basis is still SLD-resolution (switching lemma).

The literal could be moved to the right of the query to ensure that it is sufficiently instantiated at the end. However, tests should be done as soon as possible to reduce the search space. A compromise consists of *suspending* the evaluation of the literal and waking it up as soon as its variables are completely instantiated by the resolution of the other literals of the query. It is this mechanism of coroutine which is implemented in NU–Prolog, where there is a satisfactory behavior:

```
?- X + 1 =< 3, X = 1.

↷ X = 1
```

The predicate `plus` of NU–Prolog is suspended so long as one of its arguments is a variable. If both of its arguments are integers, then the third can be computed.

These evaluating predicates only play a passive role: they check without producing values themselves (the value is produced by the evaluator, the predicate tests it). The coroutine mechanism does not suffice to guarantee completeness, since the two literals of each of the queries

```
?- X =< 3, X >= 5.
?- X =< 3, 1 =< X.
```

are suspended indefinitely, even though the first query should fail *logically* and the second should succeed. One can envisage active predicates which restrict the search space of solutions: the answer to the first query should be $X \in \varnothing$, which is equivalent to a failure, without examing other literals, and the answer to the second should be $X \in [1, 3]$. To have active predicates, SLD-resolution must be extended to the resolution of equations over $\mathrm{T}[X]$, by a mechanism of resolution and constraint propagation. The only constraint that Prolog handles correctly is equality between terms. The relations $\neq$ over T_Σ, $=$ and $\leqslant$ over numbers are not treated in a 'logical' manner.

· The only applicative part of Prolog, limited to the evaluation of arithmetic expression and to a few other elementary data types, is not sufficiently powerful as a programming language. It has neither the functional λ, nor the sequential iteration. It must therefore be combined with SLD-resolution to simulate these constructions. So one writes programs such as:

```
f(0,1).
f(N,P) :- N > 0, N1 is N - 1, f(N1,P1), P is N * P1.
```

The literal M `is` N succeeds if the evaluation of N is allowed and produces a value n, and if M unifies with n. Normally, this is restricted to arithmetic expressions, so n is a number, but other types of evaluation could be possible, such as `[X,Y] is [1+1,2*3]`. Here is the resolution of the query `?- f(3,P)`, omitting the test `N > 0`:

$$\begin{array}{rcl} \mathit{1}\cdot f(3,P) & \longrightarrow & \mathit{1}\cdot N_1 \texttt{ is } 3-1, f(N_1,P_1), P \texttt{ is } 3*P_1 \\ & \longrightarrow & \mathit{1}\cdot f(2,P_1), P \texttt{ is } 3*P_1 \\ & \xrightarrow{\star} & \mathit{1}\cdot f(0,P_3), P_2 \texttt{ is } 1*P_3, P_1 \texttt{ is } 2*P_2, P \texttt{ is } 3*P_1 \\ & \longrightarrow & \mathit{1}\cdot P_2 \texttt{ is } 1*1, P_1 \texttt{ is } 2*P_2, P \texttt{ is } 3*P_1 \\ & \xrightarrow{\star} & \begin{bmatrix} P \\ 6 \end{bmatrix}\cdot \top. \end{array}$$

7.4 Negation

SLD-resolution is defined in intuitionistic logic, and can even be defined in minimal logic. Since the latter has no rule for negation, the logical interpretation of programs offers no clue for the treatment of negation. How can one, with the same programs, state, prove and use negative facts. And what kind of negation? That of minimal logic, where $\neg\varphi$ is an abbreviation for $\varphi \Rightarrow \bot$, the constant connective $\bot$ being defined by no rule? that of intuitionistic logic, which remains constructive? or that of classical logic?

7.4.1 Negation as failure

From the model point of view, the answer is easy: since each definite program P has a model At_Σ in which all the atoms are true, it is impossible to deduce $P \vdash \neg A$, whatever the concept of deduction $\vdash$ which satisfies the soundness property. To avoid this problem, two paths are possible:

- extend P into a stronger theory $\tilde{P}$, leaving the framework of definite clauses, to obtain $\tilde{P} \vdash \neg A$, where $\vdash$ is as usual;
- replace $\vdash$ by a stronger relation $\tilde{\vdash}$, leaving the usual logic, to obtain $P \tilde{\vdash} \neg A$.

Before choosing one of these paths, consider what happens in practice.

The standard implementation of negation consists of defining a predicate `not` as follows (using `X` as a literal metavariable):

```
not(X) :- X, !, fail.
not(_).
```

Given a query `not(R)`, if the query `R` succeeds, `not(R)` fails, since `fail` always fails and the `!` prevents the activation of `not(_)`. If `R` fails, then `not(R)` succeeds with the second clause. Negation is interpreted as *failure*. However, logically incomprehensible programs can result: although the query

```
?- X = a, not(X = b).
```

```
↷ X = a
```

succeeds as one would expect,

```
?- not(X = b), X = a.
```

fails after having unified `X` with `b`, traversed the `!` and encountered the `fail` (literal `X = a` is not even examined). To avoid this problem, `not` should only be applied to queries without variables or to queries in which the variables have all been instantiated beforehand, a condition which is not statically verifiable.

This interpretation of negation was formalized in database theory, where the *closed world assumption* states that any fact not explicitly affirmed is considered to be false. It is tempting to extend this assumption into an inference rule allowing the deduction of $\neg\varphi$ if φ is not deducible from the data:

$$(\mathrm{CW}): \quad \frac{\Gamma \nvdash \varphi}{\Gamma \tilde{\vdash} \neg\varphi}.$$

Up to now, for each deductive relation $\vdash$ (natural deduction, equational logic, SLD-resolution), inference rules over the formulas have allowed the construction of derivations, which are concrete objects, and $\Gamma \vdash \varphi$ stated the existence of a derivation of φ with assumptions *in* Γ. The relation was local to Γ. Here, rule (CW) cannot be used to construct the derivations of formulas. It modifies $\vdash$, which loses its local qualities, since $\Gamma, \varphi \tilde{\vdash} \neg\varphi$ should not be inferrable, even if $\Gamma \tilde{\vdash} \neg\varphi$ were obtained. Furthermore, as soon as a logical system is powerful enough, $\Gamma \nvdash \varphi$? is not even semi-decidable. An inference rule whose premises are not semi-decidable is of no *operational* interest.

This solution, which replaces classical deduction by a non-local deductive relation, can also be stated by extending the program: to prove $\neg A$, just add $\neg A$ to P! Replace P by

$$\tilde{P} \equiv P \cup \{\neg A; A \in \mathrm{At}_\Sigma \text{ and } P \nvDash A\}.$$

For every closed atom A, then either $\tilde{P} \vDash A$ or $\tilde{P} \vDash \neg A$. However, the theory $\tilde{P}$ is not recursively enumerable in general (see Chapter 7), which means it has no *declarative* value. If a recursive $\tilde{P}$ is wanted, preferably finite, then there will be negations $\neg A$ such that $P \nvDash A$ cannot be proven. The declarative reading gives three values: some atoms are true ($P \vDash A$), others are false (one can prove $\neg A$), the others are unknown. The latter class should be reduced as much as possible.

7.4.2 Decidable refutation of atoms

Let P be a definite program. It is not possible to prove $\neg A$ for all $A \notin T_P^{\uparrow\omega}$, but there is a class of atoms which can be refuted effectively: those for which there is a finitely failed search tree, i.e., a finite search tree with no solution. This class is easily found using T_P, by iterating from At_Σ instead of $\varnothing$: $T_P^{\downarrow 0} = \mathrm{At}_\Sigma$, $T_P^{\downarrow 1} = T_P(T_P^{\downarrow 0})$, ... So, if A admits a search tree without solution of depth 1 (root with no child), then A is not an instance of a clause head of P: $A \notin T_P^{\downarrow 1}$.

More generally, if A admits a tree without solution of depth h, then $A \notin T_P^{\downarrow h}$. Hence, if A has a finitely failed search tree, then $A \notin T_P^{\downarrow\omega}$. This result is easily proven in [4].

The set $T_P^{\downarrow\omega}$ has already been presented: if A allows an infinite fair derivation, then $A \in T_P^{\downarrow\omega}$ (proposition 7.8). Hence, if $A \notin T_P^{\downarrow\omega}$, every fair search tree is necessarily finite and without solution, since $T_P^{\uparrow\omega} \subseteq T_P^{\downarrow\omega}$. The following result has been proven.

Proposition 7.16 The following properties are equivalent, for $A \in \mathrm{At}_\Sigma$ and a definite program P:

1. $A \notin T_P^{\downarrow\omega}$.
2. A has a finitely failed search tree.
3. Every fair search tree for A is finitely failed.

From an operational view, one must (and can) choose a *fair* selection rule S. Given $A \in \mathrm{At}_\Sigma$, the construction determined by S of the resolution tree of A yields one of the two following results, or yields nothing at all:

- `true`, as soon as a solution is found (be the tree finite or not), i.e, $A \in T_P^{\uparrow\omega}$;
- `fail`, if the tree is finitely failed, i.e., $A \notin T_P^{\downarrow\omega}$.

The `true/fail` pair of NU–Prolog is preferable to `yes/no`, which makes the two answers appear to be symmetrical. The intermediate case is $A \in T_P^{\downarrow\omega} \setminus T_P^{\uparrow\omega}$, which corresponds to an infinite tree without solution: no interpreter can give an answer in such a case. In practice, a response such as `stack overflow` can correspond to any of the three cases, but it is quite probable that an infinite tree is being constructed which would exclude $A \notin T_P^{\downarrow\omega}$.

Note that `fail` is not sensitive to the search strategy, since it can only be generated for finite trees. This is not the case for `true`. With an unfair rule, which is standard, there can also be non-termination when $A \notin T_P^{\downarrow\omega}$: `stack overflow` no longer gives any indication.

7.4.3 Completed program

The problem is to extend the program P into a 'recursively axiomatizable' theory $\tilde{P}$ such that $\tilde{P} \vdash \neg A$ if A is effectively refutable, i.e, $A \notin T_P^{\downarrow\omega}$. Clark introduced program completion to achieve this goal.

Clause completion

Let C be a clause $p(t_1, \ldots, t_r)$:- $A_1, \ldots, A_p$ with variables $y_1, \ldots, y_m$. Its homogeneous form is clause C^h,

$$p(x_1, \ldots, x_r) \text{ :- } x_1 = t_1, \ldots, x_r = t_r, A_1, \ldots, A_p,$$

written $p(\vec{x})$:- C^{h-}, where $x_1, \ldots, x_r$ are new variables. Variables $y_1, \ldots, y_m$ all appear on the right-hand side, where they are quantified existentially. So this clause can be written as the formula

$$p(x_1, \ldots, x_r) \Leftarrow \exists \vec{y}(x_1 = t_1 \wedge \ldots \wedge x_r = t_r \wedge A_1 \wedge \ldots \wedge A_p).$$

A clause and its homogeneous form are equivalent modulo the equality axioms: $Ax_= \vdash_M C \Longleftrightarrow C^h$. Consider the homogeneous forms $p(\vec{x})$:- C_i^{h-} of all of the clauses defining p. In the following discussion, variables $\vec{x}$ are fixed and distinct from all of the variables in clauses. The *homogeneous definition* of p is the formula

$$\mathrm{hom}(p) \equiv \forall \vec{x}\, p(\vec{x}) \Leftarrow \bigvee_i C_i^{h-},$$

which is equivalent to the conjunction of the clauses defining p, since $\vdash_M (p(\vec{x}) \Leftarrow \bigvee_i C_i^{h-}) \Longleftrightarrow \bigwedge_i (p(\vec{x}) \Leftarrow C_i^{h-})$ and $Ax_= \vdash_M C_i \Longleftrightarrow C_i^h$ for each i. Hence, for the whole program

$$Ax_= \vdash_M P \Longleftrightarrow \bigwedge_p \mathrm{hom}(p).$$

This is a program transformation, analogous to the transformation of call by pattern matching in functional programming into a `case` structure inside the body of the defined function. This homogeneous form makes the non-determinism of pure logic programming explicit through a disjunction: C_i^h is chosen 'randomly' to solve a literal of predicate p. If these homogeneous forms are considered as programs, the SLD-resolution must be restricted so that is does not select any equations: $\mathit{1} \cdot R_0 \xrightarrow{\star} \mathit{1} \cdot E$ is obtained, where E is a sequence of equations, no variable of R_0 having yet been bound.

Example The homogeneous form of the predicate p defined by

```
p(U,0,U).
p(U,s(V),s(W)) :- p(U,V,W).
```

is

```
p(X,Y,Z) :- X=U, Y=0, Z=U.
p(X,Y,Z) :- X=U, Y=s(V), Z=s(W), p(U,V,W).
```

or, more simply, by using the *or* ';' of Prolog:

```
p(X,Y,Z):- (X=U, Y=0, Z=U)
         ; (X=U, Y=s(V), Z=s(W), p(U,V,W)).
```

□

If symbol p appears in the program, without any clause defining it, the disjunction $\bigvee_i C_i^{h-}$ reduces the neutral element of $\vee$ to $\bot$. Since $p \Leftarrow \bot$ is an intuitionistic tautology, one can define $\mathrm{hom}(p) \equiv \top$: for a definite program, an undefined fact, which is therefore not provable using SLD-resolution, is considered to be true. This is exactly the opposite of negation as failure! Instead of $p \Leftarrow \bot$, or $\top$, it is $p \Rightarrow \bot$, or $\neg p$, which represents the absence of information about p under negation.

More generally, since all that is known about p is in one formula, the definition is 'closed' by replacing $\Leftarrow$ (p if ...) by $\Longleftrightarrow$ (p if and only if ...). The *completed definition* of the predicate p is

$$\text{comp}(p) \equiv \forall \vec{x}(p(\vec{x}) \Longleftrightarrow \exists \vec{y}(C_1^{h-} \vee \ldots \vee C_n^{h-})).$$

Example (cont.)

$$\forall(x,y,z)\left(\begin{array}{ll} p(x,y,z) \Longleftrightarrow & \exists u(x = u \wedge y = 0 \wedge z = u) \\ & \vee \exists uv(x = u \wedge y = s(v) \wedge z = s(w) \wedge p(u,v,w)) \end{array}\right).$$

□

In the case where p has no definition in the program, then $n = 0$, which leads to $\forall \vec{x}(p(\vec{x}) \Longleftrightarrow \bot)$, or

$$\text{comp}(p) \equiv \forall \vec{x} \neg p(\vec{x}).$$

which is equivalent. Since comp(p) is the conjunction of hom(p) and of the symmetric (in $\Leftarrow / \Rightarrow$) formula, it is clear that comp(p) $\Rightarrow$ hom(p), hence

$$Ax_= \vdash_M \bigwedge_p \text{comp}(p) \Rightarrow P.$$

The formula $\forall \vec{x}(p(\vec{x}) \Rightarrow \exists \vec{y}(C_1^{h-} \vee \ldots \vee C_n^{h-}))$ is the *homogeneous converse* definition of p.

Fixpoints of T_P

Proposition 7.17 A Herbrand algebra $\mathcal{H}$ is a model of $\bigwedge_p \text{comp}(p)$ if and only if $\mathcal{H}$ is a fixpoint of T_P.

Proof Since $\mathcal{H} \models P$ if and only if $T_P(\mathcal{H}) \subseteq \mathcal{H}$, it suffices to show that $\mathcal{H}$ satisfies the homogeneous converse definitions of P if and only if $T_P(\mathcal{H}) \supseteq \mathcal{H}$.

$T_P(\mathcal{H}) \supseteq \mathcal{H}$ means that $\mathcal{H}$ is an algebra *supported* by P, i.e, for every $A \in \mathcal{H}$, there exists a closed instantiation θC of a clause C of P such that $A \equiv \theta C^+$ and $\mathcal{H} \models \theta C^-$.

$\mathcal{H} \models \forall \vec{x}(p(\vec{x}) \Rightarrow \bigvee_i \exists \vec{y}(\vec{x} = \vec{t} \wedge C_i^-))$ means that for every $M_1, \ldots, M_r \in \mathrm{T}_\Sigma$, if $\mathcal{H} \models p(\vec{M})$, then there exists an i such that $\mathcal{H} \models \exists \vec{y}(\vec{M} = \vec{t} \wedge C_i^-)$. Hence there exists an i and a closed substitution θ such that $\mathcal{H} \models \vec{M} = \theta \vec{t} \wedge \theta C_i^-$.

It is clear that the two properties are equivalent. □

Negation as omission and Clark axioms

The only way to state a negative fact in a definite program is to leave it out. For example, the predecessor over the natural integers is defined by the unit clause:

```
pred(s(X),X).
```

which is unifiable with no literal of the form `pred(0,Y)`, and the interpreter correctly answers `fail` for such a query.

It is true that comp(*pred*) $\equiv \forall xy(pred(x,y) \iff \exists u(x = s(u) \wedge y = u))$, but $Ax_= \nvdash \text{comp}(pred) \Rightarrow \neg pred(0,t)$ no matter what t is, simply because there exist models with equality of comp(*pred*) where 0 receives the same interpretation as a successor term. For example, if $(sso)^{\mathcal{A}} = 0^{\mathcal{A}}$ in a model with equality $\mathcal{A}$ of comp(*pred*), then $\mathcal{A} \models pred(0, s0)$, which shows that $\neg pred(0, s0)$ cannot be a consequence of comp(*pred*).

Since the completion of clauses is insufficient to express this negation as omission, we add a set of axioms, due to Clark, defining the theory of equality of terms:

decomposition $f(x_1, \ldots, x_r) = f(y_1, \ldots, y_r) \Rightarrow x_1 = y_1 \wedge \ldots \wedge x_r = y_r$, for each f;
consistency $f(x_1, \ldots, x_r) \neq g(y_1, \ldots, y_s)$, if $f \neq g$;
acyclicity $x \neq t$ for every term t different from x and containing x.

The particular case of consistency axioms for constants ($c \neq c'$ if c and c' are distinct constants) is well known in the framework of databases.

The *completion* comp(P) of a program P is the union of the equality axioms, the Clark axioms and the completed definitions of the predicates.

Corollary 7.18 If P is a definite program, theory comp(P) is consistent.

Proof T_P has at least one fixpoint (its least fixpoint $\mu T_P = \mathcal{M}_P$) which is a model of $\bigwedge_p \text{comp}(p)$ (proposition 7.17). Since every Herbrand algebra satisfies the Clark axioms, the fixpoint is also a model of comp(P). So comp(P) is consistent. □

Every model with equality of the Clark axioms contains a subalgebra isomorphic to T_Σ. It has already been shown (without the Clark axioms) that $\bigwedge_p \text{comp}(p) \vdash_M P$. Since every Herbrand algebra satisfies the Clark axioms, we have the following corollary.

Corollary 7.19 comp(P) $\vdash_M P$.

From proposition 7.17, we also have the following corollary.

Corollary 7.20 The Herbrand models of comp(P) are all the Herbrand algebras between μT_P and $\bar{\mu} T_P$. Hence, comp(P) $\cup \{A\}$ has a Herbrand model if and only if $A \in \bar{\mu} T_P$.

Positive consequences of comp(P)

The completion of clauses is a conservative extension of the program, with respect to positive formulas.

Proposition 7.21 Let ψ be a positive formula, i.e constructed over $\wedge$, $\vee$, $\forall$ and $\exists$. If comp(P) $\models \psi$, then $P \models \psi$.

Proof Suppose that $\text{comp}(P) \models \psi$. It will be shown that $P \Rightarrow \psi$ is a tautology. From the Herbrand–Skolem theorem, $P \Rightarrow \psi$ is a tautology if and only if its existential skolemized function $(P \Rightarrow \psi)^e \equiv P \Rightarrow \psi^e$ is satisfied by all of the Herbrand algebras of domain $\mathrm{T}_{\Sigma'}$, where Σ' follows from Σ by adding Skolem symbols. Let $\mathcal{H}$ be a Herbrand model of P of domain $\mathrm{T}_{\Sigma'}$. By the definition of $\mathcal{M}_P$, $\mathcal{M}_P \subseteq \mathcal{H} \cap \mathrm{At}_\Sigma$. Since $\mathcal{M}_P$ is a fixpoint of T_P, it is also a model of comp(P). By assumption, $\text{comp}(P) \models \psi$, hence $\mathcal{M}_P \models \psi$. Now $\models \psi \Rightarrow \psi^e$, so $\mathcal{M}_P \models \psi^e$. From exercise 10, since ψ^e is existential positive, deduce that $\mathcal{H} \models \psi^e$ from $\mathcal{M}_P \subseteq \mathcal{H}$ and $\mathcal{M}_P \models \psi^e$. Formula $P \Rightarrow \psi$ is therefore a tautology. □

Corollary 7.22 Let R be a positive query and σ a substitution.

$$\text{If comp}(P) \models \forall(\sigma R), \text{ then } P \models \forall(\sigma R).$$

Negative consequences of comp(P)

It is easy to show that if $A \notin T_P^{\downarrow\omega}$, then $\text{comp}(A) \vdash_M \neg A$, by using the homogeneous converse definitions of P and the Clark axioms. Only the base case of the proof will be considered: $A \notin T_P^{\downarrow 1}$, i.e, $A \equiv p\vec{M}$ is not an instance of a clause head. The converse definition of p is used: $p(\vec{x}) \Rightarrow \ldots$. By instantiating $\vec{x}$ in $\vec{M}$, and by supposing A, the disjunction $\exists\vec{y}(\vec{M} = \vec{t} \wedge C^-) \vee \ldots$ follows. By assumption, the equations $p\vec{M} = p\vec{t}$ between A and the clause heads defining p have no solution. The Clark axioms allow one to derive $\perp$ from each equation $\vec{M} = \vec{t}$. So $\perp$ was derived by supposing A, which proves $\neg A$ in intuitionistic logic, and even in minimal logic, since $\perp$-elimination was not used.

The converse is more delicate, since a model of the theory $\text{comp}(P) \cup \{A\}$ must be constructed when $A \in T_P^{\downarrow\omega}$. If one had the assumption $A \in \bar{\mu}T_P$, which is stronger since $\bar{\mu}T_P \subseteq T_P^{\downarrow\omega}$, proposition 7.20 would show that there exists a *Herbrand* model of $\text{comp}(P) \cup \{A\}$. But for $A \in T_P^{\downarrow\omega}$, a general model must be constructed, using the fact that there exists a successful or infinite derivation in a fair search tree of A. This construction is given in [95, §16].

Theorem 7.23 If P is a definite program, and $A \in \mathrm{At}_\Sigma$, then

- $\text{comp}(P) \vdash_M A$ is equivalent to $A \in T_P^{\uparrow\omega}$;
- $\text{comp}(P) \vdash_M \neg A$ is equivalent to $A \notin T_P^{\downarrow\omega}$.

7.4.4 General programs

For a programmer, it is easier to write negation explicitly rather than through omission in definite programs. *General* programs and queries allow the bodies of clauses and queries to be formed of arbitrary literals.

However, the extension of the logical approach to general programs is problematic.

The intersection of Herbrand models is not always a model. There does not always exist a least Herbrand model, since the T_P operator is no longer monotone. The definition of completion can be transposed to general programs, but comp(P) can be inconsistent even if P is consistent.

Example Consider the clause:

```
p :- not p.
```

The definition of p is classically equivalent to the assertion of p, but the theory comp(P) $\equiv p \Longleftrightarrow \neg p$ is inconsistent. □

Normally, a query's variables are quantified existentially. If non-closed negative atoms are allowed in queries, negation as failure can only work if universal quantifiers are introduced: the failure of a proof of $\exists x p(x)$ does not prove $\exists x(\neg p(x))$, rather $\neg(\exists x p(x))$, i.e, $\forall x(\neg p(x))$.

SLDNF-resolution The operational semantics of a general program is given by SLDNF-resolution (NF means *negation as failure*).

The same environment-query state is used, but both SLDNF-derivations and SLDNF-search trees must be defined. The nodes of a search tree are labeled by the search states. A search state is a pair $(\rho \cdot R, s)$, where $R = \top$, or R/s is a literal of R which is an atom or the negation of a *closed* atom. The search tree for the solutions of an initial query $R_0 \neq \top$ has as root the state $(\mathit{1} \cdot R_0, s_0)$. Let $(\rho \cdot R, s)$ be a search state.

- If R/s is an atom, let C be a separate variant of a clause of P, and
  - if $R/s = C^+$ has no solution, then there is no transition tagged by s, C out of $\rho \cdot R$;
  - if $R/s = C^+$ has a most general unifier θ, then $\rho \cdot R \xrightarrow{s,C} \theta\rho \cdot \theta(R[s \leftarrow C^-])$.

  The node $(\rho \cdot R, s)$ has a descendent with state $\theta\rho \cdot \theta(R[s \leftarrow C^-])$ for each such C. It is a failure node if it has no descendents.
- If R/s is the negation $\neg B$ of a closed atom B, and
  - if B has a solution, then $(\rho \cdot R, s)$ is a failure node, which has no descendents;
  - if B has finitely failed, then $\rho \cdot R \xrightarrow{s,C} \rho \cdot R \setminus s$, and the node $(\rho \cdot R, s)$ has a unique descendent, with state $\rho \cdot R \setminus s$.

Clark gave a soundness theorem for SLDNF-resolution:

Theorem 7.24 Let P be a general program and R a general query:

- if R fails, then comp(P) $\vdash_M \neg\exists(R)$;
- if ρ is an answer to R, then comp(P) $\vdash_M \forall(\rho R)$.

SLDNF-resolution proves theorems in minimal logic, as does SLD-resolution.

Prolog interpreters, such as C–Prolog, implement negation as failure, but rarely safely since they can select a non-closed negative literal: these interpreters may give incorrect answers. Suppose that the closure condition on negative literals were left out, and consider program `p(a)` and query `not p(X)`. Since there is the solution $\mathit{1} \cdot p(x) \longrightarrow \left[\begin{smallmatrix} x \\ a \end{smallmatrix}\right] \cdot \top$, `not p(X)` fails and an erroneous application of the soundness theorem leads to $\text{comp}(P) \vdash_I \neg\exists(\neg p(x))$, hence $\text{comp}(P) \vdash \forall x p(x)$, which is incorrect.

NU–Prolog is a safe implementation of negation as failure, which suspends the selection of a negative literal if it is not closed, selects another literal, and awakens the suspended literal as soon as its variables are bound to closed terms. It is possible that all of the literals of a query are negative and not closed: in that case no literal may be selected and the search *flounders*. Hence `?- not p(X)` flounders immediately. But whether a query flounders is an undecidable problem. Because of these problems, the extension of Prolog to general programs can no longer be considered as an automatic theorem prover.

Examples

- With the program `p(X) :- not q(X)`, query `p(a)` succeeds, but `p(X)` flounders: hence $p(a)$ is provable, but not $\exists x p(x)$.
- By adding `r(a)` to the preceding program, query `r(X), p(X)` succeeds but `p(X)` flounders: $\exists x(p(x) \wedge q(x))$ is provable, but not, $\exists x p(x)$.
- With the program `p :- not q`, `p` succeeds and `q` fails, while `q` succeeds and `p` fails with the equivalent program `q :- not p`.

As was stated previously, negation as failure requires that the free variables of negative sign be quantified universally: the literal `not p(X)` means $\forall x(\neg p(x))$. NU–Prolog has explicit quantifiers `some` and `all`: by default `not p(X)` is equivalent to `all X not p(X)` and to `not (some X p(X))`, while the query `?- some X not p(X)` flounders immediately. It is often useful to formulate queries, or to write $\neg(\exists\vec{x}R_0)$ in the body of a rule, where R_0 is a query and $\vec{x}$ are variables (not necessarily all those of R_0).

Example The `disjoint/2` predicate, which tests if two lists have no element in common, is defined by one of the following equivalent clauses:

```
disjoint(X,Y) :- not (some A (member(A,X), member(A,Y))).
disjoint(X,Y) :- all A not (member(A,X), member(A,Y)).
```

□

The soundness theorem shows that $\neg(\exists\vec{x}R_0)$ is erasable by constructing a finitely failed tree for R_0 if $\neg(\exists\vec{x}R_0)$ is closed, i.e., all of the variables in R_0 are bounded by the $\exists$. A coroutine mechanism allows the suspension of the selection of $\neg(\exists\vec{x}R_0)$ so long as it contains free variables. Even better, the search can begin even if there are free variables, and it will be suspended before broadcasting a binding relating to these variables. If R_0 succeeds, then $\neg(\exists\vec{x}R_0)$ fails. If R_0 has no solution, then $\neg(\exists\vec{x}R_0)$ is deleted.

In addition to the logical weaknesses of negation as failure, its major computing defect is that it computes nothing: it is a test which checks the non-existence of a solution without producing any solution. This defect has already been mentioned with respect to evaluating predicates. With a program `p(a)`, it would be useful if a literal $\neg p(x)$ were deleted and replaced by the constraint $x \neq a$. This idea is the basis for constraint programming.

7.5 Equality and resolution

The two computing paradigms presented in this book, reduction and resolution, are based on quite distinct principles. Several proposals have been made to unify functional and logic programming. These two paradigms have one point in common: equations. SLD-resolution is based on the resolution of equations in $\mathrm{T}[X]$, while functional programs can be seen as rewrite rules (oriented equations) and equality is axiomatized by equations (first-order in combinatory logic).

An interesting and useful extension would be to incorporate an equational theory $\mathcal{E}$, for example to state a recursive function definition, properties of a data structure or a domain of computation, such as Booleans, rationals or multi-sets. Although the treatment of equality poses no theoretical difficulties in logic, the complexity of automatic proof methods constitutes a real obstacle to their transformation into effective computations. In a system based on the resolution principle, the equality axioms multiply the number of equations to be solved.

Although it is more natural to define functions in an equational manner (Prolog appeared fifteen years after Lisp), Prolog programmers have become used to writing non-equational definitions. This shows that equational definitions are not irreplaceable, and that it is possible to write efficient logic programs by renouncing the ability to express equality directly.

7.5.1 Reduction by resolution

A first method consists of adding to the *program* axioms of $\mathcal{E}$ which are simple facts $M = N$, and the equality axioms in definite clause form for each of the symbols used. Here is the equational definition of addition:

```
e(X+0,X).
e(X+s(Y), s(X+Y)).
e(X,X).
e(X,Y) :- e(Y,X).
e(X,Z) :- e(X,Y), e(Y,Z).
```

```
e(s(M),s(N)) :- e(M,N).
e(M1+N1, M2+N2) :- e(M1,M2), e(N1,N2).

?- e(s(s(0))+s(s(0)),Y).
```

The interpreter supplies a great number of solutions of little interest, and in this case does not give the 'good' solution, which is `Y = s(s(s(s(0))))`:

```
↷ Y = s(s(s(0)) + s(0)) ;
  Y = s(s(0)) + s(s(0)) ;
  Y = (s(s(0)) + s(s(0))) + 0 ;
  Y = s(s(0)) + s(s(0)) ;          etc ...
```

As can be checked, this program is very sensitive to the order of clauses: the search tree has infinite branches and far too many stupid and repeated solutions. It correctly defines an equality (congruence over terms, stable under substitutions), but ignores the existence of a canonical rewriting system for this equality. In particular the interpreter is incapable of recognizing the normal form of a term, which is the only interesting response. Yet this idea is easily defined:

```
normal(0).
normal(s(M)) :- normal(M).
```

It is then possible to state the rewriting rules with an execution order using the following logic program, where the predicates eq1 and eq2 both state the equality:

```
eq1(M,M) :- normal(M).
eq1(M,X) :- not normal(M), eq2(M,Y), eq1(Y,X).
eq2(X+0,X).
eq2(X+s(Y),s(X+Y)).
eq2(s(M),s(N)) :- eq1(M,N).
eq2(M1+N1, M2+N2) :- eq1(M1,M2), eq1(N1,N2).

?- eq1(s(s(0))+s(s(0)),X).
```

```
↷ X = s(s(s(s(0)))) ;
```

The role of the transitivity rule is to introduce a new variable for each reduction step. Doing this avoids returning the results of intermediate steps. The first rule is only applied at the end, when M is normal. At each step of the reduction, the selected literal is of the form eq2(M,Y), where M is a non-normal term and Y is free. If M is of the form θP for a rule $P \to Q$, then the literal is deleted, Y receiving as value θQ. The following selected literal is eq1(θQ,X). Otherwise, M is decomposed by the other rules for eq2. The first answer obtained is the required normal form.

7.5.2 Equational programs and unification

SLD-resolution eliminates equations from programs, but incorporates them into the computing mechanism. They clearly appear in the homogeneous form of programs which introduce equations in queries. These homogeneous forms can be written using the $==$ symbol instead of $=$, which will be kept for writing the systems of equations to be unified.

Resolution (in the ordinary sense) of equations by unification can be interpreted at the level of homogeneous programs by adding the unit clause $V == V$ to the program. The unification problem $(M == N) = (V == V)$ is transformed into the equivalent problem $M = V, N = V$, then to $M = N, V = M$. The natural idea is to extend this interpretation by adding a theory (over the symbol $==$) more specific than $V == V$ to the program.

Another idea consists of not introducing equations in the programs, but to solve the equations in domains other than the algebra of terms, for example, by unification modulo $\mathcal{E}$. The difficulties are handed to unification in the cases where there is no most general unifier: the don't know non-determinism is introduced even before the choice of clause: a search state should have as many children as there are 'minimal' solutions for each unification problem, and minimal solutions do not always exist.

7.5.3 Narrowing

Narrowing is a transformation which has aspects of rewriting and resolution. While resolution solves queries using a logic program, narrowing solves equations using a functional program, which is a rewriting system. The narrowing of a term M consists of applying a smaller substitution to M, making M reducible by a rule, and of applying that rule. The narrowing of a closed term corresponds to rewriting.

The definitions are formally analogous to SLD-resolution. A narrowing state is a pair $\rho \cdot M$, where ρ is a logical environment, M a term, and $\mathrm{dom}(\rho) \cap \mathrm{var}(M) = \varnothing$. Let $\mathcal{R}$ be a rewriting system. A narrowing step is defined as follows: let $\rho \cdot M$ be a state, and $P \to Q$ a rule such that P is superposable on M at non-variable occurrence u, with most general unifier θ. Then

$$\rho \cdot M \xrightarrow{u, P \to Q} \theta\rho \cdot \theta(M[u \leftarrow Q]).$$

As in SLD-resolution, there are concepts of derivation, of successful derivation $\mathit{1} \cdot M \xrightarrow{\star} \rho \cdot \top$ and of result $\rho\lceil_{\mathrm{var}\, M}$.

This mechanism will be used to solve equations. In this case the terms are always of the form $M == N$.

Example Given the rewriting system $x+0 \to x$, $x+s(y) \to s(x+y)$, let $0+x = s(0)$ be an equation to be solved. Here are two examples of narrowing, one failed, one successful:

$$\mathit{1}\cdot 0+x == s(0) \xrightarrow{1} \mathit{1}\cdot 0 == s(0),$$

$$\mathit{1}\cdot 0+x == s(0) \xrightarrow{2} \begin{bmatrix} x \\ s(y_1) \end{bmatrix}\cdot s(0+y_1) == s(0),$$

$$\xrightarrow{1} \begin{bmatrix} x \\ s(0) \end{bmatrix}\cdot s(0) == s(0).$$

□

The soundness and completeness properties have the following form:

- if $\mathit{1}\cdot M == N \xrightarrow{\star} \rho\cdot\top$, then $\mathcal{E}\vdash \rho(M)=\rho(N)$;
- if $\mathcal{E}\vdash\sigma(M)=\sigma(N)$, there exists a derivation $\mathit{1}\cdot M == N \xrightarrow{\star} \rho\cdot\top$ such that $\sigma \geqslant \rho$.

These properties are satisfied if one has a canonical rewriting system for $\mathcal{E}$.

7.5.4 Knuth–Bendix completion

Narrowing is a particular case of completion. In fact, the narrowing state $\rho\cdot M$ can be interpreted as a *rule* $M \to \mathrm{res}(\rho(x_1),\ldots,\rho(x_n))$, where $\{x_1,\ldots,x_n\}$ are the variables of the initial query, and res is a new symbol.

The superposition of $P\to Q$ on $M == N \to \mathrm{res}(\rho(x_1),\ldots,\rho(x_n))$ at occurrence u, θ being the most general unifier of $(M == N)/u$ and $\theta(P)$, gives the equation

$$\theta((M == N)/u \leftarrow Q) = \theta(\mathrm{res}(V_1,\ldots,V_n)),$$

i.e, $\mathrm{res}(\theta\rho(x_1),\ldots,\theta\rho(x_n))$. It follows that the narrowing of $\rho\cdot M$ by the rule $P\to Q$ is equivalent to the computation of a critical pair due to a superposition of $P\to Q$ on $M == N \to \mathrm{res}(\rho(x_1),\ldots,\rho(x_n))$.

Critical pairs are computed by superposition of the 'programs' on a query' by considering the rules $P\to Q$ of the system $\mathcal{R}$ as the program, and the $M == N \to \mathrm{res}(\vec{V})$ as the queries. No superposition between rules of the program, nor between query and rule, nor between successive queries has been computed. The rewrite rule $u == u \to \top$ is added in order to halt the computation with a result.

Example

	$0+x == s(0) \to \mathrm{res}(x)$	superposed by 2
$\to$	$s(0+y_1) == s(0) \to \mathrm{res}(s(y_1))$	reduced
$\to$	$0+y_1 == 0 \to \mathrm{res}(s(y_1))$	superposed by 1
$\to$	$0 == 0 \to \mathrm{res}(s(0))$	superposed by $u == u \to \top$
$\to$	$\mathrm{res}(s(0)) \to \top$.	

□

While Knuth–Bendix completion is, like Robinson's resolution, a method of automatic theorem-proving, narrowing is, as is SLD-resolution, a method of computation.

Exercises[4]

1. Define two unary predicates for the mutually recursive types of trees and forests.

2.★ Write a *quicksort* program using D-lists.

3. Define a D-tree structure analogous to D-lists, along with a concatenation program.

4. Show that if $\mathcal{H}$ and $\mathcal{H}'$ are two Herbrand algebras, there exists a morphism of algebras from $\mathcal{H}$ to $\mathcal{H}'$ if and only if $B(\mathcal{H}) \subseteq B(\mathcal{H}')$, in which case the morphism is necessarily the identity.

5. Let P be the program

```
p1(f(X)) :- p1(X).
p2(a)    :- p1(X).
p2(f(X)) :- p2(X).
```

Compute $T_P^{\uparrow\omega}$, $T_P^{\downarrow n}$, $T_P^{\downarrow\omega}$, $T_P^{\downarrow(\omega+n)}$ and $T_P^{\downarrow(\omega+\omega)}$. Write a program P_k, for $k \in \mathrm{N}$, such that $\bar{\mu}T_{P_k} = T_{P_k}^{\downarrow\omega k}$ (the ordinal ωk represents k copies of N placed end to end).

6.★ Let $\Phi_\Sigma : \mathcal{P}(\Sigma^\star) \to \mathcal{P}(\Sigma^\star)$ be the operator associated with a signature Σ (Chapter 1). Show that $\Phi_\Sigma^{\uparrow\omega} = \Phi_\Sigma^{\downarrow\omega}$.

7.★ Let P be a definite program such that $\mathrm{var}(C^-) \subseteq \mathrm{var}(C^+)$ for every clause C of P. Show that $T_P^{\downarrow\omega} = \bar{\mu}T_P$.

8. Show that if the signature does not contain any functional symbols of arity > 1, then $T_P^{\downarrow\omega} = \bar{\mu}T_P$.

9. Prove that the selection rule which selects at each step the leftmost of the residual radicals, or if there is none, one of the introduced literals, is fair.

10. Let $\mathcal{H}$ and $\mathcal{H}'$ be two Herbrand algebras such that $B(\mathcal{H}) \subseteq B(\mathcal{H}')$, and φ be a positive existential formula, i.e, constructed with $\wedge$, $\vee$ and $\exists$ as the only logical symbols. Show that for every valuation ξ, $\varphi^{\mathcal{H}}(\xi) \leqslant \varphi^{\mathcal{H}'}(\xi)$. Is this inequality still true for an arbitrary φ?

[4]Hints to the exercises labeled with a ★ can be found at the end of the book.

11. Write a logic program to evaluate rewriting systems with lazy constructors. Treat the example of §3.3.1.

12.★ Consider the class of P-formulas defined by:

- an atomic formula is a P-formula;
- if B and C are P-formulas and if F is an arbitrary formula, then $B \wedge C$, $\forall x B$ and $F \Rightarrow B$ are P-formulas.

Show, by induction over the derivations of cut-free LJ, that if Π is a list of P-formulas, and if A and B are arbitrary formulas, then

1. if the sequent $\Pi : \exists x A(x)$ is derivable, then there exists a term t such that $\Pi : A[x := t]$ is derivable;
2. if the sequent $\Pi : A \vee B$ is derivable, then one of the two sequents, $\Pi : A$ or $\Pi : B$, is derivable.

Bibliographic notes

On the art of programming in Prolog, Sterling and Shapiro's book [141] is recommended, as is O'Keefe [115]. C–Prolog is described in [118], NU–Prolog in [149], and Prolog–II in [50, 153]. The model of activation boxes is due to L. Byrd [12]. A study of Prolog implementation , in particular of suspension mechanisms, is made in [11].

The many refinements of the resolution principle, due to Robinson [128], are developed in the standard books of Chang and Lee [13] and of Loveland [96]. The logical interpretation of Prolog is due to Kowalski [88]. The links between SLD-resolution and the sequent calculus are the focus of [47, 54]. The foundations of logic programming are given in Lloyd's book [95], which compiles the results of several essential articles [4, 17, 155].

The handling of negation, initiated by articles of [17, 127], is analyzed by Shepherdson in [134], where theorem 7.24 is given. The introduction of equality to logic programming is studied in [40, 74, 156, 157]. The writing of a rewriting system as a logic program is due to van Emden and Yukawa [157].

There are numerous extensions to logic programming: constraint programming (Jaffar and Lassez [73], van Hentenryck [61] and van Hentenryck and Deville van-Hentenryck91), Gallier and Raatz's Horn clauses [48], the higher-order logic programming of Miller and Nadathur [108], the extension to languages more general that Horn clauses [109, 107]. Shapiro's survey [133] deals with concurrent logic programming and Saraswat *et al.* [131] study constraint concurrent programming.

Chapter 8

The computable landscape

8.1 Computability

Functions have been approached from two opposing and important views. First, mathematicians made functions into banal, particular cases of binary relations, via set theory, opening the road, most remarkably, to functional analysis. Second, the logicians of the 1930s (Herbrand, Church, Kleene, Gödel, Turing, Post, etc.) succeeded in defining the concept of computable function, as well as the mechanisms of computation. Equal in expressive power since they characterize the same computable function, the different paradigms for computation, complementing each other, allow for an analysis of the different styles of programming.

8.1.1 A few encodings

Computation is a concrete activity which manipulates very real objects, typically words, which represent integers, programs, etc. It excludes everything which is not effectively given, in particular the concepts of function or set in their contemporary mathematical sense.

Since questions of computability and decidability are untouched by dimension, encodings are used to translate $\mathbf{N}^k$ into $\mathbf{N}$. Let $\langle , \rangle : \mathbf{N}^2 \to \mathbf{N}$ be a bijective function, with inverse $(\pi_2^1, \pi_2^2) : \mathbf{N} \to \mathbf{N} \times \mathbf{N}$. For each $k \geqslant 1$, there is a bijection $\langle\ \rangle_k ; \mathbf{N}^k \to \mathbf{N}$, defined by induction over k by

$$\begin{aligned}\langle m \rangle_1 &= m \\ \langle m_1, \dots, m_{k-1}, m_k \rangle_k &= \langle \langle m_1, \dots, m_{k-1} \rangle_{k-1}, m_k \rangle, \qquad k \geqslant 2,\end{aligned}$$

and using this function, a bijection $\langle\ \rangle^\star$ from $\bigcup_{k \geqslant 0} \mathbf{N}^k$ to $\mathbf{N}$:

$$\begin{aligned}\langle\rangle_0^\star &= 0 \\ \langle m_1, \dots, m_k \rangle_k^\star &= \langle m_1, \dots, m_k, k-1 \rangle_{k+1} + 1 \qquad k \geqslant 1.\end{aligned}$$

It therefore suffices to know how to code pairs of integers. The encoding that follows has been in use ever since Cantor showed that there are as many rationals as there are integers. Enumerate $\mathbf{N} \times \mathbf{N}$ along the lines $x + y = p$, for increasing p, and on each line, with increasing x (i.e, $(0,0)$, $(0,1)$, $(1,0)$, $(0,2)$, $(1,1)$, $(2,0)$, $(0,3)$, ...).

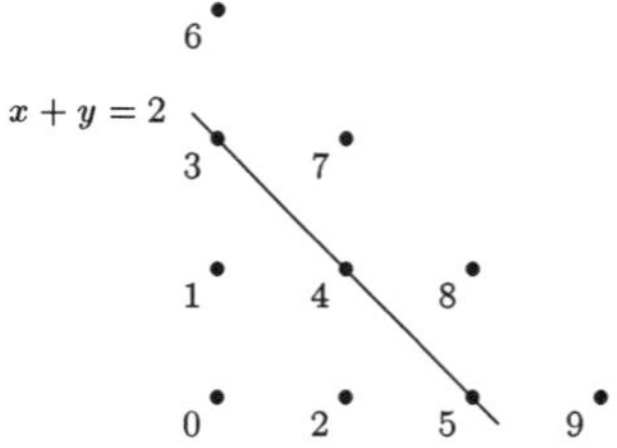

It is useful to define some auxiliary functions: if $x = \langle x_1, \ldots, x_k\rangle^\star$, let

$$\begin{aligned}
\|x\| &= k, \\
\mathrm{val}(x) &= \langle x_1, \ldots, x_k\rangle_k, \\
\mathrm{pre}(x) &= \langle x_1, \ldots, x_{k-1}\rangle^\star, \\
\mathrm{last}(x) &= x_k, \\
\mathrm{pre}(0, x) &= x, \\
\mathrm{pre}(i+1, x) &= \mathrm{pre}(\mathrm{pre}(i, x)), \\
\mathrm{last}(i, x) &= \mathrm{last}(\mathrm{pre}(i, x)), \\
cons(\langle x_0, \ldots, x_{k-1}\rangle^\star, x_k) &= \langle x_0, \ldots, x_k\rangle^\star.
\end{aligned}$$

This encoding allows several useful constructions. First, associate with each function $f : \mathbf{N} \to \mathbf{N}$ a history function

$$f^\star(x) = \langle f(0), f(1), \ldots, f(x-1)\rangle^\star.$$

A history function for $f : \mathbf{N}^k \to \mathbf{N}$ can also be defined by distinguishing one of the k arguments of f. Second, there is a bijection between unary and k-ary functions, by associating with each unary function f the function $f_k : \mathbf{N}^k \to \mathbf{N}$, defined by $f_k = f \circ \langle\rangle_k$.

8.1.2 Computable functions

The development of the concept of computable function will be presented. This concept first came to light in the 1930s and has now become well known because of computer science. A function from $\mathbf{N}^p$ to $\mathbf{N}$, for $p \geqslant 0$, is called a numerical function; write $f(\vec{n})$ instead of $f(n_1, \ldots, n_p)$. A subset A of $\mathbf{N}^k$ is a k-ary relation over $\mathbf{N}$. The notations $m \in A$, $A(m)$, $1_A(m) = 1$ and even $A(m) = 1$ are interchangeable. When A is seen as a mapping over $\mathbf{N}^k$ with Boolean values, A is called a predicate.

Church, Kleene and λ-definable functions
Recall that an integer n is coded in the λ-calculus as a *Church numeral*, a 'function iterator' $\mathbf{n} \equiv \lambda f x \,.\, f^n x$. For example, $\mathbf{3} f x =_\beta f(f(fx))$.

Definition 8.1 Let $f : \mathbf{N}^k \to \mathbf{N}$ be a numeric function. A λ-term M_f *λ-defines* f if for every $(n_1, \ldots, n_k) \in \mathbf{N}^k$, with $f(n_1, \ldots, n_k) = m$, $M_f \mathbf{n_1} \ldots \mathbf{n_k} \xrightarrow{\star} \mathbf{m}$. The function f is *λ-definable* if there exists a λ-term M_f which defines f.

The successor, addition, product and exponentiation functions are λ-definable (see Chapter 2).

A λ-definable function is computable: it suffices to compute the normal form of $M_f \mathbf{n_1} \ldots \mathbf{n_k}$ by a left reduction, which is effective and deterministic. In 1932 Kleene [79], Church's student, checked that all the numeric functions which he knew how to compute were λ-definable. In 1933 this led Church [14] to consider λ-definability as the correct concept of 'computability'. This conviction was not shared by everyone, in particular not yet by Gödel, who had just arrived at Princeton Institute.

Herbrand, Gödel and primitive recursive functions
Following Dedekind (1888 [36]) and Peano, it became known that recursion could be used to define numeric functions: the recursive definitions of addition and multiplication are a subset of the Peano axioms of arithmetic. In the form of the day, now called primitive recursive, this mechanism did not allow the computation of certain obviously 'computable' functions (Ackermann 1928 [3]). Inspiring himself by a suggestion made by Herbrand, in 1934 Gödel [57] proposed his definition of recursive functions. This definition used three operators over functions, composition, primitive recursion and minimization, as well as a closure condition.

A set $\mathbf{E}$ of numeric functions is

1. *closed under composition* if for every $h, g_1, \ldots, g_p \in \mathbf{E}$, if f is defined by

$$\boxed{f(\vec{n}) \quad = \quad h(g_1(\vec{n}), \ldots, g_p(\vec{n})),}$$

then $f \in \mathbf{E}$;

2. *closed under primitive recursion* if for every $h, g \in \mathbf{E}$, if f is defined by

$$\boxed{\begin{array}{rcl} f(0, \vec{n}) & = & g(\vec{n}), \\ f(m+1, \vec{n}) & = & h(f(m, \vec{n}), m, n) \end{array}}$$

then $f \in \mathbf{E}$;

3. *closed under (total) minimization* if for every $g \in \mathbf{E}$ such that for each $\vec{n}$, there exists p such that $g(\vec{n}, p) = 0$, if f is defined by

$$\boxed{f(\vec{n}) \quad = \quad \min\{p \in \mathbf{N}; g(\vec{n}, p) = 0\}}$$

then $f \in \mathbf{E}$. Write $f = \mu p \,.\, [g(.,p) = 0]$.

The *base functions* are

- the constant $0 : \mathbf{N} \to \mathbf{N}$;
- the successor $s : \mathbf{N} \to \mathbf{N}$; and
- the projections $pr_k^i : \mathbf{N}^k \to \mathbf{N}$, for $1 \leqslant i \leqslant k$.

Definition 8.2 The *primitive recursive functions* are the elements of the least set of numeric functions containing the base functions and closed under composition and primitive recursion. A *subset* A of $\mathbf{N}^k$ is *primitive recursive* if its characteristic function $1_A : \mathbf{N}^k \to \{0, 1\}$ is primitive recursive.

All the usual functions are primitive recursive (see exercise 4).

Example Addition is primitive recursive:

$$\begin{aligned} 0 + y &= y &&= pr_1^1(y), \\ s(x) + y &= s(x + y) &&= s(pr_3^1(x + y, x, y)). \end{aligned}$$

It was obtained by primitive recursion and composition starting from the base functions pr_1^1, pr_3^1 and s. □

The primitive recursive functions are easy to program. Here is a primitive recursive operator in ML:

```
fun rec_prim g h =
    let fun f(m,n) = if m=0 then g(n) else h(f(m-1,n),m-1,n)
    in f end ;
```

The definition of primitive recursive functions is easily extended from $\mathbf{N} \sim \mathrm{T}_{\{0,s\}}$ to any Σ-algebra of terms, without needing to code the terms by integers.

A primitive recursion definition is only a particular case of a (syntactically) recursive definition. From the sequential programming point of view, the primitive recursive programs are those in which the only iteration structure is the finite loop (Pascal's `for`). Here is the grammar of a sequential language using this structure:

$$\begin{aligned} \mathit{term} \ &::= \ 0 \\ &\ \ \mid \ \ \mathit{variable} \\ &\ \ \mid \ \ \mathit{variable} + 1 \\ \mathit{program} \ &::= \ \mathit{variable} := \mathit{term} \\ &\ \ \mid \ \ \mathit{program}\,;\, \mathit{program} \\ &\ \ \mid \ \ \textbf{do}\ \mathit{variable}\ \textbf{times}\ \mathit{program}\ \textbf{end} \end{aligned}$$

In a loop **do** x **times** π **end**, the variable x is first evaluated, and if a is that value, the program π is executed a times.

The set of primitive recursive functions is stable under certain operations using bounded sets.

If g is primitive recursive, the function f defined by

$$f(n,\vec{m}) = \sum_{p\leqslant n} g(p,\vec{m})$$

is primitive recursive (see exercise 6). The same is true for the bounded quantifiers $\forall_{\leqslant}$ and $\exists_{\leqslant}$: if A is a predicate, define

$$\begin{array}{rcl}(\exists p \leqslant n)A(p,\vec{m}) & = & \max_{p\leqslant n}[(p,\vec{m}) \in A], \\ (\forall p \leqslant n)A(p,\vec{m}) & = & \min_{p\leqslant n}[(p,\vec{m}) \in A].\end{array}$$

If A is primitive recursive, then the predicates $(\exists p \leqslant n)A(p,\vec{m})$ and $(\forall p \leqslant n)A(p,\vec{m})$ are as well. The search for an element in a bounded set is defined by the operator $\mu_{\leqslant}$, or *bounded minimization* μ:

$$\mu y \leqslant m\,.\,A(y) = \begin{cases} \min\{y \leqslant m; y \in A\} & \text{if that minimum exists} \\ m+1 & \text{otherwise.}\end{cases}$$

This operator is computable by primitive recursion:

$$\begin{array}{rcl}\mu y \leqslant 0\,.\,A(y) & = & \begin{cases} 0 & \text{if } 0 \in A \\ 1 & \text{otherwise,}\end{cases} \\ \mu y \leqslant n+1\,.\,A(y) & = & h(\mu y \leqslant n\,.\,A(y), n),\end{array}$$

where h is

$$h(x,n) = \begin{cases} x & \text{if } x \leqslant n \\ n+1 & \text{if } x = n+1 \text{ and } n+1 \in A \\ n+2 & \text{otherwise.}\end{cases}$$

The class of primitive recursive functions is also stable under the bounded maximization operator μ' (see exercise 7).

By applying this operator to the integer quotient $m \mapsto m/2$, it can be shown that the encoding of pairs $\langle\ \rangle$ is primitive recursive, as are the inverse projections π^1 and π^2 (note that $m, n \leqslant \langle m, n\rangle$). It follows that the encoding functions $\langle\rangle_k : \mathbf{N}^k \to \mathbf{N}$ are recursive, as are their inverses π_k^i.

Recall that the history $f^\star$ of a function f is defined by

$$f^\star(x) = \langle f(0), f(1), \ldots, f(x-1)\rangle^\star.$$

By using the definition of the encoding $\langle\ \rangle^\star$, it is easy to see that $f^\star$ is obtained starting from f by the following primitive recursion:

$$\begin{array}{rcl}f^\star(0) & = & 0 \\ f^\star(m+1) & = & \mathit{cons}(f^\star(m), f(m)).\end{array}$$

It follows that $f^\star$ is primitive recursive if f is, and conversely, since $f(m) = \text{last}(f^\star(m+1))$. The history of f incorporates all of the memory of f. It is therefore tempting to generalize primitive recursion (which defines $f(m+1)$ from $f(m)$) using $f^\star$, to define $f(m+1)$ from $f(0), \ldots, f(m)$.

Definition 8.3 Let $h : \mathbf{N}^{k+2} \to \mathbf{N}$. The function $f : \mathbf{N}^{k+1} \to \mathbf{N}$ is defined by *complete* primitive recursion from h if

$$f(m, \vec{n}) = h(f^\star(m, \vec{n}), m, \vec{n}).$$

Given the primitive recursive function h, define g by

$$\begin{aligned} g(0, \vec{n}) &= 0, \\ g(m+1, \vec{n}) &= cons(g(m, \vec{n}), h(g(m, \vec{n}), m, \vec{n})). \end{aligned}$$

It is a primitive recursive definition of g, and it is clear that $g = f^\star$. All that is left to do is to define f by composition:

$$f(m, \vec{n}) = \text{last}(g(m+1, \vec{n})),$$

which proves that f is primitive recursive.

The definition of primitive recursive functions is an inductive definition. Since each of the operators has only a finite number of arguments, this definition can be reformulated 'from below': for every primitive recursive function f, there exists a sequence of functions $f_0, \ldots, f_n$, with $f_n = f$, each function f_i being a base function or a function defined from preceding functions by composition or primitive recursion. Inductive definition can also be seen as an inference system: the base functions correspond to the axioms and the composition operator to the rule:

$$\frac{h, g_1, \ldots, g_p \ \text{ primitive recursive}}{h(g_1, \ldots, g_p) \ \text{ primitive recursive}}.$$

Thus it can be formally proven that a given function is primitive recursive, as a derivation (a term) in this inference system (more precision would be needed, since the preceding rule using '...' is not very formal). Note that a derivation of 'f primitive recursive' is almost a method to compute f. Furthermore, an arbitrary derivation (a term) of this inference system evidently defines a function, since its definition appears in the conclusion.

Recursive functions

It is easy to write programs which do not compute primitive recursive functions, ones which do not always terminate (they compute *partial* functions). It is difficult to write an interesting program which always terminates but which does not compute a primitive recursive function.

Example Ackermann defined a function A in 1928 [3]; see Chapter 2. Here is a slight variant:

$$\begin{aligned} A(0,y) &= y+1, \\ A(x+1,0) &= A(x,1), \\ A(x+1,y+1) &= A(x,A(x+1,y)). \end{aligned}$$

This function is easily computable by a recursive program—although proof of its termination is not easy—but it is not primitive recursive. It can be shown that $A(n,n)$ increases faster than any primitive recursive function. □

Here is another example, which appears to be more artificial. It is possible to enumerate all of the derivations (or constructions) of primitive recursive functions effectively. To the n-th derivation, associate a primitive recursive function ϕ_n, computable by following the derivation step by step. Define $\phi(n) = \phi_n(n) + 1$. This function is computable: to compute $\phi(5)$, write the 5-th derivation of the primitive recursive functions, which constitutes a procedure for ϕ_5, compute $\phi_5(5)$ and add 1. However, if ϕ were primitive recursive, there would exist an integer p such that $\phi = \phi_p$. But $\phi(p) = \phi_p(p)+1 \neq \phi_p(p)$. There therefore exist computable functions which are not primitive recursive. The third operation, minimization, must be introduced.

Definition 8.4 (Herbrand, Gödel) The *recursive functions* are the elements of the least set of numeric functions containing the base functions and closed under composition, primitive recursion and total minimization.

Recursive functions can be programmed in any language allowing arbitrary recursive calls or having a 'while' iteration structure. In ML, the minimization operator can be defined by:

```
fun mu g n = let fun h c = if g(n,c)=0 then c else h(c+1)
             in h 0 end ;
```

Making this closure condition explicit implies that, for every recursive function f, there exists a sequence of functions $f_0, \ldots, f_n$, with $f_n = f$, each function f_i either being a base function or defined from preceding functions using one of the three operators. This sequence of definitions is called a *Herbrand–Gödel system*. However, it is not natural to consider this inductive definition as an inference system, because of the condition to be checked before 'inferring' that a function obtained through minimization is recursive. Rules containing validity conditions have already been shown. This is the case for the $\forall$-introduction rule in natural deduction: there, the condition (that a variable is not free in the assumptions) is easy to check; here, the existence of a zero is of another order of difficulty: it is an 'undecidable' problem.

The minimization operator prescribes a search for a value in an *unbounded* set. This characteristic of computing procedures is essential. It is missing in primitive recursion, which limits the usable resources. The λ-calculus terms constructed using β-reduction can be of arbitrary length (one cannot predict ahead before starting the computation how much paper and ink are necessary to attain a normal form). And Turing machines, it will be shown, use an 'infinite' tape.

The minimization operator is more powerful than primitive recursion. It will be shown that a function defined by *complete* primitive recursion can be obtained by compositions and minimizations. This is particularly true for functions defined by ordinary primitive recursion. This result of recursion 'elimination' will be the subject of a much more general theorem in the case of partial recursive functions (theorem 8.18).

Let f satisfy $f(m) = h(f^\star(m), m)$ for every $m \geqslant 0$ (omit the parameters to simplify). Then $f(m-i) = h(f^\star(m-i), m-i)$ for every i, $0 \leqslant i \leqslant m$, which can be stated by

$$\text{last}(i-1, f^\star(m)) = h(\text{pre}(m-i, f^\star(m)), m-i).$$

Hence, $x = f^\star(m)$ is a solution of the system of equations:

$$\begin{aligned} \text{last}(i-1, x) &= h(\text{pre}(m-i, x), m-i) \qquad (1 \leqslant i \leqslant m), \\ \|x\| &= m. \end{aligned}$$

Define $\bar{f}$ by $\bar{f}(m) = \mu x \,.\, h'(x, m) = 1$, where h' is the primitive recursive function defined by composition and bounded minimization of primitive recursive functions:

$$h'(x, m) \;=\; \|x\| = m \,\wedge\, \forall i (1 \leqslant i \leqslant m) \text{last}(i-1, x) = h(\text{pre}(m-i, x), m-i).$$

Then $\bar{f}$ is primitive recursive, since it is defined explicitly by minimization of h' and satisfies $\bar{f} = f^\star$. And f is defined by composition of $\bar{f}$ and last:

$$f(m) = \text{last}(\bar{f}(m+1)).$$

It is remarkable that the introduction of minimization allows the generation of all computable functions. The following theorem states the equivalence between Herbrand–Gödel and Church's definitions.

Theorem 8.1 For $f : \mathbf{N}^p \to \mathbf{N}$ the following properties are equivalent:

1. f is λ-definable;
2. f is recursive.

Proof It will be shown that recursive functions are λ-definable. The proof is quite simple, but long. A 'compiler' must be created; it translates a Herbrand–Gödel system into λ-calculus terms. This process is realistic, since the source language is of sufficiently high-level, and almost immediately translatable into usual programming languages. The proof uses induction, with cases for base functions, composition, primitive recursion and minimization.

For the base functions, the following terms are used:

$$\begin{aligned}\mathbf{0} &\equiv \lambda fx\,.\,x,\\ \mathbf{succ} &\equiv \lambda nfx\,.\,f(nfx),\\ \mathbf{pr_n^k} &\equiv \lambda x_1\ldots x_n\,.\,x_k.\end{aligned}$$

First consider composition. Let $h : \mathbf{N}^p \to \mathbf{N}$ and $g_i : \mathbf{N}^k \to \mathbf{N}$ $(1 \leqslant i \leqslant p)$, functions whose respective λ-definitions are terms H and G_i. If

$$f(n_1,\ldots,n_k) = h(g_1(n_1,\ldots,n_k),\ldots,g_p(n_1,\ldots,n_k)),$$

let

$$F \equiv \lambda x_1\ldots x_k\,.\,H(G_1x_1\ldots x_k)\ldots(G_px_1\ldots x_k).$$

Then F defines f.

For primitive recursion, suppose that $g : \mathbf{N}^k \to \mathbf{N}$ and $h : \mathbf{N}^{k+2} \to \mathbf{N}$ are λ-defined by terms G and H respectively and let

$$\begin{aligned}f(0,\vec{n}) &= g(\vec{n}),\\ f(m+1,\vec{n}) &= h(f(m,\vec{n}),m,\vec{n}).\end{aligned}$$

This definition of f can be interpreted as a fixpoint equation, by writing it in the form

$$f(m,\vec{n}) = \begin{cases} g(\vec{n}) & \text{if } m = 0\\ h(f(m-1,\vec{n}),m-1,\vec{n}) & \text{otherwise.}\end{cases}$$

It suffices to know the λ-terms **pred** and **zero** which define the functions $m \mapsto m-1$ and $m,u,v \mapsto$ 'if $m = 0$ then u else v' to apply a fixpoint combinator $\mathbf{Y}$ and deduce from it that

$$F = \mathbf{Y}\big(\lambda fx\vec{y}\,.\,\mathbf{zero}(G\vec{y})(H(f(\mathbf{pred}\,x)\vec{y})(\mathbf{pred}\,x)\vec{y})\big).$$

In this case, using a fixpoint combinator is a sledgehammer solution that a good compiler would not use. In fact, this is not an arbitrary syntactic recursive definition but a very particular computation scheme, for which it is known, for example, that it always terminates in m (the first argument) steps: it is preferable that F be normalizable. The solution is to construct a primitive recursion combinator $\mathbf{R}$, normalizable, so that F can be defined from $\mathbf{R}$, G and H.

The idea is to replace primitive recursion by iteration, a simpler operation, over pairs. In fact, for a given k-tuple $\vec{n}$, $f(m,\vec{n})$ is computed by iterating the mapping

$$(x,y) \mapsto (h(x,y,\vec{n}),y+1)$$

m times from the pair $(g(\vec{n}),0)$. There is also a combinator, $\mathbf{cons} \equiv \lambda xym\,.\,mxy$, which produces pairs. Write $[X,Y]$ instead of $\mathbf{cons}\,XY$. The iteration over pairs can be defined by the term

$$\lambda p\,.\,[H(p\mathbf{T})(p\mathbf{F})\vec{n},\mathbf{succ}(p\mathbf{F})],$$

where p is a 'pair' variable, $\mathbf{T}$ and $\mathbf{F}$ are the Booleans *true* and *false*. Let

$$Q \equiv \lambda vp\,.\,[v(p\mathbf{T})(p\mathbf{F}), \mathbf{succ}(p\mathbf{F})].$$

The function Qv is iterated m times, starting from $[u, \mathbf{T}]$ (u will be $G\vec{n}$ and v will be $\lambda xy\,.\,Hxy\vec{n}$):

$$m(Qv)[u, \mathbf{T}].$$

It is a pair whose first component is applied to $\mathbf{T}$:

$$m(Qv)[u, \mathbf{T}]\mathbf{T}.$$

Abstraction is made over m and v to make a function f:

$$\lambda vm\,.\,m(Qv)[u, \mathbf{T}]\mathbf{T},$$

and then over u to obtain the primitive recursion combinator $\mathbf{R}$:

$$\mathbf{R} \equiv \lambda uvm\,.\,m(Qv)[u, \mathbf{T}]\mathbf{T}.$$

It remains to λ-define the function f by the term:

$$F \equiv \lambda m\vec{n}\,.\,\mathbf{R}(G\vec{n})(\lambda xy\,.\,Hxy\vec{n})m.$$

For minimization, suppose that $g : \mathbf{N}^{k+1} \to \mathbf{N}$ satisfies the total minimization condition and is λ-defined by term G. Let $f(\vec{n}) = \mu m\,.\,[g(\vec{n}, m) = 0]$. Essentially, a `while` loop is to be translated into the λ-calculus. First a combinator $\mathbf{W}$ is constructed such that

$$\mathbf{W}U\mathbf{n} = \begin{cases} \mathbf{n} & \text{if } U\mathbf{n} = \mathbf{0} \\ \mathbf{W}U(\mathbf{succ}\,\mathbf{n}) & \text{if } U\mathbf{n} \text{ is a numeral } \neq \mathbf{0}. \end{cases}$$

How does one obtain $\mathbf{W}$? $\mathbf{W}U\mathbf{n}$ must contain a term V which takes as argument $U\mathbf{n}$ and which tests if it is null. Use the combinator $\mathbf{D}$,

$$\mathbf{D} \equiv \lambda xym\,.\,m(\mathbf{K}y)x.$$

The pair $\mathbf{D}XY$ is a functional object, which serves as conditional operator testing if its argument is null:

$$\mathbf{D}XY\mathbf{p} = \begin{cases} X & \text{if } p = 0 \\ Y & \text{if } p > 0. \end{cases}$$

Take for V a pair $\mathbf{D}AB$. If $U\mathbf{n} = \mathbf{0}$, A must return $\mathbf{n}$ and forget everything else. Otherwise, B must continue the computation where it was, with V and $\mathbf{n}$ as arguments, by incrementing $\mathbf{n}$. It is therefore possible to have $\mathbf{W}U\mathbf{n} = V(U\mathbf{n})V\mathbf{n}$, with $V = \mathbf{D}AB$ and to take $A \equiv \lambda vn\,.\,n$ so that $AVn = n$, and $B \equiv \lambda vn\,.\,v(U(\mathbf{succ}\,n))v(\mathbf{succ}\,n)$, so that $BVn = \mathbf{W}U(\mathbf{succ}\,n) = V(U(\mathbf{succ}\,n))$

$V(\mathbf{succ}\, n)$. Since B depends on U, define $V = TU$, defining first T, then $\mathbf{W}$ using T:

$$\begin{aligned} T &\equiv \lambda u \,.\, \mathbf{D}(\lambda vn \,.\, n)(\lambda vn \,.\, v(u(\mathbf{succ}\, n))v(\mathbf{succ}\, n)), \\ \mathbf{W} &\equiv \lambda un \,.\, Tu(un)(Tu)n. \end{aligned}$$

The function f is then λ-defined by the term

$$F = \lambda \vec{n} \,.\, \mathbf{W}(G\vec{n})\mathbf{0}.$$

The converse (every λ-definable function is recursive) will be deduced from general results on inference systems. □

Turing machines

Neither λ-definable functions nor recursive functions are particularly convincing models of computability. Turing undertook to analyze the (human, mental) computation process and proposed a new model in 1936 [151]. His machines are automata, such as finite state push-down automata, which preceded the von Neumann machine by ten years.

A fixed 'blank' symbol, □, will be used below.

Definition 8.5 A *Turing machine* is specified by three finite sets Γ, Q and R:

- Γ is the tape alphabet, where $\square \notin \Gamma$;
- Q is the set of states;
- R is the set of rules, which are 5-tuples (q, s, q', s', d) where $q, q' \in Q$, $s, s' \in \Gamma \cup \{\square\}$ and $d \in \{-, 0, +\}$.

It is useful to have a 'physical' description of a Turing machine. There are three components:

- an infinite tape divided into cells, each capable of holding a symbol from Γ or a blank symbol;
- a read/write head on the tape which moves at most one cell to the left or right at each step;
- a control unit which contains the 'program' R and which, at each step, can be found in one of the states of Q.

As for any automaton, a Turing machine works by transitions between configurations, starting from an initial configuration and working towards a possible final configuration. It is supposed that in the initial configuration, the tape contains a word from $\Gamma^\star$, with blank symbols to the left and to the right. Each transition rule can trigger the writing of an element of $\Gamma \cup \{\square\}$ on the tape. After a finite number of transitions, the tape can only contain a finite number of elements of Γ, with, perhaps, inter-spaced blanks. The contents of the tape can therefore be represented by a word in $(\Gamma \cup \{\square\})^\star$, by omitting all the prefix and suffix □'s.

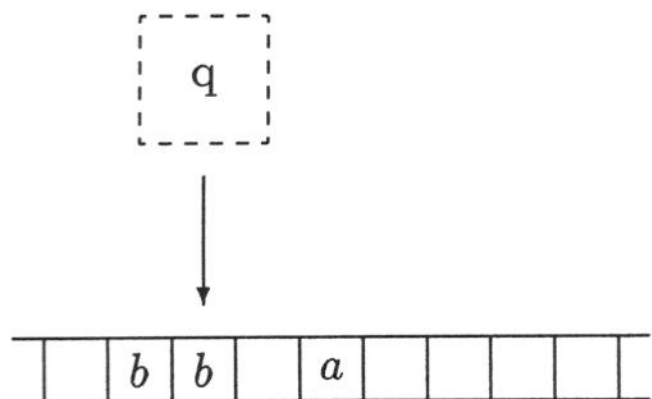

Example The tape contains $bb \,\Box\, a \in (\Gamma \cup \{\Box\})^{\star}$. $\Box$

To represent a machine other than by a picture, the concept of configuration must be formalized by incorporating the current state. A *configuration* is a word of the form $\alpha q \beta$ where $q \in Q$, α, $\beta \in (\Gamma \cup \{\Box\})^{\star}$, with α (respectively β) not starting (respectively finishing) with $\Box$. The configuration $\alpha q \beta$ states that the tape contains the word $\alpha\beta$, that the state is q and that the head points to the beginning of β. In the preceding example, the configuration is $bqb \,\Box\, a$.

Here is the definition of the transitions of a Turing machine (Γ, Q, R): if, in state q, the head points to the symbol $s \in \Gamma \cup \{\Box\}$ and if $\rho = (q, s, q', s', d) \in R$, the ρ triggers the following transition:

- the state becomes q';
- the head writes s' to replace s and then moves one cell to the left if $d = -$ or to the right if $d = +$.

Write this transition $c \xrightarrow{\rho} c'$ as a relation between configurations c (before) and c' (after). The transition rules can be interpreted as rewrite rules *of words* over configurations:

$$aqs \longrightarrow \begin{cases} as'q' & \text{if } d = + \\ aq's' & \text{if } d = 0 \\ q'as' & \text{if } d = -. \end{cases}$$

A *halting* configuration is a configuration for which no rule is applicable (it is a word that is irreducible by rewriting). A *computation* is a sequence of transitions which is either infinite, or finite and terminating by a halting configuration:

$$C_0 \xrightarrow{\rho_1} C_1 \xrightarrow{\rho_2} \dots \xrightarrow{\rho_p} C_p.$$

The sequence $\rho_1, \dots, \rho_p$ is the *trace* of the computation.

A Turing machine has two uses: as an automaton recognizing a language, and as a transducer. This last use will allow the definition of Turing-computable functions. The definition of the computations of a Turing machine leaves, however, the choice of rule to apply to a given configuration, which does not correspond to the normal idea of a program (but does correspond to the activity of a human being[1]). A *deterministic* Turing machine (DTM) is a machine (Γ, Q, R) such that for every

[1]Neither of us believes in predestiny.

$(q, s) \in Q \times \Gamma \cup \{\Box\}$, there exists at most one rule $(q, s, \dots)$ in R. In other words, a DTM is specified by a function $(q, s) \mapsto (q', s', d)$ defined by a subset of $Q \times (\Gamma \cup \{\Box\})$. It is supposed that an initial state q_0 is designated so that an initial configuration will be of the form $q_0\alpha$, with $\alpha \in \Gamma^\star$. With a DTM, for each initial configuration there is only one possible computation.

Let $f : \Gamma^\star \to \Gamma^\star$ be a function. A DTM (Γ, Q, R) computes f if for every $\alpha \in \Gamma^\star$ the computation starting from the initial configuration $q_0\alpha$ terminates in a configuration $q\beta$ and if $f(\alpha) = \beta$ (it is not necessary to impose that the head points to the first letter of β). The function f is *Turing-computable* if there exists a DTM computing f.

This concept of computability, applicable to words, can be transposed to the integers by encoding the integers and k-tuples of integers. Take as alphabet $\Gamma = \{0, 1\}$, and encode the integer n by $n+1$ letters '1' and the k-tuple $(n_1, \dots, n_p)$ by separating the codes for n_i, n_{i+1} with a '0': 1110111101 encodes the triple $(2, 3, 0)$. A function $f : \mathbf{N}^k \to \mathbf{N}$ therefore defines a function $\tilde{f} : \Gamma^\star \to \Gamma^\star$ associating with the encoding of a k-tuple the encoding of its image by f. Note that the image of $\tilde{f}$ is contained in the language $\{1\}^+$.

A numeric function $f : \mathbf{N}^k \to \mathbf{N}$ is *Turing-computable* if, after encoding in Γ, the function $\tilde{f} : \Gamma^\star \to \Gamma^\star$ is Turing-computable by a DTM.

Example The successor function, $s : \mathbf{N} \to \mathbf{N}$, is Turing-computable by the following DTM:

$$\begin{aligned} \Gamma &= \{1\}, \\ Q &= \{q_0, q_1\}, \\ R &= \{(q_0, 1, q_0, 1, +), (q_0, \Box, q_1, 1, 0)\}. \end{aligned}$$

□

Turing showed that all the recursive functions are Turing-computable. His proof, by induction over the construction of recursive functions, consists of compiling the Herbrand–Gödel system into a formalism resembling a macro-assembler and is not particularly interesting. Conversely, all the Turing-computable functions are recursive: the proof of this property is not immediate, but it will follow from more general properties.

If Turing machines give a convincing description of computation, their influence on the design of high-level languages is very limited. From the theoretical point of view, many constructions and proofs are done using Turing machines. Furthermore, they allow a precise definition of the number of 'steps' and 'cells' necessary for a computation, hence of the temporal and spatial complexities of an algorithm. Finally, the Turing machine model is often extended (several tapes, several heads, ...) to facilitate the handling of certain problems, without the expressive power being changed (this is proven for each extension): it is called a 'robust' model.

8.1.3 Church's thesis

This 'thesis', formulated by Church in 1936 [15], is the result of theoretical results and practical experience. The theoretical results are the 'equivalence theorems' between recursive functions, λ-definable functions and Turing-computable functions. Each of these definitions offers a different vantage over the same class of functions. What is the interest of these diverse characterizations?

The usefulness of the λ-calculus for functional programming has already been indicated.

The Herbrand–Gödel definition of recursive functions is the closest to typical programming languages. It highlights the class of primitive recursive functions, which contains all of the usual arithmetic functions. In terms of sequential programming, the primitive recursive functions can be characterized as those functions which can be computed using finite (`for`) loops. Furthermore, their equational definition can be seen as a noetherian and confluent system.

Finally, the description of computation given by Turing seems to be fine enough to cover any imaginable computation.

In addition, it is a result of experience that all known computable functions have been shown to be recursive. And contemporary practice has only confirmed this experience.

Church thesis | 'Every effectively computable function is recursive.' |

In this formulation, only the word *recursive* has a rigorous definition, while *effectively computable* refers to the 'physical' concept of computation: effective, mechanical, automatic, etc. It is certainly not a theorem, but a postulate of experimental origin. The thesis has practical consequences:

- Do not try to write a program to compute a function which has been shown to be non-recursive.
- A visibly computable function can, during a proof, be considered to be recursive to avoid a tedious proof (but one must be sure!), and one can then deduce, for example, the possibility of a Turing machine computing it without having to construct one explicitly.

One normally refers to computable functions, for recursive, λ-definable or Turing-computable functions. The list of equivalent characterizations of computable functions can still be completed:

- canonical systems (Post 1943 [123]);
- theory of Markov algorithms (1951 [99]);
- elementary formal systems (Smullyan 1961 [138]);
- and in 1981 Kleene himself [82] gave a new characterization, originating in the recursion theorem for computable functionals.

The Church thesis, as given here, deals with the class of computable functions as extensional objects, inside the class of functions, and not as intensional computational processes. The equivalence theorems state that this class is well defined. It can be obtained by using several more or less useful definitions, and it is the multiplicity of approaches which makes this class particularly interesting and which motivates the Church thesis.

The proofs of equivalence theorems prove a lot more than the simple equality between sets of functions. They are constructive proofs which transform a Herbrand–Gödel definition into a λ-term, and into a Turing machine, and conversely. The same concept of *algorithm* underlies all these constructions. These remarks justify the following statement:

Church (strong) thesis ‘If a function f is computable by an algorithm, then this algorithm can be effectively transformed into a Turing machine computing f.’

One can then freely refer to an algorithm, in the same manner that the first version of the Church thesis allows one to refer to a computable function—without specifying the model used. In what follows, algorithms will be designated using the notations which define them: for example a Turing machine will be referred to as an algorithm, and when an encoding of these machines is introduced, an integer n will represent an algorithm executed by the n-th Turing machine.

Logicians have never been able to define the concept of algorithm as generally as the concept of function. For Turing machines, one could state that an algorithm is defined by a Turing machine. For the λ-calculus, a reduction strategy must be associated with a term. For recursive functions, it is a system of Herbrand–Gödel equations. An algorithm always uses a notation (a triplet (Γ, Q, R), a term or a system of equation), which is only usable as ‘instructions’ when more general constructions (a word rewriting relation, β-reduction or an inference system) indicate how these instructions are executed. These notations do not in themselves represent the computing procedures, but do suffice to define them.

For us, an algorithm is a computation procedure which is practised without really looking for a formal definition. A major contribution of logic is to have given a mathematical status to ‘computation’ by defining convincing models. Logic should then be seen as an experimental science, just like mechanics, which succeeded in defining the concept of force from that of interaction of bodies. It is this aspect, and not that of metamathematics or that of theory of theories, which is at the origin of its natural links with computer science.

8.1.4 Kleene and partial functions

The preceding processes sought the computable functions from $\mathbf{N}^k$ to $\mathbf{N}$ and how to compute them. By doing this, computing methods—leftmost β-reduction of a

λ-term and rewriting of configurations in a Turing machine—and notations were introduced to specify these processes: λ-terms, deterministic Turing machines or Herbrand–Gödel systems. If f is computable, there exists a term M_f, a Herbrand–Gödel system and a DTM which compute f.

Conversely, what should be done with an arbitrary term, system or DTM? The same computing processes can always be used, but in general, they will compute *partial functions*.

A partial function from A to B, written $f : A \rightharpoonup B$, is a (total) function from a subset $D(f) \subseteq A$, the domain of f, to B. The expression $f(a)\downarrow$ means that the partial function f is defined in a, i.e, $a \in D(f)$. What is the meaning of a partial function and its domain? From a denotational point of view, $a \notin D(f)$ states simply that $f(a)$ is not defined. From an operational point of view, either the computation of $f(a)$ never terminates or there is no instruction to compute it, in which case an error is detected and the computation cannot continue. It will be shown that it is not possible in general to decide in advance if a computation halts: the concept of domain of a partial function is not effective. The first undecidability problems appear with the general use of these computation processes.

Hence, if M is a λ-term, $M\mathbf{n_1}\ldots\mathbf{n_k}$ is not always normalizable and its normal form, it it exists, is not necessarily a Church numeral.

In the case of a deterministic Turing machine (Γ, Q, R) with initial state $q_0 \in Q$, consider the initial configuration $q_0\alpha$, where $\alpha \in \Gamma^\star$. The unique computation starting in this configuration does not necessarily halt. If it halts in configuration $\beta q \gamma$, let $f_R(\alpha) = \beta\gamma$. The function $f_R : \Gamma^\star \rightharpoonup \Gamma^\star$ is the partial function computed by (Γ, Q, R).

Consider numerical functions. The problem is no longer to find a DTM computing a given function, rather to start from an arbitrary DTM over the alphabet $\Gamma = \{0, 1\}$ and to make it run. Let $k \geqslant 1$ be fixed, and consider the initial configurations which correspond to encodings of k-tuples of integers: they are of the form $q_0\alpha$, where $\alpha = 1^{m_1}0\ldots01^{m_k}$. As in the case of λ-terms, the computation from one of these initial configurations, when it halts, does not necessarily produce a configuration containing the encoding of an integer. When the final configuration is $\beta q\gamma$, with $\beta\gamma = 1^{m+1}$, let $f_R^k(m_1, \ldots, m_k) = m$. The function f_R^k is the k-ary partial function computed by R. For a given R over the alphabet $\{0, 1\}$, the preceding constructions yield partial functions $f_R : \Gamma^\star \rightharpoonup \Gamma^\star$, $f_R^1 : \mathbf{N} \rightharpoonup \mathbf{N}$, $f_R^2 : \mathbf{N}^2 \rightharpoonup \mathbf{N}$, etc. The f_R^k are, by definition, the *Turing-computable partial functions*.

Consider also systems of recursive functions. Each of the three Herbrand–Gödel operations is 'computable' by a sequential program. Minimization is very different from the others: the existence of zeros over g must be satisfied before one can define $f(x) = \mu y\,.\,[g(x, y) = 0]$. If minimization is applied without checking this condition, a program might not halt and will then only compute a partial function. If it halts, the number of iterations is not known in advance, while in primitive recursion exactly k iterations are used to compute $f(k, \vec{n})$). The existence of such zeros is not necessarily decidable: the only general way to check and obtain the minimum is to compute $g(\vec{m}, 0)$, $g(\vec{m}, 1)$, $g(\vec{m}, 2)$, ... successively. Furthermore, there is no way to extend effectively a partial function f obtained under minimization to a total function. A definition such as

$$f(\vec{n}) = \begin{cases} \min\{k \in \mathbf{N}; g(\vec{n}, k) = 0\} & \text{if this minimum exists} \\ 0 & \text{otherwise} \end{cases}$$

is not effective, even if g is computable.

Therefore, the class of *total* recursive functions is not well adapted to programming. It must be extended to partial functions, which are the objects naturally computed by a computation process. Kleene studied them systematically starting in 1938 [80]. If partial functions are allowed, the total minimization condition is no longer necessary.

There is worse to come: the minimization operation, applied as a *partial function*, is itself not necessarily computable.

Example Let h be a computable partial function such that the problem $m \in D(h)$? is undecidable. Let g be the partial function defined by

$$g(m, n) = 0 \quad \text{if } n = 1, \text{ or if } n = 0 \text{ and } m \in D(h),$$

and f by $f(m) = \mu n \,.\, [g(m, n) = 0]$. Then

$$f(m) = \begin{cases} 0 & \text{if } m \in D(h) \\ 1 & \text{otherwise.} \end{cases}$$

It is clear that f is not computable since it is a definition by cases over an undecidable condition. However, if this definition is translated into a program, the latter will compute f', which equals 0 over $D(h)$ and is not defined elsewhere, since if $m \notin D(h)$, the program computing $h(m)$ does not halt, so the decision as to whether $f(m) = 1$ is never made. □

This example shows that the class of computable partial functions is not closed under minimization and that when there are partial functions the operational interpretation of a definition can differ from its (mathematical) denotational interpretation.

A precise definition for the class of *partial recursive* functions, the same as the classes of partial Turing-computable and λ-definable functions, will be given later on. Practice allows the statement of a version of the Church thesis for partial computable functions:

Church–Kleene thesis 'An effectively partial computable function is recursive.'

The use of this version is sometimes dangerous, as was illustrated by the above example of non-effective minimization.

The fundamental results of recursive function theory, due to S. Kleene, are presented in the framework of computable partial functions. Recursive function theory is an important branch of logic, and several applications to metaprogramming will be presented. Before defining the partial recursive functions rigorously, the Church thesis will be used to equate partial recursive Turing-computable functions.

It is time to distinguish the different uses of the word 'recursive': the logical, syntactical and algorithmic points of view.

In logic, 'recursive' refers to recursive functions and, by extension, to the study of computability; the quality of being recursive is recursiveness.

In computer science, there are several uses of 'recursive'. A definition of an object is called recursive when this object is defined in terms of itself. Hence an inductive definition is recursive: this use corresponds better to *impredicativity* in logic. Syntactic recursivity refers to when the same symbol appears in the two sides of a definition. For example, the rule in a context-free grammar $Z \longrightarrow aZ$ is recursive since the non-terminal Z appears to the left *and* to the right of the $\longrightarrow$. A definition of the form

```
datatype nat = zero | succ of nat ;
```

is recursive since `nat` appears on both sides of the `=`. Similarly, one refers to recursive definitions of functions and procedures. These functions are obviously computable, hence recursive, in the logical sense. These definitions generally have a fixpoint equation as semantics. However a mathematician would never call the differential equation $y' = y$ 'recursive'.

In programming, the resources required to execute a task distinguish recursive from iterative. A procedure is recursive if its execution requires the repeated allocation of memory, in practice, the growth of an unbounded stack. In general, a procedure whose definition is syntactically recursive is executed in a recursive manner. A procedure is iterative if its execution can be done in constant space. This is the case of a program which only uses iterative control structures. It it also the case for procedures using tail-recursion, which is syntactically recursive, if the compiler knows how to handle them correctly: it is unnecessary to preserve the stack when tail-recursion is being used.

When recursive functions of logic are to be computed by a program, a Herbrand–Gödel sytem can be translated into recursive function definitions—primitive recursion is already in this form, and minimization calls for computing $g(0)$, where $g(n)$ calls $g(n+1)$. An iterative procedure can also be constructed: a `for` loop for primitive recursion and a `while` loop for minimization.

These three uses are all distinct.

8.1.5 Computing in a logical system

First-order logic and arithmetic

Consider first-order logic with the NK system, and let T be a theory with equality. Since what are of interest are numeric functions, suppose that the language L_Σ of T contains a constant (or a closed term) $\mathbf{n}$, called *numeral*, for each integer n, and that $T \vdash \neg(\mathbf{n} = \mathbf{m})$ as soon as n and m are distinct integers. This is the case for elementary arithmetic where $\mathbf{n}$ is $s^n 0$ (see exercise 26).

Definition 8.6 Let $f : \mathbf{N}^k \to \mathbf{N}$. The function f is *T-defined* by formula $\varphi \in L_\Sigma[x_1, \ldots, x_k, y]$ if: for every $(n_1, \ldots, n_k, m) \in \mathbf{N}^{k+1}$, $f(n_1, \ldots, n_k) = m$ if and only if

$$\boxed{T \vdash \forall y(\varphi(\mathbf{n_1}, \ldots, \mathbf{n_k}, y) \iff y = \mathbf{m}).}$$

f is then *T-definable*. Note that this condition implies in particular the theorem $\varphi(\mathbf{n_1}, \ldots, \mathbf{n_k}, \mathbf{m})$, if $f(n_1, \ldots, n_k) = m$.

The problem is to determine which theories define all the recursive functions. It is clear that if a theory T is suitable, every extension of T is suitable as well. However, a restriction T' of T (restriction of the language, axioms or inference rules) might not suffice. It is therefore interesting to know the minimal theories.

Theorem 8.2 (Gödel definability) A numeric function is recursive if and only if it is definable in Peano arithmetic.

The proof of this system consists precisely of a compilation of a Herbrand–Gödel system into a finite set of formulas of arithmetic. The proof can also be seen as the synthesis of a program from its specification if a definition of a recursive function is seen as a specification and the formulas of arithmetic as programs. Here are some details. The reasoning uses induction over the Herbrand–Gödel definitions of recursive functions; there are three cases. That of composition is easy (exercise 10).

For minimization, let $f = \mu y \,.\, [g(.,y) = 0]$. If g is defined by the formula $\varphi(x, y, z)$, it is easy to prove that f is defined by the formula

$$\psi(x, u) \equiv \forall y(y < u \Rightarrow \neg\varphi(x, y, \mathbf{0})) \wedge \varphi(x, u, \mathbf{0}).$$

It is not as easy for primitive recursion. It was shown that if f is defined by complete primitive recursion from h, i.e.,

$$f(m, \vec{n}) = h(f^\star(m, \vec{n}), m, \vec{n}),$$

then the history $f^\star$ of f is obtained by minimization of a function h' constructed explicitly (i.e, by composition and bounded minimization) from h and a few functions linked to the encoding of n-tuples of integers. Hence f is obtained directly by composition: $f(m) = \text{last}(f^\star(m+1))$. Apart from the other two cases, it suffices to prove that the few encoding functions are definable, which will not be done here.

It would have been more astute to omit primitive recursion from the definition of recursive functions, by adding the predicate $<$ over $\mathbf{N}^2$ to the base functions. It would then suffice to show that the encoding functions can be obtained by (bounded) minimization and compositions from the base functions. But the primitive recursion scheme is quite natural in programming. The difficulty of this case shows that definability in arithmetic is not well adapted to programming. It would lead to a programming style avoiding 'recursion' and favoring imperative structures, such as `while`, which translates directly into an arithmetic formula, and systematic encodings of data structures by integers. Peano arithmetic appears to be a very low-level language to describe typical programs. Other concepts of definability, with other inference systems, can be used more naturally.

This result has been extended to elementary arithmetic, which is finitely axiomatizable, and it will be used later on. However, Presburger arithmetic, i.e. with only addition, is insufficient.

Equational logic

Instead of working in a fixed theory, one can, for each recursive function f, construct a theory which defines f: the representation is in the inference system rather than in the theory.

Kleene (1952 [81]) showed that partial recursive functions are definable in equational logic (see Chapter 6, section 6.1.3). It is clear that the construction by composition and primitive recursion is equational in itself. The case for minimization is less simple. Introduce two new binary symbols a and b, along with the equations:

$$\begin{aligned} a(u, sv) &= u, \\ b(u, 0) &= u. \end{aligned}$$

Consider f to be defined so that $f(x) = \mu y \,.\, [g(x, y) = 0]$, where g has already been defined by a set of equations. Introduce a new symbol $\tilde{g}$ with the same arity as g, along with the equations,

$$\begin{aligned} \tilde{g}(x, 0) &= g(x, 0) \\ \tilde{g}(x, sy) &= a(g(x, sy), \tilde{g}(x, y)), \end{aligned}$$

and then the equation defining f:

$$f(x) = b(u, \tilde{g}(x, u)).$$

Write E for the set of introduced equations (including those defining g, a and b). Then $E \vdash f(\mathbf{n}) = \mathbf{m}$ if and only if $f(n) = m$.

Logic programs

In 1977 Tärnlund ([147]) proved that all partial recursive functions are computable by a definite logic program. This can be done by simulating Turing machines, or in the sense of §8.1.5. As in the equational case, only minimization is not obvious.

The integers are coded by terms over the signature $\{0, s\}$. A k-ary partial function f is defined by a program P with a $k+1$-ary predicate $\mathtt{f}$ associated with f such that: $f(\vec{n}) = m$ if and only if $\mathtt{f}(s^{n_1}(0), \ldots, s^{n_k}(0), s^m(0)) \in \mathcal{M}_P$, the minimal model of P.

To define $f(x) = \mu y\,.\,(g(x,y) = 0)$, suppose g is defined by the program P_0 and the $k+2$-ary predicate `g`. Introduce the $(k+1)$-ary symbols `h` and `f`, and form the program P starting from P_0 and the clauses (here $k = 1$):

```
h(X,0).
h(X,s(Y)) :- h(X,Y), g(X,Y,s(Z)).
f(X,Y) :- h(X,Y), g(X,Y,0).
```

Then the goal `f(`s^n`(0),Y)` has a solution `Y =` $s^m(0)$ if and only if $f(n) = m$.

The λ-calculi
The pure λ-calculus is formalized by an inference system whose judgments are the $M \triangleright N$, where M and N are terms. The very first definition which defined computable functions was that of functions definable in this inference system.

The typing rules for terms of a typed λ-calculus involve judgments of the form $C \vdash M : \tau$ and restrict the formation of terms. Since there are fewer terms, there are fewer numeric functions definable by a typed λ-calculus. In the case of simple types, only 'polynomials with conditional' can be defined, which is insufficient for programming. The addition of constants, along with their contraction rules, or of a more powerful typing system, allows the definition of more numeric functions. For example, Gödel's system T (a typed λ-calculus with atomic types for integers and Booleans, and for constants zero, successor, recursors and conditionals) allows the definition of all functions whose termination is provable in Peano arithmetic.

8.2 Decidability

This section studies *problems* from the decidability point of view. The logicians of the 1930s, dealing with the new methodological requirements of mathematicians, the foundation problem, discovered unthought of phenomena which would have profound implications. By creating a real 'effective' laboratory, computer science illustrates these phenomena concretely.

8.2.1 Problems and encodings

A problem is specified by the set of its *instances* and by a *query*.

Example
Given: an context-free grammar (Z, A, P, z_0).
Problem: is the generated language $L(z_0)$ non-empty? □

The queries considered here allow *yes* or *no* as answers: a query determines the set of instances for which the answer is positive. From an abstract point of view, these are decidability problems of the form $x \in A$?, for a subset A of the set of instances.

To define a decision problem to an algorithm, the statement of each instance must be made precise. It is natural to suppose that the instances are 'finite' objects which are uniformly and effectively representable by words over a finite alphabet or by integers.

Example (cont.) A context-free grammar can be given by

- the number n of its non-terminal symbols, designated by $z_0, \dots, z_{n-1}$;
- the number t of its terminal symbols, designated by $a_0, \dots, a_{t-1}$;
- a finite sequence of production rules, each being a pair formed of a non-terminal and a word over the alphabet $\{z_0, \dots, z_{n-1}, a_0, \dots, a_{t-1}\}$; this sequence can be represented by a word over the alphabet, extended by a symbol, e.g., '#', separating the left- and right-hand sides of rules, along with the rules themselves.

This representation is not quite satisfactory, since it requires an infinite alphabet to allow for an arbitrary number of grammar symbols. These symbols must be coded in a finite alphabet using, for example, the three letters '`N`' (non-terminal), '`T`' (terminal) and '`1`': z_0 is represented by `N`, z_1 by `N1`, z_2 by `N11`, etc., and a_0 by `T`, a_4 by `T1111`, etc. A grammar is therefore represented by a word over the alphabet $\{\mathtt{N}, \mathtt{T}, \mathtt{1}, \mathtt{\#}\}$. □

It is this kind of encoding which must be produced when the decidability of a problem is to be studied. What follows does not deal with problems which cannot be encoded explicitly. In particular, the results of this section will not be applied to any arbitrary problem. However, in many domains there are quite natural encodings: programming practice ensures that normal objects such as programs or data such as files, floating point numbers and graphs are obviously encoded: it will be rare that the encoding of a problem will be given. An encoding for k-tuples of integers has already been given in §8.1.1, but the construction of the Gödel encoding of logic will not be given.

So decision problems will be studied in the abstract form $x \in A$?, for $A \subseteq \mathbf{N}^k$ $k \geqslant 1$, or sometimes, $A \subseteq X^\star$, with a finite alphabet X. A decision algorithm is equivalent to the computation of the characteristic function $1_A : \mathbf{N} \to \{0, 1\}$, of the set A (of encodings) of the instances of the problem which have *yes* as answer. The semi-characteristic function of a subset A of $\mathbf{N}^k$ will also be used: it is the partial function $(1/2)_A$ of domain A where its value is 1.

8.2.2 Recursive and recursively enumerable sets

Recursive sets are defined from recursive functions.

Definition 8.7

- A subset A of $\mathbf{N}^k$ is called *recursive* if its characteristic function $1_A : \mathbf{N}^k \to \{0, 1\}$ is recursive.

- Let $A \subseteq \mathbf{N}^k$. The problem $n \in A$? having as instances the $n \in \mathbf{N}^k$ is called *decidable* if A is recursive.

Note a few properties of recursive sets which follow from the same properties of recursive functions.

Proposition 8.3 Let A, B be recursive subsets of $\mathbf{N}^k$. Then $A \cap B$, $A \cup B$, $A \setminus B$ and $\mathbf{N}^k \setminus A$ are recursive.

Proof Since $1_{A\cap B} = \min(1_A, 1_B)$, $1_{A\cup B} = \max(1_A, 1_B)$ and $1_{\mathbf{N}^k\setminus A} = 1 - 1_A$, the proposition follows from the recursiveness of functions min, max and $t \mapsto 1 - t$. □

Conversely, recursive functions can be defined from recursive sets. This characterization cannot be extended to partial functions.

Proposition 8.4 A function $f : \mathbf{N}^k \to \mathbf{N}$ is recursive if and only if its graph

$$G_f = \{(n_1, \dots, n_k, f(n_1, \dots, n_k)); n_i \in \mathbf{N}\}$$

is a recursive subset of $\mathbf{N}^{k+1}$.

Proof This equivalence is proven by the two relations

$$\begin{aligned} 1_{G_f}(n_1, \dots, n_k, p) &= 1_{\{0\}}(f(n_1, \dots, n_k) \dot{-} p), \\ f(n_1, \dots, n_k) &= \mu p \,.\, [1_{G_f}(n_1, \dots, n_k, p) = 1]. \end{aligned}$$

□

The class of recursive sets is closed under union, intersection and complement. However, it is clear that there exist non-recursive sets, using a simple cardinality argument, and they can be obtained by other operations. The simplest construction is projection: if $A \subseteq \mathbf{N}^2$ is recursive, a projection from A to $\mathbf{N}$ is not necessarily recursive, just as the projection of a measurable set of $\mathbf{R}^2$ is not necessarily measurable. The projections of recursive sets can also be obtained by operations significant to computer science, such as the set of results of a program or the set of inputs for which a program halts.

Definition 8.8

- A subset A of $\mathbf{N}^k$ is *semi-recursive* if it is the projection over $\mathbf{N}^k$ of a recursive subset of $\mathbf{N}^{k+1}$.
- A subset A of $\mathbf{N}$ is *recursively enumerable (r.e.)* if it is empty or is the image of a recursive function.

It will be shown using the Church thesis that these two classes of sets are the same.

Let $A \subseteq N^k$ be a semi-recursive set, i.e. a projection over $\mathbf{N}$ of a recursive set $B \subseteq \mathbf{N}^{k+l}$, $l \geqslant 1$. Then the semi-characteristic function of A,

$$(1/2)_A(\vec{m}) = \exists n\,.\,[1_B(\vec{m}, n) = 1],$$

is computable since it is obtained by minimization of a recursive function, and A is the domain of $(1/2)_A$. Hence a semi-recursive set is the domain of a partial computable function.

Let $f : \mathbf{N}^k \rightharpoonup \mathbf{N}$ be a partial computable function and M an algorithm (a λ-term or a Turing machine) computing f. Associate with M a recursive set G_M which is an 'effective hull' of the graph of f. The set G_M is formed of $(k+2)$-tuples $(\vec{m}, n, p) \in \mathbf{N}^k \times \mathbf{N} \times \mathbf{N}$ such that the computation of $f(\vec{m})$ halts in less than p steps and produces n. It is computable: the computation of $1_{G_M}(\vec{m}, n, p)$ consists of making M run at most the p first steps of the computation of $f(\vec{m})$. If it halts at a step $p' \leqslant p$, then $1_{G_M}(\vec{m}, n, p) = 1$, otherwise $1_{G_M}(\vec{m}, n, p) = 0$. This proof can be formalized to prove rigorously that G_M is recursive. It is a typical example where one can invoke the Church thesis. Note that the image of f is the projection from G_M to the $(k+2)$-th component, that the domain of f is the projection from G_M to the k first components, and that the graph of f is the projection from G_M to the $k+1$ first components. Hence, if f is partial computable, its image, its domain and its graph are semi-recursive sets.

It remains to be shown that the non-empty domain of a partial computable function can be enumerated by a recursive function. Since all of the considered functions take values in $\mathbf{N}$, suppose that $A = D(f) \neq \varnothing$, with $f : \mathbf{N} \rightharpoonup \mathbf{N}$. An algorithm M computing f will be used, along with the encoding $\langle\ \rangle$ of $\mathbf{N} \times \mathbf{N}$ to produce a sequence of integers, i.e., a function from $\mathbf{N}$ to $\mathbf{N}$, enumerating A by interleaving the computations of $f(0), f(1), \ldots$. For $n \in \mathbf{N}$, the $(\pi^2(n))$-th step of the execution of $f(\pi^1(n))$ is executed, if it has not halted previously, i.e, the first step of $f(0)$ is executed, then the second step, then the first of $f(1)$, the third of $f(0)$, the second of $f(1)$, the first of $f(2)$, etc. If it terminates at instant n, $\pi^1(n)$ is produced as the new element of the list. This function must be shown in total, i.e., this enumeration must not halt. Since $A \neq \varnothing$, there exists at least one integer m in the domain of f, hence an integer p such that the computation of $f(m)$ halts in p steps. Then m will be produced at instant n, with $n = \langle m, p\rangle$, and an infinite number of times subsequently.

To refer to a recursively enumerable set of dimension $k \geqslant 2$, the recursive bijection $n \mapsto (\pi^1(n), \ldots, \pi^k(n))$ between $\mathbf{N}$ and $\mathbf{N}^k$ is used. It is standard to use 'recursively enumberable' (r.e.) rather than 'semi-recursive' no matter what the dimension or the mode of definition (projection, domain, graph or image) is. The following proposition has been shown:

Proposition 8.5 Let $A \subseteq \mathbf{N}^k$. The following properties are equivalent:

1. A is the projection of a recursive subset of $\mathbf{N}^{k+1}$.
2. A is the image of a recursive function.

3. A is the image of a partial computable function.
4. A is the domain of a partial computable function.
5. A is the graph of a partial computable function.
6. The semi-characteristic function $(1/2)_A$ is partial computable.

In the preceding discussion, the Church thesis was used to handle partial computable functions. It is time to give a precise and simple definition of partial recursive functions.

Definition 8.9

- A *partial recursive* function is a partial function whose graph is r.e.
- Let $A \subseteq \mathbf{N}^k$. The problem $n \in A$? having the $n \in \mathbf{N}^k$ as instances is called *semi-decidable* if A is r.e.

Every recursive set is r.e. The construction of non-recursive r.e. sets is not easy. This is the technical version of undecidability results: the best-known examples are the set of provable formulas of arithmetic and the set of normalizable terms of the λ-calculus. The following proposition explains the term 'semi-decidable', which comes from 'semi-recursive':

Proposition 8.6

- $A \subseteq \mathbf{N}^k$ is recursive if and only if A and $\mathbf{N}^k \setminus A$ are semi-recursive.
- The problem $n \in A$? is decidable if and only if the problems $n \in A$? and $n \notin A$? are semi-decidable.

Proof

(1) If A is recursive, then A and $\mathbf{N}^k \setminus A$ are recursive, hence semi-recursive.

(2) Conversely, if A and $\mathbf{N}^k \setminus A$ are semi-recursive, let B, $C \subseteq \mathbf{N}^{k+1}$ be recursive projections A, $\mathbf{N}^k \setminus A$, respectively, over $\mathbf{N}^k$. Let $f(x) = \inf\{n \in \mathbf{N}; (x, n) \in B \cup C\}$, $x \in \mathbf{N}^k$. Under minimization, f is a recursive function from $\mathbf{N}^k$ to $\mathbf{N}$ and $A = \{x \in \mathbf{N}^k; (x, f(x)) \in B\}$. Hence, A is recursive. □

The equivalence between some characterizations of r.e. sets, hence semi-decidable problems, was given. The justification of this equivalence is constructive, where, starting from an algorithm computing a partial function of domain A, an algorithm effectively enumerating A is constructed. The 'declarative' definition must be distinguished from its operational interpretations: how is a semi-decision procedure for the problem $x \in A$ to be otained from a definition of A?

If A is the projection of a recursive set B, the procedure is to compute $1_B(0)$, $1_B(1)$, $1_B(2)$, ..., until, possibly, $1_B(x) = 1$.

If A is the image of a recursive function f, the sequence $f(0)$, $f(1)$, $f(2)$, ..., is computed, until, possibly, x is in the sequence. Note that the number of $f(p)$ to be computed is not known ahead of time, while the resolution of the decidable problem $n \in A$? requires a single computation, that of $1_A(n)$. Thus a problem is semi-decidable is there exists a program which, given an instance of the problem, halts with 'yes' as result if the answer is 'yes'.

This procedure is no longer valid if A is the image of a *partial* recursive function f. The naïve operational interpretation, which consists of successively computing $f(0)$, $f(1)$, ... , need not enumerate A completely. It suffices that the computation of $f(0)$ does not terminate for this procedure to only enumerate the empty set—even if $f(\mathbf{N}) = A$. An analogous situation is that of a tree traversal having infinite branches, as is the case in logic programming: a breadth-first search is 'complete', since all the leaves are reached, while a left-to-right depth-first search is 'incomplete': if the left branch is infinite, no leaf will be enumerated. The algorithm presented here does not enumerate the results of the computations of the $f(p)$, but enumerates the computations themselves by 'interleaving' them: hence the non-termination of the computation of $f(p)$ will not block the enumeration of the other results. Programming this interleaving requires coroutine mechanisms and the interruption of computations. This is an uncommon programming structure, similar to the *engines* of Scheme. Among the programming mechanisms which present themselves naturally in this chapter, such as recursion and compilation, it is the only one which is not yet implemented in a satisfying manner in languages. These techniques are now being introduced in logic and functional programming.

Summary

- The recursive functions are the functions obtained from the base function using composition, primitive recursion and total minimization.
- The recursive sets are the sets whose characteristic function is recursive.
- The recursively enumerable (r.e.) sets are the projections of recursive sets.
- The partial recursive functions are the partial functions whose graph is r.e.
- The problem $x \in A$ is decidable (respectively semi-decidable) if A is recursive (respectively r.e.).

8.2.3 Definition of recursive sets in arithmetic

In first-order logic, a subset A of $\mathbf{N}^k$ is *defined* in theory T by a formula $\varphi \in \mathrm{L}_\Sigma[x_1, \ldots, x_k]$ if:

(1) $(n_1, \ldots, n_k) \in A$ implies $T \vdash \varphi(\mathbf{n_1}, \ldots, \mathbf{n_k})$;
(2) $(n_1, \ldots, n_k) \notin A$ implies $T \vdash \neg\varphi(\mathbf{n_1}, \ldots, \mathbf{n_k})$.

If T is consistent, 'implies' can be replaced by 'if and only if'. If the function 1_A is T-definable, then A is also T-definable (see exercise 13). Gödel's theorem (8.2) allows the statement of the following proposition, which will only be extended to r.e. sets in a weaker form, by only requiring part (1) of the preceding definition.

Proposition 8.7 Every recursive set is definable in elementary arithmetic.

8.2.4 The decidability problem in logic

Let Σ be a signature, $\Gamma \subseteq \mathrm{L}_\Sigma$ a set of axioms and T the theory generated by Γ. The decidability problem in T is: $\varphi \in T$? i.e, $\Gamma \vdash \varphi$? To study the recursiveness of T, an encoding must be constructed. This was done by Gödel in 1931 [55] and it will not be described in detail here. In the following discussion, it is supposed that the signature Σ and the set X of variables are countable. All syntactic objects, i.e., elements of Σ, X, $\mathrm{T}_\Sigma[X]$, $\mathrm{L}_\Sigma[X]$ and $\mathrm{D}_\Sigma[X]$ (derivations) are encoded by integers. The symbol $\#k$ will denote the integer encoding object k, be it a term, a formula or something else. What is important is that the sets of encodings of each of the sets be recursive and that the usual operations over these objects be themselves encoded by recursive functions: inductive constructions of terms, formulas, derivations and substitutions of a variable by a term in a formula. The classical construction gives sets of encodings and operations which are *primitive recursive*: this fact will be needed. In particular, the substitution functions s^k satisfying

$$s^k(m, n_1, \ldots, n_k) = \begin{cases} p & \text{if } m = \#\varphi, \text{ for } \varphi \in \mathrm{L}_\Sigma[x_1, \ldots, x_k] \\ & \text{and } p = \#\varphi[x_1 := \mathbf{n_1}, \ldots, x_k := \mathbf{n_k}] \\ m & \text{otherwise} \end{cases}$$

will be used. The symbol $\ulcorner k \urcorner$ will stand for term $\mathbf{n}$ if $\#k = n$. It is important to distinguish the integer n from the term $\mathbf{n}$, and the syntactic object k from its numeric encoding $\#k$ and its syntactic encoding $\ulcorner k \urcorner$.

T is *recursively axiomatized* if (the set of encodings of) Γ is a recursive subset of $\mathbf{N}$. Intuitively, this means that one can effectively decide if a proposition is an axiom. This assumption is reasonable and is satisfied in particular if Γ is finite.

Peano arithmetic, which contains an infinity of axioms, because of the recursion axiom scheme, is recursively axiomatized. This is also the case of the theory of algebraically closed fields, of fields of characteristic zero and of ZF set theory. The completion of a logic program is also a recursively axiomatized theory. However, consider complete arithmetic, a theory formed of all the true formulas in $\mathbf{N}$. All those formulas could be taken as axioms, but it will be shown that this set is not recursive, which makes those 'axioms' rather limited in their utility since they would be unrecognizable.

If Γ is recursive, it can be proven that the set der_Γ of pairs $(\#d, \#\varphi)$ where d is a derivation of φ with assumptions in Γ is recursive: this means that one can recognize a derivation, that the formulas in the leaves are axioms and that the conclusion is φ. It follows that the set th_Γ of $\#\varphi$ such that $\Gamma \vdash \varphi$, which is the second projection of der_Γ, is recursively axiomatizable. Hence:

Proposition 8.8 The decision problem of a recursively axiomatizable theory is semi-decidable.

In the case where $\Gamma = \varnothing$, we obtain: the decision problem for the predicate calculus is semi-decidable.

Definition 8.10 A theory is *decidable* if the set (of encodings) of its theorems is recursive. Otherwise, the theory is called *undecidable*. A theory T is called *essentially undecidable* if every extension T' of T, over the same signature, is undecidable.

Examples

- Group theory was proven undecidable by Tarski in 1946 [148]. It is easy to generate decidable extensions: the theory of commutative groups, with the additional axiom $\forall xy(xy = yx)$), is decidable (Wanda Szmielew 1948 [146]). There are also decidable fragments of group theory: this is the case for the equational theory since, from the axioms, a noetherian and confluent rewriting system can be generated (6.3.3).
- Ring theory and the theory of commutative rings are undecidable. The theory of algebraically closed fields of characteristic zero is decidable.
- Elementary arithmetic is essentially undecidable. Presburger arithmetic is decidable (1929 [126]).
- If a signature Σ contains a binary symbol, then the predicate calculus over Σ is undecidable (Church 1936 [15]). If Σ is formed only of unary relational symbols, then the predicate calculus over Σ is decidable.

Proposition 8.9 Let T be a complete theory. Then T is decidable if and only if T is recursively axiomatizable.

Proof If T is complete, for every proposition φ, $T \vdash \varphi$ or $T \vdash \neg\varphi$. If T is recursively axiomatizable, there is a semi-decision procedure. It is applied to φ and $\neg\varphi$ to solve the problems, $T \vdash \varphi$? and $T \vdash \neg\varphi$?. By running the two procedures in parallel and by halting when one of them halts, which is guaranteeed since T is complete, a decision procedure is produced for $T \vdash \varphi$? This proof can be written more formally by using properties of encodings of formulas and the characterization of a recursive set by the semi-recursiveness of itself and its complement. Conversely, if T is decidable, the recursive set of theorems can be taken as the set of axioms. □

Proposition 8.10 Let T be a consistent theory, recursively axiomatizable and satisfying the conditions of §8.1.5. Then every numeric function (respectively every subset of $\mathbf{N}^k$) which is T-definable is recursive.

Proof Take $k = 1$ to simplify and consider the case of a function $f : \mathbf{N} \to \mathbf{N}$. Suppose that f is defined in T by the formula $\varphi \in \mathrm{L}_\Sigma[x, y]$. Since T is recursively axiomatizable, its theorems can be obtained by an enumeration procedure. If $n \in \mathbf{N}$, $m = f(n)$ must be computed. The enumeration procedure is initiated, and halted as soon as a formula $\varphi(\mathbf{n}, \mathbf{p})$ is obtained. It will stop since by assumption on f and φ, $T \vdash \varphi(\mathbf{n}, \mathbf{m})$. So let p be the integer upon which the procedure is halted. Since T is consistent and $T \vdash \neg\varphi(\mathbf{n}, \mathbf{p})$ if $p \neq m$, $p = f(n)$ must hold. □

From this proof, it should be remembered that under the assumptions of the proposition, an enumeration procedure of the theorems can be transformed into a computation procedure. This (too) general idea makes the design of logic(al) programming theoretically possible. It will be shown in a similar manner that the λ-definable and Turing-computable functions are recursive.

8.2.5 Undecidability of arithmetic

The discovery of the phenomenon of undecidability was one of the major discoveries of this century, particularly in the case of arithmetic, one of the best-known subjects. The undecidability of this theory is a result of its power of computation in the sense of §8.1.5. Here is a technical lemma allowing a syntactic manipulation necessary for the proof of the theorem. Recall that $\ulcorner\psi\urcorner$ is the term $\mathbf{n}$ if $\#\psi = n$.

Lemma 8.11 Let T be a theory such that the recursive functions are definable in T. Let x be a variable and $\varphi \in \mathrm{L}_\Sigma[x]$. There exists a proposition ψ such that

$$T \vdash \psi \Longleftrightarrow \varphi(\ulcorner\psi\urcorner).$$

Proof Consider the diagonal function $d : \mathbf{N} \to \mathbf{N}$ defined by:

$$d(n) = \begin{cases} n' & \text{if } n = \#\chi \text{ and } n' = \#(\chi[x := \mathbf{n}]) \text{ for } \chi \in \mathrm{L}_\Sigma[X] \\ n & \text{otherwise.} \end{cases}$$

d is recursive, hence is defined in T by a formula $\delta \in \mathrm{L}_\Sigma[x, y]$. In fact, d is the only recursive function for which one supposes T-definability. Hence

$$T \vdash \forall y(\delta(\mathbf{n}, y) \Longleftrightarrow y = d(\mathbf{n}))$$

for each $n \in \mathbf{N}$. Let

$$\begin{aligned} m &= \#\big(\forall y(\delta(x, y) \Rightarrow \varphi(y))\big), \\ m' &= d(m), \\ \psi &= \forall y(\delta(\mathbf{m}, y) \Rightarrow \varphi(y)). \end{aligned}$$

By the definition of d, $\#\psi = d(m)$. From the derivation of δ, we obtain the two following derivations:

$$(E_\Rightarrow) : \frac{(E_\forall) : \dfrac{\forall y(\delta(\mathbf{m}, y) \Rightarrow \varphi(y))}{\delta(\mathbf{m}, \mathbf{m}') \Rightarrow \varphi(\mathbf{m}')} \qquad \dfrac{(E_\forall) : \dfrac{\forall y(\delta(\mathbf{m}, y) \Longleftrightarrow y = \mathbf{m}')}{\delta(\mathbf{m}, \mathbf{m}') \Longleftrightarrow \mathbf{m}' = \mathbf{m}'} \qquad \dfrac{}{\mathbf{m}' = \mathbf{m}'}}{\delta(\mathbf{m}, \mathbf{m}')}}{\varphi(\mathbf{m}')},$$

$$(I_\forall) : \frac{(E_\Rightarrow) : \dfrac{(E_\Longleftrightarrow) : \dfrac{\forall y(\delta(\mathbf{m}, y) \Longleftrightarrow y = \mathbf{m}')}{\delta(\mathbf{m}, y) \Longleftrightarrow y = \mathbf{m}'} \qquad (=) : \dfrac{\varphi(\mathbf{m}')}{y = \mathbf{m}' \Rightarrow \varphi(y)}}{\delta(\mathbf{m}, y) \Rightarrow \varphi(y)}}{\forall y(\delta(\mathbf{m}, y) \Rightarrow \varphi(y))},$$

which show that $T \vdash \psi \iff \varphi(\mathbf{m}')$. Since $m' = \#\psi$, it follows that $T \vdash \psi \iff \varphi(\ulcorner\psi\urcorner)$. □

Theorem 8.12 If T is a consistent theory such that all of the recursive functions are T-definable, then T is essentially undecidable.

Proof Suppose that T is decidable and let $\theta \in \mathrm{L}_\Sigma[x]$ be the formula defining in T the set th_T (of encodings) of its theorems. Then:

- $n \in th_T$ if and only if $T \vdash \theta[x := \mathbf{n}]$;
- $n \notin th_T$ if and only if $T \vdash \neg\theta[x := \mathbf{n}]$.

The 'if and only if' comes from the consistency of T. Apply the lemma to $\neg\theta$ instead of φ. A proposition ψ satisfying $T \vdash \psi \iff \neg\theta(\ulcorner\psi\urcorner)$ is obtained, and hence the following contradiction: $T \vdash \psi$ if and only if $T \vdash \neg\theta(\ulcorner\psi\urcorner)$ by the construction of ψ, if and only if $T \nvdash \psi$ by the definition of θ. □

Once this general theorem is proven, the following important and well-known results easily follow.

Theorem 8.13 (Church undecidability) Elementary arithmetic is essentially undecidable.

Proof Follows from theorems 8.2 and 8.12. Hence Peano arithmetic is undecidable. □

Theorem 8.14 (Gödel incompleteness) Every consistent and recursively axiomatizable extension of elementary arithmetic is incomplete.

Proof Follows from theorem 8.13 and proposition 8.9. □

There is no hope of completing such a theory, e.g., Peano arithmetic, by adding a recursive set of axioms, i.e., by adding to T one of φ or $\neg\varphi$ if $T \nvdash \varphi$ and $T \nvdash \neg\varphi$, since the extended theory would remain recursively axiomatizable and consistent. Theorem 8.14 means that there exist propositions φ, necessarily true or false in $\mathbf{N}$, but for which $AP \nvdash \varphi$ and $AP \nvdash \neg\varphi$. Rosser constructed such formulas, but artificially, as for lemma 8.11. In 1976 Paris and Harrington [116] gave a statement in the area of combinatorics, true in $\mathbf{N}$ but unprovable in Peano arithmetic. There are now several 'significant' statements which fall in the same case: results on Goodstein sequences (Kirkby–Paris 1982 [78], Cichon 1983 [16]) and a finite form of Kruskal's theorem (Friedman 1981, unpublished), among others. See also the remarks on the second incompleteness theorem (next page). The following theorem states the failure of 'axiomatism'.

Theorem 8.15 Complete arithmetic is not recursively axiomatizable.

Proof Follows from theorem 8.13 and proposition 8.9. □

8.2.6 The second incompleteness theorem

Gödel's second incompleteness theorem is even more famous than his first because of its epistemological interpretations. It deals with the consistency of arithmetic. Everyone knows that it is consistent, since **N** is a model. The proof of this fact is not done in arithmetic, but in general mathematics—it is obvious that all the theorems in this book were proven using mathematics—and to be formalist, in set theory. However, the consistency of arithmetic can be stated by an arithmetic formula, for example $\neg\exists x\, der(x, \ulcorner \mathbf{0} = \mathbf{1} \urcorner)$. If this formula were provable *in* arithmetic, it would guarantee it an autonomous position, independent of the rest of mathematics. For the rest of mathematics, this problem is even more important. This question was crucial at the beginning of the century, to ensure a solid foundation for mathematics, which studies objects well beyond our intuition of concrete objects. The objective of Hilbert's program was a consistency proof of formalized mathematics—by finite means and mechanically verifiable. This stress on finite means recalls the old debate on the purity of methods and restricted means: in geometry, restricting constructions to the use of a ruler and a compass or banning coordinates from proofs, and in number theory, not using analytical means, among others.

Gödel's second incompleteness theorem states that a formula stating the consistency of Peano arithmetic is not derivable from its axioms. Its proof consists of formalizing the derivation of the first incompleteness theorem inside arithmetic. For computer science, the interest of this statement is not direct. It resides in an analysis of 'minimal' consistency proofs and particularly in proofs of other significant properties such as the normalization of the typed λ-calculus. What are the necessary supplementary means?

For example, the consistency of elementary arithmetic is provable in Peano arithmetic, thanks to induction over the integers. It is not necessary to use all of set theory to prove the consistency of Peano arithmetic. Gentzen's proof uses induction over the ordinals up to ϵ_0. Induction up to ω is one of the axioms of Peano arithmetic. It is sufficiently powerful to prove some *transfinite* induction principles $TI(\alpha)$, i.e., up to an infinite ordinal α. Precisely, it allows the proof of $TI(\alpha)$ for every ordinal $\alpha < \epsilon_0$, but not $TI(\epsilon_0)$. Yet Gentzen used $TI(\epsilon_0)$ to prove the consistency of arithmetic.

8.3 Metaprogramming

The general results of the study of recursive functions can be interpreted naturally in programming terms, and even in metaprogramming, the area of computer science dealing with general algorithmic constructions such as interpretation, compilation, verification and tests.

8.3.1 Universal predicates T, U

First consider the case of Turing machines. For each partial Turing-computable function, there exists a Turing machine (Γ, Q, R) computing f, i.e. computing $f(n_1, \ldots, n_k)$ if $n_1, \ldots, n_k$ is placed on the tape. In computer science, since von Neumann it has been known that there 'exists' a unique machine, programmed (in the Turing machine sense) once and for all to which one passes as data both the instructions of the computation and the initial values of the variables $n_1, \ldots, n_k$: every compiler is an implementation for it. The same is true for Turing machines (Turing 1936 [151]): there exist universal Turing machines. What follows will proceed differently in order to remain inside logic's inference systems.

Recall that, by definition, a partial function $f : \mathbf{N}^k \rightharpoonup \mathbf{N}$ is recursive if and only if its graph $G(f)$ is r.e., hence the projection of a recursive subset A of $\mathbf{N}^{k+2}$. By Gödel's theorem, A is definable by a formula $\varphi \in \mathrm{L}_\Sigma[x_1, \ldots, x_{k+2}]$ in Peano arithmetic. Let $e = \#\varphi$ its numeric encoding.

By the definition of the graph $G(f)$, of A, of φ, of the relation *der* and of the encoding e of φ: $f(\vec{m}) = n$ if and only if there exists c (a 'computation') such that $(\vec{m}, n, c) \in A$; if and only if there exists c such that $AP \vdash \varphi(\vec{\mathbf{m}}, \mathbf{n}, \mathbf{c})$; if and only if there exist c and d (a 'derivation') such that $der(\#(\varphi(\vec{\mathbf{m}}, \mathbf{n}, \mathbf{c})), d)$; and if and only if there exist c and d such that $der(s^{k+2}(e, \vec{m}, n, c), d)$. A substitution function s^{k+2} is used.

By encoding the pair (c, d) by an integer p, it follows that there exists p such that $der(s^{k+2}(e, \vec{m}, n, \pi^1(p)), \pi^2(p))$. Recall that der, s^{k+2}, π^1 and π^2 are recursive primitive. So, to compute $f(\vec{m})$ it suffices to find the smallest n—there is only one since f is a partial function—and the smallest p satisfying the preceding relation. All that has to be done is to encode the pair (n, p) by an integer q, to minimize in q and deduce $n = \pi^1(q)$. Let

$$\begin{aligned} T^k(e, \vec{m}, q) &= der(s^{k+2}(e, \vec{m}, \pi^1(q), \pi^1(\pi^2(q))), \pi^2(\pi^2(q))) \\ U(q) &= \pi_1(q). \end{aligned}$$

Functions T^k and U are primitive recursive. The operator T^k depends only on the encoding used for the syntactic operations of arithmetic. Only f remains to be defined, using T^k, U and the integer e: it has been shown that for every k-ary partial recursive function f, there exists an integer e such that

$$f(\vec{m}) = U(\mu q\,.\,[T^k(e, \vec{m}, q) = 1]).$$

This is Kleene's *normal form* of partial recursive functions: the integer e is called an *index* of f.

From Kleene, write $\{e\}^k$ for the partial function $\vec{m} \mapsto U(\mu q\,.\,[T^k(e,\vec{m},q) = 1]$. Its domain is written W_e^k. It is the set of $\vec{m}$ such that there exists q (a derivation or a computation) satisfying $T^k(e,\vec{m},q)$. In the definition of $\{e\}^k(\vec{m})$, the integers e and $\vec{m}$ are all arguments of T^k: the $(k+1)$-ary partial function $u : (e,\vec{m}) \mapsto \{e\}^k(\vec{m})$ is therefore recursive. It is a *universal* partial recursive function in the sense that, for every partial recursive function $f : \mathbf{N}^k \rightharpoonup \mathbf{N}$, there exists e such that $f(\vec{m}) = u(e,\vec{m})$. Note that for a given integer e, there is a partial function $\{e\}^k$ for each k. The k indicating the arity is often omitted in practice, in particular when $k = 1$. The following theorem has been proven:

Theorem 8.16 (Kleene universality) There exists a primitive recursive function $U : \mathbf{N} \to \mathbf{N}$, and for each $k \geqslant 1$, a primitive recursive set $T^k : \mathbf{N}^{k+2} \to \mathbf{N}$ such that if one defines

$$\{e\}^k(\vec{m}) = U(\mu q\,.\,[T^k(e,\vec{m},q) = 1]),$$

then, for every partial recursive function $f : \mathbf{N}^k \rightharpoonup \mathbf{N}$, there exists an integer e such that $f = \{e\}^k$.

The recursive functions were first defined using the composition, primitive recursion and (total) minimization operations. Then those operations were tentatively used to obtain—without success—the partial computable functions, since minimization is not effective over partial functions. The existence of a Kleene normal form shows that every partial recursive function is obtained from a primitive recursive function by a single application of minimization, then by composition with a primitive recursive function. In particular, it is not necessary to minimize partial functions.

This proof could have been written using Turing machines. First a universal Turing machine would have been constructed, hardly an interesting exercise. What plays the role of universal machine here is Peano arithmetic, already constructed, or any other theory of equal expressiveness. However, an encoding of arithmetic and of the substitution functions was used, and it was not presented explicitly. The approach chosen shows the links between logic and computation better.

One must understand the intuitive content of T and U. In the logical approach $T^k(e,\vec{m},q)$ states that e is the encoding of a formula φ, $\vec{m}$ a k-tuple of values, and that q encodes a derivation in Peano arithmetic, whose conclusion is of the form $\varphi(\vec{\mathbf{m}},\mathbf{n},\mathbf{c})$: n is the value of $f(\vec{m})$. In the Turing machine approach, $T^k(e,\vec{m},q)$ states that e is the code of a Turing machine, $\vec{m}$ a k-tuple of values, and that q encodes a computation executed on a universal machine with e and $\vec{m}$ on the tape in the initial position: at the end of the computation the tape contains the value $f(\vec{m})$.

The mapping $n \mapsto \{n\}$ associating with an algorithm a function which computes it is obviously not injective. This corresponds to the idea that a function can be computed by an infinite number of different algorithms. The integer n should be seen as encoding an algorithm as a real, finite object, while the partial function $\{n\}$ is an ideal, mathematical object. The equality $n = m$ of two algorithms corresponds to the *intensional* equality of functions $\{n\}$ and $\{m\}$. The equality $\{n\} = \{m\}$ of computed functions corresponds to the *extensional* equality of algorithms: they compute the same function 'in extension'.

Everything here can immediately be transcribed to r.e. sets: if A is an r.e. set, an *index* of A is an integer e such that $A = W_e$. In particular, an index of the semi-characteristic partial function $(1/2)_A$ is an index of A. For each k, the recursive set defined by T^k is *universal*: for every r.e. set $A \subseteq \mathbf{N}^k$, there exists an integer e such that

$$A = W_e^k = \{\vec{m} \in \mathbf{N}^k; (\exists q \in \mathbf{N})T^k(e, \vec{m}, q)\}.$$

This result allows the enumeration of the partial recursive functions, and, consequently, of the r.e. sets: W_0, W_1, W_2, ...

Indices allow the statement of some problems. Let f be a partial recursive function, defined by a given algorithm e. The formula $\forall m \exists n T(e, m, n)$ states that it is total. It will be shown that the set of e for which this formula is true is not recursively enumerable. The problem here is: for a given e, for example 6867, to know if the formula is provable. If this is the case $\{e\}$ is called *total provable* in Peano arithmetic.

There are recursive functions which are not total provable—since the proof of their termination must use an induction scheme stronger than the induction over integers. These functions are rarely encountered in programming, and a language which computed only the provably total functions would be sufficient: this is the case for certain typed λ-calculi such as Gödel's system T. The simplest example of a function which is not provably total is obtained by induction up to ϵ_0, by iterating the construction which leads to Ackermann's function, itself already complex enough.

8.3.2 Semantics of languages—interpreters

Let L be a programming language computing all of the partial recursive functions, which is the case for usual languages. A way to give a semantics to L is to associate with each program π its interpretation $\pi^{\mathbf{N}}$ in $\mathbf{N}$. Since every program contains only a finite set of variables, $\pi^{\mathbf{N}} : \mathbf{N}^\star \rightharpoonup \mathbf{N}$ is the partial function computed by π.

This idea applies to an arbitrary L. The universality theorem transposes itself to L: there is an encoding of programs of L by integers and the semantics $\pi \mapsto \pi^{\mathbf{N}}$ is a partial function S_L (or simply S) : $\mathbf{N} \to (\mathbf{N}^\star \rightharpoonup \mathbf{N})$, or $\mathbf{N} \times \mathbf{N}^\star \rightharpoonup \mathbf{N}$

$$(p, (m_1, \ldots, m_k)) \mapsto S_L(p, (m_1, \ldots, m_k)).$$

This is the result of the p-th program with $m_1, \ldots, m_k$ as initial values for the variables. This function S_L gives the execution of a program of L in a 'machine-language L'. Actually, programs must be executed in a 'universal machine' U, for which Kleene's notation $\{\ \}$ will be reserved, instead of S_U. From the universality theorem, since $(p, m_1, \ldots, m_k) \mapsto S_L(p, (m_1, \ldots, m_k))$ is computable (partial recursive), there exists an integer i such that

$$S_L(p,(m_1,\dots,m_k)) = \{i\}^{k+1}(p,m_1,\dots,m_k).$$

This integer i encodes a program in machine-language U which, executed on this machine with $p, m_1, \dots, m_k$ as input, yields the same result as the program p of L executed on L with $m_1, \dots, m_k$ as input.

In programming, i is *interpreted* from L to U. The universality theorem therefore states the existence of interpreters.

8.3.3 Partial evaluation and compilation

A natural problem is that of partial evaluation of a computable function. Let p be an integer and consider the function $\{p\}^{k+l}$. Fix the first k arguments $m_1, \dots, m_k$. It is clear that the function

$$(m_{k+1},\dots,m_{k+l}) \mapsto \{p\}^{k+l}(m_1,\dots,m_k,m_{k+1},\dots,m_{k+l})$$

is computable. The universality states the existence of p' such that this function is $\{p'\}^k$. Program p' is computable as a function of p: a text editor such as Emacs computes p' from p by substituting values to the identifiers. But a better partial evaluation of `if m>0 then A else B` with $m = 4$ gives `A` instead of `if 4>0 then A else B`: this is the sort of thing done by a good optimizing compiler.

Theorem 8.17 (s–m–n) Let k and l be two integers $\geqslant 1$. There exists a primitive recursive function $\rho_k^l : \mathbf{N}^{l+1} \to \mathbf{N}$ such that

$$\{s\}^{k+l}(m_1,\dots,m_{k+l}) = \{\rho_k^l(s,m_1,\dots,m_k)\}^l(m_{k+1},\dots,m_{k+l}).$$

Proof The notation used for the universality theorem will be used here:

$$T^k(s,\vec{m},q) = der(s^{k+2}(s,\vec{m},\pi^1(q),\pi^1(\pi^2(q))),\pi^2(\pi^2(q))).$$

The proof relies on a general property of Gödel's encoding concerning substitutions: for every substitution σ, the function $\#\varphi \mapsto \#(\sigma(\varphi))$ is recursive primitive. To each formula $\varphi \in \mathrm{L}_\Sigma[x_1,\dots,x_{k+l+2}]$ is associated $\psi \in \mathrm{L}_\Sigma[x_1,\dots,x_{l+2}]$ by substitution,

$$\psi \equiv \varphi[x_1 := \mathbf{m_1},\dots,x_k := \mathbf{m_k}, x_{k+1} := x_1,\dots,x_{k+l+2} := x_{l+2}],$$

and let $\rho^{k,l}$ be the function associating $\#\psi$ with $(\#\varphi, m_1, \dots, m_k)$. Then

$$\varphi(\mathbf{m_1},\dots,\mathbf{m_k}) \equiv \psi(\mathbf{m_{k+1}},\dots,\mathbf{m_{k+l}}),$$

from which

$$der(s^{k+l+2}(s,m_1,\dots,m_{k+l},\pi^1(q),\pi^1(\pi^2(p))),\pi^2(\pi^2(p))) =$$
$$der(s^{l+2}(\rho^{k,l}(s,m_1,\dots,m_k),\pi^1(\pi^2(p)),\pi^1(q),m_{k+1},\dots,m_{k+l}),\pi^2(\pi^2(p))).$$

It follows that

$$\begin{aligned} T^{k+l}(s, m_1, \ldots, m_{k+l}, q) &= T^k(\rho^{k,l}(s, m_1, \ldots, m_k), m_{k+1}, \ldots, m_{k+l}, q), \\ \{s\}^{k+l}(m_1, \ldots, m_{k+l}) &= \{\rho^{k,l}(s, m_1, \ldots, m_k)\}^k(m_{k+1}, \ldots, m_{k+l}). \end{aligned}$$

□

Apply the universality theorem again. Let e be an index of ρ:

$$\rho(s, m_1, \ldots, m_k) = \{e\}(s, m_1, \ldots, m_k).$$

Program e is called a *partial evaluator*.

This result, important in recursive function theory, forms the basis for powerful methods of metaprogramming. Let e be a partial evaluator, i.e.,

$$\{\{e\}(p, m)\}\}(m') = \{p\}(m, m').$$

Define c by partially evaluating e with the interpreter i, of L over U, as input:

$$c = \{e\}(e, i).$$

Let s be a source program in L. Compute

$$\begin{aligned} o &= \{c\}(s) \\ &= \{\{e\}(e, i)\}(s) \\ &= \{e\}(i, s), \\ \{o\}(m) &= \{\{e\}(i, s)\}(m) \\ &= \{i\}(s, m) \\ &= S_L(s, m). \end{aligned}$$

It follows that c is a *compiler* from L to U and that o is the object form of s. Now c can be obtained from i by letting

$$\begin{aligned} g &= \{e\}(e, e), \\ c &= \{g\}(i), \end{aligned}$$

since

$$\begin{aligned} c &= \{\{e\}(e, e)\}(i) \\ &= \{e\}(e, i). \end{aligned}$$

So g is therefore a *compiler generator*, which produces a compiler from an interpreter. This method of compiler generation is used in several software engineering projects, in particular Danish and Japanese. The difficult aspects lie in the construction of a good partial evaluator.

8.3.4 Recursivity and reflexivity

The following theorem is one of the best known in recursive function theory. In programming, a syntactically recursive definition is of the form

```
f(x) = ...  f ...
```

How does one formalize this syntactic recursivity when it is interpreted effectively? It is in fact an equation that can be written

$$f(x) = \Phi(f, x),$$

where Φ is a 'functional', which must be computable. But the computable functionals have not been defined. The idea is to replace Φ by a partial recursive function g and the argument f of Φ by an index n of f, so that this equation is transformed into

$$\{n\}(x) = g(n, x).$$

This equation is in n, but once a solution n has been found, the function $f = \{n\}$ is defined explicitly by $f(x) = g(n, x)$: there is no more 'recursivity'. In particular, g can be defined using a partial recursive function $(n, x) \mapsto \{n\}(x)$, yielding the standard case of recursion where g contains 'recursive calls' $f(\dots)$.

This formulation has larger scope: g can call f not only extensionally, by using *values* $f(a)$ for a given integer a, but also intensionally by using the algorithm computing f globally, represented by its index n, i.e, by the code of the program. For example, a solution n of the equation $\{n\}(x) = n$ is a program which gives as output its own text, whatever the input.

Its proof uses the s–m–n theorem.

Theorem 8.18 (Kleene recursivity) Let g be a partial recursive function from $\mathbf{N}^k$ to $\mathbf{N}$. There exists $n \in \mathbf{N}$ such that $\{n\}^k(x_1, \dots, x_k) = g(n, x_1, \dots, x_k)$ for every $\vec{x}$.

Proof Let $\rho : \mathbf{N}^2 \to \mathbf{N}$ be the recursive function given by theorem 8.17. Consider the partial recursive function:

$$(t, x_1, \dots, x_k) \mapsto g(\rho(t, t), x_1, \dots, x_k).$$

From the universality theorem, there exists $m \in \mathbf{N}$ such that

$$g(\rho(t, t), x_1, \dots, x_k) = \{m\}^{k+1}(t, x_1, \dots, x_k).$$

By the definition of ρ,

$$\{m\}^{k+1}(t, x_1, \dots, x_k) = \{\rho(m, t)\}^k(x_1, \dots, x_k).$$

Let $n = \rho(m, m)$. Then $g(n, x_1, \dots, x_k) = \{n\}^k(x_1, \dots, x_k)$. □

This theorem evidently applies to recursive definitions of functions, but also to reflexive constructions, common in functional and object-oriented programming, where data structures belonging to the interpreter are accessible at the program level.

8.3.5 Undecidability of β-conversion

This is the first undecidability result, proven in 1936 by Church [15], from which he proved the undecidability of the predicate calculus. A more general version due to D. Scott (1963, unpublished) will be presented here. Its proof closely resembles those of the Rice and Church theorems, which also use a fixpoint lemma and an encoding.

Suppose there is an encoding $M \mapsto \#M$ of λ-terms, with recursive functions $app : \mathbf{N}^2 \to \mathbf{N}$ and $num : \mathbf{N} \to \mathbf{N}$ such that $\#(MN) = app(\#M, \#N)$ and $\#\mathbf{n} = num(n)$. Let $\ulcorner M \urcorner = \mathbf{m}$ if $m = \#M$.

Lemma 8.19 For every combinator G, there exists X such that $G\ulcorner X \urcorner = X$, i.e, if $\#X = p$, then $G\mathbf{p} = X$.

Example For $G \equiv \mathbf{I}$, there exists X such that $\ulcorner X \urcorner = X$. □

Proof The functions *app* and *num* are λ-definable by terms app and num. Hence $\text{num}\ulcorner M \urcorner = \ulcorner\ulcorner M \urcorner\urcorner$ and $\text{app}\ulcorner M \urcorner \ulcorner N \urcorner = \ulcorner MN \urcorner$.

Let $\omega \equiv \lambda x \,.\, G(\text{app}\, x(\text{num}\, x))$ and $X \equiv \omega\ulcorner \omega \urcorner$. Then

$$\begin{aligned} X \equiv \omega\ulcorner\omega\urcorner &\to G(\text{app}\ulcorner\omega\urcorner(\text{num}\ulcorner\omega\urcorner)) \\ &= G(\text{app}\ulcorner\omega\urcorner\ulcorner\ulcorner\omega\urcorner\urcorner) \\ &= G(\ulcorner\omega\ulcorner\omega\urcorner\urcorner) \\ &= X. \end{aligned}$$

□

Definition 8.11 Two subsets A_0, A_1 of $\mathbf{N}$ are recursively separable if there exists a recursive function $f : \mathbf{N} \to \{0, 1\}$ separating A_0 and A_1, i.e, such that $f(n) = i$ if $n \in A_i$, for $i = 0, 1$.

Two sets of terms are recursively separable if their sets of encodings are recursively separable.

Theorem 8.20 Two sets of terms, non-empty and stable under β-conversion, are not recursively separable.

Proof Suppose A_0 and A_1 are separated by the recursive function f, and let F be a term defining f. If A_0 and A_1 are non-empty, let $M_0 \in A_0$ and $M_1 \in A_1$. Let $G \equiv \lambda x \,.\, \mathbf{D}M_1M_0(Fx)$: by the definition of the combinator $\mathbf{D}$, if $M \in A_0$, $G\ulcorner M \urcorner = M_1$ and if $M \in A_1$, then $G\ulcorner M \urcorner = M_0$. By lemma 8.19, there exists X such that $G\ulcorner X \urcorner = X$. If $X \in A_0$, then $X = G\ulcorner X \urcorner \in A_1$, by the stability of A_1. If $X \in A_1$, then $X = G\ulcorner X \urcorner \in A_0$, by the stability of A_0. Contradiction! □

Corollary 8.21
- For fixed A, the set $\{M; M = A\}$ is not recursive.
- The relation $M = N$ is not recursive.
- The set of normalizable terms is not recursive.

8.3.6 Decidability problems in programming

It is still assumed that the programming language is sufficiently powerful to compute all the partial computable functions.

The halting problem
Given: a pair (p, m) formed of a program p and data m.
Problem: does program p, with m as input, halt?

Intuitively, it should be clear that the problem is semi-decidable: it suffices to run the program, and give a positive answer when it halts. The set of its solutions is

$$\{(p, m); \{p\}(m)\downarrow\} = \{(p, m); m \in W_p\}.$$

A more particular problem is obtained though diagonalization. The set of its solutions is

$$K = \{n; \{n\}(n)\downarrow\} = \{n; n \in W_n\}.$$

Since the partial function $n \mapsto \{n\}(n)$ is recursive, the set K is r.e. It will be shown that it is not recursive, using a standard diagonalization method. Let

$$f(n) = \begin{cases} \{n\}(n) + 1 & \text{if } n \in K \\ 0 & \text{otherwise.} \end{cases}$$

This function is total and is not recursive if K is. Furthermore it is distinct from all the partial recursive functions since $f(n) \neq \{n\}(n)$ for every n. Hence K cannot be recursive.

Rice's theorem follows easily from the Kleene recursion theorem.

Theorem 8.22 (Rice) Let $I_E = \{x \in \mathbf{N}; \{x\} \in E\}$, where E is a set of partial recursive functions of one variable. The set I_E is recursive if and only if E is empty or equal to the set of partial recursive functions.

Proof Suppose I_E is recursive. Suppose that E is non-trivial. There exist partial recursive functions $f, g : \mathbf{N} \rightharpoonup \mathbf{N}$ such that $f \in E, g \notin E$. Let

$$h(x, y) = \begin{cases} g(y) & \text{if } x \in I_E \\ f(y) & \text{if } x \notin I_E. \end{cases}$$

g, f and I_E being recursive, h is partial recursive. From theorem 8.18, there exists $n \in \mathbf{N}$ such that $\{n\}(y) = h(n, y)$ for every y.

If $n \in I_E$, then $\{n\}(y) = g(y)$, so $n \notin I_E$.

If $n \notin I_E$, then $\{n\}(y) = f(y)$, so $n \in I_E$.

Contradiction. □

Examples This theorem proves the non-existence of decision procedures for the following problems involving programs p, q of one variable:

- Does p halt for all input?
- Does p not halt for any input?
- Does p compute a primitive recursive function?
- Does p compute a function of finite domain?
- Do p and q compute the same function?

The problems of partial (respectively total) correctness of a program with respect to a condition (φ, ψ) are also undecidable since certain particular cases appear among the above problems. □

Rice's theorem can also be stated for sets instead of functions, with E a set of r.e. subsets of $\mathbf{N}$, and $I_E = \{n; W_n \in E\}$.

The characterization of r.e. sets is more delicate. It uses an effective enumeration of *finite* sets: for example, to the finite set $\{a_1, \ldots, a_n\}$ is associated the integer $2^{a_1} + \ldots + 2^{a_n}$. It is a bijection from the set of finite subsets of $\mathbf{N}$ to $\mathbf{N}$. Write D_n for the finite set encoded by the integer n and $\mathcal{U}(D_n)$ for the set of r.e. sets containing D_n. It is an open neighborhood of D_n in an appropriate topology. The following theorem is stated without proof.

Theorem 8.23 (Rice–Shapiro) Let E be a set of r.e. subsets of $\mathbf{N}$. The set I_E of its indices is r.e. if and only if E is empty or there exists a recursive function f such that $E = \bigcup_{n \in \mathbf{N}} \mathcal{U}(D_{f(n)})$. The set E is an effective union of open neighborhoods.

In other words, $A \in E$ if and only if there exists n such that $A \supseteq D_{f(n)}$. In particular, it is necessary that if $A \in E$ and $A \subseteq B$, then $B \in E$. It is easy not to satisfy this condition.

Example Let $E = \{\varnothing\}$: then $I_E = \{n; W_n = \varnothing\}$ is not r.e. and the problem $A = \varnothing$ is not semi-decidable. □

8.3.7 Equation resolution

If $P(X_1, \ldots, X_k, Y_1, \ldots, Y_l)$ is a polynomial with coefficients in $\mathbf{Z}$, the set

$$R = \{(x_1, \ldots, x_k) \in \mathbf{N}^k; \exists (y_1, \ldots, y_l) \in \mathbf{N}^l, P(x_1, \ldots, x_k, y_1, \ldots, y_l) = 0\}$$

is called a *diophantine set*. As a projection of the recursive set $P = 0$, R is recursively enumerable.

Among the famous problems stated by Hilbert in 1900 [63], the tenth was called the 'Determining the solvability of a diophantine equation'. It proposed to 'find a process under which one can determine in a finite number of operations if the equation is solved for rational integers'. That was in 1900, and before Gödel and Turing, the first explicit statement of a decidability problem. It was solved in 1970 by Matjasevič [105] by a surprising theorem:

Theorem 8.24 (Matjasevič) Each r.e. set is diophantine.

Its proof is difficult, the key step consisting of proving that the relation $x = y^z$ is diophantine. Since there exist r.e. sets, there exist non-recursive diophantine sets, hence undecidable diophantine equations.

Unification The unification problem for first-order terms is decidable. This is an important result which shows the importance of the first-order to computation. In 1981 Goldfarb [58] showed that the unification problem for second-order terms is undecidable, where second-order terms are formed of function variables which can be substituted for terms.

Unification modulo an equational theory $\mathcal{E}$ was mentioned in Chapter 5: find σ such that $\sigma(M_1) =_{\mathcal{E}} \sigma(M_2)$. Here are examples of different behaviors which also show the difficulty of equational unification:

- The associative–commutative case is decidable and leads to a complete, finite set of unifiers (Stickel 1975 [142]).
- The associative case is decidable, but there does not exist a complete, finite set of unifiers.
- The distributive–associative case is undecidable (Szabo 1982 [145]).

8.3.8 Computable reals

Recursive function theory, condemned to handle only finite objects such as integers and words, seems inadequate for one of the most useful applications of computer science, numerical computation. Of course, the 'floating point numbers' of programming are only decimals, almost integers, finitely representable by a machine. However, the mathematical methods using numerical computations are those of analysis, not of arithmetic! So what is the effective content of the approximations of the continuous by the discrete?

To characterize the computable reals, the basic concrete objects are the rational numbers, which are easily coded by integers. To the integer n is associated the rational $(\pi_3^1(n) - \pi_3^0(n))/\pi_3^3(n) + 1$. The rational $p/q + 1$ (with $p \in \mathbf{Z}$, $q \in \mathbf{N}$) is coded by the integer $\langle p, 0, q\rangle$ if $p \geqslant 0$, or by $\langle 0, -p, q\rangle$ otherwise. If such an encoding were adopted, one could refer to recursive functions from $\mathbf{N}$ to $\mathbf{Q}$ or to $\mathbf{Q} \times \mathbf{Q}$.

Definition 8.12 The recursive functions $r, s : \mathbf{N} \to \mathbf{Q}$ define a *recursive real* if the sequence of intervals $[r(n), s(n)] \subseteq \mathbf{Q}$ is decreasing and $\lim_{n\to\infty} |s(n) - r(n)| = 0$.

Even though it is not necessary to construct $\mathbf{R}$ beforehand, the recursive reals that these functions define are the intersections of the intervals $[r(n), s(n)]$ *in* $\mathbf{R}$.

A recursive real is an abstract object if it it compared to all reals. It is much more concrete than the latter from an operational point of view, as is the program computing the interval $[r(n), s(n)]$. It is easy to show that the sum and the product of two recursive reals and the inverse of a non-null recursive real are recursive. It follows that the set $\mathbf{R}_{rec}$ of recursive reals is a subfield of $\mathbf{R}$. It contains $\mathbf{Q}$ and is countable, like the set of recursive functions. It also contains the algebraic reals (real roots of polynomial equations with coefficients in $\mathbf{Q}$) and even certain transcendental numbers such as e and π for which approximation methods are known. By working a little harder, one can show that every positive recursive real has a square root in $\mathbf{R}_{rec}$ and that every polynomial of odd degree with coefficients in $\mathbf{R}_{rec}$ has a root in $\mathbf{R}_{rec}$. These properties mean that $\mathbf{R}_{rec}$ is a *closed real field*, just as $\mathbf{R}$ and the field of algebraic reals are. One of Tarski's famous theorems states that the theory of closed real fields is complete. It follows that $\mathbf{R}$ and $\mathbf{R}_{rec}$ are elementarily equivalent, i.e., satisfy exactly the same propositions. The recursive reals are therefore indistinguishable from the reals from the point of view of global algebraic properties. Unfortunately, these reals are not well suited to effective analysis; see exercise 21 on page 361, which shows that $\mathbf{R}_{rec}$ is not complete.

Example The equality of two recursive reals is not decidable, which is not surprising, since equality here is an extensional concept. Suppose that the equality were decidable; this would imply the decidability of the halting problem. A 'counter' c is used, initialized to 1 and capable of holding a rational. At each step of the program, $c/2$ is added to c, which produces a Cauchy sequence defining a recursive real $c^\star$. The program does not halt if and only if $c^\star = \sum_{n=0}^{+\infty} 1/2^n = 2$, which *reduces* the halting problem to the equality problem in $\mathbf{R}_{rec}$. The method of problem reduction will be studied more below. □

8.4 A classification of problems

How does one rate the difficulty of a problem? Decidability is a problem, but it is not the only one. Among decidable problems, the last section of this chapter will study the analysis of the *complexity* of decision algorithms. Among undecidable problems, it is also useful to classify them. It is useful to know not only the level of difficulty of a problem, but also to have general ways of measuring this level.

8.4.1 The arithmetic hierarchy

The decidable and semi-decidable problems have already been distinguished. More generally, the arithmetic hierarchy allows one to classify problems according to global criteria. Recall that the set of recursive sets is closed under intersection, union and complement, but not under projection. By definition, the r.e. are projections of recursives. The class of r.e. is closed under intersection, union and projection, but not under complement. The complements of r.e. sets are called *co-r.e.*.

The two projection and complement operations can be iterated alternately to generate two sequences of classes of sets, Σ_n and Π_n:

$$\begin{aligned}
\Sigma_0 &= \Pi_0 = \text{class of recursive sets,}\\
\Sigma_{n+1} &= \text{class of projections of elements of } \Pi_n,\\
\Pi_n &= \text{class of complements of elements of } \Sigma_n \text{ (i.e. of the co-}\Sigma_n\text{),}\\
\Delta_n &= \Sigma_n \cap \Pi_n.
\end{aligned}$$

In particular, Σ_1 is the class of r.e., Π_1 of the co-r.e. A set is *arithmetic* if it belongs to one of these classes.

Recall that the recursive sets are definable in Peano arithmetic. It follows that they are definable in $\mathbf{N}$. A subset A of N^k is *defined in the interpretation* $\mathbf{N}$ by φ if

$$A = \{(m_1, \ldots, m_k) \in \mathbf{N}^k; \varphi^{\mathbf{N}}(m_1, \ldots, m_k) = 1\}.$$

This semantic definability of the recursives extends to the r.e. and beyond. It is easily proven that every arithmetic set is definable in $\mathbf{N}$ by a first-order formula: each arithmetic set can be defined in the form

$$\{(x_1, \ldots, x_k) \in \mathbf{N}^k; Q_0 y_0 \ldots Q_p y_p R(x_1, \ldots, x_k, y_0, \ldots, y_p)\},$$

where R is recursive, $Q_1 \ldots Q_p$ is a sequence of alternating quantifiers, the first being $\exists$ if the set is Σ_p, and $\forall$ if it is Π_p. For example, the sets Π_2 are of the form $\forall \exists \forall R$. Using standard coding techniques, a sequence of n identical quantifiers can be reduced to a single quantifier over an n-tuple. For example, $\exists y_0 \exists y_1 R(y_0, y_1)$ becomes $\exists y R'(y)$, where $R(y_0, y_1) = R'(\langle y_0, y_1\rangle_2)$. Furthermore, the quantifiers are naturally introduced by projection and complement. Let $A \subseteq \mathbf{N}^k$, φ be a formula which $\mathbf{N}$-defines the projection of a set $B \subseteq \mathbf{N}^{k+1}$, and $\bar{A} = \mathbf{N}^k \setminus A$: from

$$B = \{(\vec{n}, p) \in \mathbf{N}^{k+1}; \varphi^{\mathbf{N}}(\vec{n}, p) = 1\},$$

deduce

$$A = \{\vec{n} \in \mathbf{N}^k; \exists p\, \varphi^{\mathbf{N}}(\vec{n}, p) = 1\}.$$

Hence A is $\mathbf{N}$-defined by $\exists x_{k+1}\varphi$, and

$$\bar{A} = \{\vec{n} \in \mathbf{N}^k; \forall p\, \varphi^{\mathbf{N}}(\vec{n}, p) = 0\},$$

hence $\bar{A}$ is $\mathbf{N}$-defined by $\forall x_{k+1}(\neg\varphi)$.

By the Matjasevič theorem, every r.e. set is the set of integer solutions to a diophantine equation, so a finer result is obtained: a set is Σ_n if and only if it is definable by a prenex formula, starting with $\exists$, and containing n alternations of quantifiers.

The important result is Kleene's hierarchy theorem, which follows by diagonalization from the universality result of exercise 24:

Theorem 8.25 (Kleene hierarchy) Beyond Δ_1 all inclusions are strict.

Proof Let $U_n \subseteq \mathbf{N}^2$ be a universal set Σ_n, $n \geqslant 1$. Define $A = \{m \in \mathbf{N}; (m, m) \in U_n\}$. Then A is Σ_n, hence $\bar{A}$ is Π_n. If $\bar{A}$ were also Σ_n, then by universality there would exist e such that $\bar{A} = \{m \in \mathbf{N}; (e, m) \in U_n\}$. Then $e \in A$ if and only if $e \notin A$, a contradiction. Hence $\Sigma_n \neq \Pi_n$ and $\Delta_n \neq \Sigma_n$. □

There is therefore a hierarchy of arithmetic sets which transposes to a hierarchy of problems. The problem $n \in A$? is decidable, semi-decidable or co-semi-decidable if A is recursive, r.e. (or Σ_1), or co-r.e. (or Π_1). Beyond, the problem $n \in A$? cannot be touched by machines. Concrete examples of these problems will be given in §8.4.4. Are there non-arithmetic sets? A simple example is given by the set (of encodings) of true arithmetic formulas in $\mathbf{N}$: this is Tarski's theorem (see exercise 25).

To place a set in the arithmetic hierarchy is generally quite easy.

Example Let $R = \{p; W_p = \mathbf{N}\}$ be the set of programs computing recursive functions. It suffices to write that

$$\begin{aligned} R &= \{p; \forall m(m \in W_p)\} \\ &= \{p; \forall m \exists q T(p, m, q)\}. \end{aligned}$$

Since T is recursive and its prefix is $\forall\exists$, deduce that R is Π_2. It is more difficult to show that R is not Σ_2. □

Natural problems are found among the first levels of the hierarchy, for example those mentioned around Rice's theorem. However, even though the hierarchy is strict, practically no 'natural' problems are found above level four. This seems to be linked to the difficulty of formulating (and intuitively understanding) properties containing a high number of alternating quantifiers. Mathematicians generally introduce definitions to designate important properties, e.g., continuity for $\forall\epsilon\,\exists\eta$, or even new objects to 'reset to 0' the number of alternations to facilitate future developments.

While the $\mathcal{M}_P$ model of a definite logic program is r.e., after an encoding of At_Σ, Apt and Blair have shown that the standard model of a general program 'with n strata' is Σ_n: this result shows the gap between a definite and even a restricted class of general programs.

8.4.2 Reductions

There are other ways to measure the difficulty between problems, by studying the relative difficulty of two problems. Science often progresses when a solution to a difficult problem is discovered. What more is known when a problem is solved? One can measure the importance of a solution by the class of other problems which become solved. The relations between problems are formalized by the concepts of reduction. There are two kinds of reduction. One allows the transformation of the *solution* of a problem into the solution of other (easier) problems. The other allows an extension of the available *means*, such as the use of a solution of instances of a solved problem as a supplementary means.

To reduce a problem to another is standard practice: a geometry problem is often reduced to an algebra problem. Each concept of reduction is defined as a preorder $\leqslant$ over a set of subsets of $\mathbf{N}$, $A \leqslant B$ meaning that the problem $x \in A$? reduces to $x \in B$?: one can consider that $x \in B$? is more difficult, or more general than $x \in A$?. Associate with the preorder $\leqslant$ the equivalence relation $\sim$ that it generates, i.e. $A \sim B$ if and only if $A \leqslant B$ and $B \leqslant A$: $A \sim B$ states that the problems $x \in A$? and $x \in B$? each reduce to the other, and hence are equivalent problems and of the same difficulty. Furthermore, the maximal elements of this preorder are useful reference points. More precisely, let $\leqslant$ be a reduction relation and $\mathcal{E}$ a set of problems. A problem is called *$\mathcal{E}$-difficult* for $\leqslant$ if it is an upper bound for all the elements of $\mathcal{E}$, and *$\mathcal{E}$-complete* for $\leqslant$ (or $\leqslant$-complete in $\mathcal{E}$) if it belongs to $\mathcal{E}$ and is $\mathcal{E}$-difficult. This last concept is due to Post.

Proposition 8.26

1. If A and B are $\mathcal{E}$-complete, then $A \sim B$.
2. If A is $\mathcal{E}$-complete and if $B \in \mathcal{E}$, then B is $\mathcal{E}$-complete if and only if $A \leqslant B$.

These reductions are useful, since one can easily prove that a problem is undecidable by showing that an undecidable problem can be reduced to it. It is therefore useful to have a catalogue of already classified problems which serve as a reference point.

8.4.3 Turing reduction (or relative reducibility)

Since the beginning of the chapter it has been assumed that the objects being handled in a computation are all given effectively: integers, words, etc. The addition of two integers is an effective operation: if a and b are two given integers (for example 36 and 5), then $a+b$ can be computed effectively. If f and g are two given functions from $\mathbf{N}$ to $\mathbf{N}$, consider the function $h = f + g : n \mapsto f(n) + g(n)$. To compute $h(5)$, the values of $f(5)$ and $g(5)$ must first be obtained, then they must be added, this last operation being effective.

In a certain sense, the addition of two integers and the addition of two functions are equally effective operations, even though the first handles finite objects, integers, and the second handles infinite objects, functions. However, note that the way in which elements are 'given' must be made precise. There are not many ways to give an integer. There are, however, many ways, standard in programming, to give a function f: as the text of a program computing it, or as the call to an 'external routine' producing the value $f(n)$ on demand, when n is supplied. In one case, the (source) text of the program is a finite object and it is known how f is computed. In the other, only a finite number of useful values is known for f, and these are supplied by the external routine: a library, or in an interactive manner by a user. In the latter case, it is not even necessary that f be effectively computable. So effective operations can be done on non-effective objects: computation is a transformation, and the effectiveness is relative to what is given. Turing was the first to approach computability in a relative manner by allowing the use of an 'oracle' during a computation.

The relation $A \leqslant_T B$ states that A is recursive relative to B: during the computation of the function 1_A, the use of an *oracle* is allowed: it gives a value of $1_B(x_0)$ when the latter is needed for an arbitrary value of x_0. No effectiveness assumption is made on this oracle, hence the name, but it can only be called a finite number of times, at certain steps of a computation. Turing reducibility can be stated in terms of elementary instructions of a machine. If the problem $x \in A$? can be solved, and if one adds to a Turing machine the test $x \in A$? as elementary instruction, then a new class of problems $x \in B$? can be solved: the class of $B \leqslant_T A$.

Example A set is always recursive relative to itself and to its complement. □

If B is recursive, this oracle can be replaced by a DTM, in which case A is recursive. The converse is used for undecidability proofs.

The concept of oracle is intuitive but may not appear rigorous. The precise definition of relative reducibility uses functionals rather than functions. Let $\mathcal{N}^{k,l} = \mathbf{N}^k \times (\mathbf{N}^\mathbf{N})^l$. A partial function from $\mathcal{N}^{k,l}$ to $\mathbf{N}$ is called a partial functional. To simplify, suppose $l = 1$ and write $\mathcal{N}^k$ instead of $\mathcal{N}^{k,1}$:

- A subset A of $\mathcal{N}$ is *semi-recursive* if there exists a recursive subset (in the ordinary sense) B of $\mathbf{N}^{k+2}$, such that $(m_1, \ldots, m_k, f) \in A$ if there exists $p \in \mathbf{N}$ such that $(m_1, \ldots, m_k, p, f^\star(p)) \in B$, where $f^\star$ is the history of f.
- A partial *functional* $F : \mathcal{N} \rightharpoonup \mathbf{N}$ is *recursive* if its graph, which is a subset of $\mathcal{N}^{k+1,1}$, is semi-recursive.
- A *subset* of $\mathcal{N}$ is *recursive* if its characteristic function is recursive.
- Finally a function $f : \mathbf{N}^k \to \mathbf{N}$ is recursive *relative* to a function $g \in \mathbf{N}^k \to \mathbf{N}$, written $f \leqslant_T g$, if $f(\vec{m}) = F(\vec{m}, g)$ for a partial recursive functional F. If A and B are two subsets of $\mathbf{N}^k$, A is called recursive relative to B, written $A \leqslant_T B$, if $1_A \leqslant_T 1_B$.

Now that these definitions have been given, it has to be shown that they cover the intuitive approach to computations using oracles. Consider the problem $(\vec{m}, f) \in$

A?, where $A \subseteq \mathcal{N}$ is a semi-recursive subset. To solve it, use $B \subseteq \mathbf{N}^{k+2}$, recursive. For $p \geqslant 0$, the p-th step consists of obtaining the value of $f^\star(p)$ from an 'oracle', then to compute $1_B(m_1, \ldots, m_k, p, f^\star(p))$. If the computed value is 1, return 1, otherwise start over with $p+1$.

This reduction is very important in logic. The associated equivalence classes are called *degrees of unsolvability* (Kleene and Post 1954 [83]). The ordered set of these degrees has surprising properties, which have been the object of original methods (Post, Friedberg, Muchnik, Sacks, etc.). The set of degrees has a least element $\mathbf{0}$ formed of all the recursive sets (it is in particular the degree of $\varnothing$ and of $\mathbf{N}$, these being $\leqslant_T$ to any set). Another degree is well known: the degree $\mathbf{0}'$ of the r.e. set K of the halting problem. It was thought for a long time that every r.e. set was of degree $\mathbf{0}$ (if it was recursive) or $\mathbf{0}'$. In fact all the undecidability proofs consisted of showing $K \leqslant_T A$ (or even, which is stronger, $K \leqslant_m A$): in 1943 this led Post [123] to try to construct intermediate degrees between $\mathbf{0}$ and $\mathbf{0}'$. This problem was solved in 1956 by Muchnik [112] and Friedberg [44] independently by original techniques, thereby inaugurating a specialized new branch of logic. There exists a non-recursive r.e. set E such that $K \not\leqslant_T E$, hence whose degree is strictly between $\mathbf{0}$ and $\mathbf{0}'$. There even exists an uncountable number of r.e. degrees, incomparable under $\leqslant_T$. In 1964 Sacks [130] showed that the order over r.e. degrees between $\mathbf{0}$ and $\mathbf{0}'$ is dense. From these theorems, there exists an infinite number of 'distinct' semi-decidable problems in the Turing reduction sense, while all the usual problems can be found in the degrees $\mathbf{0}$ and $\mathbf{0}'$. The moral of the story is that it is very difficult to imagine complicated problems.

8.4.4 Strong reduction

In practice, to reduce a problem to another one is to transform its statement. The idea is to let $A \leqslant B$ if there is a sufficiently simple transformation f such that $x \in A$ if and only if $f(x) \in B$. Since this general definition depends on the class of accepted transformations, there are several concepts of strong reduction: recursive transformation, injective recursive transformation and polynomial transformation.

Definition 8.13 The preorder $\leqslant_m$ is defined by: $A \leqslant_m B$ if there exists a recursive function f such that $x \in A$ if and only if $f(x) \in B$.

This reduction is stronger than that of Turing, since $A \leqslant_m B$ implies $A \leqslant_T B$: the computation of $1_A(x)$ is reduced to that of $f(x)$ by a DTM and a call to an oracle for $1_B(f(x))$. If B is r.e, then A is as well: this property is not true if one only has $A \leqslant_T B$.

Consider once again the set

$$K = \{n; \{n\}(n)\downarrow\} = \{n; n \in W_n\}$$

of the halting problem.

Theorem 8.27 1. K is r.e.

2. K is not recursive.
3. K is $\leqslant_m$-complete for the class of r.e.

Proof (1) $n \mapsto \{n\}(n)$ is partial recursive, hence by proposition 8.5, its domain is r.e.

(2) If K were recursive, $\mathbf{N}\backslash K$ would be r.e. (proposition 8.6) from which $\mathbf{N}\backslash K = \{x \in \mathbf{N}; \{p\}(x)\downarrow\}$ for a $p \in \mathbf{N}$ (universality theorem). Then $p \in K$ if and only if $p \notin K$. Contradiction!

(3) Let A be an r.e. set. Then $A = \{x \in \mathbf{N}; f(x)\downarrow\}$, where f is partial recursive. Introduce the partial recursive function $g(x, y) = f(x)$.

There exists $m \in \mathbf{N}$ such that $g(x, y) = \{m\}(x, y)$ (universality). From the s–m–n theorem: $\{m\}(x, y) = \{\rho(m, x)\}(y)$. The function $x \mapsto \rho(m, x)$ is recursive and:

$$A = \{x \in \mathbf{N}; \rho(m, x) \in K\},$$

from which $A \leqslant_m K$. □

It can be shown that

- $\{n \in \mathbf{N}; W_n = \mathbf{N}\}$ is Π_2-complete;
- $\{n \in \mathbf{N}; W_n$ finite $\}$ is Σ_2-complete;
- $\{n \in \mathbf{N}; W_n$ infinite $\}$ is Π_2-complete;
- $\{n \in \mathbf{N}; W_n$ cofinite $\}$ is Σ_3-complete;
- $\{n \in \mathbf{N}; W_n$ recursive $\}$ is Σ_3-complete.

A set is strictly Π_n (respectively Σ_n) if it belongs to the class Π_n (respectively Σ_n) and does not belong to the class Σ_n (respectively Π_n). The hierarchy theorem implies that each of the preceding sets constitutes an example of a set strictly in its class.

Example It has already been proven that $R = \{n \in \mathbf{N}; W_n = \mathbf{N}\}$, the set of indices of recursive functions, is Π_2. We show that it is Π_2-complete for $\leqslant_m$. Let A be an arbitrary Π_2 set, hence of the form $\{x; \forall n((x, n) \in B)\}$, where B is a r.e. Let e be an index of B: $(x, n) \in B$ if and only if $\{e\}(x, n)\downarrow$. By the s–m–n theorem, $\{e\}(x, n) = \{\rho(e, x)\}(n)$. Let $f(x) = \rho(e, x)$. Then $(x, n) \in B$ is equivalent to $n \in W_{f(x)}$. Deduce that $x \in A$ is equivalent to $\forall n(n \in W_{f(x)})$, i.e, $W_{f(x)} = \mathbf{N}$, or else $f(x) \in R$. This shows that $A \leqslant_m R$, using f. □

The proof of the other properties is not immediate. Hence, these sets are neither r.e. nor co-r.e. There does not even exist a semi-decision procedure to determine whether a program terminates (or not). Such problems, crucial in programming, are the subject of human activity, as is most of mathematics.

This is also the case for programming, where one generally defines an 'algorithm' as a procedure which always terminates. If one accepts this definition, then one must accept the fact that this set of algorithms is Π_2, hence the simple problem of determining if something is an algorithm has no effective solution. Algorithm analysis is not a theory of algorithms in the same way as arithmetic is a theory of numbers. This analysis studies interesting algorithms by developing general methods such as construction methods and complexity analysis, and restricted classes of algorithms of which none attain the heights of complexity that this hierarchy allows. In fact, the ceiling is more that of efficiency, which will be studied in the final section.

The situation is analogous to that in mathematics, where one looks for interesting theorems, and not for everything which can be deduced from the axioms. Besides, the axioms never have an extraordinary syntactic complexity: they are simply there because they serve to model useful objects.

These concepts of reduction often allow the 'transfer' of undecidability results. For example, in 1955 Novikov proved (in an article [114] of 153 pages!) that the word problem for groups is undecidable. In 1958 Markov [98] showed how to associate with each finite presentation of a group G a topological manifold $V(G)$ of dimension 4 whose homotopy group is G, in such a way that G_1 is a group isomorphic to G_2 if and only if the manifolds $V(G_1)$ and $V(G_2)$ are homeomorphic. It follows that the problem of homeomorphy of manifolds of dimension 4 is not decidable: here is a transfer from algebra to topology.

It is *via* algebra that mathematicians look for substantial aid from computer science, e.g., in the use of computer algebra for algebraic geometry or algebraic topology. Yet, it is also *via* algebra that many undecidable problems branched out into other domains of mathematics. The use of algebra allows a transfer of problems from topology towards algebra, but also a transfer, in the reverse direction, of undecidability results.

8.4.5 Creative problems

It has been seen that the complement $\bar{A}$ of a non-recursive r.e. set A cannot be r.e. Yet $\bar{A}$ contains many r.e. sets, for example all of its finite subsets. But every r.e. W_n is distinct from $\bar{A}$: there exists an element a_n in $\bar{A} \setminus W_n$ or in $W_n \setminus \bar{A}$. In particular, there does not exist an enumeration procedure which is both 'sound' and 'complete' for $\bar{A}$: if a procedure enumerates only elements of $\bar{A}$ (it is sound), it will necessarily forget some, and if it does not forget any (it is complete), it will also produce elements of A. If the counterexample a_n can be computed effectively from n, the following definition is reached:

Definition 8.14 A recursive function f is a *productive* function for a set P if for every n, $f(n) \in P \setminus W_n$ or $f(n) \in W_n \setminus P$. A set P is *productive* if there exists a productive function for P. A set is *creative* if it is r.e. and if its complement is productive.

The productive problems are not semi-decidable, and *unfoolable*. Each time that someone pretends to have a decision procedure for P, P productive, by supplying an algorithm of index n such that $P = W_n$, only $f(n)$ need be computed to put it to rest.

Example The set K of the halting problem is creative, since it is r.e. and its complement $\bar{K}$ is productive. In fact, $\bar{K}$ has the identity as production function. For every n, $n \in \bar{K} \setminus W_n$, by the definition of K. □

It turns out that a large number of known r.e. are creative, for example the set K, and the set of formulas provable in arithmetic: it is this last example which gives these sets the qualifier of 'creative', due to Post, since they illustrate the creative character of mathematical activity. The incompleteness of arithmetic is explained in this framework. The set of theorems is creative, hence the set P of non-provable formulas is productive. The set of negations of theorems is r.e., like the set of theorems, hence it is a certain W_n, and a subset of P. If f is a production function for P, then $f(n) \in P \setminus W_n$, i.e, $f(n)$ is the encoding of a non-provable formula φ, such that $\neg\varphi$ is equally non-provable.

The creative problems are exactly the Σ_1-complete problems for the reduction $\leqslant_m$ (Myhill theorem).

8.5 Complexity

Even though computability was developed well before computer science, it is the former which largely motivated the study of the complexity of algorithms.

There is an 'abstract theory' of complexity, due to M. Blum (1967 [10]) who, starting from minimal assumptions about what a complexity measure should be, proves remarkable general phenomena (gap and speedup properties). It is in the same tradition of the fundamental results of Kleene, such as universality and recursivity. Complexity will not be studied here in this manner, but, rather, algorithmically.

Computability does not deal with the resources needed for a computation. It just supposes a Turing machine with an infinite tape. Algorithm analysis studies algorithms to determine the resources needed for them to run. There are several kinds of resources:

- Two are theoretically unlimited, though in practice finite and expensive:
  - time and
  - space (main memory).
- Others are available in a fixed, limited number, and the optimal allocation problem can be very difficult:
  - registers, stacks, ... ,
  - processors, and

- peripherals (secondary memory, random or sequential, input-output, etc.).

Only the space and time problems will be studied here. Their study is necessary to produce *efficient* algorithms.

8.5.1 Complexity measures

Before looking at the general context of partial recursive functions and Turing machines, here is a simple example of a class of functions and the programs computing them.

Example It is easy to see that the primitive recursive functions are the functions computed by a restricted class of iterative programs, the finite loop programs defined in §8.1.2.

There are two ways to measure the complexity of such a program. First consider the depth $d(\pi)$ of a program π defined by the nesting depth of **do** ... **end** loops. For example, the parenthesization $(()(()))$ has depth 3. Second, let $x_1, \dots, x_q$ be variables of π. Introduce the function $T_\pi : \mathbf{N}^q \to \mathbf{N}$ measuring for $a_1, \dots, a_q \in \mathbf{N}$ the number of assignments executed by π with $a_1, \dots, a_q$ as initial values for variables $x_1, \dots, x_q$. The function T_π is a temporal complexity of dynamic character since it is computable by simulating the execution of the program, while $d(\pi)$ is a static complexity, i.e., computable upon reading the program.

What is interesting about this class of programs is that T_π and $d(\pi)$ are related. Introduce the functions $f_n : \mathbf{N} \to \mathbf{N}$ defined by:

$$\begin{aligned} f_0(m) &= \begin{cases} m+1 & \text{if } m = 0 \text{ or } 1 \\ m+2 & \text{otherwise,} \end{cases} \\ f_{n+1}(m) &= (f_n^{(m)}(1)). \end{aligned}$$

These functions grow very quickly and can set bounds on T_π. It can be proved that $d(\pi) = p$ if and only if there exists k such that

$$T_\pi(a_1, \dots, a_k)^k \leqslant f_p(\max_i a_i).$$

Hence the complexity analysis of finite loop programs has a satisfactory solution. But this language is not powerful enough, since it only computes the primitive recursive functions, and particularly since its programs halt. □

In the case of a programming language computing all of the partial recursive function, there is no analogous result. There are programs whose execution time, compared to their textual length, are as long as one wishes, just as in logic, there are trivial theorems with gigantic derivations. These phenomena are due to the undecidability of the corresponding problems.

There are, however, static measures of complexity, i.e., depending only on the text of the program, much used in software engineering: number of lines or identifiers, or diverse characteristics of the calling graph for procedures. In a given class of applications for which standard values are known, these measures enable a superficial understanding of the readability of programs, but not of their complexity, be it of the logic of the design or of the efficiency of execution. These typical values are mainly a good source of information to determine the most *useful* compiler optimizations.

In the absence of a static measure of complexity, a dynamic measure must be given, which will take into account all the possible executions over all the possible data.

Among the models of computation described in this chapter, those of deterministic Turing machines (DTM) are the best adapted. The length of a computation $c_0 \to c_1 \to \ldots \to c_n$ is n.

Definition 8.15

1. Let (Γ, Q, R) be a DTM and S and T be functions from $\mathbf{N}$ to $\mathbf{N}$. The DTM (Γ, Q, R) is of *temporal complexity* (less than) T if the length of the computation of every word of length n is at most $T(n)$. It is of *spatial complexity* (less than) S if, during the computation of every word of length n, it traverses at most $S(n)$ cells.
2. Let L be a subset of $\Gamma^\star$. L, or the problem $x \in L$? is of complexity T if there exists a DTM of complexity T accepting L, i.e. such that every computation from a word of L halts.
3. Let $f : \Gamma^\star \to \Gamma^\star$. The function f is of complexity T if there exists a DTM of complexity T computing f.

In practice, the evaluation of the complexity of an algorithm solving a problem U is done as follows:

- The set I_U of of U's instances is provided with a 'size' function $|\;| : I_U \to \mathbf{N}$, for example the length of a word or the dimension of a problem.
- A maximal temporal complexity $T : \mathbf{N} \to \mathbf{N}$ is defined such that the execution of the algorithm with $x \in I_U$ as data takes at most $T(|x|)$ time units, this time being computed by associating with each instruction of the algorithm an execution time and by simulating its execution.

For example, computing the maximum of n integers requires $(n-1)$ comparisons. It runs in $O(n)$, n being the size of the problem. Recall the usual asymptotic notations: let f be a function from $\mathbf{N}$ to $\mathbf{R}^\star$.

$$O(f) = \{g : \mathbf{N} \to \mathbf{R}^\star \,;\, \exists c \in \mathbf{R}^{+\star}\ \exists n_0 \in \mathbf{N}\ \forall n \geqslant n_0\ (g(n) \leqslant cf(n))\}.$$

Write $g(n) = O(f(n))$ instead of $g \in O(f)$.

To justify such a practice in the framework of DTM, the instances of a problem must first be encoded in an alphabet. In the case of numeric data it will always be supposed that the encoding of integers is binary (or decimal, but never unary) and that the encoding of sets of integers can be done by listing their separate elements by a supplementary alphabet. This allows one to refer directly to the complexity of $A \subseteq \mathbf{N}^k$ (or of the problem $x \in A$?) and of a function $f : \mathbf{N}^k \to \mathbf{N}$. The algorithm then has to be translated into a DTM, which is far from a normal translation.

This is why algorithm analysis uses the *random access machine* (RAM) model. RAMs realistically simulate the behavior of a computer with an assembler-like language. The logarithmic cost assumption is used, i.e., the time to execute instruction 'INST n' is proportional to the logarithm of n.

The complexity computations in different models, DTM or RAM, obviously give different results. The model and the assumptions about the cost of instructions must therefore be made precise. The DTM model, more primitive, is well adapted to the theoretical study of complexity. The RAM model, more realistic, applies to concrete algorithms of computer science. There are nevertheless simple relations between the computed complexities in each of the models.

8.5.2 Polynomial problems

In the second section of the chapter, problems were divided into two classes: decidable and undecidable. The decidable problems are not all equally tractable. Many standard algorithms, such as searches, sorts and polynomial evaluations have asymptotic complexities of $O(\log n)$, $O(n)$, $O(n \log n)$, $O(n^2)$, $O(n^3)$, ... , hence are bounded by a polynomial. As they are standard, they are of course tractable and increasing the computation speed of machines significantly increases the size of allowable data: a machine 100 times faster than the old one allows one to handle data ten times larger for an algorithm in $O(n^2)$. There are also algorithms which theoretically solve problems, by exhaustive search in a finite set which can be a powerset, for example, but which take lots of time $(2^n, n^n, n!, \dots)$. They are executable for only very small values of n (for example $50! \approx 3.10^{23}$) and increasing the power of a machine has little effect on the size of possible data.

Definition 8.16 An algorithm is *polynomial* if there exists $k \in \mathbf{N}$ such that its complexity is $O(n^k)$. The problem $x \in A$?, or the set A, is polynomial (or *in class* **P**) if there exists a polynomial algorithm solving it. A function f is a *polynomial transformation* (or is in class **P**) if there exists a polynomial algorithm computing it (A is in **P** if and only if 1_A is in **P**).

Example The unification of two terms is decidable. Although the usual algorithm is exponential, there are polynomial algorithms, and even linear ones. The unification problem is therefore in **P**. □

The definition of class **P** is *robust.* For example, it is not affected if DTMs with several tapes are used. The same class is achieved with RAMs of logarithmic cost. A RAM program of complexity $O(n)$ will have a complexity $O(n^5)$ once simulated by a DTM but will remain polynomial. There are therefore good reasons to state Cook's thesis:

Cook thesis 'The *tractable* problems are those in class **P**.'

Like Church's thesis, Cook's thesis is of experimental origin, sustained by convincing theoretical results, and provides a definition of the word *tractable.*

Remark The tractable/non-tractable delimitation is less drastic than that of decidability. It is sometimes more convenient to use a non-polynomial algorithm for a problem in **P**, because it is easy to write and is of reasonable execution speed for the normal instances of the problem. This is the case for linear programming in which the non-polynomial simplex algorithm (Dantzig 1951 [32]) is used rather than the complicated and inefficient polynomial algorithm of Khachiyan (1979 [77]); there is a new efficient polynomial algorithm, proposed by Karmarkar in 1984 [75]. The role of constants is ignored when tractability is studied; it is possible to prefer an algorithm in 2^n rather than one in $10^{100}n$. So efficiency involves more criteria than just tractability. □

There are also several important problems for which no polynomial algorithm is known. In order to solve them by preserving the polynomial computation time, the concept of algorithm will be extended by introducing faster machines than DTM. They cannot solve all problems in polynomial time but only those whose solution, once given, can be checked in polynomial time.

Example The existence problem of a Hamiltonian cycle in a graph is of this kind: it is immediate—hence polynomial!—to check if a given cycle is Hamiltonian. □

Definition 8.17 $A \subseteq \Gamma^\star$ is in class **NP** (*non-deterministic polynomial*) if there exists $B \subseteq \Gamma^\star \times \Delta^\star$ in **P** and a polynomial p such that

$$A = \{x \in \Gamma^\star; \exists y \in \Delta^\star, |y| \leqslant p(|x|) \text{ and } (x, y) \in B\}.$$

It is the existential quantifier $\exists y$ which introduced non-determinism. To solve $x \in A$?, one must

- generate y such that $|y| \leqslant p(|x|)$;
- compute $1_B(x, y)$.

The complexity of the first operation is at most $p(|x|)$. The second is polynomial in $|x| + |y|$ by assumption on B, hence from the condition on y it is polynomial in x. This procedure can be executed by a (non-deterministic) Turing machine:

Definition 8.18 Let $L \subseteq \Gamma^\star$. A Turing machine (Γ, Q, R) *accepts* L in time T over a subset Q_0 of Q, if for every $x \in L$ there exists a computation from x, ending in a state in Q_0 and of length less than $T(|x|)$.

Proposition 8.28

(1) L is in **NP** if and only if there exists a Turing machine accepting L in polynomial time.
(2) If L is accepted by a Turing machine in time $T(n) = O(n^k)$, then there exists a constant c and a DTM of complexity $O(c^{n^k})$ solving the problem $x \in L?$.

Part (1) is often taken as the definition of **NP**. Part (2) states that the problems in **NP** have at most exponential complexity. The essential idea of the proof (of the converse of (1) and (2)) is as follows. Let d be the degree of non-determinism of a Turing machine, i.e, the maximum number of rules $(q, a, \dots)$ starting with the same pair (q, a); it is the choice between these rules which creates non-determinism. For a DTM, $d = 1$. Then for $x \in \Gamma^\star$, there are at most $d^{T(|x|)}$ different computations of length $\leqslant T|x|$ from x. By exploring the computations, a DTM accepting L can be constructed and one proves that L is in **NP**.

The class **NP** does not cover all non-polynomial problems. In fact **NP** *does not* abbreviate Not Polynomial. For example, the decidability problem in Presburger arithmetic is decidable in a non-deterministic time of at least 2^{k^n}, where k is constant: this problem is not even in **P**.

It is clear that $\mathbf{P} \subseteq \mathbf{NP}$. But is it clear that the power of computation has been increased by allowing non-determinism? No, since, for the moment, no example of a problem in **NP** has been shown to not be in **P**. However, there are good reasons to believe that $\mathbf{P} \neq \mathbf{NP}$.

8.5.3 Strong polynomial reduction

The definition of this reduction is directly inspired from the strong reduction under a recursive transformation: $A \leqslant_P B$ if there exists a polynomial transformation f such that $x \in A$ if and only if $f(x) \in B$. If one can solve $x \in B?$, then, given extra polynomial time, $x \in A?$ can be solved. This reduction is particularly interesting outside of **P**. Here is a first example of polynomial reduction.

Hamiltonian cycle (HC)
Given: an undirected graph (S, A).
Problem: does (S, A) have a Hamiltonian cycle?

Traveling Salesperson (TS)
Given: a distance $d : C \times C \to \mathbf{N}$ and an integer M.
Problem: is there a tour of C of length at most M ?

(HC) $\leqslant_P$ (TS) is proven by constructing an instance (C, d, M) of (TS) from an instance (S, A) of (HC). Let:

$$\begin{aligned} C &= S, \\ M &= |S|, \\ d(x,y) &= \begin{cases} 1 & \text{if } (x,y) \in A \\ 2 & \text{otherwise.} \end{cases} \end{aligned}$$

This transformation is polynomial, after a suitable encoding in an alphabet describing the problems and (S, A) has a Hamiltonian cycle if and only if there is a tour of C of length at most M.

Definition 8.19 An **NP**-complete problem is a problem in **NP**, complete under polynomial reduction $\leqslant_P$.

The following properties are immediate ($\sim_P$ is the equivalence generated by $\leqslant_P$):

Proposition 8.29

1. If A and B are **NP**-complete, then $A \sim_P B$.
2. If A is **NP**-complete and in **P**, then $\mathbf{P} = \mathbf{NP}$.
3. If A is **NP**-complete and if B is in **NP**, then B is **NP**-complete if $A \leqslant_P B$.

Remarks

1. states that all **NP**-complete problems are as difficult as each other;
2. explains why the study of these problems is crucial if one wishes to show that $\mathbf{P} = \mathbf{NP}$ or $\mathbf{P} \neq \mathbf{NP}$;
3. gives the standard method for proving that a problem (B) is **NP**-complete if this is already known for (A).

□

Several hundred **NP**-complete problems have been catalogued in many domains: operations research, graph theory, robotics, number theory, etc. For each of these problems, no polynomial algorithm has been found, despite intensive effort. This is why, because of property (2), $\mathbf{P} \neq \mathbf{NP}$ is likely, even though it has never been proven that it is impossible to obtain polynomial solutions to these problems.

Conjecture $\boxed{\mathbf{P} \neq \mathbf{NP}}$

SAT (satisfiability in propositional logic) Let X be a finite countable set.
Given: a conjunctive formula $\varphi \in \mathrm{Prop}[X]$.
Problem: is φ satisfiable?

Theorem 8.30 (Cook) (SAT) is **NP**-complete.

This was the first problem shown to be **NP**-complete, in Cook's initial article defining polynomial reduction, **NP** and NP-complete problems, in 1971 [24]. Let $A \subseteq \Gamma^\star$ be accepted in polynomial time by a Turing machine (Γ, Q, R). The proof, long and tedious, explicitly constructs for each $x \in \Gamma^\star$ a propositional formula φ_x which is satisfiable if and only if $x \in A$. From (SAT), one can prove the **NP**-completeness of other problems (proposition (3)). Karp gave twenty or so in 1972 [76], and Garey and Johnson more than 300 in 1979 [49].

Before presenting a few examples, note the following parallel:

recursive	class **P**
r.e.	class **NP**
$\leqslant_m$	$\leqslant_P$
Church thesis	Cook thesis

In particular, the definition of **NP** by existential quantification from **P** is similar to the definition of the r.e. from the recursives, with the added restriction that there be a polynomial bound on the quantifier. However, many results in recursive function theory do not have equivalents in complexity theory.

8.5.4 Examples of NP-complete problems

The first problem (3-SAT) essentially plays a technical role, as it is easier to use than (SAT) to initiate the list of NP-complete problems. Recall that if (S, A) is a unoriented graph, a subset R of S is a cover of S if every arc has at least one end in R.

3-SAT
Given: a conjunctive formula φ of *Prop*$[X]$, in which each clause contains exactly *three* literals.
Problem: is φ satisfiable?

Here is the transformation allowing the reduction of (SAT) to (3-SAT). Let $C = \{l_1, \ldots, l_k\}$ be a clause containing $k \geqslant 4$ literals. Introduce $k-3$ new propositional variables $y_1, \ldots, y_{k-3}$ and replace clause C by the $k-2$ clauses:

$$\begin{gathered}\{l_1, l_2, y_1\},\\ \{l_3, \neg y_1, y_2\},\\ \{l_4, \neg y_2, y_3\},\\ \cdots\cdots\\ \{l_{k-2}, \neg y_{k-4}, y_{k-3}\},\\ \{l_{k-1}, l_k, \neg y_{k-3}\}.\end{gathered}$$

A two-literal clause $\{l_1, l_2\}$ is replaced by the two clauses $\{l_1, l_2, y\}$, $\{l_1, l_2, \neg y\}$, and a one-literal clause $\{l_1\}$ by the four clauses $\{l_1, y_1, y_2\}$, $\{l_1, \neg y_1, y_2\}$, $\{l_1, y_1, \neg y_2\}$, and $\{l_1, \neg y_1, \neg y_2\}$.

Vertex cover (VC)
Given: a graph (S, A) and an integer k.
Problem: does (S, A) allow a cover R of S with $|R| \leqslant k$?

(3-SAT) $\leqslant_P$ (VC) is proven by associating with each instance $\varphi = C_1 \wedge \ldots \wedge C_n$ of 3-SAT a graph having $3n$ vertices, one for each literal of each clause, by linking the three vertices associated with a clause, and between clauses, the vertices corresponding to opposite literals $(l, \neg l)$. Let $k = 2n$. For example, for

$$\varphi = (x_1 \vee \neg x_3 \vee x_4) \wedge (\neg x_1 \vee x_2 \vee x_4),$$

the graph obtained is:

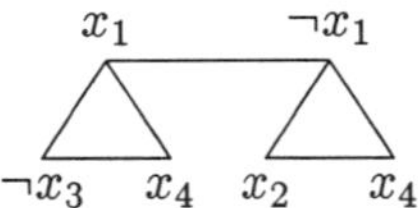

Some other **NP**-complete problems of graph theory: k-coloring, clique and Hamiltonian cycle—but the edge cover and Euler cycle problems are in **P**. For network problems: traveling salesperson, maximal path between two vertices and various flow problems are **NP**-complete—but the minimal path problem is in **P**. In computer science, many optimal code generation, data compression or dynamic memory allocation problems are **NP**-complete. Scheduling problems, for one or more processors, with priority relations between tasks, with time or resource constraints, etc., are easily **NP**-complete. Here are some examples:

Partition
Given: a finite set A and a size function $s : A \rightarrow \mathbf{N}^\star$.
Problem: does there exist a subset $A' \subseteq A$ such that $\sum_{a \in A'} s(a) = \sum_{a \in A \setminus A'} s(a)$?

Conjunctive queries

Given: a finite set D, a family $R_1, \ldots, R_p$ of r_i-ary relations over D, and a (closed formula) query $\rho : \exists x_1 \ldots x_n(\alpha_1 \wedge \ldots \wedge \alpha_m)$, where each α_i is an atom with parameters in D.

Problem: $D \models \rho$?

Multiprocessor ordering (MO)

Given: a finite set T of tasks with a length function $l : T \to \mathbf{N}^\star$, a number $m > 0$ of processors and a deadline $d > 0$.

Problem: is there a scheduling compatible with the data?

(Partition) $\leqslant_p$ (MO) easily, by restriction, since (Partition) is the particular case of (MO) when $m = 2$ and $d = 1/2 \sum_{a \in T} l(a)$.

8.5.5 Solving NP-complete problems

For a given **NP**-complete problem, it is possible that only a restricted class of problems is presented and that this class is in **P**. For example, (2-SAT), analogous to 3-SAT, and the 2-coloring of graphs are in **P**. The priority scheduling problem is in **P** if the priority graph is a tree or if there are only two processors. (SAT) restricted to Horn clauses is polynomial.

A non-polynomial algorithm can have an average polynomial temporal complexity, and quite satisfactory behavior in practice, if the worst cases are 'rare'. The probabilistic study of algorithms is not far advanced.

It is also possible that an algorithm may not be polynomial in the size of its data but polynomial, say, in the largest integer appearing in the data; obviously, n is linear in n, but not in $\log n$. This is the case when an exhaustive search is being simplified, for example by dynamic programming. For the (Partition) problem, this method leads to an algorithm in $O(nk)$ where $n = |A|$, $k = \sum_{a \in A} s(a)$. This algorithm is not linear in the size of its data, which is $O(n \log k)$. However, in typical cases, k is in $O(n)$, so a polynomial algorithm is obtained. Problems of this kind are called pseudo-polynomial and are typically solved by dynamic programming or 'branch and bound' methods. But not all **NP**-complete problems are pseudo-polynomial. This is the case for the (TS), for which the complexity does not come from possibly large distances, since the reduction of (HC) to (TS) only uses distances of 1 or 2, and (HC) itself is **NP**-complete.

If the known algorithms are not really polynomial, even with the probabilistic or pseudo variants, a polynomial algorithm can be sought to *approximate* a solution to an *optimization* problem. There are many good approximating algorithms. For (TS), one can find a tour of at most twice the optimal length in polynomial time. For some problems, it is possible to obtain a solution ϵ away from optimal for an arbitrary positive ϵ. For others, such as (TS), this is impossible unless $\mathbf{P} = \mathbf{NP}$.

Exercises[2]

1. Show that $\langle\ \rangle_k$ is a bijection from $\mathbf{N}^k$ to $\mathbf{N}$, and find its inverse $(\pi_k^1, \ldots, \pi_k^k) : \mathbf{N} \to \mathbf{N}^k$.

2. Show that N^2 is enumerated by the formula:

$$\langle x, y\rangle = 1/2((x+y)^2 + 3x + y).$$

3. Define $\|x\|$, $\mathrm{val}(x)$, $\mathrm{pre}(i,x)$, $\mathrm{last}(i,x)$ and *cons* using $\langle\ \rangle$ and the projections π^1 and π^2.

4. Show that the following functions are primitive recursive:

1. the predecessor p ($p(0) = 0$, $p(x+1) = x$);
2. (total) substraction $\dot{-}$ (defined by $x \dot{-} y = x - y$ if $x \geqslant y$, 0 otherwise);
3. the relations $x \leqslant y$, $x = y$;
4. the distance $|x - y|$.

5.★ Show that if g, h and P are primitive recursive, then f defined by

$$f(n) = \begin{cases} g(n) & \text{if } n \in P \\ h(n) & \text{otherwise} \end{cases}$$

is primitive recursive.

6.★ Show that if g is primitive recursive, then f defined by

$$f(m, \vec{n}) = \sum_{p \leqslant m} g(p, \vec{n})$$

is also primitive recursive. Similar result for $\prod$, max and min instead of $\sum$.

7. Show that the class of primitive recursive functions is stable under the bounded maximization operator μ' (to be defined).

8. Show that addition is Turing-computable.

9. Show that the successor and addition are respectively defined in Peano arithmetic by the formulas $y = sx$ and $z = x + y$.

10. Show that if $f(n) = h(g_1(n), g_2(n))$ and if h, g_1 and g_2 are definable in arithmetic, then f is as well.

11.★ Show that finite sets are recursive.

[2]Hints to the exercises labeled with a ★ can be found at the end of the book.

12. Show that if A is the domain of a computable partial function and is infinite, then it is the image of an injective recursive function. Show that if $f : \mathbf{N} \to \mathbf{N}$ is recursive increasing, then $f(\mathbf{N})$ is recursive.

13. Let $A \subseteq \mathbf{N}^k$. Show that if the function 1_A is T-definable, then the set A is also T-definable.

14. Deduce from the s–m–n theorem that for every r.e. set $A \subseteq \mathbf{N}^2$, there exists a recursive function f such that $A = \{(x, y); x \in W_{f(y)}\}$.

15.★ Solve for X in the equations: (1) $X = \ulcorner AX \urcorner$; (2) $X = \ulcorner XA \urcorner$.

16. Show that each of the problems on page 340 is undecidable.

17. Show that the mapping associating the integer $2^{a_1} + \ldots + 2^{a_n}$ with the finite set $\{a_1, \ldots, a_n\}$ defines a bijection from the set of finite subsets of $\mathbf{N}$ to $\mathbf{N}$.

18.★ Apply the Rice–Shapiro theorem to show that the following properties of r.e. sets are not semi-decidable:

1. A is recursive;
2. A is not recursive;
3. $A = \mathbf{N}$.

19.★ Apply the Rice–Shapiro theorem to show that the following properties are semi-decidable:

1. $33 \in A$;
2. $A \neq \varnothing$.

20.★ Deduce from Matjasevič's theorem and from the universality theorem the existence of a polynomial P such that for every r.e. subset $A \subseteq \mathbf{N}$, there exists an integer i such that

$$A = \{x \in \mathbf{N}; \exists (y_1, \ldots, y_k) \in \mathbf{N}^k P(i, x, y_1, \ldots, y_k) = 0\}.$$

21.★ Let f be an injective recursive function from $\mathbf{N}$ to $\mathbf{N}$, and s be the limit of the sequence $s_n = \sum_{k=0}^{n} 1/2^{f(k)}$. Show that s is a recursive real if and only if $f(\mathbf{N})$ is a recursive set.

22. Show the following inclusions: $\Sigma_n \cup \Pi_n \subseteq \Delta_{n+1}$ for every $n \geqslant 0$.

23. Prove that the interpretation in $\mathbf{N}$ of a formula of arithmetic is an arithmetic set (start with the interpretation of terms, then reason by induction over formulas).

24.★ By using the universality theorem for r.e. sets, show that for every $n \geqslant 1$, there exists a subset $U_n \subseteq \mathbf{N}^2$, which is Σ_n and universal for the subsets Σ_n of $\mathbf{N}$.

25.★ Show, using contradiction and lemma 8.11 that the set of true formulas in $\mathbf{N}$ is not definable in $\mathbf{N}$ by a formula of arithmetic (it is a theorem of Tarski's).

26.★ Show that $A = \{p; W_p \text{ is recursive}\}$ is Σ_3.

27. Prove proposition 8.26.

28. Show that K and its complement $\overline{K}$ are incomparable for $\leqslant_m$.

29. Write a program with finite loops π to compute the product of two integers, and compute $T_\pi(a, b)$ and $d(\pi)$.

30. Using the transformations given on page 358, prove that (SAT) $\leqslant_P$ (3-SAT).

Bibliographic notes

Rogers' book [129] is a standard reference for recursive function theory. [139] is a recent book treating these questions from a programming point of view. Programs with finite loops are introduced by Meyer and Ritchie [106]. See also [34, 139]. The encoding of logic is made explicit in several books on logic, including [135]. On proofs of soundness and the strength of theories, see [53]. The notion of reflexivity in programming is analyzed in, among others, [45]. Shoenfield's little book [136] is an introduction to the study of degrees of unsolvability. Garey and Johnson's book [49] is the reference for **NP**-complete problems.

Hints to selected exercises

Chapter 2

2. For the lexical order over $A^\star$, if $i < j$ in A, every word ujv has as lower bounds the ui^n, for $n \geqslant 0$.

6. These properties are proved by induction. Consider (3) for a term M by supposing it is true for its immediate subterms. If $M = c$, $\mathrm{ar}(c) = 0$ and $u \in \mathcal{O}(M)$, then $u = \epsilon$, then $u1 \notin \mathcal{O}(M)$. If $M = fM_1 \ldots M_p$ and $u \in \mathcal{O}(M)$, then $u = i \cdot v$ with $1 \leqslant i \leqslant p$ and $v \in \mathcal{O}(M_i)$, hence $M(u) = M_i(v)$; by applying the inductive hypothesis to M_i, if $\mathrm{ar}(M_i(v)) = n$, then $v \cdot 1, \ldots, v \cdot n \in \mathcal{O}(M_i)$ and $u \cdot (n+1) \notin \mathcal{O}(M_i)$; hence $i \cdot v \cdot 1, \ldots, i \cdot v \cdot n \in \mathcal{O}(M)$ and $i \cdot u \cdot (n+1) \notin \mathcal{O}(M)$.

11. Proof by induction over λ-terms. For $M \equiv PQ$, distinguish depending on P's form.

12. The set of free variables of a de Bruijn term is given by

$$\begin{aligned}
\mathrm{var}(i) &= \{v_i\},\\
\mathrm{var}(MN) &= \mathrm{var}(M) \cup \mathrm{var}(N),\\
\mathrm{var}(\lambda(M)) &= \{v_{i-1}; i > 0 \text{ and } v_i \in \mathrm{var}(M)\}.
\end{aligned}$$

Chapter 3

2. The graph is a parallelopiped of $2 * 3 * 3$ vertices.

3. If $p = m+1$, then $M\mathbf{pq} = M'\mathbf{qm}$, where $M' \equiv \lambda pq\,.\,pF(\mathbf{K}b)(qF(\mathbf{K}a))$, and $M\mathbf{0q} = a$. Hence $M\mathbf{pq} = a$ if $p \leqslant q$, and $= b$ otherwise. The length of the reduction is proportional to $\min(p, q)$.

6. The types of the combinators **S**, **B** and **C** are

```
↷ val S = fn : ('a -> 'b -> 'c) -> ('a -> 'b) -> 'a -> 'c
  val B = fn : ('a -> 'b) -> ('c -> 'a) -> 'c -> 'b
  val C = fn : ('a -> 'b -> 'c) -> 'b -> 'a -> 'c
```

11. Show that $M = \pi^1(M \wedge N)$ and $N = \pi^2(M \wedge N)$ by induction over $M \wedge N$.

14. Iterators over trees and forests:

```
fun Tree_it u v c (Union(x,f)) = u(x, Forest_it u v c f)
and Forest_it u v c =
    let fun F Empty = v
          |  F (Cons(a,f)) = c(Tree_it u v c a, F f)
    in F end;
```

```
↷ val Tree_it = fn : ('a * 'b -> 'c) ->
                      'b ->
                      ('c * 'b -> 'b) ->
                      'a Tree -> 'c

  val Forest_it = fn : ('a * 'b -> 'c) ->
                        'b ->
                        ('c * 'b -> 'b) ->
                        'a Forest -> 'b
```

Chapter 4

1. $\Sigma'_0 = \{\bot\} \cup \mathrm{At}_\Sigma$, $\Sigma'_1 = \{\forall x; x \in X\} \cup \{\exists x; x \in X\}$, $\Sigma'_2 = \{\wedge, \vee, \Rightarrow\}$.

3. $\lambda z^{\varphi\vee\psi}\lambda f^{\varphi\Rightarrow\psi}\,.\,\mathbf{case}\,(z, [x^\varphi].fx, [y^\psi].y) :$
$(\varphi \vee \psi) \Rightarrow ((\varphi \Rightarrow \psi) \Rightarrow \psi)$

$\lambda f^{\varphi\Rightarrow(\psi\Rightarrow\chi)}\lambda x^{\varphi\wedge\psi}\,.\,f(\pi^1 x)(\pi^2 x) :$
$(\varphi \Rightarrow (\psi \Rightarrow \chi)) \Rightarrow (\varphi \wedge \psi) \Rightarrow \chi$

$\lambda z^{(\varphi\Rightarrow\psi)\vee(\varphi\Rightarrow\chi)}\lambda x^\varphi\,.\,\mathbf{case}\,(z, [f^{\varphi\Rightarrow\psi}].\iota^1(fx), [g^{\varphi\Rightarrow\chi}].\iota^2(gx)) :$
$((\varphi \Rightarrow \psi) \vee (\varphi \Rightarrow \chi)) \Rightarrow \varphi \Rightarrow (\psi \vee \chi)$

In Standard ML, the types and the terms would be:

```
fun A (one x) f = f x
|   A (two y) f = y
and uncurry f (y,z) = f y z
and disjunct (one f) x = one (f x)
|   disjunct (two g) x = two (g x) ;
```

```
↷ val A = fn : ('a,'b) union -> ('a -> 'b) -> 'b
```

```
val uncurry = fn : ('a -> 'b -> 'c) -> 'a * 'b -> 'c
val disjunct = fn : (('a -> 'b),('a -> 'c)) union ->
                    'a -> ('b,'c) union
```

4. Rules of introduction and elimination for equivalence:

$$(I_{\Leftrightarrow}) : \frac{\begin{array}{cc}[\varphi] & [\psi] \\ \psi & \varphi\end{array}}{\varphi \Leftrightarrow \psi}, \qquad (E^1_{\Leftrightarrow}) : \frac{\varphi \Leftrightarrow \psi \quad \varphi}{\psi}, \quad (E^2_{\Leftrightarrow}) : \frac{\varphi \Leftrightarrow \psi \quad \psi}{\varphi}.$$

11. The η rule for conjunction is obtained by combining an elimination, then an introduction of $\wedge$,

$$\frac{\dfrac{\varphi \wedge \psi}{\varphi} \qquad \dfrac{\varphi \wedge \psi}{\psi}}{\varphi \wedge \psi},$$

and by considering that if x is a derivation of $\varphi \wedge \psi$, the previous combination, with x above each assumption, can be simplified to x:

$$(\pi^1(x), \pi^2(x)) \longrightarrow_\eta x.$$

For implication, the rule $\lambda x \,.\, Mx \longrightarrow_\eta M$ is obtained, if $x \notin \text{var}(M)$.

14. For contradiction, suppose $\neg\exists x(\neg\varphi)$. Then, (by $\vdash_M$) $\forall x(\neg\neg\varphi)$, then ($\vdash_C$) $\forall x\varphi$. $\perp$ is obtained by supposing $\neg\forall x\varphi$. By (C), $\exists x(\neg\varphi)$ follows.

17. Proof in LK of Łukasiewicz's axiom

$$\dfrac{\dfrac{\dfrac{\dfrac{\dfrac{\dfrac{\varphi : \varphi, \chi}{: \varphi, \varphi \Rightarrow \chi} \qquad \psi : \psi}{(\varphi \Rightarrow \chi) \Rightarrow \psi : \varphi, \psi} \qquad \varphi, \omega : \varphi}{(\varphi \Rightarrow \chi) \Rightarrow \psi, \psi \Rightarrow \varphi, \omega : \varphi, \varphi}}{(\varphi \Rightarrow \chi) \Rightarrow \psi, \psi \Rightarrow \varphi : \omega \Rightarrow \varphi}}{(\varphi \Rightarrow \chi) \Rightarrow \psi : (\psi \Rightarrow \varphi) \Rightarrow (\omega \Rightarrow \varphi)}}{((\varphi \Rightarrow \chi) \Rightarrow \psi) \Rightarrow ((\psi \Rightarrow \varphi) \Rightarrow (\omega \Rightarrow \varphi))}.$$

21. At least three elements: $\exists x_1x_2x_3(x_1 \neq x_2 \wedge x_2 \neq x_3 \wedge x_3 \neq x_1)$. At most three elements: $\exists x_1x_2x_3\forall x(x = x_1 \vee x = x_2 \vee x = x_3)$. Exactly three: the conjunction of the two formulas.

23. Associativity is defined in $\Sigma'' = \{e, m\}$ by

$$\forall xyzstu\Big(m(x,y,s) \wedge m(s,z,u) \wedge m(y,z,t) \Rightarrow m(x,t,u)\Big).$$

26. If $m > n$, prove $AE \vdash \neg(s^m(0) = s^n(0))$ by induction over n. If $n > m$, $\neg(s^m(0) = s^n(0))$ follows from $\neg(s^n(0) = s^m(0))$ by contraposition. $AE \vdash s^m(0) + s^n(0) = s^p(0)$, if $m + n = p$, is proven by induction over n.

Chapter 5

2. Let $f : \mathbf{2}^2 \to \mathbf{2}$ define a binary connective $\mathbf{c}$ with which every connective can be defined. By considering negation, show first that $f(0,0) = 1$ and $f(1,1) = 0$. If $f(x,y) \neq f(y,x)$, then $f(x,y) = 1 - x$ or $1 - y$, and so f cannot be used to express disjunction. To be proven by induction over terms formed with $\mathbf{c}$.

3. Show that for every $f : \mathbf{2} \to \mathbf{2}$ defined using these connectives, $f(1) = 1$.

8. For $(p \Rightarrow q) \vee (q \Rightarrow p)$, use $U = \{a, b, c\}$, with $a < b$, $a < c$, and $I(b) = \{p\}$, $I(c) = \{q\}$. For $(p \Rightarrow (q \vee r)) \Rightarrow ((p \Rightarrow q) \vee (p \Rightarrow r))$, use the same U with $I(b) = \{p, q\}$, $I(c) = \{p, r\}$. For $(p \Rightarrow q) \Rightarrow (\neg p \vee q)$, use $U = \{a, b, c, d\}$, with $a < b < d$, $a < c < d$ and $I(b) = \{q\}$, $I(c) = \{p, q\}$.

10. The theory of fields of characteristic zero (TC_0) is axiomatized by adding the axioms $1 + \ldots + 1\,(p \text{ times}) \neq 0$, for all $p \geqslant 2$, to the theory of fields (TC). If $TC_0 \models \varphi$, then $TC_0 \cup \{\neg\varphi\}$ is inconsistent, so a finite subset of $TC_0 \cup \{\neg\varphi\}$ is inconsistent, by compactness. Let p be the largest i such that the axiom $i.1 \neq 0$ appears in this finite subset. Then φ is a consequence of TC and of the axioms $i.1 \neq 0$ for $i \leqslant p$.

14. If $\mathcal{E}$ and $\mathbf{Alg}_\Sigma \setminus \mathcal{E}$ are axiomatized by Γ and Δ, respectively, then $\Gamma \cup \Delta$ is inconsistent, hence by compactness, a finite subset $\Gamma' \cup \Delta'$ is inconsistent, with $\Gamma' \subseteq \Gamma$ and $\Delta' \subseteq \Delta$. Show that $\mathcal{E} = \mathbf{Mod}(\Gamma')$.

15. Let $\mathcal{A}$ and $\mathcal{B}$ be two models of Γ. If Γ is complete, let φ be a proposition. Then, either $\Gamma \vdash \varphi$, hence by the soundness of $\vdash$, $\mathcal{A} \models \varphi$ and $\mathcal{B} \models \varphi$; or $\Gamma \vdash \neg\varphi$, hence $\mathcal{A} \not\models \varphi$ and $\mathcal{B} \not\models \varphi$. Therefore, $\mathcal{A} \equiv \mathcal{B}$. Conversely, if all of the models of Γ are elementarily equivalent for every proposition φ, φ or $\neg\varphi$ is true in all the models of Γ, i.e, $\Gamma \models \varphi$ or $\Gamma \models \neg\varphi$. Hence, by the completeness of $\vdash$, $\Gamma \vdash \varphi$ or $\Gamma \vdash \neg\varphi$.

20. With a unary Skolem function D, one obtains:

$$S(\varphi) \;=\; \left\{ \begin{array}{l} \forall \varepsilon x(\neg(\varepsilon > 0) \vee D(\varepsilon) > 0), \\ \forall \varepsilon x(\neg(\varepsilon > 0) \vee |x - a| < D(\varepsilon) \vee |f(x) - f(a)| < \varepsilon) \end{array} \right\}.$$

23. A subset E of $\mathcal{F}$ is consistent if for every x, the set of $f(x)$, for $f \in E$ and $x \in D(f)$, has at least one element.

25. Every function is a solution of $f(x) = f(x)$. Every constant function over $\mathbf{N}$ is a solution of $f(x) = f(x+1)$. In both cases the least solution is the constant $\perp$.

Chapter 6

11. One can go from one presentation to the next by $u = ab^2$, $v = b$ and $a = (uv)^2$, $b = v$. The defined group is that of permutations of three elements, where u and v are the transpositions (1 2) and (1 3).

12. Modify the algebra of colored integers by taking three colors instead of two, addition conserving the 'best' color of its two arguments.

13. These properties of $\mathrm{T}_{\Sigma/\Sigma^0}$ follow only from the universal property. For composition, consider a Σ_0-algebra $\mathcal{X}$ and a Σ_0-morphism ξ of $\mathcal{X}$ in a Σ'-algebra. Extend ξ to a Σ-morphism defined over $\mathrm{T}_{\Sigma/\Sigma^0}[\mathcal{X}]$, then to a Σ'-morphism over $\mathrm{T}_{\Sigma'/\Sigma}[\mathrm{T}_{\Sigma/\Sigma^0}[\mathcal{X}]]$, ... Then $\mathrm{T}_{\Sigma/\emptyset}[X] = \mathrm{T}_\Sigma[X]$, $\mathrm{T}_{\Sigma/\Sigma}[\mathcal{X}] = \mathcal{X}$, and $\mathrm{T}_{\Sigma/\Sigma^0}[\mathrm{T}_{\Sigma^0}] = \mathrm{T}_\Sigma$.

15. The completion of $\mathcal{E}_0 = \{x^{-1}(xy) = y\}$ gives three rules $x^{-1}(xy) \to y$, $(x^{-1})^{-1}y \to xy$ and $x(x^{-1}y) \to y$.

16. The complete system is $ee \to e$, $(xy)z \to x(yz)$ and $e(ez) \to ez$.

17. If $E \cup \{M_i \neq N_i\}$ has a solution σ_i, it is in particular a solution of E, hence an instance $\mu_i\theta$ of a most general unifier θ of E . Therefore, $\sigma(M_i) \not\equiv \sigma(N_i)$ implies $\theta(M) \not\equiv \theta(N)$. Hence θ is a solution of $E \cup \{M_1 \neq N_1, \ldots, M_p \neq N_p\}$.

18. Take $M = \lambda n \lambda f \lambda x \,.\, f(nfx)$. Associate a type variable to each bound variable and each application node of the term's syntactic tree,

$$M = (\lambda n : \alpha)(\lambda f : \beta)(\lambda x : \gamma) \,.\, @^\delta f^\beta (@^\epsilon @^\zeta n^\alpha f^\beta x^\gamma),$$

by writing the constraints in these nodes,

$$\begin{aligned} \beta &= \epsilon \to \delta, \\ \zeta &= \gamma \to \epsilon, \\ \alpha &= \beta \to \zeta, \end{aligned}$$

and by synthesizing the types of the abstractions, at the root, obtain: $\alpha \to \beta \to \gamma \to \delta$. Solving the constraints gives

$$((\epsilon \to \delta) \to (\gamma \to \epsilon)) \to (\epsilon \to \delta) \to \gamma \to \delta.$$

Chapter 7

2. D-lists are useful to concatenate lists without using `append`. Define a predicate `quicksort_dl` whose first argument is a list and whose second is a D-list, to call `quicksort_dl(X,Y-[])`:

```
quicksort_dl([],X-X).
quicksort_dl([H|T],S-V) :-
   partition(T,H,Littles,Bigs),
   quicksort_dl(Littles,S-[H|U]),
   quicksort_dl(Bigs,U-V).
```

6. Show by induction over $n \geqslant 1$ that if $u \in \Phi_\Sigma^{\downarrow n}$ and $|u| < n$, then $u \in \Phi_\Sigma^{\uparrow n}$.

7. Let $A \in T_P^{\downarrow\omega}$. Since the set of clauses of P is *finite*, from 'for every n, there exists a clause such that ... ', deduce that 'there exists a clause such that for an infinity of n ... '. Use the assumption on the variables to work with a single substitution σ for every n, and use the monotonicity of T_P.

12. Of the seventeen cases to examine (four structural rules of LJ, two times six logical rules, identity), certain rules are not applicable because of the structure of P-formulas or of the consequent. There remain only eight rules: L_C, L_A, L_P, $L_\wedge$, $L_\Rightarrow$, $L_\forall$, $R_\vee$ and $R_\exists$.

Chapter 8

5. $f(n) = P(n)g(n) + (1 - P(n))h(n)$.

6. To prove that f is primitive recursive, it is defined in terms of g:

$$\begin{aligned} f(0,\vec{n}) &= g(0,\vec{n}), \\ f(m+1,\vec{n}) &= f(m,\vec{n}) + g(m+1,\vec{n}). \end{aligned}$$

11. Prove by induction over n that the singleton sets $\{n\}$ are recursive, and that a set with p elements is recursive by induction over p.

15. Define $\ulcorner MN \urcorner$ using app and apply lemma 8.19.

18. A set containing a recursive (respectively non-recursive) set is not necessarily recursive (respectively non-recursive). For example, $\varnothing \subseteq K \subseteq \mathbf{N}$. The set $\mathcal{U}(D_m)$ always being infinite, it is impossible that the indices of a finite class of sets be r.e.

19. $33 \in A$ if and only if $A \in \mathcal{U}(\{33\})$. $A \neq \varnothing$ if and only if $A \in \bigcup_{n \geqslant 0} \mathcal{U}(\{n\})$, and $\{n\} = D_{2^n}$.

20. Since the r.e. set $U = \{(e,x); \exists q T(e,x,q)\}$ is diophantine (Matjasevič's theorem), there exists a polynomial P such that $U = \{(e,x); \exists \vec{y} P(e,x,\vec{y}) = 0\}$. Hence $W_e = \{x; \exists \vec{y} P(e,x,\vec{y}) = 0\}$.

21. If $f(\mathbf{N})$ is recursive, the binary development of s is computable: the p-th numeral is a 1 if and only if $p \in f(\mathbf{N})$, which is decidable. Conversely, if s is recursive, the binary development of s constitutes an *increasing* recursive enumeration of $f(\mathbf{N})$, which must then be recursive.

24. Let, for example, $n = 3$:

$$U_3 = \{(e, x) \in \mathbf{N}^2; \exists x_1 \forall x_2 \exists x_3 T^3(e, x, x_1, x_2, x_3)\}$$

is universal for the Σ_n. In fact, if A is Σ_3, then

$$A = \{x \in \mathbf{N}; \exists x_1 \forall x_2 \exists x_3 R(x, x_1, x_2, x_3)\}$$

for a recursive relation R. Apply the universality theorem to the r.e. set $\exists x_3 R$.

25. If θ is a formula of arithmetic with one variable x such that $\mathbf{N} \models \varphi$ if and only if $\mathbf{N} \models \theta(\ulcorner \varphi \urcorner)$, apply lemma 8.11 to $\neg\theta$.

26. An r.e. set is recursive if and only if its complement is r.e., and therefore has an index. On the other hand, $x \in W_p$ if and only if $\exists q T(p, x, q)$, from which $x \notin W_p$ if and only if $\forall q \neg T(p, x, q)$. Hence $x \in A$ if and only if $\exists e \forall x (\forall q \neg T(p, x, q) \iff \exists q T(e, x, q))$. By putting this formula in prenex form, and by contracting the successive quantifiers of the same strength, a prefix $\exists \forall \exists$ is obtained, which proves that A is Σ_3.

Bibliography

[1] H. Abelson, G. J. Sussman and J. Sussman. *Structure and Interpretations of Computer Programs.* The MIT Press, Cambridge, MA, 1984.

[2] S. Abramsky and C. Hankin, editors. *Abstract Interpretations of Declarative Languages.* Ellis Horwood, Chichester, UK, 1987.

[3] W. Ackermann. Zum hilbertschen Aufbau der reellen Zahlen. *Math. Ann.*, 99:118–133, 1928.

[4] K. R. Apt and M. H. van Emden. Contributions to the theory of logic programming. *Journal of the ACM*, 29(3):841–862, July 1982.

[5] A. Asperti and G. Longo. *Categories, Types and Structures.* The MIT Press, 1991.

[6] L. Bachmair, N. Dershowitz and J. Hsiang. Orderings for equational proofs. In *Proc. 1st Symp. on Logic in Computer Science*, pages 346–357, Cambridge, June 1986.

[7] J. Backus. Can programming be liberated from the von Neumann style? A functional style and its algebra of programs. *Journal of the ACM*, 21(8):613–641, 1978.

[8] H. P. Barendregt. *The Lambda-Calculus, its Syntax and Semantics.* North-Holland, Amsterdam, revised edition, 1984.

[9] J. G. P. Barnes. *Programming in Ada.* Addison-Wesley, New York, 1982.

[10] M. Blum. A machine-independent theory of the complexity of recursive functions. *Journal of the ACM*, 14:322–336, 1967.

[11] P. Boizumault. *Prolog, l'implantation.* Masson, Paris, 1988.

[12] L. Byrd. Understanding the control flow of programs. In S.ÅTärnlund, editor, *Logic Programming Workshop*, pages 127–138, Debrecen, Hungary, 1980.

[13] C. L. Chang and R. C. T. Lee. *Symbolic Logic and Mechanical Theorem Proving.* Academic Press, New York, 1973.

[14] A. Church. A set of postulates for the foundation of logic (second paper). *Ann. of Math.*, 33:839–864, 1933.

[15] A. Church. A note on the Entscheidungsproblem. *J. Symbolic Logic*, 1:40–41, 101–102, 1936.

[16] E. A. Cichon. A short proof of two recently discovered independence results using recursion theoretic methods. *Proc. Amer. Math. Soc.*, 87(4):704–706, 1983.

[17] K. L. Clark. Negation as failure. In H. Gallaire and J. Minker, editors, *Logic and Data Bases*, pages 293–322. Plenum Press, New York, 1978.

[18] D. Clément, J. Despeyroux, T. Despeyroux, L. Hascoët and G. Kahn. Natural semantics on the computer. Rapport de Recherche 416, INRIA, Sophia-Antipolis, 1985.

[19] D. Clément, J. Despeyroux and G. Kahn. A simple applicative language: Mini-ML. In *Proc. 1986 ACM Symp. on Lisp and Functional Programming*, pages 13–27, 1986.

[20] W. Clinger and J. Rees. Revised[4] report on the algorithmic language Scheme. *SIGPLAN Notices*, 26(3):1–51, March 1991.

[21] P. J. Cohen. *Set Theory and the Continuum Hypothesis.* Benjamin, New York, 1966.

[22] P. M. Cohn. *Universal Algebra.* Reidel, Dordrecht, second edition, 1981.

[23] A. Colmerauer, H. Kanoui, P. Roussel and R. Pasero. Un système de communication homme-machine en français. Technical report, Groupe de Recherche en Intelligence Artificielle, Université d'Aix-Marseille, Marseille, 1973.

[24] S. A. Cook. The complexity of theorem-proving procedures. In *Proc. 3rd Annual ACM Symp. on the Theory of Computing*, pages 151–158, New York, 1971. ACM.

[25] T. Coquand. Metamathematical investigations of a calculus of constructions. In P. Odifreddi, editor, *Logic and Computer Science*, pages 91–122. Academic Press, 1990.

[26] J. Corbin and M. Bidoit. A rehabilitation of Robinson's unification algorithm. In *Information Processing 83*, pages 909–914. North-Holland, Amsterdam, 1983.

[27] B. Courcelle. Fundamental properties of infinite trees. *Theoretical Computer Science*, 25:95–169, 1983.

[28] G. Cousineau. An algebraic definition for control structures. *Theoretical Computer Science*, 12:175–192, 1980.

[29] G. Cousineau, P. L. Curien and M. Mauny. The categorical abstract machine. In J. P. Jouannaud, editor, *Functional Programming Languages and Computer Architecture*, pages 50–64. Springer-Verlag LNCS 201, Berlin, 1985.

[30] P. Cousot and R. Cousot. Abstract interpretations: a unified lattice model for static analysis of programs by construction of approximation of fixpoints. In *4th ACM Symp. on Principles of Programming Languages*, pages 238–252, 1977.

[31] P. L. Curien. *Categorical Combinators, Sequential Algorithms and Functionnal Programming.* Pitman, London, 1986.

[32] G. B. Dantzig. Maximization of a linear function of variables subject to linear inequalities. In T. C. Koopmans, editor, *Activity Analysis of Production and Allocation*, pages 339–347. Wiley, New York, 1951.

[33] M. A. Davis. Eliminating the irrelevant from mechanical proofs. In *Proc. Symp. on Applied Mathematics*, pages 15–30, Providence, RI, 1963.

[34] M. D. Davis and E. J. Weyuker. *Computability, Complexity and Languages.* Academic Press, 1983.

[35] N. G. de Bruijn. Lambda-calculus notation with nameless dummies, a tool for automatic formula manipulation, with application to the Church-Rosser theorem. *Indag. Math*, 34(5):381–392, 1972.

[36] R. Dedekind. *Was sind und was sollen die Zahlen?* Braunschweig, 1888.

[37] M. Dehn. Über die Topologie des dreidimensionalen Raumes. *Math. Ann.*, 69:137–168, 1910.

[38] N. Dershowitz. A note on simplification orderings. *Information Processing Letters*, 9(5):212–215, 1979.

[39] N. Dershowitz and J. P. Jouannaud. Rewrite systems. In J. van Leeuwen, editor, *Handbook of Theoretical Computer Science*, volume B, chapter 6, pages 243–320. Elsevier, Amsterdam, 1990.

[40] M. Dincbas and P. van Hentenryck. Extended unification algorithm for the integration of functional programming into logic programming. *Journ. of Logic Programming*, 4:199–227, 1987.

[41] G. Escalada-Imaz and M. Ghallab. A practically efficient and almost linear unification algorithm. *Artificial Intelligence*, 36(2):249–263, September 1988.

[42] M. Felleisen. λ_v-CS: an extended λ-calculus for Scheme. In *Proc. 1988 ACM Symp. on Lisp and Functional Programming*, pages 72–85, 1988.

[43] A. J. Field and P. G. Harrison. *Functional Programming.* Addison-Wesley, New York, 1988.

[44] R. M. Friedberg. Two recursively enumerable sets of incomparable degrees of unsolvability (solution of Post's problem 1944). *Proc. Nat. Acad. Sci. U.S.A.*, 43:236–238, 1957.

[45] D. P. Friedman and M. Wand. Reification: reflexion without metaphysics. In *Proc. 1984 ACM Symp. on Lisp and Functional Programming*, pages 348–355, 1984.

[46] D. P. Friedman and D. S. Wise. Cons should not evaluate its arguments. In S. Michaelson and R. Milner, editors, *Automata, Languages and Programming*, pages 257–284. Edinburgh University Press, 1976.

[47] J. Gallier. *Logic for Computer Science.* Harper & Row, 1986.

[48] J. H. Gallier and S. Raatz. Hornlog: a graph-based interpreter for general Horn clauses. *Journ. of Logic Programming*, 4:119–155, 1987.

[49] M. R. Garey and D. S. Johnson. *Computers and Intractability, a Guide to the Theory of NP-completeness.* Freeman, San Francisco, CA, 1979.

[50] F. Giannesini, H. Kanoui, R. Pasero and M. van Caneghem. *Prolog.* InterEditions, Paris, 1985.

[51] J. Y. Girard. *Interprétation fonctionnelle et élimination des coupures dans l'arithmétique d'ordre supérieur.* PhD thesis, Université Paris VII, 1972.

[52] J. Y. Girard. The system F of variable types, fifteen years later. *Theoretical Computer Science*, 45:159–192, 1986.

[53] J. Y. Girard. *Proof Theory and Logical Complexity.* Bibliopolis, Naples, 1987.

[54] J.Y. Girard, Y. Lafont and P. Taylor. *Proofs and Types.* Cambridge University Press, 1989.

[55] K. Gödel. Über formal unentscheidbare Sätze der Principia Mathematica und verwandter Systeme. *Monatsh. Math. Phys.*, 38:173–198, 1931.

[56] K. Gödel. *The Consistency of the Continuum Hypothesis.* Princeton University Press, Princeton, NJ, 1944.

[57] K. Gödel. On undecidable propositions of formal mathematical systems. In M. Davis, editor, *The undecidable, basic papers on undecidable propositions, unsolvable problems and computable functions*, pages 39–74. Raven Press, Hewlett, NY, 1965.

[58] W. D. Goldfarb. Note on the undecidability of the second-order unification problem. *Theoretical Computer Science*, 13:225–230, 1981.

[59] I. Guessarian. *Algebraic Semantics.* Springer-Verlag, Berlin, 1981.

[60] L. Henkin. The completeness of first-order functional calculus. *J. Symbolic Logic*, 14:159–166, 1949.

[61] P. van Hentenryck. *Constraint Satisfaction in Logic Programming.* The MIT Press, Cambridge, MA, 1989.

[62] J. Herbrand. Investigations in proof theory. In Van Heijenoort, editor, *From Frege to Gödel*, pages 529–581. Harvard University Press, Cambridge, MA, 1967. (translation).

[63] D. Hilbert. Mathematical problems. *Bull. of the Amer. Math. Soc.*, 8:437–479, 1902. (translation).

[64] J. R. Hindley and J. P. Seldin. *Introduction to Combinators and λ-Calculus.* Cambridge University Press, 1986.

[65] J. E. Hopcroft and J. D. Ullman. *Introduction to Automata Theory, Languages and Computation.* Addison-Wesley, New York, 1979.

[66] W. A. Howard. The formulae-as-types notion of construction. In J. P. Seldin and J. R. Hindley, editors, *To H. B. Curry: Essays on Combinatory Logic, Lambda-Calculus and Formalism*, pages 479–490. Academic Press, 1980.

[67] G. Huet. *Résolution d'Equations dans des Langages d'Ordre 1,2, ... , ω.* PhD thesis, Université Paris VII, 1976.

[68] G. Huet. Confluent reductions: abstract properties and applications to term rewriting systems. *Journal of the ACM*, 27(4):797–821, October 1980.

[69] G. Huet. A complete proof of correctness of the Knuth–Bendix completion algorithm. *Journal of Computer and System Sciences*, 23:11–21, 1981.

[70] G. Huet, editor. *Logical Foundations of Functional Programming.* Addison-Wesley, New York, 1990.

[71] G. Huet and D. Oppen. Equations and rewrite rules: a survey. In R. Book, editor, *Formal Languages: Perspectives and Open Problems.* Academic Press, 1980.

[72] R. J. M. Hughes. Super combinators — a new implementation method for applicative languages. In *Proc. 1982 ACM Symp. on Lisp and Functional Programming*, pages 1–10, Pittsburgh, PA, 1982.

[73] J. Jaffar and J. L. Lassez. Constraint logic programming. In *14th ACM Symp. on Principles of Programming Languages*, pages 111–119, Munich, January 1987.

[74] J. Jaffar, J. L. Lassez and M. J. Maher. A theory of complete logic programs with equality. *Journ. of Logic Programming*, 1:211–223, 1984.

[75] N. Karmarkar. A new polynomial-time algorithm for linear programming. In *Proc. 16th Annual ACM Symp. on Theory of Computing*, pages 302–311, Washington, DC, 1984. ACM.

[76] R. M. Karp. Reducibility among combinatorial problems. In R. E. Miller and J. Thatcher, editors, *Complexity of Computer Computations*, pages 85–104. Plenum Press, New York, 1972.

[77] L. G. Khachiyan. A polynomial algorithm in linear programming. *Dokl. Akad. Nauk SSSR*, 244:1093–1096, 1979. (in Russian).

[78] L. Kirkby and J. Paris. Accessible independence results for Peano arithmetic. *Bull. of the London Math. Soc.*, 14:285–293, 1982.

[79] S. C. Kleene. A theory of positive integers in formal logic. *Amer. J. Math*, 57:153–173, 219–244, 1935.

[80] S. C. Kleene. On notation for ordinal numbers. *J. Symbolic Logic*, 3:150–155, 1938.

[81] S. C. Kleene. *Introduction to Metamathematics*. North-Holland, Amsterdam, 1952.

[82] S. C. Kleene. The theory of recursive functions, approaching its centennial. *Bulletin of the AMS*, 5(1):43–61, July 1981.

[83] S. C. Kleene and E. L. Post. The upper semilattice of degrees of recursive unsolvability. *Ann. of Math.*, 59:379–497, 1954.

[84] J. W. Klop. *Combinatory Reduction Systems*. PhD thesis, Mathematisch Centrum Amsterdam, 1980.

[85] J. W. Klop and R. C. de Vrijer. Unique normal forms for lambda-calculus with surjective pairings. *Information and Control*, 80(2):97–113, February 1989.

[86] K. Knight. Unification: a multidisciplinary survey. *ACM Computing Surveys*, 21(1):93–124, March 1989.

[87] D. Knuth and P. Bendix. Simple word problems in universal algebras. In J. Leech, editor, *Computational Problems in Abstract Algebra*. Pergamon, Oxford, UK, 1970.

[88] R. A. Kowalski. Predicate logic as a programming language. In *Information Processing 74*, pages 569–574, Stockholm, 1974. North-Holland, Amsterdam.

[89] J. L. Krivine. Un algorithme non typable dans le système F. *Notes C.R. Acad. Sciences, Série I*, 304(5):123–126, 1987.

[90] J. L. Krivine. *Lambda-calcul, Types et Modèles*. Masson, Paris, 1990.

[91] J. B. Kruskal. Well-quasi-orderings, the tree theorem and Vazsonyi's conjecture. *Trans. Amer. Math. Soc.*, 95:210–225, May 1960.

[92] P. Landin. The mechanical evaluation of expressions. *Computer Journal*, 6:308–320, 1964.

[93] P. Landin. A correspondance betwwen ALGOL-60 and Church's lambda notation. *Communications of the ACM*, 8:89–101, 158–165, 1965.

[94] P. Landin. The next 700 programming languages. *Communications of the ACM*, 9:157–166, March 1966.

[95] J. W. Lloyd. *Foundations of Logic Programming*, second edition. Springer-Verlag, Berlin, 1987.

[96] D. W. Loveland. *Automated Theorem Proving: a Logical Basis*. North-Holland, New York, 1978.

[97] Z. Manna and S. Ness. On the termination of Markov algorithms. In *Proc. 3rd Hawaii Intern. Conf. on System Science*, pages 789–792, Honolulu, HI, 1970.

[98] A. A. Markov. Unsolvability of the problem of homeomorphy. In *Proc. Intern. Congress of Mathematicians*, pages 300–306, London, 1958. Cambridge University Press. (in Russian).

[99] A. A. Markov. Theory of algorithms. *Amer. Math. Soc. Transl.*, 2(15):1–14, 1960.

[100] A. Martelli and U. Montanari. An efficient unification algorithm. *ACM Transactions on Programming Languages and Systems*, 4(2):258–282, April 1982.

[101] A. Martelli and G. Rossi. Efficient unification with infinite terms in logic programming. In ICOT, editor, *Proc. Intern. Conf. on Fifth Generation Computer Systems*, pages 203–209. North-Holland, Amsterdam, 1984.

[102] U. Martin and T. Nipkov. Unification in Boolean rings. In *Proc. 8th Intern. Conf. on Automated Deduction*, pages 506–513, Oxford. Springer Verlag, Berlin, 1986.

[103] P. Martin-Löf. An intuitionistic theory of types: predicative part. In H. E. Rose and J. C. Shepherdson, editors, *Proc. Logic Colloquium '73*, pages 73–118, Amsterdam, 1973. North-Holland, Amsterdam.

[104] P. Martin-Löf. *Intuitionistic Type Theory*. Bibliopolis, Napoli, 1984.

[105] Y. Matjasevich. Enumerable sets are diophantine. *Dokl. Akad. Nauk SSSR*, 191:354–357, 1970.

[106] A. R. Meyer and D. M. Ritchie. The complexity of loop programs. In *Proc. ACM Nat. Meeting*, pages 465–469, 1967.

[107] D. Miller. A logical analysis of modules in logic programming. *Journ. of Logic Programming*, 6(1 & 2):79–108, 1989.

[108] D. A. Miller and G. Nadathur. Higher-order logic programming. In E. Shapiro, editor, *Proc. 3rd Intern. Conf. on Logic programming*, pages 448–462, London. Springer-Verlag LNCS 225, Berlin, 1986.

[109] D. A. Miller, G. Nadathur and A. Scedrov. Hereditary Harrop formulas and uniform proof systems. In *Proc. 2nd Symp. on Logic in Computer Science*, pages 98–105, 1987.

[110] R. Milner. A theory of type polymorphism in programming. *Journal of Computer and System Sciences*, 17:348–375, 1978.

[111] R. Milner, M. Tofte and R. Harper. *The Definition of Standard ML*. The MIT Press, Cambridge, MA, 1990.

[112] A. A. Muchnik. On the unsolvability of the problem of reducibility in the theory of algorithms. *Dokl. Akad. Nauk SSSR*, 108:194–197, 1956. (in Russian).

[113] A. Mycroft. The theory and practice of transforming call-by-need into call-by-value. In *International Symposium on Programming*. Springer-Verlag LNCS 83, 1980.

[114] P. S. Novikov. On the algorithmic unsolvability of the word problem in group theory. *Dokl. Akad. Nauk SSSR Mathematičeskii Institut Trudy*, 44, 1955. (in Russian).

[115] R. O'Keefe. *The Craft of Prolog*. The MIT Press, Cambridge, MA, 19990.

[116] J. Paris and L. Harrington. A mathematical incompleteness in Peano arithmetic. In J. Barwise, editor, *Handbook of Mathematical Logic*, chapter D.8, pages 1133–1142. North-Holland, Amsterdam, 1977.

[117] M. S. Paterson and M. N. Wegman. Linear unification. *Journal of Computer and System Sciences*, 16:158–167, 1978.

[118] F. Pereira and C. Tweed. *C-Prolog User's Manual*. University of Edimburgh, GB, 1987. Versions 1.5 and 1.5+.

[119] S. Peyton-Jones. *The Implementation of Functional Programming Languages*. Prentice-Hall, London, 1987.

[120] S. Peyton-Jones and D. Lester. *Implementating Functional Languages*. Prentice-Hall, London, 1987.

[121] G. Plotkin. Building in equational theories. *Machine Intelligence*, 7:73–90, 1972.

[122] G. D. Plotkin. A structural approach to operational semantics. DAIMI FN 19, Computer Science Departement, Aarhus University, Aarhus, Denmark, 1981.

[123] E. L. Post. Formal reductions of the general combinatorial decision problem. *Amer. J. Math*, 65:197–215, 1943.

[124] D. Prawitz. An improved proof procedure. *Theoria*, 26:102–139, 1960.

[125] D. Prawitz. *Natural Deduction.* Almqvist and Wiksell, Stockholm, 1965.

[126] M. Presburger. Über die Vollständigkeit eines gewissen Systems der Arithmetik ganzer Zahlen, in welchem die Addition als einzige Operation hervortritt. In *Comptes Rendus du* 1er *Congrès des Mathématiciens des Pays Slaves*, pages 92–101, Warszawa, 1929.

[127] R. Reiter. On closed world data bases. In H. Gallaire and J. Minker, editors, *Logic and Data Bases*, pages 55–76. Plenum Press, New York, 1978.

[128] J. A. Robinson. A machine-oriented logic based on the resolution principle. *Journal of the ACM*, 12(1):23–41, January 1965.

[129] H. Rogers Jr. *Theory of Recursive Functions and Effective Computability.* The MIT Press, Cambridge, MA, 1987 (first edition, McGraw-Hill 1967).

[130] G. E. Sacks. The recursively enumerable degrees are dense. *Ann. of Math.*, 80:300–312, 1964.

[131] V. Saraswat, M. Rinard and P. Panangaden. The semantic foundation of concurrent constraint programming. In *18th ACM Symp. on Principles of Programming Languages*, pages 333–352, Orlando, January 1991. ACM Press.

[132] M. Schönfinkel. On the building blocks of logic. In Van Heijenoort, editor, *From Frege to Gödel*, pages 355–366. Harvard University Press, Cambridge, MA, 1967. (translation).

[133] E. Shapiro. The family of concurrent logic programming languages. *ACM Computing Surveys*, 21(3):413–510, September 1989.

[134] J. C. Shepherdson. Negation in logic programming. In J. Minker, editor, *Foundations of Deductive Databases and Logic Programming*, chapter 1, pages 19–88. Morgan Kaufmann, Los Altos, 1987.

[135] J. Shoenfield. *Mathematical Logic.* Addison-Wesley, New York, 1967.

[136] J. Shoenfield. *Degrees of Insolvability.* North-Holland, Amsterdam, 1971.

[137] J. H. Siekmann. Unification theory. *Journ. of Symbolic Computation*, 7:207–274, 1989.

[138] R. M. Smullyan. *Theory of Formal Systems*, volume 47 of *Annals of Math. Studies.* Princeton University Press, Princeton, NJ, 1961.

[139] R. Sommerhalder and S. C. van Westrhenen. *The Theory of Computability: Programs, Machines, Effectiveness and Feasibility.* Addison-Wesley, New York, 1988.

[140] G. L. Steele Jr. *Common LISP, The Language.* Digital Press, Cambridge, MA, second edition, 1990.

[141] L. Sterling and E. Shapiro. *The Art of Prolog.* The MIT Press, Cambridge, MA, 1986.

[142] M. E. Stickel. A complete unification algorithm for associative-commutative functions. In *Proc. 4th Intern. Joint Conf. on Artificial Intelligence*, pages 71–82, Tblisi, 1975.

[143] J. E. Stoy. *Denotational Semantics: the Scott–Strachey Approach to Programming Languages.* The MIT Press, Cambridge, MA, 1977.

[144] M. E. Szabo, editor. *The Collected Works of Gerhard Gentzen.* North-Holland, Amsterdam, 1969.

[145] P. Szabo. *Theory of first-order unification.* PhD thesis, Universität Karlsruhe, 1982. (in German).

[146] W. Szmielew. Elementary properties of Abelian groups. *Fund. Math.*, 41:203–271, 1954.

[147] S. Å. Tärnlund. Horn clause computability. *BIT*, 1977.

[148] A. Tarski, A. Mostowski and R. M. Robinson. *Undecidable Theories.* Studies in Logic. North-Holland, Amsterdam, 1968.

[149] J. A. Thom and J. Zobel. *NU-Prolog Reference Manual.* University of Melbourne, Australia, 1988. Version 1.3.

[150] A. S Troelstra and D. van Dalen. *Constructivism in Mathematics, an Introduction.* North-Holland, Amsterdam, 1988.

[151] A. Turing. On computable numbers, with an application to the Entscheidungsproblem. *Proc. London Math. Soc.*, 2(42):230–265, 1936.

[152] D.A. Turner. A new implementation technique for applicative languages. *Software and Experience*, 9:31–49, 1979.

[153] M. van Caneghem. *L'anatomie de Prolog.* InterEditions, Paris, 1986.

[154] D. van Dalen. *Logic and Structure.* Springer-Verlag, Berlin, second edition, 1983.

[155] M. H. van Emden and R. A. Kowalski. The semantics of predicate logic as a programming language. *Journal of the ACM*, 23(4):733–742, October 1976.

[156] M. H. van Emden and J. W Lloyd. A logical reconstruction of Prolog II. *Journ. of Logic Programming*, 2:143–149, 1984.

[157] M. H. van Emden and K. Yukawa. Logic programming with equations. *Journ. of Logic Programming*, 4:265–288, 1987.

[158] P. Weis. *The CAML Reference Manual*, Version 2.6. INRIA, Le Chesnay, France, 1989.

[159] Å. Wikström. *Functional Programming using Standard ML.* Prentice-Hall, London, 1987.

Index